Physical and Computational Aspects of Convective Heat Transfer

Tuncer Cebeci
Peter Bradshaw

Physical and Computational Aspects of Convective Heat Transfer

With 180 Figures

Springer-Verlag
New York Berlin Heidelberg
London Paris Tokyo

Tuncer Cebeci
Douglas Aircraft Company
3855 Lakewood Boulevard
Long Beach, California 90846
U.S.A.

and

Department of Aerospace Engineering
California State University, Long Beach
Long Beach, California 90840
U.S.A.

Peter Bradshaw
Department of Mechanical Engineering
Stanford University
Stanford, CA 94305
U.S.A.

Springer study edition based on corrected
second printing of original hard cover
(© 1984 by Springer-Verlag New York Inc).

Library of Congress Cataloging-in-Publication Data
Cebeci, Tuncer.
    Physical and computational aspects of convective heat transfer /
Tuncer Cebeci, Peter Bradshaw.
        p. cm.
    "Springer study edition based on corrected second printing of
original hard cover (© 1984 by Springer-Verlag, New York, Inc.)"–
–Verso of t.p.
    Includes bibliographies.

    1. Heat—Convection.   2. Fluid dynamics.   I. Bradshaw, P.
(Peter), 1935–   .   II. Title.
TJ260.C35 1988
621.402′2—dc19   88-24924
                                                        CIP
Printed on acid-free paper.

9 8 7 6 5 4 3 2 1

ISBN-13: 978-0-387-96821-6   e-ISBN-13: 978-1-4612-3918-5
DOI: 10.1007/978-1-4612-3918-5

In memory of our fathers

Ömer Zihni Cebeci

and

Joseph William Newbold Bradshaw

Preface

This volume is concerned with the transport of thermal energy in flows of practical significance. The temperature distributions which result from convective heat transfer, in contrast to those associated with radiation heat transfer and conduction in solids, are related to velocity characteristics and we have included sufficient information of momentum transfer to make the book self-contained. This is readily achieved because of the close relationship between the equations which represent conservation of momentum and energy: it is very desirable since convective heat transfer involves flows with large temperature differences, where the equations are coupled through an equation of state, as well as flows with small temperature differences where the energy equation is dependent on the momentum equation but the momentum equation is assumed independent of the energy equation.

The equations which represent the conservation of scalar properties, including thermal energy, species concentration and particle number density can be identical in form and solutions obtained in terms of one dependent variable represent those of another. Thus, although the discussion and arguments of this book are expressed in terms of heat transfer, they are relevant to problems of mass and particle transport. Care is required, however, in making use of these analogies since, for example, identical boundary conditions are not usually achieved in practice and mass transfer can involve more than one dependent variable.

The book is intended for senior undergraduate and graduate students of aeronautical, chemical, civil and mechanical engineering, as well as those studying relevant aspects of the environmental sciences. It should also be of value to those engaged in design and development. The format has been

arranged with this range of readers in mind. Thus, it is readily possible to
find answers to specific heat-transfer problems by reference to correlations
or to results presented in graphical form. Those wishing to understand the
physical processes which form the basis of the results and their limitations
should read the corresponding derivations and explanations.

Many of the results presented in graphical form have been obtained with
the numerical procedure described in Chapters 13 and 14. This stems from
many years of research and is presented in a manner suitable, even for those
with little training in the numerical solution of equations, for the solution of
problems described by boundary-layer equations. This aspect of the book is
of particular significance in view of the rapidly increasing application of
numerical techniques and the present method is described in Chapter 13 in
sufficient detail for the reader to use it. These two chapters can also serve as
a basis for an independent graduate course on the numerical solution of
parabolic linear and nonlinear partial-differential equations with emphasis
on problems relating to fluid mechanics and heat transfer.

We have used Chapters 13 and 14, and various combinations of the
material of other chapters, as the basis for graduate courses at California
State University, Long Beach and at Imperial College, and have learned
that, with access to computers which range from personal microprocessors
to mainframe machines, students are able to reinforce their understanding
of classical and practical material of previous chapters and to apply the
numerical procedures to the solution of new problems. For simplicity, the
descriptions are confined to two-dimensional equations and, for turbulent
flow, to simple turbulence and heat-transfer models. These limitations can
readily be removed, but the treatment as it stands is likely to be enough for
most undergraduate and graduate courses. The computer programs described in Chapters 13 and 14 are available together with sample calculations and, in some cases, additional detail. Requests should be addressed to
the first author at the California State University, Long Beach.

Some teachers of undergraduate and graduate courses may prefer to
exclude the use of numerical methods due to lack of available time. In this
case, Chapters 1–8 will form the basis for a useful one semester course for
senior undergraduates and Chapters 1–9 a comprehensive and longer course
for senior undergraduates or graduate students. The derivations of the
equations provided in Chapter 2 may be omitted in an undergraduate
course and emphasis placed on the boundary-layer equations of Chapter 3,
their solution and related applications.

In the preparation of this text, we have benefited from the advice and
assistance of many people. We are especially grateful to Professors Keith
Stewartson and Jim Whitelaw for their constant advice and helpful contributions. The book could not have been written without the considerable
assistance of Nancy Barela, Sue Schimke, Kalle Kaups, A. A. Khattab and
K. C. Chang. Finally, it is a pleasure to acknowledge the help rendered by
our families.

TUNCER CEBECI
PETER BRADSHAW

Contents

CHAPTER 1

Introduction

The simplest configuration for a flow with heat transfer is a uniform external flow over a flat surface, part or all of which is at a temperature different from that of the oncoming fluid (Fig. 1.1). In slightly more complicated cases the surface may be curved and the external-flow velocity u_e may be a function of the longitudinal coordinate x, but in a large number of practical heat-transfer problems the variation of u_e with y in the external flow is negligibly small compared with the variation of velocity in a region very close to the surface. Within this region, called the *boundary layer*, the x-component velocity u rises from zero at the surface to an asymptotic value equal to u_e; in practice one defines the thickness of this layer as the value of y at which u has reached, say, $0.995u_e$. The *temperature* also varies rapidly with y near the surface, changing from the surface value T_w (subscript w means "wall") to the external-flow value T_e, which, like u_e, can often be taken independent of y. This region of large temperature gradient is called the *thermal boundary layer*; if the fluid has high thermal conductivity, it will be thicker than the hydrodynamic (velocity) boundary layer, and if conductivity is low, it will be thinner than the hydrodynamic boundary layer. Later we shall be more precise about the meanings of "high" and "low," which involves comparing the conduction of heat by molecular motion with the "conduction" of momentum—that is, the viscosity of the fluid, which controls the rate of growth of the velocity boundary layer. The reason for labeling the external flow in Fig. 1.1 "inviscid" is not that the viscosity vanishes but that the velocity in the external flow changes so slowly that viscous stresses are negligibly small compared with, say, the surface shear stress. The surface shear stress represents the rate at which momentum is

1

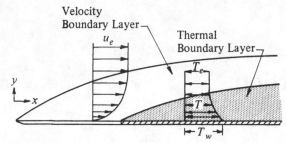

Figure 1.1. Boundary layer on partly heated plate. Shaded area denotes the heated portion.

being extracted from the flow as the boundary layer of slower-moving fluid grows. The rate at which heat is being extracted from the flow (if $T_w < T_e$) or transferred to it (if $T_w > T_e$) is the main subject of this book.

Figure 1.2 illustrates a few more examples of fluid flow with heat transfer. Heat transfer in circular pipes or noncircular ducts [Fig. 1.2(a)] behaves in much the same way as in boundary layers, even though a uniform stream can be distinguished only in the entry region before the growing boundary layers have met on the centerline of the duct. Heat transfer in jets like that shown in Fig. 1.2(b) is not of great interest in itself, since heat is merely being spread out into a wide region of fluid rather than being exchanged with a solid surface, but we need to calculate the temperature distribution in the jet of Fig. 1.2(c), which is being used to heat—or cool—a solid surface. Figure 1.2(c) could show either a blowtorch flame or the exhaust of a rocket motor; in both cases buoyancy effects may be important, as they are in the problem of dispersion of pollutants in a smokestack plume or even the transfer of heat between a heating or air-conditioning system and the room whose temperature it controls. Mass or pollutant transfer is governed by equations that are nearly the same as the heat-transfer equations. Figure 1.2(d) shows an extreme example of heat transfer from a fluid flow to a solid surface—a reentering spacecraft. This book will not enable the reader to predict the heat transfer to a spacecraft, but it will enable the reader to understand the problem qualitatively and to produce quantitative results for simpler problems using the formulas, charts, and computer programs presented later in the book.

Because convective heat transfer is by definition associated with a flow field, and because the equations governing heat or pollutant transfer are very similar to those that govern flow fields (momentum transfer), this book contains a discussion of momentum transfer, sufficiently detailed for the book to be self-contained. A more extensive treatment of momentum transfer is given in Cebeci and Bradshaw [1]. The present book extends the coverage of [1] to flows with large density differences, caused either directly by large temperature differences or by large pressure differences (leading to large temperature differences) in high-speed flows such as those over aircraft

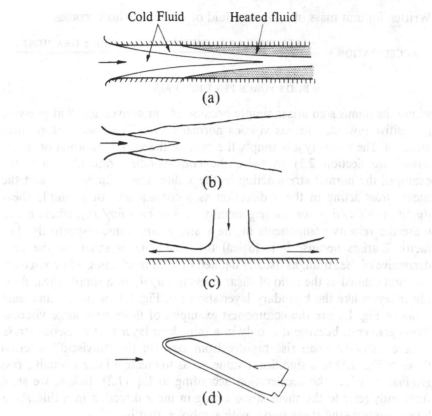

Figure 1.2. Examples of flow with heat transfer. (a) Pipe flow with unheated entry length. (b) Mixing layers merging to form a jet. (c) Jet impinging on a solid surface. (d) Orbiter reentry; boundary layer and separated flow with heat transfer to the solid surface.

or spacecraft. In these cases momentum transfer is affected by heat transfer, because the density appears in the momentum-transfer equations; so the two must be discussed together. In most cases, fluid flows in boundary layers, ducts, or jets are turbulent—a blowtorch roars because turbulence makes the rate of combustion unsteady and the resulting density and pressure fluctuations lead to the radiation of noise. Turbulence generally causes large increases in heat transfer and momentum transfer, and we discuss it at length in this book.

1.1 Momentum Transfer

Newton's second law of motion, applicable to a solid mass or a fluid mass, is

RATE OF CHANGE OF MOMENTUM = APPLIED FORCE. (1.1)

Written for unit mass of a viscous fluid of density ρ, this becomes

$$\text{ACCELERATION} = \frac{-(\text{PRESSURE GRADIENT}) + \text{VISCOUS STRESS GRADIENTS}}{\rho}$$

$$+ \text{BODY FORCE PER UNIT MASS},\qquad(1.2)$$

where the minus sign arises simply because of our convention that pressure is positive inward, whereas viscous normal ("direct") stresses are positive outward. The viscosity μ is simply the ratio of the viscous (normal or shear) stress (see Section 2.1) to twice the corresponding rate of strain; for example, the normal stress acting in the x direction is $2\mu\,\partial u/\partial x$, and the shear stress acting in the x direction as a consequence of a simple shear $\partial u/\partial y$ is $\mu\,\partial u/\partial y$—or, more generally, $\mu(\partial u/\partial y + \partial v/\partial x)$, where u and v are the velocity components in the x and y directions, respectively. The factor 2 arises because it is logical to define rate of strain as the time derivative of the strain, as usually defined in solid mechanics, while viscosity was first defined as the ratio of shear stress to $\partial u/\partial y$ in a simple shear flow. Shear layers like the boundary layer shown in Fig. 1.1 or the jet and duct flows of Fig. 1.2 are the commonest examples of flows with large viscous-stress gradients, because it is only in a thin shear layer that a viscous stress —here $\mu\,\partial u/\partial y$—can rise rapidly from zero (in the "inviscid" external flows of Fig. 1.1) to a significant value, so as to create a large enough stress *gradient* to affect the acceleration according to Eq. (1.2). Below, we shall frequently refer to the shear stress acting in the x direction in a thin shear layer simply as the *shear stress*, with symbol τ, that is,

$$\tau = \mu \frac{\partial u}{\partial y}.\qquad(1.3)$$

The viscosity μ and kinematic viscosity $\nu \equiv \mu/\rho$ depend on temperature (Figs. 1.3 and 1.4), and if there is a temperature gradient in the y direction, the gradient in the y direction of the viscous shear stress $\mu\,\partial u/\partial y$ is affected. An adequate approximation for the viscosity μ in *air* is Sutherland's law; in terms of the metric system and British units it is given by

$$\mu = \begin{cases} 1.45 \times 10^{-6} \dfrac{T^{3/2}}{T+110} \text{ kg m}^{-1}\text{s}^{-1} & (1.4a) \\[2ex] 2.270 \times 10^{-8} \dfrac{T^{3/2}}{T+198.6} \text{ lb}_f\text{-sec/ ft}^2 & (1.4b) \end{cases}$$

with T expressed in degrees Kelvin ($^\circ$K) and in degrees Rankine ($^\circ$R), respectively. Equation (1.4) can also be approximated as $\mu \propto T^{0.76}$ for temperatures near atmospheric (15°C).

In turbulent flows (see Section 2.3) momentum transfer by the unsteady, eddying motion is added to (viscous) momentum transfer by molecular collisions. Analogies between eddy motion and molecular motion can be highly misleading, but it is traditional and convenient to regard the extra

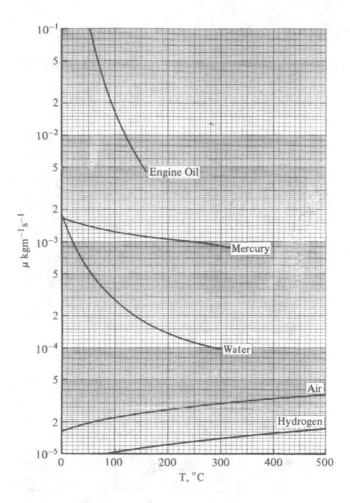

Figure 1.3. Variation of dynamic viscosity of several fluids with temperature: even for gases, μ is nearly independent of pressure.

momentum transfer rate, averaged over time, as equivalent to extra apparent stresses, so that the second law of motion for the time-averaged motion in turbulent flow satisfies Eq. (1.2) if turbulent stress gradients are added to viscous stress gradients. Turbulent stresses are simply related to the time-average products of the fluctuating parts of the velocity components, and equations for turbulent stresses can be obtained as weighted time averages of the original equations of motion; however, the turbulent stress equations contain more complicated kinds of time averages of the fluctuations, and the only way to achieve a closed set of equations (number of equations equal to number of unknowns) is to use empirical data about the time-averaged behavior of the turbulence. Thus the calculation of turbulent flow requires approximate empirical data, in addition to any approximations that may be made for computational convenience—as they might be

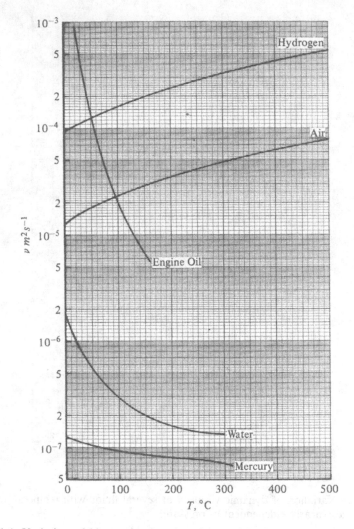

Figure 1.4. Variation of kinematic viscosity of several fluids with temperature: for gases $\nu \equiv \mu/\rho$ is directly proportional to pressure.

even in purely-viscous laminar "flow." Time averaging is a computational convenience, to which we are forced because current computers are not powerful enough to solve the time-dependent equations of motion for the extremely complicated velocity-fluctuation patterns of turbulence. Most fluid flows of engineering interest are turbulent; so we need calculation methods that give reasonable accuracy at reasonable cost—the definition of the word "reasonable" varies according to the problem concerned. Because turbulent-flow calculations are, in effect, laminar-flow calculations with additional contributions to the stresses, it is convenient to discuss laminar-flow problems first and then proceed to turbulent flows.

1.2 Heat and Mass Transfer

"Heat is that which transfers from one system to a second system at lower temperature, by virtue of the temperature difference, when the two are brought into communication.... Heat, like work, is a transitory quantity; it is never contained *in* a body."

This quotation, from Keenan's *Thermodynamics* [2], explains why "heat" appears in the study of heat transfer only as a name. The properties for which we write our quantitative "conservation" equations, analogous to the momentum-conservation equations, are the thermal internal energy per unit mass, e, and the enthalpy per unit mass, h. Heat transfer to a body increases its thermal energy, just as doing work on a body increases its momentum and kinetic energy. Although e and h have the dimensions of energy per unit mass, it is the analogy between heat (thermal energy) transfer and *momentum* transfer—rather than kinetic-energy transfer—that is most valuable. Note that it is customary to refer to the conservation equation from which the fluid temperature is calculated as the "energy" equation—usually it is in fact the enthalpy equation, or the "total enthalpy" equation for the sum of the enthalpy and the kinetic energy, rather than the internal energy equation. In low-speed flows with constant specific heat, enthalpy is just internal energy times the ratio of specific heats, the kinetic energy contribution to total enthalpy is negligible, and the equations for the three quantities are effectively the same.

The first law of thermodynamics, applicable to a solid mass or a fluid mass, is

$$
\boxed{\begin{array}{c}\text{RATE OF CHANGE}\\ \text{OF ENERGY}\end{array}} = \boxed{\begin{array}{c}\text{RATE OF WORK DONE}\\ \text{BY EXTERNAL FORCES}\end{array}} + \boxed{\begin{array}{c}\text{RATE OF}\\ \text{HEAT ADDITION.}\end{array}}
$$

$$(1.5)$$

Here, "energy" includes potential energy and kinetic energy as well as thermal internal energy (for the moment we ignore energy changes due to combustion or other chemical reaction). An equation for thermal internal energy alone can be obtained by subtracting the equations for kinetic energy and the trivial "hydrostatic" equation for potential energy. In this book we shall ignore potential energy. Also, it will be shown below that in low-speed flow with heat transfer, changes in kinetic energy are negligible compared with changes in thermal internal energy, and the work term in Eq. (1.5) is usually negligible compared with the heat addition term. Written for unit mass of a conducting fluid of density ρ, Eq. (1.5) becomes

$$
\boxed{\begin{array}{c}\text{RATE OF CHANGE}\\ \text{OF SPECIFIC}\\ \text{INTERNAL ENERGY}\end{array}} = \boxed{\begin{array}{c}\text{RATE OF WORK}\\ \text{DONE PER}\\ \text{UNIT MASS}\end{array}} - \frac{1}{\rho} \boxed{\begin{array}{c}\text{CONDUCTIVE}\\ \text{HEAT-FLUX}\\ \text{GRADIENTS}\end{array}}, \quad (1.6)
$$

where the minus sign arises before the last term because a positive heat-flux gradient in the upward vertical direction (say) means that more heat is being lost from the top of a body than is being gained at the bottom. The thermal conductivity k is the ratio of the conductive heat-flux rate to (minus) the corresponding temperature gradient; for example, the heat-flux rate in the y direction, $\dot{q}_y$ in the usual notation, is given for any simple fluid by the *thermal conductivity* (or *heat-conduction*) *law*

$$\dot{q}_y = - k \frac{\partial T}{\partial y}. \tag{1.7}$$

Heat fluxes have only one direction associated with them, whereas stresses have two (the direction of the stress and the direction of the normal to the surface on which it acts), but the final heat-flux terms in Eq. (1.6) are algebraically similar to the viscous-stress terms in Eq. (1.2), as we shall see in Chapter 2.

The mass-transfer equation corresponding to (1.6) is

RATE OF CHANGE OF MASS FRACTION OF SPECIES	=	RATE OF CREATION OF SPECIES PER UNIT MASS	$-\dfrac{1}{\rho}$	DIFFUSIVE MASS-FLUX GRADIENTS

$$\tag{1.8}$$

where the species is the contaminant substance whose mass is being considered (e.g., dye in water or tracer gas in air) and mass fraction is measured as mass of species per unit mass of mixture. The mass diffusivity D is simply the ratio of the diffusive mass-flux rate to (minus) the corresponding gradient of mass fraction: it depends on the contaminant *and* on the host fluid. Creation of species can take place by chemical reaction, and since reaction rates usually depend on the concentrations of both reactants and product, several equations like Eq. (1.8) may have to be solved, coupled by the reaction equation; some reactions, notably combustion, release heat, so that a heat-transfer equation has to be solved as well. Equation (1.8) is rather similar in appearance to Eq. (1.6) except for obvious differences of notation.

The time-average equations for heat or mass transfer in turbulent flow contain extra apparent transfer rates, simply related to the mean product of the temperature fluctuation and the velocity fluctuation in the direction of the transfer. As in the case of turbulent momentum transfer, exact equations for the transfer rates can be derived from weighted time averages of the original equations, but again empirical data are needed to obtain a closed, solvable system of equations.

It is important to recognize that the temperature-fluctuation field in heated turbulent flow is entirely a *consequence* of the velocity-fluctuation field, which mixes the fluid. If all velocity fluctuations were somehow eliminated at time $t = 0$, the spatial distribution of temperature would thereafter remain unaltered except for the effects of thermal conductivity and the mean motion.

1.3 Relations between Heat and Momentum Transfer

According to the simplest version of the kinetic theory of gases, heat and momentum are transferred or "diffused" in exactly the same way by molecular collisions. The quantity analogous to the viscosity μ is k/c_p; both have the dimensions mass/(length × time), so that $\mu c_p/k$ is dimensionless and, according to the simple kinetic theory of gases, equal to *unity*. The dimensionless quantity $\mu c_p/k$ is called the *Prandtl number* Pr:

$$\mathrm{Pr} \equiv \frac{\mu c_p}{k} = \frac{\nu}{k/\rho c_p}, \tag{1.9}$$

where ν is the *kinematic* viscosity μ/ρ and $k/(\rho c_p) \equiv \kappa$ is the *thermometric conductivity*, or *thermal diffusivity*. The actual value of Pr for diatomic gases is about 0.7, almost independent of temperature; the difference from unity is due to the importance of the vibrational and rotational modes of molecular motion, neglected in simple kinetic theory. The specific heat of a gas is also nearly constant; k varies with temperature in nearly the same way as μ (Fig. 1.5). In liquids, Pr falls rapidly with increasing temperature (see Fig. 1.5) because μ falls rapidly while k and c_p are nearly constant. In liquid metals Pr can be as small as 0.02 because metals, even when liquid, have high thermal conductivity; in water Pr falls from about 7 at room temperature to

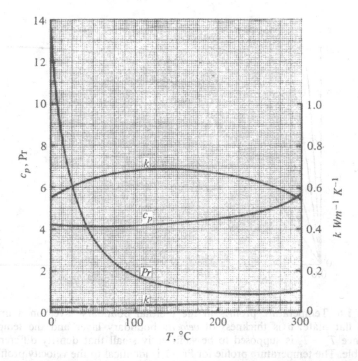

Figure 1.5. Variation with temperature of k for air (dashed line) and water, and of c_p and Pr for water. See Table 1.1 for units of k.

1.7 at the boiling point, whereas in lubricating oils, which have high viscosity, Pr can be 1000 or more.

Since Pr is the ratio of a fluid's ability to diffuse momentum to its ability to diffuse heat, the wide range of Pr found in common substances leads to a wide range of possible temperature distributions in a given flow configuration, according to the fluid being used. For example, as we shall see in Section 4.2, the thickness of a heated region of fluid (the thermal boundary layer) in constant-pressure laminar flow over a uniformly heated flat plate—a special case of the partly heated plate shown in Fig. 1.1—is roughly $Pr^{-0.34}$ times the thickness of the velocity boundary layer. Some numerical results are shown in Fig. 1.6. In this (very special) case, if, and only if, Pr = 1, the profile of the normalized temperature difference $(T_w - T)/(T_w - T_e)$ is identical with the profile of the normalized velocity u/u_e; that is,

$$\frac{T_w - T}{T_w - T_e} = \frac{u}{u_e} \tag{1.10}$$

for all y. Here subscripts w and e refer to the surface (wall) and the external flow, respectively. This very simple relation is found, even if Pr = 1, only in a few special cases where the boundary conditions on the momentum equation are analogous to those on the thermal energy equation. We see that there are important similarities, and important differences, between molecu-

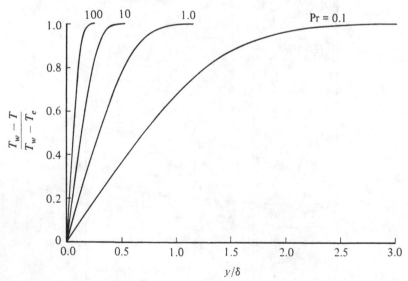

Figure 1.6. Temperature profiles in the thermal boundary layer on a uniformly heated flat plate: δ is thickness of *velocity* boundary layer and the temperature difference $T_w - T_e$ is supposed to be sufficiently small that density differences are negligible. The temperature profile for Pr = 1 is identical to the velocity profile u/u_e which is the same for any Pr if density differences are negligible.

lar transfer of heat and that of momentum. In turbulent flow the mechanisms of heat transfer and of momentum transfer are again *roughly* similar. (In Chapter 6 we shall define a "turbulent Prandtl number" that is almost always of order 1 and is at most only weakly dependent on the molecular Prandtl number Pr except near solid surfaces.)

As mentioned in Section 1.2, heat transfer and mass transfer obey equations that are identical, in simple cases, with thermal energy replaced by mass fraction and viscosity ν by diffusivity D. The analogy between mass and momentum transfer is therefore identical, or at least close, to the analogy between heat and momentum transfer. The ratio ν/D, analogous to the Prandtl number $\nu/(k/\rho c_p)$, is called the *Schmidt number* Sc. It depends on the host fluid and on the contaminant, whose mass is many times that of the fluid molecules (being about 700 for the diffusion of salt in water and of order 10^3–10^4 for the diffusion of smoke or dye particles) and which is therefore only weakly affected by collisions with the fluid molecules. The equations do not hold for particles so large that they fail to follow the macroscopic motion of the fluid, and therefore the study of large particles is outside the scope of this book (see Lumley [3] for an introductory discussion). However, nearly all the discussions of heat transfer in the following chapters can be applied directly to mass transfer, and the further extensions required in the case of two or more contaminants (such as fuel and combustion product in an oxidizing host fluid) are conceptually obvious, if computationally complicated.

The presence of pressure terms in Eq. (1.2) and other equations describing the velocity field and their absence from Eq. (1.6) and other equations describing the temperature field—except for work terms, which are small in low-speed flow—implies that an exact correspondence between momentum transfer and heat transfer can be found only in constant-pressure flow. Since the pressure in a turbulent flow is always a function of time, this means a restriction to *laminar* constant-pressure flow. An additional necessary condition is that the Prandtl number be unity; this condition is satisfied sufficiently by common gases (Pr ≈ 0.7), so that the case Pr $=1$ is of semiquantitative interest. The most restrictive condition of all is that the boundary conditions on the velocity and temperature must be analogous. For example, even if Pr $=1$, the velocity and temperature profiles in the laminar boundary layer on a heated flat plate, with freestream velocity u_e independent of x, are of the same shape *only* if the surface temperature is also independent of x (see Fig. 1.1 and Section 2.3). A few other laminar-flow cases of practical interest yield exact correspondences like Eq. (1.10). A laminar jet of slightly heated fluid with Pr $=1$, whose initial profiles of velocity and excess temperature are geometrically similar, will continue like this if the temperature of the stationary fluid into which the jet emerges is uniform. Note that although the shape of the velocity profiles depends on the ratio of jet speed to external-stream speed, the excess-temperature profile shape does not depend on the ratio of jet temperature to external-stream temperature as long as the latter ratio is near unity. This is because, if density and conductivity are constant, the thermal energy equations become linear and solutions can be added or arbitrarily factored.

Although the analogy between momentum transfer and heat transfer is usually not exact, it is always close enough to be of qualitative use, even in turbulent flow, especially in choosing methods for solving the heat-transfer equations. It is frequently close enough for a dimensionless property of the temperature field to be equated to the corresponding dimensionless property of the velocity field multiplied by an "analogy factor," to be found from experiment. The most common example, the *Reynolds analogy factor*, is the ratio of the surface heat-transfer parameter, called the *Stanton number* St, defined by

$$St = \frac{\dot{q}_w}{\rho c_p (T_w - T_e) u_e},$$ (1.11)

to half the local skin-friction coefficient defined by

$$c_f = \frac{\tau_w}{\frac{1}{2} \rho u_e^2}.$$ (1.12)

Here $\dot{q}_w$ and τ_w are the surface heat-transfer rate and surface shear stress, respectively. As a rule, the Reynolds analogy factor is expected to be fairly close to unity.[1] This suggests that the most difficult part of the task of predicting a given property of the temperature field is predicting the analogous property of the velocity field. This is a useful general rule: some exceptions exist (for example, the heat-transfer rate at a stagnation point, or at a point of reattachment downstream of a region of separated flow, is usually larger than elsewhere, although the surface shear stress is zero). However, the success of the analogy approach in most cases has discouraged the development of more refined methods of calculating heat transfer because the improvements in accuracy are thought not to justify the effort–which is sometimes, but not always, true.

In "constant-property" flow—nominally constant ρ and μ (see Section 1.4)—the velocity field can be calculated by the methods used in truly isothermal flows without considering the temperature field [1]. Then, the temperature field can be calculated by solving the heat-transfer equations, taking the velocity components as known. The linearity of the heat-transfer equations for constant-property flow permits solutions to be superimposed to satisfy the required boundary conditions. [In practice the velocity and temperature calculations proceed side by side to minimize computer storage space (Chapter 13).] The numerical methods used on the enthalpy equation, often referred to as the *energy* equation, are usually slight adaptations of those used on the momentum equation.

---

[1]Actually it is close to the reciprocal of the molecular or turbulent Prandtl number (empirically, $Pr^{-0.66}$ in laminar flow).

1.4 Coupled and Uncoupled Flows

In some heat-transfer problems, temperature differences are small compared with the absolute temperature (and pressure changes are small compared with the absolute pressure). Therefore, the changes in density, viscosity, and conductivity produced by the temperature differences are small enough to be *neglected* in the momentum and thermal energy equations, even though we are trying to calculate the thermal energy change produced by the same temperature differences. This is the *uncoupled-flow*, or *constant-property*, approximation, using the word "property" in its correct thermodynamic sense. It is used in the solutions shown in Fig. 1.6. Also, in buoyancy problems (see Chapter 9) we can often neglect density changes everywhere except in the buoyancy-force term itself; this is called the *Boussinesq approximation*, named for its originator. Degenerate versions of these approximations can also appear; in liquid flows, for example, the variation of viscosity with temperature may have to be taken into account, whereas changes in density and conductivity are usually negligible.

Note that significant variation of fluid properties with temperature means that the flow and heat transfer cannot be described entirely by the Reynolds number, Prandtl number, etc., evaluated at any single reference temperature. Consider boundary-layer flows of two different liquids over flat plates as in Fig. 1.1, with the same Reynolds number and Prandtl number based on fluid properties in the freestream (for example, benzene at room temperature has about the same Prandtl number as water at $50°C$). If the plates are heated, the Reynolds number and Prandtl number based on *surface* conditions will be different for the two different liquids, so that the two temperature fields cannot be completely similar, and simple results like Eq. (1.10) and its extension for $Pr \neq 1$ will no longer hold. If the constant-property approximation is not valid, the variation of fluid properties with absolute temperature must be specified in any calculation.

In heat transfer with large temperature differences, the temperature-field equations become nonlinear and are "coupled" to the velocity-field equations, because the viscosity (and in a gas flow, the density) depends on temperature. However, the coupling is usually weak and, in fact, the numerical solution techniques used are similar to those for linear equations. We shall see below that modern numerical methods for solving the heat- and momentum-transfer equations are easy to extend to coupled problems, although extra empirical data may be needed for turbulent flows. Therefore, the treatment of coupled problems is much easier than it used to be in the days when heat-transfer calculations were based on wholly empirical correlations of test data with a crude power-law dependence on absolute-temperature ratio.

The discussions of laminar and turbulent flows in later chapters begin with uncoupled problems and then pass on to coupled ones. A reader who has studied the chapters on uncoupled problems should have no difficulty with the coupled-flow chapters, whether the coupling arises because of "kinetic heating" in high-speed flows or because of large temperature differences in flows at low Mach number. It is because high-speed gas flows always involve internal heat transfer, even if solid surfaces are insulated that high-speed shear layers are discussed in this book in conjunction with strongly heated low-speed flows. In high-speed flow two extra coupling terms appear in the enthalpy equation, one being the enthalpy equivalent of the work done on the fluid by pressure gradients and the other the rate of dissipation of kinetic energy (mean or turbulent) into thermal internal energy by viscous stresses. Both of these are enthalpy sources within the fluid and have no analogies in the velocity-field equations (other than improbable distributions of body force). Combustion and other chemical reactions also lead to enthalpy sources within the fluid as well as to changes in fluid properties resulting from changes in chemical composition.

1.5 Units and Dimensions

Throughout the book we use metric units and adopt the SI system (Système International d'Unitès), which is based on the meter, kilogram, second, and kelvin, with the newton (N) as the unit of force and the newton-meter, or joule (J), as the unit of work, energy, and heat. The conversion factors needed to express units in the British system (BS) based on the foot, pound-mass (lb), pound-force (lb_f), second, and Fahrenheit degree, with the British thermal unit (Btu) as the unit of heat and Joule's equivalent 778 ft lb_f Btu^{-1} are given in Table 1.1. A more extensive table is given in Appendix A.[1] The tables provided in Appendix B indicate the physical properties for a representative sample of gases, liquids, and solids. Although this book is concerned with convection, information on the properties of solids is provided since the engineer frequently has to deal with heat transfer in a gas or liquid flowing over a solid body.

Table 1.1 Conversion factors for BS and SI units[a]

Quantity	BS	SI	Conversion factor
Specific heat, c_p	Btu $(lb \cdot °F)^{-1}$	kJ $(kg \cdot K)^{-1}$	4.1868
Thermal conductivity, k	Btu $(hft \cdot °F)^{-1}$	W $(m \cdot K)^{-1}$	1.7307
Dynamic viscosity, μ	lb_f s ft^{-2}	kg ms^{-1}	0.0462
	lb $(ft \cdot s)^{-1}$	kg ms^{-1}	1.4882
Kinematic viscosity, ν	$ft^2 s^{-1}$	$m^2 s^{-1}$	0.09290
Density, ρ	lb ft^{-3}	kg m^{-3}	16.023

[a]To convert from BS to SI units, multiply by the conversion factor.

[1] The authors are grateful to Prof. N. Ozisik for providing this appendix and Table B-3. The other tables in Appendix B are taken from Ref. 4.

1.6 Outline of the Book

Chapter 2 is a presentation and discussion of the partial differential equations governing fluid flow and heat transfer. These equations are complicated algebraically but are based on the principles of conservation of mass, momentum, and energy for a fluid with a well-behaved viscosity and thermal conductivity. After derivation of the equations for a general, variable-property coupled flow, we present special cases for uncoupled flows.

The number of possible configurations of flows with heat transfer is very large because:

The different kinds of boundary conditions for the temperature field are additional to, and at least as numerous as, the different kinds of boundary conditions for the velocity field.

Sources of heat may be present within the fluid, as a result of chemical reaction or kinetic heating, and their distributions can be much more complex than the distributions of momentum "sources" such as body forces or pressure gradients.

Buoyancy forces, due to density changes in the presence of a gravitational field, can themselves create flows that have no analog in constant-density cases.

In engineering practice, flows in which heat transfer is the primary interest often have complicated boundary conditions for the velocity field; examples include heat exchangers with fins or tubes protruding into the flow, and the several forms of injection-cooling device.

Fortunately, regions of high heat transfer in a fluid flow are usually fairly thin layers, associated with the regions of high momentum transfer that we know as *shear layers*. By far the most common kind of shear layer in heat-transfer problems is the boundary layer, because the engineer's interest is usually in the rate of heat transfer to or from a solid boundary. Therefore, the multiplicity of configurations found in heat-transfer problems does not imply a multiplicity of different phenomena; a good way of approaching the subject is by a study of convective heat transfer in boundary layers and other shear layers, and this is the subject of the present book. As well as this conceptual simplification, thin shear layers offer the mathematical and computational advantage that the conservation equations derived in Chapter 2 can be simplified by neglecting small terms. The resulting boundary-layer, or thin-shear-layer, equations are introduced in Chapter 3, and the remainder of the book is concerned largely with their solution for boundary conditions of practical importance. For those who wish to bypass the detailed derivations of the conservation equations presented in Chapter 2, a summary of the general equations is presented in Section 3.1 and forms the starting point for the derivation of the boundary-layer forms of the continuity, momentum, and thermal energy equations, which again are presented in full only for two-dimensional and axisymmetric flows, for simplicity. The chapter ends with a discussion of the ordinary differential equations obtained by integrating the boundary-layer equations across the shear layer.

Chapter 4, on laminar uncoupled boundary layers, begins with a discussion of the special cases in which velocity and/or temperature profiles at different downstream positions are geometrically similar. Although these cases are mostly rare in practice, they are useful in simple discussions of flow behavior, and the coordinate transformation that eliminates streamwise variations in "similar" flows can still be used to reduce streamwise variations in nonsimilar flows, which is a help in numerical solutions. After a discussion of nonsimilar flows in general, simple calculation methods for the velocity and temperature profile parameters are presented; they are based on the ordinary differential equations in x derived in Chapter 3 and are best regarded as empirical correlations of the properties of "exact" numerical solutions for laminar flow.

Chapter 5 discusses heat transfer in uncoupled laminar duct flows, i.e., duct flows in which the shear layers on the duct walls interact.

Chapter 6, on turbulent uncoupled boundary layers, explores the possibilities of limited profile similarity in this case and then discusses methods of predicting the apparent extra rates of transfer of momentum and heat by the turbulence. The simplest way of doing this is to define an apparent extra viscosity and conductivity, which is mathematically legitimate, and to correlate experimental data for these quantities in a form that can be used over a worthwhile range of flows.

Chapter 7 discusses heat transfer in uncoupled turbulent duct flows, and Chapter 8 discusses heat transfer in laminar and turbulent free shear flows.

Chapter 9 is concerned with buoyant flows. On the one hand, buoyancy may exert a small influence on, say, a boundary layer, leaving it still qualitatively recognizable as a boundary layer; on the other hand, buoyancy is entirely responsible for the creation of *free-convection* flows such as the plume that rises from a heat source in still air. As before, a qualitative discussion can be based on laminar flows, with an empirical extension to turbulent flows. Some buoyant flows, such as those created by free convection in confined spaces, have a very complicated form for which numerical solution is difficult even if the flow is laminar. However, a discussion of free-convection shear layers is still useful as an introduction to the subject.

Chapter 10 considers coupled laminar flows (in the absence of buoyancy) where the variation in fluid properties results from large changes in temperature, due either to large rates of heat transfer in low-speed flows or to kinetic heating in high-speed flows where the rate of dissipation of kinetic energy into thermal internal energy is appreciable. In special cases the equations and their solutions can be reduced, by transformation of variables, to the constant-property form, which is useful for exposition, but more general solutions involve simultaneous numerical solution of the full equations for the velocity and temperature fields. However, the same numerical methods can be used in low-speed flows and high-speed flows.

Chapter 11 treats coupled turbulent flows, on the assumption—usually satisfied—that density fluctuations do not directly affect the turbulence behavior. This enables the similarity analysis and empirical formulas of Chapter 6 to be extended, and although high-speed flows and low-speed flows with large temperature differences now differ in principle because of the presence of large pressure fluctuations in the former, no distinction need be made in practice because the effect of the density fluctuations associated with the pressure fluctuations is negligible.

Chapter 12 discusses heat transfer in laminar and turbulent coupled duct flows, and finally Chapters 13 and 14 present details of the numerical methods that can be used to solve the velocity-field and heat-field equations in full partial differential form. In the latter two chapters, it will be shown that the same method can be used effectively for coupled and uncoupled flows and that at least with some models of turbulence behavior the same method is applicable to laminar and turbulent flows if the extra turbulent transfer rates are included.

Students concerned only with uncoupled flows can omit Chapters 9 to 12, but they should realize that flows with buoyancy effects or other forms of coupling between the velocity and temperature fields are an important part of engineering practice. All students should realize that numerical solution techniques like those presented in Chapters 13 and 14 are now an essential part of engineering practice and have displaced many of the simplified empirical formulas that had to be used in the past to predict momentum and heat transfer.

Problems

1.1 Calculate the Prandtl number of fluids with the following properties (a) a liquid metal with $\rho = 879$ kg m^{-3}, $\mu = 3.3 \times 10^{-4}$ kg m^{-1} s^{-1}, $\kappa = 66.5 \times 10^{-6}$ m^2 s^{-1}; (b) a gas with $c_p = 0.2415$ Btu lb^{-1}°F^{-1}, $\mu = 5.193 \times 10^{-2}$ lb ft^{-1}h^{-1}, $k = 1.806 \times 10^{-2}$ Btu h^{-1} ft^{-1}°F^{-1}; (c) a liquid with $\rho = 900$ kg m^{-3}, $c_p = 1.902$ kJ kg^{-1} °K^{-1}; $\nu = 9 \times 10^{-4}$ m^2 s^{-1}, $k = 0.143$ W m^{-1} °K^{-1}.

1.2 Calculate the Reynolds number for the following cases: (a) air with a freestream velocity of 100 ft s^{-1}, a pressure of 14.7 lb in^{-2}, and a temperature of 150°F flowing over a 6-inch long flat plate; (b) water with an average velocity of 12 m s^{-1} and an average temperature of 25°C flowing in a circular duct of 0.25 m in diameter; (c) glycerin at 30°C with a velocity of 10 m s^{-1} at a distance of 1 m from the leading edge of a flat plate.

1.3 In problem 1.2(a), if the local skin-friction coefficient c_f is given by the formula,

$$c_f = \frac{0.664}{\sqrt{R_x}}$$

where $R_x = u_e x / \nu$, find the (a) wall shear, τ_w, (b) wall heat flux, $\dot{q}_w$. Take $T_w - T_e = 10$°F.

References

[1] Cebeci, T. and Bradshaw, P.: *Momentum Transfer in Boundary Layers*. Hemisphere, Washington, DC, 1977.

[2] Keenan, J. H.: *Thermodynamics*. John Wiley, New York, 1941.

[3] Lumley, J. L.: In *Topics in Applied Physics*, **12**, Turbulence (P. Bradshaw, ed.). Springer-Verlag, Berlin, 1977.

[4] Eckert, E. R. G. and Drake, R. M.: *Analysis of Heat and Mass Transfer*. McGraw-Hill, New York, 1972.

CHAPTER 2

Conservation Equations for Mass, Momentum, and Energy

This chapter presents derivations of the differential equations that, with corresponding boundary conditions, describe convective heat transfer processes. Since convective heat transfer always involves transfer of mass and momentum, the derivations of the corresponding equations are also presented and serve as an introduction to the heat-transfer equations, which are conceptually rather similar. The derivations in Sections 2.1 to 2.3 lead to equations that represent conservation of mass, momentum, and energy—including thermal energy—for unsteady two-dimensional flows. These derivations use control-volume analysis, together with the laws for heat- and momentum-flux rates in a viscous conducting fluid that were introduced in Chapter 1. The equations for three-dimensional flows contain extra terms but no new principles (see Problems 2.1 and 2.4). Since most practical cases of convective heat transfer involve turbulent flow, the usual decomposition of the velocity and fluid properties into mean and fluctuating quantities, with subsequent averaging of the equations, is described in Section 2.4.

The equations representing conservation of mass ("continuity"), momentum, and energy (or enthalpy) equations can be obtained by analyzing the properties of the flow into and out of an infinitesimal control volume (CV) such as that of Fig. 2.1. The lengths of the sides dx and dy are taken small enough for quantities of order dx^2 or dy^2 to be neglected. In two-dimensional flow the value of dz is immaterial. The conservation principle for any conserved quantity Q can be written for the CV in the form

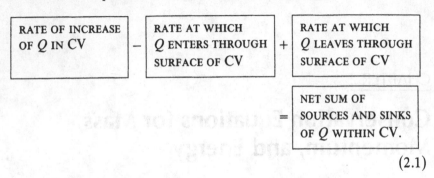

$$(2.1)$$

2.1 Continuity Equation

To obtain the continuity equation from Eq. (2.1), we choose the control volume as shown in Fig. 2.1 and set $Q = \text{mass} \equiv \text{density} \times \text{volume}$. Since there are no sources or sinks of mass, the last term in Eq. (2.1) is zero, and the equation states that the net rate at which mass is entering the CV equals the rate at which the mass of fluid within the CV is increasing. The mass of the fluid in the CV is $\rho\,dx\,dy\,dz$, and its rate of increase with respect to time is

$$\frac{\partial \rho}{\partial t}\,dx\,dy\,dz,$$

which is of course zero in constant-density flow. The rate of flow of mass through the face normal to the x direction whose center is P is equal to the product of the density, the component of velocity normal to the face, and the area of the face, namely $\rho u\,dy\,dz$. The corresponding rate of flow of mass out of the parallel face whose center is Q is $(\rho u + \partial(\rho u)/\partial x \cdot dx)\,dy\,dz$, and so the *net* rate of flow of mass out of the CV through the faces whose centers are P and Q is

$$\frac{\partial \rho u}{\partial x}\,dx\,dy\,dz.$$

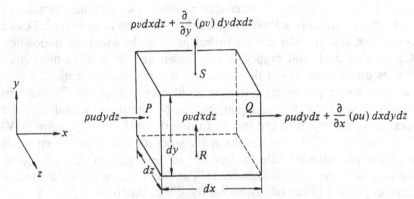

Figure 2.1. Infinitesimal control volume for derivation of conservation of mass.

Similarly the net rate of flow of mass out of the CV through the faces normal to the y axis is

$$\frac{\partial \rho v}{\partial y} dx\, dy\, dz.$$

Inserting these expressions into Eq. (2.1), we get

$$\frac{\partial \rho}{\partial t} dx\, dy\, dz = -\left(\frac{\partial \rho u}{\partial x} + \frac{\partial \rho v}{\partial y}\right) dx\, dy\, dz.$$

Dividing both sides by the volume of the CV, we obtain the mass-conservation equation (*continuity equation*) for a two-dimensional unsteady compressible flow,

$$\frac{\partial \rho}{\partial t} + \frac{\partial \rho u}{\partial x} + \frac{\partial \rho v}{\partial y} = 0. \qquad (2.2a)$$

Equation (2.2a) shows that if the density is constant, the continuity equation reduces to

$$\frac{\partial u}{\partial x} + \frac{\partial v}{\partial y} = 0 \qquad (2.2b)$$

in steady *or* unsteady flow. For a steady compressible flow, Eq. (2.2a) reduces to

$$\frac{\partial \rho u}{\partial x} + \frac{\partial \rho v}{\partial y} = 0. \qquad (2.2c)$$

2.2 Momentum Equations

The momentum-conservation equations, known as the *Navier – Stokes equations*, can be derived by similar use of CV concepts. Since momentum is a vector, there are three momentum equations. To derive (say) the x component of the momentum equation, we make use of Newton's second law of motion (the principle of conservation of momentum) and for the CV of Fig. 2.1 write, in the formalism of Eq. (2.1),

RATE OF INCREASE OF x-COMPONENT MOMENTUM OF FLUID IN CV	−	RATE OF FLOW OF x-COMPONENT MOMENTUM INTO CV	+	RATE OF FLOW OF x-COMPONENT MOMENTUM OUT OF CV	=	SUM OF x-COMPONENTS OF FORCES APPLIED TO FLUID IN CV.

$$(2.3)$$

The x component of momentum of the fluid in the CV is just u times the mass of the fluid in the CV, that is, $\rho u\, dx\, dy\, dz$. Therefore,

$$\boxed{\begin{array}{l}\text{RATE OF INCREASE}\\ \text{OF } x\text{-COMPONENT}\\ \text{MOMENTUM OF FLUID}\\ \text{IN CV}\end{array}} = \frac{\partial}{\partial t}(\rho u)\, dx\, dy\, dz. \tag{2.4}$$

The rate of flow of x-component momentum into the CV through the face whose center is R (perpendicular to the y direction) is the x-component momentum per unit mass, u, times the rate of mass flow through the face, $\rho v\, dx\, dz$, that is,

$$\rho uv\, dx\, dz.$$

Note the distinction between the direction of the momentum component and the direction of the component of mass flow rate that transports the momentum, x and y, respectively, in this case. The rate of flow of x-component momentum out of the opposite face (center S) is

$$\left[\rho uv + \frac{\partial}{\partial y}(\rho uv)\right] dx\, dy\, dz,$$

and so the net rate of flow of x-component momentum out of the CV via this pair of faces is

$$\frac{\partial}{\partial y}(\rho uv)\, dx\, dy\, dz.$$

We can also write expressions similar to this for the net rate of flow of x-component momentum out of the CV via the pair of faces containing the points P or Q. The sum of the net rates of outflow of x-component momentum is

$$\boxed{\begin{array}{l}\text{RATE OF FLOW}\\ \text{OF } x\text{-COMPONENT}\\ \text{MOMENTUM OUT}\\ \text{OF CV}\end{array}} - \boxed{\begin{array}{l}\text{RATE OF FLOW}\\ \text{OF } x\text{-COMPONENT}\\ \text{MOMENTUM INTO}\\ \text{CV}\end{array}}$$

$$= \left[\frac{\partial}{\partial x}(\rho u^2) + \frac{\partial}{\partial y}(\rho uv)\right] dx\, dy\, dz. \tag{2.5}$$

The forces acting on the fluid in the CV are of two types, body forces and surface forces. The simplest example of a body force is gravity; the fluid in the CV experiences a gravitational force equal to g times its mass, $\rho g\, dx\, dy\, dz$. Clearly, a body force is a vector; for a two-dimensional flow in the xy plane it has two components, which we call f_x, f_y per unit mass. Surface forces (i.e., forces on the imaginary surfaces of the CV) arise because of molecular stresses in the fluid. One kind of molecular stress is the pressure, present even in a fluid at rest, and it is normal to the surface on which it acts. If a fluid element changes shape or size with time, viscosity leads to the presence of further stresses that may act normal to a surface or tangentially (shear stress). We define the different components of normal stress and shear stress, *excluding* the pressure, as shown in Fig. 2.2: the first subscript to the symbol σ represents the direction of the stress, and the second the direction of the surface normal. By convention, an *outward* normal stress acting on the fluid in the CV is positive, and the shear stresses are taken as positive on the faces furthest from the origin of the coordinates. Thus σ_{xy} acts in the *positive* x direction on the visible (upper) face perpendicular to the y axis; a corresponding shear stress acts in the *negative* x direction on the invisible (lower) face perpendicular to the y axis. For a variable-density "Newtonian" viscous fluid, the normal viscous stresses σ_{xx}, σ_{yy} and shear stresses σ_{xy}, σ_{yx} are given by the rather complicated formulas:

$$\sigma_{xx} = 2\mu\frac{\partial u}{\partial x} + \left(\beta - \frac{2}{3}\mu\right)\left(\frac{\partial u}{\partial x} + \frac{\partial v}{\partial y}\right), \tag{2.6a}$$

$$\sigma_{yy} = 2\mu\frac{\partial v}{\partial y} + \left(\beta - \frac{2}{3}\mu\right)\left(\frac{\partial u}{\partial x} + \frac{\partial v}{\partial y}\right), \tag{2.6b}$$

$$\sigma_{xy} = \sigma_{yx} = \mu\left(\frac{\partial u}{\partial y} + \frac{\partial v}{\partial x}\right). \tag{2.6c}$$

Here β is the bulk viscosity, producing a viscous normal stress in any direction proportional to the rate of dilatation (rate of increase of volume) of the fluid. The factor $-2\mu/3$ is inserted in σ_{xx}, σ_{yy}, and σ_{zz} so that the sum of these three stresses is zero; by convention, the sum of normal stresses due to molecular motion is taken as part of the pressure and not part of the viscous stress. In a constant-density fluid, because of Eq. (2.2b), normal stresses reduce to

$$\sigma_{xx} = 2\mu\frac{\partial u}{\partial x}, \qquad \sigma_{yy} = 2\mu\frac{\partial v}{\partial y}. \tag{2.6d}$$

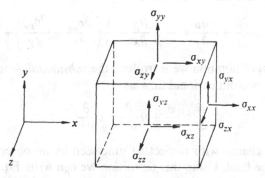

Figure 2.2. Definitions of stress components (excluding pressure) applied to faces of control volume by surrounding fluid. Force components are stress components multiplied by areas of corresponding faces.

However, the shear-stress formulas remain the same.

The net force in the x direction applied to the CV of Fig. 2.1 by the pressure is the difference in pressure between the face containing the point P and that containing Q, multiplied by the area of either face, namely

$$-\left(\frac{\partial p}{\partial x} dx\right) dy\, dz,$$

the minus sign arising because, by definition, a positive pressure acts inward. Only a pressure *gradient*, not a constant pressure, produces a net force on an elementary CV. The faces with centers R and S do not contribute because the pressure on them is normal to the x direction.

By the same argument as that just used for the pressure, the net force in the x direction applied by the normal stresses σ_{xx} and σ_{xy} is found to be

$$\left(\frac{\partial \sigma_{xx}}{\partial x} + \frac{\partial \sigma_{xy}}{\partial y}\right) dx\, dy\, dz.$$

Therefore,

$$\boxed{\begin{array}{l}\text{SUM OF } x \text{ COMPONENTS}\\ \text{OF FORCES APPLIED}\\ \text{TO FLUID IN CV}\end{array}} = \left(-\frac{\partial p}{\partial x} + \frac{\partial \sigma_{xx}}{\partial x} + \frac{\partial \sigma_{xy}}{\partial y} + \rho f_x\right) dx\, dy\, dz.$$

$$(2.7)$$

Inserting Eqs. (2.4), (2.5), and (2.7) into Eq. (2.3) and dividing by $dx\, dy\, dz$, we get the *x-component momentum equation* for unsteady two-dimensional variable-density flow:

$$\frac{\partial \rho u}{\partial t} + \frac{\partial}{\partial x}\rho u^2 + \frac{\partial}{\partial y}\rho uv = -\frac{\partial p}{\partial x} + \frac{\partial \sigma_{xx}}{\partial x} + \frac{\partial \sigma_{xy}}{\partial y} + \rho f_x. \qquad (2.8)$$

Multiplying the continuity equation, Eq. (2.2a), by u and subtracting it from Eq. (2.8) gives, after dividing by the density ρ, an *alternative* version of the x-component momentum equation:

$$\frac{\partial u}{\partial t} + u\frac{\partial u}{\partial x} + v\frac{\partial u}{\partial y} = -\frac{1}{\rho}\frac{\partial p}{\partial x} + \frac{1}{\rho}\left(\frac{\partial \sigma_{xx}}{\partial x} + \frac{\partial \sigma_{xy}}{\partial y}\right) + f_x. \qquad (2.9)$$

For compactness of notation we introduce the *substantial derivative* defined for an unsteady two-dimensional flow as

$$\frac{d}{dt} \equiv \frac{\partial}{\partial t} + u\frac{\partial}{\partial x} + v\frac{\partial}{\partial y}. \qquad (2.10)$$

It is the rate of change with respect to time seen by an observer following the motion of the fluid. Using this definition we can write Eq. (2.9) as

$$\frac{du}{dt} = -\frac{1}{\rho}\frac{\partial p}{\partial x} + \frac{1}{\rho}\left(\frac{\partial \sigma_{xx}}{\partial x} + \frac{\partial \sigma_{xy}}{\partial y}\right) + f_x. \qquad (2.11)$$

Note that Eqs. (2.9) and (2.11) are still valid for variable- or constant-density flow. Note also that du/dt is the x component of the acceleration, the rate of change of velocity of a small element of moving fluid. Any equation whose left-hand side can be written as $d\phi/dt$ is called a *transport* equation for the quantity ϕ. As shown in Problem 2.1, the continuity equation, Eq. (2.2a), is a transport equation for ρ.

A similar equation for the y component of momentum can be derived in the same way as Eq. (2.8) (see Problem 2.2), and the resulting expression can be written in a form similar to Eq. (2.11),

$$\frac{dv}{dt} = -\frac{1}{\rho}\frac{\partial p}{\partial y} + \frac{1}{\rho}\left(\frac{\partial \sigma_{yx}}{\partial x} + \frac{\partial \sigma_{yy}}{\partial y}\right) + f_y. \tag{2.12}$$

When the general symbol σ is used for stresses, the above equations are valid for any fluid whatever its stress-strain relationship. The σ terms for a fluid of variable viscosity and density are complicated, but in constant-property flows Eqs. (2.11) and (2.12) become, using the continuity equation,

$$\frac{du}{dt} = -\frac{1}{\rho}\frac{\partial p}{\partial x} + \frac{\mu}{\rho}\left(\frac{\partial^2 u}{\partial x^2} + \frac{\partial^2 u}{\partial y^2}\right) + f_x, \tag{2.11a}$$

$$\frac{dv}{dt} = -\frac{1}{\rho}\frac{\partial p}{\partial y} + \frac{\mu}{\rho}\left(\frac{\partial^2 v}{\partial x^2} + \frac{\partial^2 v}{\partial y^2}\right) + f_y. \tag{2.12a}$$

In two-dimensional flow the third equation is $dw/dt = 0$.

2.3 Internal Energy and Enthalpy Equations

Internal (Kinetic and Thermal) Energy Equation

To derive the internal energy equation of fluid flow, we start with the first law of thermodynamics. Recalling that, if we neglect potential energy for simplicity, the internal energy per unit mass of the fluid consists of the sum of the kinetic energy $(u^2 + v^2)/2$ and the *thermal* internal energy $e \equiv c_v T$, we can write the first law as

$$dE = dQ + dW. \tag{2.13}$$

Here dE is the increment in internal (kinetic plus thermal) energy of the system, dQ is the heat transfer *to* the system, and dW is the work done *on* the system.

We next write Eq. (2.13) in substantial derivative form, with d/dt defined by Eq. (2.10), so that it applies to transport of E by a moving system,

$$\frac{dE}{dt} = \frac{dQ}{dt} + \frac{dW}{dt}, \tag{2.14}$$

and apply it to a fluid element (a small mass of moving fluid which can legitimately be regarded as a thermodynamic system), so that, in words,

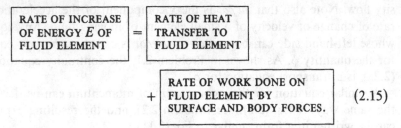

$$(2.15)$$

Also for convenience, we consider a fixed control volume rather than a moving fluid element [in effect, this merely involves replacing the substantial derivative d/dt by its definition, Eq. (2.10)] and write Eq. (2.15) as

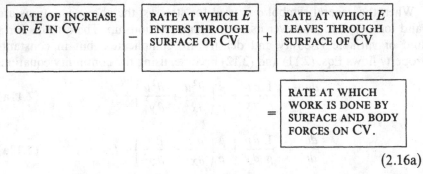

$$(2.16a)$$

Equation (2.16a) is a form of Eq. (2.1) with E as the conserved quantity. In deriving the momentum equation, we adopted the usual convention that viscous effects are equivalent to stresses; however, they are in fact net rates of transport of momentum by molecular agencies and, so regarded, would appear on the left-hand side of Eq. (2.3) instead of in the force terms on the right. On the other hand, molecular transport of heat (the effect of thermal conductivity) is always regarded as transport and is therefore at first allocated to the second and third terms on the left-hand side of Eq. (2.16a). Just as in the derivation of the momentum equation, it is convenient to distinguish transport by the *flow* from transport by molecular agencies, and we therefore (1) refer to conductive transport of heat as *heat transfer* and (2) allocate it terms of its own in Eq. (2.16a), which becomes

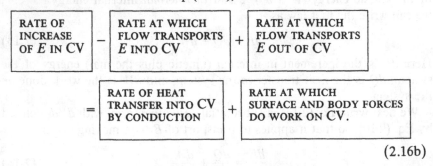

$$(2.16b)$$

We now consider the terms in Eq. (2.16b), in order, for the CV shown in Fig. 2.3.

The derivation of the left-hand side of Eq. (2.16b) follows procedures similar to those used in the derivation of the left-hand side of the x-component momentum equation, Eq. (2.8). Taking E as the energy per unit mass, we have

$$\boxed{\begin{array}{c} \text{RATE OF INCREASE} \\ \text{OF } E \text{ IN CV} \end{array}} = \frac{\partial \rho E}{\partial t} dx\, dy\, dz \qquad (2.17)$$

and the net rate of transport of energy out of the CV by the flow (see Fig. 2.3) is

$$\boxed{\begin{array}{c} \text{RATE AT WHICH} \\ \text{FLOW TRANSPORTS} \\ E \text{ OUT OF CV} \end{array}} - \boxed{\begin{array}{c} \text{RATE AT WHICH} \\ \text{FLOW TRANSPORTS} \\ E \text{ INTO CV} \end{array}}$$

$$= \left[\frac{\partial}{\partial x}(\rho u E) + \frac{\partial}{\partial y}(\rho v E) \right] dx\, dy\, dz. \qquad (2.18)$$

The left-hand sides of Eqs. (2.17) and (2.18) sum to give the left-hand side of Eq. (2.16b); with the use of the continuity equation, Eq. (2.2a), and with the definition of d/dt in Eq. (2.10), this left-hand side can be written as

$$\rho\, dx\, dy\, dz\, \frac{d}{dt}\left(e + \frac{u^2 + v^2}{2} \right). \qquad (2.19)$$

The first term on the right of Eq. (2.16b) represents heat transfer into the control volume by conduction. If $\dot{\mathbf{q}}$ denotes the heat flux per unit area per unit time, with components $\dot{q}_x$ and $\dot{q}_y$ in the x and y directions, the *net* rate of heat transfer into the control volume can be derived, applying standard

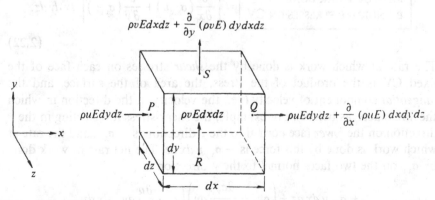

Figure 2.3. CV showing rate of transport of energy through the faces normal to the x- and y-axes.

CV arguments used previously for the σ stresses, as

$$\boxed{\begin{array}{l} \text{RATE OF HEAT} \\ \text{TRANSFER TO CV} \end{array}} = -\left(\frac{\partial \dot{q}_x}{\partial x} + \frac{\partial \dot{q}_y}{\partial y}\right) dx\, dy\, dz, \qquad (2.20)$$

where the minus sign arises because heat transfer is counted as positive in the positive coordinate direction.

We now consider the surface and body force terms of Eq. (2.16).

The rate at which the pressure does work on one side of a *moving* flat surface in the fluid is the product of the pressure, the area of the surface, and the normal component of velocity—a special case of the general rule *rate of doing work = force $\times$ velocity*. In control-volume terms, the rate at which work is done on fluid that enters the control volume through the left-hand face normal to the x direction is $pu\,dy\,dz$, the rate at which work is done on fluid that leaves by the other face normal to the x direction is

$$-\left(p + \frac{\partial p}{\partial x} dx\right)\left(u + \frac{\partial u}{\partial x} du\right) dy\, dz,$$

and (to first order in small quantities) the net pressure work associated with flow through the two faces normal to the x direction is

$$-\frac{\partial}{\partial x}(pu)\, dx\, dy\, dz.$$

Adding the corresponding expression for the net pressure work on the other two pairs of faces gives

$$\boxed{\begin{array}{l} \text{RATE OF WORK DONE} \\ \text{BY PRESSURE FORCES ON CV} \end{array}} = -\left[\frac{\partial}{\partial x}(pu) + \frac{\partial}{\partial y}(pv)\right] dx\, dy\, dz.$$

$$(2.21)$$

The rate of work done by the normal stresses σ_{xx} and σ_{yy} can be obtained similarly, remembering that σ is positive outward from the CV (tensile stress: see Fig. 2.2) while p is positive inward. It is

$$\boxed{\begin{array}{l} \text{RATE OF WORK DONE} \\ \text{BY NORMAL STRESSES ON CV} \end{array}} = \left[\frac{\partial}{\partial x}(\sigma_{xx}u) + \frac{\partial}{\partial y}(\sigma_{yy}v)\right] dx\, dy\, dz.$$

$$(2.22)$$

The rate at which work is done by the *shear* stresses on each face of the fixed CV is the product of the stress, the area of the surface, and the *tangential* component of velocity (i.e., the velocity in the direction in which the stress acts, as before). For example, the shear-stress force acting in the x direction on the *lower* face normal to the y direction is $-\sigma_{xy}$, and the rate at which work is done by this force is $-\sigma_{xy}u\,dx\,dz$. The net rate of work done by σ_{xy} on the two faces normal to the y direction is

$$-\sigma_{xy}u\,dx\,dz + \left(\sigma_{xy} + \frac{\partial \sigma_{xy}}{\partial y} dy\right)\left(u + \frac{\partial u}{\partial y} dy\right) dx\, dz$$

or, to first order in small quantities,

$$\frac{\partial}{\partial y}(\sigma_{xy}u)\,dx\,dy\,dz.$$

The other term contributing to shear-stress work in two-dimensional flow, which is attributable to the stress σ_{yx} acting on the pair of faces whose area is $dy\,dz$, is

$$\frac{\partial}{\partial x}(\sigma_{yx}v)\,dx\,dy\,dz.$$

Finally the rate at which work is done by the components of the body force $\mathbf{f}$ per unit mass is, applying once more the rule rate of doing work = force × velocity,

$$\boxed{\begin{array}{l}\text{RATE OF WORK DONE}\\ \text{BY BODY FORCES ON CV}\end{array}} = (uf_x + vf_y)\rho\,dx\,dy\,dz. \qquad (2.23)$$

We can now rewrite the internal energy "transport" equation (2.16b) by adding up the terms given by Eqs. (2.19)–(2.23) and dividing by $\rho\,dx\,dy\,dz$ to get

$$\frac{d}{dt}\left[e + \frac{1}{2}(u^2 + v^2)\right] = -\frac{1}{\rho}\left(\frac{\partial \dot{q}_x}{\partial x} + \frac{\partial \dot{q}_y}{\partial y}\right) - \frac{1}{\rho}\left[\frac{\partial}{\partial x}(pu) + \frac{\partial}{\partial y}(pv)\right]$$

$$+ uf_x + vf_y + \frac{1}{\rho}\frac{\partial}{\partial x_j}(\sigma_{ij}u_i), \qquad (2.24)$$

where the tensor notation of the last term can be expanded (with $i, j = 1, 2$ for two-dimensional flow) as the sum of the terms derived above,

$$\frac{\partial}{\partial x_j}\sigma_{ij}u_i \equiv \frac{\partial}{\partial x}(\sigma_{xx}u) + \frac{\partial}{\partial y}(\sigma_{xy}u) + \frac{\partial}{\partial x}(\sigma_{yx}v) + \frac{\partial}{\partial y}(\sigma_{yy}v). \quad (2.25)$$

Note that the complicated expression given by Eq. (2.24) is just an expression of the first law of thermodynamics, Eq. (2.13), for a fluid with internal stresses and heat transfer represented by the σ and q terms.

To obtain an equation for *thermal* internal energy e alone, we have to subtract the kinetic energy equation, whose left-hand side is $d/dt[\frac{1}{2}(u^2 + v^2)]$, from Eq. (2.24) for the thermal-plus-kinetic energy. The kinetic energy equation can be obtained from the momentum equations by straightforward algebra without further physical arguments: the derivation is set as Problem 2.6, and the kinetic energy equation is given as Eq. (P2.3)—see problems at end of chapter. Subtraction of Eq. (P2.3) from Eq. (2.24) gives the transport equation for *thermal* internal energy

$$\frac{de}{dt} = -\frac{1}{\rho}\left(\frac{\partial \dot{q}_x}{\partial x} + \frac{\partial \dot{q}_y}{\partial y}\right) - \frac{p}{\rho}\left(\frac{\partial u}{\partial x} + \frac{\partial v}{\partial y}\right) + \frac{\Phi}{\rho}, \qquad (2.26)$$

which is somewhat less complicated than Eq. (2.24). Here the dissipation term Φ is given by

$$\Phi = \sigma_{ij}\frac{\partial u_i}{\partial x_j} = \sigma_{xx}\frac{\partial u}{\partial x} + \sigma_{xy}\frac{\partial u}{\partial y} + \sigma_{yx}\frac{\partial v}{\partial x} + \sigma_{yy}\frac{\partial v}{\partial y} \qquad (2.27a)$$

and, for a variable-property Newtonian viscous fluid,

$$\Phi = 2\mu\left(\frac{\partial u}{\partial x}\right)^2 + 2\mu\left(\frac{\partial v}{\partial y}\right)^2 + \mu\left(\frac{\partial u}{\partial y} + \frac{\partial v}{\partial x}\right)^2 + \left(\beta - \frac{2}{3}\mu\right)\left(\frac{\partial u}{\partial x} + \frac{\partial v}{\partial y}\right)^2,$$
$$(2.27b)$$

and the terms are products of the viscous stresses and the corresponding rates of strain. The quantity Φ is the rate, per unit *volume*, at which work is done against viscous stresses by distortion of the fluid; it is another special case of the rule *rate of doing work = force × velocity*, the velocity of one face of a CV with respect to the opposite face being equal to the rate of strain times the distance between the two faces.

Using the heat-conduction law discussed in Chapter 1, giving $\dot{q}_x = -k\,\partial T/\partial x$ and $\dot{q}_y = -k\,\partial T/\partial y$, Eq. (2.26) becomes

$$\frac{de}{dt} = \frac{1}{\rho}\left[\frac{\partial}{\partial x}\left(k\frac{\partial T}{\partial x}\right) + \frac{\partial}{\partial y}\left(k\frac{\partial T}{\partial y}\right)\right] - \frac{p}{\rho}\left(\frac{\partial u}{\partial x} + \frac{\partial v}{\partial y}\right) + \frac{\Phi}{\rho}. \qquad (2.28)$$

Enthalpy Equation

We define the enthalpy per unit mass to be

$$h \equiv e + \frac{p}{\rho}. \qquad (2.29)$$

To obtain an equation for dh/dt, we simply add $d(p/\rho)/dt$ to Eq. (2.26). Now

$$\frac{d(p/\rho)}{dt} = \frac{1}{\rho}\frac{dp}{dt} - \frac{p}{\rho^2}\frac{d\rho}{dt}. \qquad (2.30)$$

If we rewrite the continuity equation (2.2a) as

$$\frac{d\rho}{dt} + \rho\left(\frac{\partial u}{\partial x} + \frac{\partial v}{\partial y}\right) = 0 \qquad (2.31)$$

and substitute for $d\rho/dt$ in Eq. (2.30), Eq. (2.26) gives

$$\frac{dh}{dt} \equiv \frac{d}{dt}\left(e + \frac{p}{\rho}\right) = -\frac{1}{\rho}\left(\frac{\partial\dot{q}_x}{\partial x} + \frac{\partial\dot{q}_y}{\partial y}\right) + \frac{1}{\rho}\left(\frac{\partial p}{\partial t} + u\frac{\partial p}{\partial x} + v\frac{\partial p}{\partial y}\right) + \frac{\Phi}{\rho}. \qquad (2.32)$$

Assuming a fluid with constant specific heat, so that

$$h = c_p T = e + (c_p - c_v)T = e + RT, \qquad (2.33)$$

and using the heat-conduction law for $\dot{q}_x$ and $\dot{q}_y$, Eq. (2.32) can be written as

$$\frac{dh}{dt} \equiv c_p \frac{dT}{dt} = c_p \left(\frac{\partial T}{\partial t} + u \frac{\partial T}{\partial x} + v \frac{\partial T}{\partial y} \right)$$

$$= \frac{1}{\rho} \left[\frac{\partial}{\partial x} \left(k \frac{\partial T}{\partial x} \right) + \frac{\partial}{\partial y} \left(k \frac{\partial T}{\partial y} \right) + \left(\frac{\partial p}{\partial t} + u \frac{\partial p}{\partial x} + v \frac{\partial p}{\partial y} \right) + \Phi \right]. \quad (2.34)$$

Now if the ratio of a typical pressure difference to the absolute pressure is small compared with the ratio of a typical temperature difference to the absolute temperature, the perfect gas law $p = \rho R T$ or $dp/p = d\rho/\rho = dT/T$ shows that the effect of pressure changes on the temperature is small; this is the case in "low-speed" flows (i.e., flows at low values of the Mach number M). Also at low speeds, the dissipation term Φ will be small, since it is proportional to the square of a typical velocity (see Section 3.3). Therefore, for low-speed flows we can neglect the pressure work and dissipation terms in Eq. (2.34).

If we had added $d(p/\rho)/dt$ to Eq. (2.24) instead of Eq. (2.26), we would have obtained an equation for $d[h + \frac{1}{2}(u^2 + v^2)]/dt$, where $h + \frac{1}{2}(u^2 + v^2)$ $\equiv H$ is called the *total enthalpy* and H/c_p is called the *total temperature* T_0 (a useful concept only if c_p is constant). A simplified version of the total enthalpy equation is derived in Section 3.3, and the full derivation is set as Problem 2.7.

Finally, if temperature differences in the (low-speed) flow are small enough for k to be assumed constant, Eq. (2.34) reduces to

$$\frac{\partial T}{\partial t} + u \frac{\partial T}{\partial x} + v \frac{\partial T}{\partial y} = \kappa \left(\frac{\partial^2 T}{\partial x^2} + \frac{\partial^2 T}{\partial y^2} \right), \quad (2.35)$$

where $\kappa \equiv k/\rho c_p$ is the thermometric conductivity introduced in Chapter 1. In mass-diffusion problems with constant diffusion coefficient D (see Chapter 1), Eq. (2.35) applies if κ is replaced by D and T is understood to stand for mass concentration.

2.4 Conservation Equations for Turbulent Flow

The equations derived in the previous sections for the conservation of mass, momentum, and energy in variable-density flows also apply to turbulent flows provided the values of fluid properties and dependent variables are replaced by their instantaneous values. A direct approach to the turbulence problem then consists of solving the equations for a given set of *mean* boundary or initial values but with different initial values of the fluctuations and of computing mean values over the ensemble of solutions. Even for the most restricted cases this is a difficult and extremely expensive computing problem because the unsteady, eddying motions of turbulence appear in a very wide range of sizes. For an introduction to the physical behavior of turbulence, see [1]; a quantitative discussion is given in Chapter 6, but we do not attempt to treat eddy statistics in detail. Thus, the standard procedure is to average over the equations rather than over the solutions. In flows, with steady boundary conditions, simple time averages can be used—in variable-density ("coupled") flow we can use either the conventional time-averaging procedure (as is done here) or the "mass-weighted" time-averaging procedure ("Favre averaging") discussed in Cebeci and Smith [2].

In order to obtain the conservation equations for turbulent flows, we replace the instantaneous quantities in the equations by the sum of their mean and fluctuating parts. For example, the velocity and density are written in the following forms:

$$u = \bar{u} + u', \qquad \rho = \bar{\rho} + \rho', \qquad (2.36)$$

where $\bar{u}$ and $\bar{\rho}$ are the time averages of the velocity and density, and u' and ρ' are the superimposed velocity and density fluctuations. Figure 2.4 shows the notation for velocity. Density fluctuations are of course negligible in constant-property (uncoupled) flows; even when mean density variations are large, $\rho'/\bar{\rho}$ is usually of no greater order than $u'/\bar{u}$. Estimates of density fluctuations in thin shear layers are presented in Section 3.2. The time average of a time-dependent quantity $f(t)$ is denoted by $\bar{f}$ and defined by

$$\bar{f} = \lim_{T \to \infty} \frac{1}{T} \int_0^T f(t)\, dt. \qquad (2.37)$$

Here for simplicity we shall omit the overbars on the basic time-averaged variables u, v, p, and T and on the molecular properties such as ρ, k, and μ. For derived quantities we shall denote time averages by overbars; for example $\overline{u'^2}$ and $\overline{u'v'}$ are the time averages of u'^2 and $u'v'$. The time averages of the fluctuations, such as u', are zero by definition, but $\overline{u'^2}$ is obviously positive since $u'^2 \geq 0$ always (see Problem 2.8).

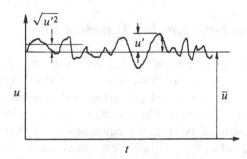

Figure 2.4. Turbulent velocity variation with time: $\bar{u} = $ mean velocity, $u' = $ fluctuations about mean. For simplicity we omit the "overbar" on mean velocity when the meaning is clear from the context.

Continuity and Momentum Equations

Let us first consider the two-dimensional unsteady continuity equation for a compressible flow given by Eq. (2.2a). Replacing ρ, u, and v by the instantaneous (i.e., mean plus fluctuating) values $\rho + \rho'$, $u + u'$, and $v + v'$, we obtain

$$\frac{\partial}{\partial t}(\rho + \rho') + \frac{\partial}{\partial x}\left[(\rho + \rho')(u + u')\right] + \frac{\partial}{\partial y}\left[(\rho + \rho')(v + v')\right] = 0.$$

(2.38)

The mean of $(\rho + \rho')(u + u')$, defined as in Eq. (2.37) and written $\overline{(\rho + \rho')(u + u')}$, is $\rho u + \overline{\rho' u'}$. (The mean of $\rho u'$ is zero because the mean of u' is zero and the mean density ρ is independent of time. The same principle applies for $\rho' u$.) By applying the same arguments to the other terms, we can write the continuity equation for a two-dimensional steady compressible turbulent flow as

$$\frac{\partial}{\partial x}\left(\rho u + \overline{\rho' u'}\right) + \frac{\partial}{\partial y}\left(\rho v + \overline{\rho' v'}\right) = 0.$$ (2.39)

We can apply the same procedure to the momentum equation. Replacing the instantaneous values in the same manner, we can write the x-momentum equation, Eq. (2.8), as

$$\frac{\partial(\rho + \rho')(u + u')}{\partial t} + \frac{\partial(\rho + \rho')(u + u')^2}{\partial x} + \frac{\partial(\rho + \rho')(u + u')(v + v')}{\partial y}$$

$$= -\frac{\partial(p + p')}{\partial x} + (\rho + \rho')(\bar{f}_x + f_x') + \text{viscous-stress gradients,}$$

(2.40)

with a similar equation for the y component. Here the viscous stresses are supposed to be the *instantaneous* values obtained by substituting $u + u'$, $\mu + \mu'$, etc., into Eqs. (2.6a) and (2.6c). The fluctuating parts of the complicated viscous terms disappear entirely on time averaging, and we ignore them. On time averaging, we get the x component of the time-averaged Navier–Stokes equation—sometimes called the *Reynolds equation*—for a two-dimensional steady compressible turbulent flow,

$$\frac{\partial \rho u^2}{\partial x} + \frac{\partial \rho u v}{\partial y} + \frac{\partial \overline{\rho u'^2}}{\partial x} + \frac{\partial \overline{\rho u' v'}}{\partial y} + \frac{\partial}{\partial x}\overline{\rho' u'^2}$$

$$+ \frac{\partial}{\partial y}\overline{\rho' u' v'} + \frac{\partial}{\partial x}\left(2u\overline{\rho' u'}\right) + \frac{\partial}{\partial y}\left(u\overline{\rho' v'} + v\overline{\rho' u'}\right)$$

$$= -\frac{\partial p}{\partial x} + \frac{\partial \sigma_{xx}}{\partial x} + \frac{\partial \sigma_{xy}}{\partial y} + \rho \bar{f}_x + \overline{\rho' f_x'},$$ (2.41)

where σ_{xx} and σ_{xy} are the *molecular* stresses, defined in Eq. (2.6) for a simple viscous fluid. Subtracting u times the continuity equation, Eq. (2.39), from Eq. (2.41) and rearranging, we get another form of Eq. (2.41),

$$\left(\rho u + \overline{\rho'u'}\right)\frac{\partial u}{\partial x} + \left(\rho v + \overline{\rho'v'}\right)\frac{\partial u}{\partial y} = -\frac{\partial p}{\partial x} + \frac{\partial \sigma_{xx}}{\partial x} + \frac{\partial \sigma_{xy}}{\partial y} + \rho \bar{f}_x + \overline{\rho'f'_x}$$

$$+ \left\{\frac{\partial}{\partial x}\left[-\left(\rho\overline{u'^2} + \overline{\rho'u'^2} + u\overline{\rho'u'}\right)\right]\right\}$$

$$+ \left\{\frac{\partial}{\partial y}\left[-\left(\rho\overline{u'v'} + \overline{\rho'u'v'} + v\overline{\rho'u'}\right)\right]\right\}$$

$$\tag{2.42}$$

The extra terms in the first line arise from the contributions of the fluctuations to the mean mass flow rate, for example, $\overline{(\rho + \rho')(u + u')} \equiv \rho u + \overline{\rho'u'}$. The terms which are enclosed in curly brackets can be regarded as gradients of extra apparent stresses (the quantities in brackets) resulting from turbulent transfer of momentum. These additions to the viscous stresses are called the *Reynolds stresses* and, except in the so-called viscous sublayer close to a solid surface, they greatly exceed the viscous stresses. Engineering solutions for turbulence problems require formulas—or more complicated differential equations—for the Reynolds stresses and, less importantly, for terms like $\overline{\rho'u'}$. We can see that the gradients of viscous stress and turbulent stress in the y-direction can be written together as

$$\frac{\partial}{\partial y}\left[\sigma_{xy} - \left(\rho\overline{u'v'} + \cdots\right)\right]$$

so that, like σ_{xy}, the turbulent terms can be regarded as a stress acting in the x direction on a plane normal to the y direction (see Section 2.1). As a simple rule, $-\rho\overline{u'_i u'_j}$ is a stress acting in the x_i direction on a plane normal to the x_j direction.

The y component of the momentum equation for a two-dimensional compressible turbulent flow can be written, following the same steps as for the x-component equation, as

$$\left(\rho u + \overline{\rho'u'}\right)\frac{\partial v}{\partial x} + \left(\rho v + \overline{\rho'v'}\right)\frac{\partial v}{\partial y} = -\frac{\partial p}{\partial y} + \frac{\partial \sigma_{yx}}{\partial x} + \frac{\partial \sigma_{yy}}{\partial y} + \rho\bar{f}_y + \overline{\rho'f'_y}$$

$$-\left\{\frac{\partial}{\partial x}\left(\rho\overline{u'v'} + \overline{\rho'u'v'} + u\overline{\rho'v'}\right)\right\}$$

$$-\left\{\frac{\partial}{\partial y}\left(\rho\overline{v'^2} + \overline{\rho'v'^2} + v\overline{\rho'v'}\right)\right\}. \tag{2.43}$$

In a constant-density flow *all* the terms containing ρ' disappear, and Eq. (2.42) becomes, on dividing by ρ,

$$u\frac{\partial u}{\partial x} + v\frac{\partial u}{\partial y} = -\frac{1}{\rho}\frac{\partial p}{\partial x} + \frac{1}{\rho}\left(\frac{\partial \sigma_{xx}}{\partial x} + \frac{\partial \sigma_{xy}}{\partial y}\right) - \frac{\partial}{\partial x}\overline{u'^2} - \frac{\partial}{\partial y}\overline{u'v'} + f_x,$$

$$\tag{2.44}$$

and use of the relations given in Eqs. (2.6c) and (2.6d) allows the viscous-stress gradients to be written, as in laminar flow,

$$\frac{1}{\rho}\left(\frac{\partial \sigma_{xx}}{\partial x} + \frac{\partial \sigma_{xy}}{\partial y}\right) = \nu\left(\frac{\partial^2 u}{\partial x^2} + \frac{\partial^2 u}{\partial y^2}\right) \tag{2.45}$$

if ν, like ρ, is constant.

Similarly Eq. (2.43) can be written for constant-property flow as

$$u\frac{\partial v}{\partial x} + v\frac{\partial v}{\partial y} = -\frac{1}{\rho}\frac{\partial p}{\partial y} + \nu\left(\frac{\partial^2 v}{\partial x^2} + \frac{\partial^2 v}{\partial y^2}\right) - \frac{\partial}{\partial x}\overline{u'v'} - \frac{\partial}{\partial y}\overline{v'^2} + \bar{f}_y.$$

$$\tag{2.46}$$

Enthalpy Equation

In turbulent flow, extra terms also appear in the enthalpy equation, Eq. (2.34). If we multiply its left-hand side by ρ and substitute $u + u'$, $v + v'$, $\rho + \rho'$, and $T + T'$ for the corresponding mean quantities, we get

$$(\rho + \rho')c_p\frac{\partial(T + T')}{\partial t} + (\rho + \rho')(u + u')c_p\frac{\partial(T + T')}{\partial x}$$

$$+ (\rho + \rho')(v + v')c_p\frac{\partial(T + T')}{\partial y}.$$

Now making the same substitutions for velocity and density in the continuity equation, Eq. (2.2a), and multiplying by $c_p(T + T')$ give

$$c_p(T + T')\left(\frac{\partial(\rho + \rho')}{\partial t} + \frac{\partial(\rho + \rho')(u + u')}{\partial x} + \frac{\partial(\rho + \rho')(v + v')}{\partial y}\right) = 0.$$

$$\tag{2.47}$$

Adding this zero quantity to the above expression and time-averaging the resulting expression we can write, after rearranging,

$$c_p\left[\frac{\partial \rho u T}{\partial x} + \frac{\partial \rho v T}{\partial y} + \frac{\partial \rho \overline{T'u'}}{\partial x} + \frac{\partial \rho \overline{T'v'}}{\partial y} + \frac{\partial}{\partial x}\overline{(\rho'T'u')}\right.$$

$$\left. + \frac{\partial}{\partial y}\overline{(\rho'T'v')} + \frac{\partial}{\partial x}\overline{(T\rho'u' + u\rho'T')} + \frac{\partial}{\partial y}\overline{(T\rho'v' + v\rho'T')}\right], \tag{2.48}$$

which is in a form analogous to that of the left-hand side of the x-component Reynolds equation for turbulent flow, Eq. (2.41).

The right-hand side of the enthalpy equation for turbulent flow contains a number of mean products of fluctuations, as can be seen by making all the above substitutions in the right-hand side of Eq. (2.34), but only those that contribute to the time-average dissipation term Φ are significant in practice. Since turbulent heat-flux rates usually greatly exceed the conductive heat transfer, correlations between thermal-conductivity fluctuations and temperature fluctuations can be neglected, like correlations between viscosity and velocity. Where conductive heat transfer is significant, i.e., in the "conductive sublayer" very close to a solid surface (see Section 4.2), the fluctuation terms are small anyway. The mean products of fluctuations arising from the pressure terms are of the order of minor terms in the conservation equation for the kinetic energy of the *turbulence* defined as $(\overline{u'^2} + \overline{v'^2} + \overline{w'^2})/2$ per unit mass: they are negligible compared with the other terms in the mean-flow energy equations.

If we insert only the mean velocity components into the dissipation term Φ in Eq. (2.27), then in turbulent flow it represents the viscous dissipation of (mean-flow) kinetic energy into thermal internal energy; the extra turbulent terms, which will be denoted by $\rho\varepsilon$, represent the dissipation of *turbulent* kinetic energy into thermal internal energy. The turbulent kinetic energy per unit mass is rarely more than a few percent of the mean-flow kinetic energy, but its rate of dissipation due to viscous forces acting on the smallest turbulent eddies is much larger than the rate of dissipation of mean-flow kinetic energy, because the fluctuating rates of strain in the smallest turbulent eddies are very large compared with the mean-flow rate of strain.

The final form of the enthalpy equation for turbulent flow, analogous to the x-component momentum equation (2.42), is obtained by transferring some of the terms in the expression (2.48) from the left-hand side to the right. After rearranging, we can write the enthalpy equation for a two-dimensional compressible turbulent flow as

$$c_p\left[\left(\rho u + \overline{\rho' u'}\right)\frac{\partial T}{\partial x} + \left(\rho v + \overline{\rho' v'}\right)\frac{\partial T}{\partial y}\right]$$

$$= \frac{\partial}{\partial x}\left(k\frac{\partial T}{\partial x}\right) + \frac{\partial}{\partial y}\left(k\frac{\partial T}{\partial y}\right)$$

$$- c_p\left[\frac{\partial}{\partial x}\left(\rho\overline{u'T'} + \overline{\rho'u'T'} + u\overline{\rho'T'}\right) + \frac{\partial}{\partial y}\left(\rho\overline{v'T'} + \overline{\rho'v'T'} + v\overline{\rho'T'}\right)\right]$$

$$+ u\frac{\partial p}{\partial x} + v\frac{\partial p}{\partial y} + \Phi + \rho\varepsilon. \tag{2.49}$$

We note that, as in the momentum equation, Eq. (2.42), additional terms appear on the right-hand side of the enthalpy equation. These terms, which are the thermal analogs of the Reynolds-stress gradients in Eq. (2.42), are called the *turbulent heat-flux gradients*. Again, empirical formulas for the turbulent heat fluxes are needed to make the time-averaged equations solvable.

In Eq. (2.49) the turbulent energy dissipation rate ε is given by an expression analogous to the form of Φ for a simple viscous fluid, Eq.

(2.27b):

$$\rho\varepsilon = 2\mu\left[\overline{\left(\frac{\partial u'}{\partial x}\right)^2} + \overline{\left(\frac{\partial v'}{\partial y}\right)^2} + \overline{\left(\frac{\partial w'}{\partial z}\right)^2}\right]$$

$$+ \mu\left[\overline{\left(\frac{\partial u'}{\partial y} + \frac{\partial v'}{\partial x}\right)^2} + \overline{\left(\frac{\partial v'}{\partial z} + \frac{\partial w'}{\partial y}\right)^2}\right.$$

$$\left. + \overline{\left(\frac{\partial w'}{\partial x} + \frac{\partial u'}{\partial z}\right)^2}\right]$$

$$+ \left(\beta - \frac{2\mu}{3}\right)\overline{\left(\frac{\partial u'}{\partial x} + \frac{\partial v'}{\partial y} + \frac{\partial w'}{\partial z}\right)^2}. \qquad (2.50)$$

Note that even if the mean motion is two-dimensional ($w = 0$), w' is of the same order as u' and v'.

For an incompressible flow, Φ, ε, and the density fluctuations are negligible, and Eq. (2.49) reduces to

$$u\frac{\partial T}{\partial x} + v\frac{\partial T}{\partial y} = \kappa\left(\frac{\partial^2 T}{\partial x^2} + \frac{\partial^2 T}{\partial y^2}\right) - \left(\frac{\partial \overline{T'u'}}{\partial x} + \frac{\partial \overline{T'v'}}{\partial y}\right), \qquad (2.51)$$

which differs from Eq. (2.35) by the appearance of turbulent heat-flux gradients.

2.5 Equations of Motion: Summary

The equations discussed in this chapter are all consequences of the principles of conservation of mass, momentum, and (thermal and/or kinetic) energy for a fluid with a simple, linear dependence of internal stress on rate of strain and of heat-flux rate on temperature gradient. The partial differential equations (PDEs) are, in general, coupled and nonlinear. See Refs. 3 and 4 for a general discussion of PDEs and their boundary conditions, and see Section 3.5 for a discussion of boundary conditions for shear layers. The PDEs describing fluid flow can have steady (non-time-dependent) laminar solutions, but they also admit very complicated time-dependent solutions, which we call *turbulence*; these solutions are extremely difficult and expensive to obtain numerically, and although turbulent flows are very important in engineering and in the earth sciences, we are forced to time-average the equations to make them numerically tractable. Time averaging discards information, and in order to obtain a "closed" set of equations for turbulent flow, we have to use empirical data. As we shall see in Chapter 6, a simple and not-too-restrictive way of incorporating empirical data about turbulent flows is to define a turbulent "eddy" viscosity as the ratio of the apparent internal stress to the corresponding time-average rate of strain. Turbulent stresses, like molecular stresses, are actually the result of momentum transfer; viscous stresses are the result of molecular interaction, and turbulent stresses are the result of interaction between turbulent motions (eddies), and there is no close relation between the two, but it is still mathematically legitimate to define an eddy viscosity provided that one accepts that it will, in general, be a complicated function of the statistical properties of the turbulence rather than a property of the fluid. If an eddy-viscosity formula, and a corresponding formula for "eddy conductivity" (also to be defined in Chapter 6) to express the transport of heat by turbulent eddies, can be devised as a correlation of experimental data, then the time-averaged equations of motion for turbulent flow can be reduced to the same form as the steady laminar flow equations for a viscous fluid. This enables us to treat the laminar and turbulent problems together and justifies spending rather more time on laminar flows, in the rest of this book, than is strictly justified by their appearance in real life. Even the more advanced turbulence models that nominally provide explicit equations for the apparent turbulent stresses and heat-flux rates can often be reformulated to yield eddy viscosity and eddy conductivity for simplicity of computation.

Problems

2.1 (a) Extend the control volume analysis used to derive the continuity equation to three-dimensional unsteady flows and show that Eq. (2.2a) can be written as

$$\frac{\partial \rho}{\partial t} + \frac{\partial}{\partial x}(\rho u) + \frac{\partial}{\partial y}(\rho v) + \frac{\partial}{\partial z}(\rho w) = 0. \qquad (P2.1)$$

(b) Deduce from Eq. (P2.1) the mass conservation equations for three-dimensional steady or unsteady constant-density flow and for three-dimensional steady, variable-density flows corresponding to Eqs. (2.2b) and (2.2c) for two-dimensional flows.

(c) Show that Eq. (P2.1) can be rewritten as a transport equation for ρ, using the three-dimensional version of the d/dt operator defined in Eq. (2.10).

2.2 By repeating the arguments used to derive the x-component momentum equation, Eq. (2.11), show that the y-component momentum equation is as given in Eq. (2.12). Check your answer by "rotating" the coordinates in the x-component momentum equation (i.e. changing x to y and y to x, u to v and v to u, throughout).

2.3 Show that

$$\frac{\partial}{\partial t} + u\frac{\partial}{\partial x} + v\frac{\partial}{\partial y} \equiv \frac{d}{dt}$$

represents the rate of change, with respect to time, as seen by an observer following the motion of a fluid element.

2.4 By extending the arguments used to derive the x-component momentum equation for two-dimensional flow, Eq. (2.9), show that in three-dimensional flow the equation becomes

$$\frac{\partial u}{\partial t} + u\frac{\partial u}{\partial x} + v\frac{\partial u}{\partial y} + w\frac{\partial u}{\partial z} = -\frac{1}{\rho}\frac{\partial p}{\partial x} + \frac{1}{\rho}\left(\frac{\partial \sigma_{xx}}{\partial x} + \frac{\partial \sigma_{xy}}{\partial y} + \frac{\partial \sigma_{xz}}{\partial z}\right) + f_x$$

where in a Newtonian fluid

$$\sigma_{xz} = \frac{\partial u}{\partial z} + \frac{\partial w}{\partial x}.$$

2.5 Show that the term $\partial(\sigma_{ij}u_i)/\partial x_j$ in Eq. (2.25) can be expanded, in two-dimensional flow, to

$$\frac{\partial \sigma_{xx}u}{\partial x} + \frac{\partial \sigma_{yx}v}{\partial x} + \frac{\partial \sigma_{xy}u}{\partial y} + \frac{\partial \sigma_{yy}v}{\partial y}$$

and hence, using Eqs. (2.6) for a Newtonian fluid, to

$$\mu\left(\frac{\partial^2 u^2}{\partial x^2} + \frac{\partial^2 v^2}{\partial y^2}\right) + \frac{\partial u^2}{\partial x}\frac{\partial \mu}{\partial x} + \frac{\partial v^2}{\partial y}\frac{\partial \mu}{\partial y} + \frac{d}{dt}\left[\left(\beta - \frac{2}{3}\mu\right)\nabla u\right]$$

$$+ \left(\beta - \frac{2}{3}\mu\right)(\nabla u)^2 + \mu e_{xy}^2 + e_{xy}\left[v\frac{\partial \mu}{\partial x} + u\frac{\partial \mu}{\partial y}\right] + \mu v\frac{\partial e_{xy}}{\partial x} + \mu u\frac{\partial e_{xy}}{\partial y}$$

where

$$\nabla u = \partial u/\partial x + \partial v/\partial y, \qquad e_{xy} = \partial v/\partial x + \partial u/\partial y$$

and d/dt is the transport operator defined in Eq. (2.10).

2.6 If we multiply the x-component of the momentum equation, Eq. (2.9) by u, and use the definition of the substantial-derivative symbol d/dt given by Eq. (2.10), we get

$$\frac{d}{dt}\left(\frac{1}{2}u^2\right) = -\frac{u}{\rho}\frac{\partial p}{\partial x} + \frac{u}{\rho}\left[\frac{\partial \sigma_{xx}}{\partial x} + \frac{\partial \sigma_{xy}}{\partial y}\right] + uf_x. \qquad (P2.2)$$

Adding the corresponding equation for d/dt $(1/2v^2)$, which can be most easily derived by changing the variables in Eq. (P2.2) in cyclic order, show that the resulting expression, with V denoting the resulting velocity, can be written as

$$\frac{d}{dt}\left(\frac{1}{2}V^2\right) = -\frac{1}{\rho}\left(u\frac{\partial p}{\partial x} + v\frac{\partial p}{\partial y}\right) + \frac{u}{\rho}\left(\frac{\partial \sigma_{xx}}{\partial x} + \frac{\partial \sigma_{xy}}{\partial y}\right) + \frac{v}{\rho}\left(\frac{\partial \sigma_{yx}}{\partial x} + \frac{\partial \sigma_{yy}}{\partial y}\right)$$

$$+ uf_x + vf_y. \qquad (P2.3)$$

Equation (P2.3) is known as the kinetic energy equation. Its left-hand side represents the rate of increase of kinetic energy per unit mass of the fluid, as the fluid moves along a streamline. The terms on the right-hand side which can be written in tensor notation as

$$\frac{u_i}{\rho}\left(-\frac{\partial p}{\partial x_i} + \frac{\partial \sigma_{ij}}{\partial x_j} + \rho f_i\right)$$

represent, respectively, the rates at which work is done on unit mass of the fluid by the pressure, by the viscous or turbulent σ-stresses, and by the body force per unit mass, $\mathbf{f}$.

2.7 By adding $d(p/\rho)/dt$ to the total-energy equation for $e + 1/2(u^2 + v^2)$, derive an equation for the total enthalpy $H \equiv h + 1/2(u^2 + v^2)$. Hint: Follow the derivation of the *static*-enthalpy equation from the *internal*-energy equation, and recall that d/dt is the transport operator defined in Eq. (2.10).

2.8 Find the mean parts u and T, the fluctuating (time-dependent) parts u' and T', the mean-square fluctuations $\overline{u'^2}$ and $\overline{T'^2}$, and the velocity-temperature mean product ("covariance") $\overline{u'T'}$ for the following variations of instantaneous velocity and temperature with time.

(a) $u + u' = a + b\sin\omega t$, $T + T' = c + d\sin(\omega t - \phi)$

(b) $u + u' = a + b\sin^2\omega t$, $T + T' = c + d\sin^2(\omega t - \phi)$

Note that the time average can be taken over just one cycle, $0 < \omega t < 2\pi$.

2.9 By repeating the arguments used to derive the time-average x-component momentum equation for turbulent flow, Eq. (2.42), show that the time-average y-component equation is

$$u\frac{\partial v}{\partial x} + v\frac{\partial v}{\partial y} = -\frac{1}{\rho}\frac{\partial p}{\partial y} + \frac{1}{\rho}\left(\frac{\partial \sigma_{xy}}{\partial x} + \frac{\partial \sigma_{yy}}{\partial y}\right) - \frac{1}{\rho}\left[\frac{\partial}{\partial x}\left(\rho\overline{u'v'} + \frac{\partial}{\partial y}\left(\overline{\rho v'^2}\right)\right)\right]$$

where terms in the density fluctuation ρ have been neglected and where the σ terms represent the viscous contributions only.

References

[1] Bradshaw, P.: Turbulence, in *Science Progress*, **67**.: 185 Oxford, 1981.

[2] Cebeci, T. and Smith, A. M. O.: *Analysis of Turbulent Boundary Layers*. Academic, New York, 1974.

[3] Hildebrand, F. B.: *Advanced Calculus for Applications*. Prentice-Hall, Englewood Cliffs, NJ, 1962.

[4] Cebeci, T. and Bradshaw, P.: *Momentum Transfer in Boundary Layers*. Hemisphere, Washington, DC, 1977.

CHAPTER 3

Boundary-Layer Equations

The momentum and heat-transfer problems solved in this text are represented by the so-called boundary-layer (or thin-shear-layer) equations, which are approximate forms of the exact conservation equations derived in Sections 2.1 to 2.4. Here we use the term "thin shear layer" to denote any member of the family of boundary layers, jets, wakes, and flows in long slender ducts. The boundary layer itself is the simplest example for qualitative discussion. The following three sections apply the boundary-layer assumptions for uncoupled (incompressible, constant-density) and coupled (compressible, variable-density) flow, respectively, and lead to the thin-shear-layer equations for steady two-dimensional and axisymmetric flows.

The equations considered in Sections 3.1 and 3.3 are partial differential equations (PDEs). The ordinary differential equations (ODEs) obtained by integrating the PDEs across the thickness of a shear layer have simple physical interpretations as equations of conservation of momentum or enthalpy for the shear layer as a whole. In the past they were the basis for nearly all calculation methods but have now been superseded in many cases by PDE-based methods. In general, these "integral" equations (strictly, ODEs) contain less information than partial differential equations, but to allow comparison and related discussion, they are derived in Section 3.4. The chapter ends with a section on the boundary conditions for which the equations must be solved—that is, the typical problems discussed in detail in later chapters.

41

3.1 Uncoupled Flows

The boundary-layer assumptions stem from the observation that, in flows at high-Reynolds-number, gradients of stress or heat flux are usually significant only in relatively thin layers, nearly parallel with the general flow direction. Thus, in the simplest case of a two-dimensional flow in the x direction, the shear stress $\mu \, \partial u / \partial y$ is significant only in a limited range of y —say, the region of the boundary layer close to a solid surface in the plane $y = 0$. Outside this region, $\partial u / \partial y$ is of the same order as $\partial u / \partial x$ (say), but within the shear layer we have

$$\frac{\partial u}{\partial y} \gg \frac{\partial u}{\partial x} \tag{3.1}$$

and the shear-layer thickness δ, by definition of a "thin" shear layer growing in the x direction, satisfies

$$\frac{d\delta}{dx} \ll 1. \tag{3.2}$$

Provided the Prandtl number (Section 1.3) is not too small, similar statements apply to enthalpy gradients in, say, the boundary layer on a *heated* surface in the plane $y = 0$; for simplicity, we just consider temperature gradients. Thus

$$\frac{\partial T}{\partial y} \gg \frac{\partial T}{\partial x} \tag{3.3}$$

and

$$\frac{d\delta_t}{dx} \ll 1, \tag{3.4}$$

where now δ_t is the thickness of the *thermal* boundary layer [Fig. 3.1(a)]. These statements can be used to perform an order-of-magnitude analysis of the terms in the equations that are derived in Chapter 2 and summarized below, and lead to the boundary-layer equations. Briefly, these equations

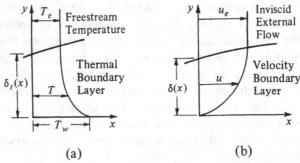

(a)   (b)

Figure 3.1. Temperature (a) and velocity (b) profiles on a solid surface.

differ from the Navier–Stokes equations in the omission of small terms from the x-component momentum equation and the almost complete collapse of the y-component equation to the statement $\partial p / \partial y = 0$; in other words, the *neglect* of pressure changes through the thickness of the shear layer.

The steady two-dimensional uncoupled forms of the conservation equations derived in Sections 2.1 to 2.4 may be written as

Continuity:

$$\frac{\partial u}{\partial x} + \frac{\partial v}{\partial y} = 0. \tag{3.5}$$

x-component momentum:

$$u\frac{\partial u}{\partial x} + v\frac{\partial u}{\partial y} = -\frac{1}{\rho}\frac{\partial p}{\partial x} + \nu\left(\frac{\partial^2 u}{\partial x^2} + \frac{\partial^2 u}{\partial y^2}\right) - \frac{\partial}{\partial x}\left(\overline{u'^2}\right) - \frac{\partial}{\partial y}\left(\overline{u'v'}\right) + f_x.$$

$$\tag{3.6}$$

y-component momentum:

$$u\frac{\partial v}{\partial x} + v\frac{\partial v}{\partial y} = -\frac{1}{\rho}\frac{\partial p}{\partial y} + \nu\left(\frac{\partial^2 v}{\partial x^2} + \frac{\partial^2 v}{\partial y^2}\right) - \frac{\partial}{\partial y}\left(\overline{v'^2}\right) - \frac{\partial}{\partial x}\left(\overline{u'v'}\right) + f_y.$$

$$\tag{3.7}$$

Energy:

$$u\frac{\partial T}{\partial x} + v\frac{\partial T}{\partial y} = \frac{k}{\rho c_p}\left(\frac{\partial^2 T}{\partial x^2} + \frac{\partial^2 T}{\partial y^2}\right) - \frac{\partial}{\partial x}\left(\overline{T'u'}\right) - \frac{\partial}{\partial y}\left(\overline{T'v'}\right). \tag{3.8}$$

They represent any steady two-dimensional uncoupled flow, including the thin shear layers introduced above and also flows where the regions of significant viscous stress are far from being thin layers parallel to the general flow direction, for example, the recirculating flow immediately downstream of a step. They allow the possibility of pressure gradients and diffusion in two directions perpendicular to each other, and a consequence is that in the formulation of any related problem, boundary conditions are required for each dependent variable, or its gradient, on all four sides of the solution domain—supposed, for simplicity, to be rectangular.

As mentioned above, the most spectacular result of the order-of-magnitude analysis of the terms in the Navier–Stokes equations for a thin shear layer is that the pressure can be approximated by its value at the edge of the shear layer and therefore ceases to be a variable in the equations. This, together with the neglect of the x-wise stress gradients (x-wise diffusion of momentum by viscosity or turbulence), implies that a disturbance introduced into a shear layer at some streamwise position x_0 does not significantly affect the flow upstream of x_0 (assuming that u is positive) unless it indirectly disturbs the pressure *outside* the shear layer. This lack of "upstream influence" of a disturbance means that the thin-shear-layer equations can be solved numerically by a "marching" process, starting at an upstream position and proceeding downstream once and for all. In contrast, the Navier–Stokes equations do permit upstream influence and must be solved by repeating the downstream-marching process many times until the solutions converge. Mathematically this means that the steady Navier–Stokes equations are "elliptic" whereas the thin-shear-layer equations are "parabolic". We shall see below that the enthalpy equation also changes from elliptic to parabolic and its numerical solution is thus simplified.

We begin the detailed discussion of the thin-shear-layer equations with the simplest case, laminar uncoupled flow.

Laminar Flow

A typical value for the velocity gradient $\partial u/\partial y$ in a boundary layer or other thin shear layer of thickness δ with an edge velocity u_e is u_e/δ [see, for example, Fig. 3.1(b)]; a typical value of $\partial u/\partial x$ is u_e/x, where x is the distance from the origin of the flow. Thus $(\partial u/\partial x)/(\partial u/\partial y)$ is δ/x or, more conveniently, $d\delta/dx$. From the continuity equation, Eq. (3.5) $\partial v/\partial y$ is also u_e/x, so that v near the edge of a boundary layer is $u_e\delta/x$ or $u_e d\delta/dx$. By repeating the argument used to find $(\partial u/\partial x)/(\partial u/\partial y)$, we can show that $(\partial^2 u/\partial x^2)/(\partial^2 u/\partial y^2)$ is $(\delta/x)^2$ or $(d\delta/dx)^2$ (see Problem 3.1).

We now apply these order-of-magnitude results to the x-component momentum equation, Eq. (3.6), ignoring the turbulent-stress gradients. Since u itself is of the same order as u_e, we see that both terms on the left-hand side of Eq. (3.6) are of order u_e^2/x. As we have just shown, $\partial^2 u/\partial x^2$ is of order $(\delta/x)^2$ times $\partial^2 u/\partial y^2$; the latter is of order u_e/δ^2. We therefore neglect $\partial^2 u/\partial x^2$ compared with $\partial^2 u/\partial y^2$. In general we expect that since Bernoulli's equation is valid outside the shear layer, $(-1/\rho)\,\partial p/\partial x$ will be of the same order as $u_e\,du_e/dx$ (the two are, of course, equal throughout the shear layer if p does not vary significantly with y) and therefore of order u_e^2/x. The part of f_x that is not balanced by a "hydrostatic" pressure gradient cannot be of higher order than the acceleration, that is, u_e^2/x. Thus, by neglecting the x-wise stress gradient $\partial^2 u/\partial x^2$, the x-component momentum equation for a laminar uncoupled thin shear layer with δ/x or $d\delta/dx \ll 1$ can be simplified from Eq. (3.6) (without the turbulent-stress terms) to

$$u\frac{\partial u}{\partial x} + v\frac{\partial u}{\partial y} = -\frac{1}{\rho}\frac{\partial p}{\partial x} + v\frac{\partial^2 u}{\partial y^2} + f_x \qquad (3.9)$$

Now, $v\,\partial^2 u/\partial y^2$, which is of order vu_e/δ^2, cannot be of larger order than the other terms (that is, u_e^2/x), and if it is of *smaller* order, viscous stresses are unimportant anyway. Therefore vu_e/δ^2 is of order u_e^2/x in a thin shear layer, and it follows from this that δ is of order $\sqrt{vx/u_e}$. Equivalently, $d\delta/dx$ is of order $(u_e x/v)^{-1/2}$ or $(u_e\delta/v)^{-1}$; the Reynolds number $u_e\delta/v$

can be written as $\rho u_e^2/(\mu u_e/\delta)$, the ratio of a typical dynamic pressure to a typical viscous stress.

The analysis of the y-component equation given by Eq. (3.7), again with turbulence terms neglected, proceeds similarly. The first term on the left-hand side of Eq. (3.7) is $u\,\partial v/\partial x$, which is of order $u_e v/x$ or $u_e^2\delta/x^2$, as is the second term $v\,\partial v/\partial y$. The larger of the viscous-stress terms, $\partial^2 v/\partial y^2$, is of order $u_e^2\delta/x^2$ also, since v is of the same order as $\delta^2/u_e x$. This leads to the result that the pressure term in the equation, $(-1/\rho)\,\partial p/\partial y$, is at most of order $u_e^2\delta/x^2$ and (neglecting the effect of body forces as previously discussed) the change in pressure across the shear layer of thickness δ is of order $\rho u_e^2(\delta/x)^2$. That is, the change in $\partial p/\partial x$ across the shear layer is negligibly small.

As stated above, $(-1/\rho)\,\partial p/\partial x$ outside the shear layer—and therefore *within* it, by the result just obtained—can be replaced by $u_e\,du_e/dx$, giving

$$u\frac{\partial u}{\partial x} + v\frac{\partial u}{\partial y} = u_e\frac{du_e}{dx} + v\frac{\partial^2 u}{\partial y^2} + f_x, \qquad (3.10a)$$

or

$$u\frac{\partial u}{\partial x} + v\frac{\partial u}{\partial y} = -\frac{1}{\rho}\frac{dp}{dx} + v\frac{\partial^2 u}{\partial y^2} + f_x \qquad (3.10b)$$

with the unaltered continuity equation (3.5), as two equations for the two variables u and v in a steady two-dimensional uncoupled laminar boundary layer.

The application of the thin-shear-layer approximation to the enthalpy equation is quite straightforward. By exactly the same arguments as above, applied to the *thermal* boundary layer or other thin shear layer with $d\delta_t/dx \ll 1$, we neglect $\partial^2 T/\partial x^2$ compared with $\partial^2 T/\partial y^2$ in Eq. (3.8), so that for steady two-dimensional constant-property laminar flow Eq. (3.8) reduces to

$$u\frac{\partial T}{\partial x} + v\frac{\partial T}{\partial y} = \frac{k}{\rho c_p}\frac{\partial^2 T}{\partial y^2} \equiv \frac{v}{\mathrm{Pr}}\frac{\partial^2 T}{\partial y^2}, \qquad (3.11)$$

which is closely analogous to Eq. (3.10) without the pressure and body-force terms. Note that neglect of $\partial^2 T/\partial x^2$ implies that Eq. (3.11) is not valid immediately downstream of a step in surface temperature. In real heat exchangers, however, longitudinal conduction in the solid surface will usually smooth out the change in temperature over a sufficiently large distance for Eq. (3.11) to be adequate; as will be seen below, it is also convenient to smooth out temperature steps in calculations.

The analysis used above to show that δ is of order $\sqrt{vx/u_e}$ can be easily adapted to estimate δ_t. If a typical temperature difference across the shear layer is ΔT, then $u\,\partial T/\partial x$ is of order $u_e\Delta T/x_t$, where x_t is measured from the origin of the *thermal* layer. If this term is to be of the same order as $(v/\mathrm{Pr})\,\partial^2 T/\partial y^2$, which is of order $(v/\mathrm{Pr})\Delta T/\delta_t^2$, it follows that δ_t is of order $\sqrt{[(v/\mathrm{Pr})x_t/u_e]}$.

Turbulent Flow

In turbulent flow we can use exactly the same arguments as in laminar flow to show that the term $\nu\,\partial^2 u/\partial x^2$ in the x-component momentum equation, Eq. (3.6), can be neglected compared with $\nu\,\partial^2 u/\partial y^2$. In the case of the turbulent-stress terms it is found from experiment that $\overline{u'^2}$ is of the same order as $\overline{u'v'}$, so that $\partial\overline{u'^2}/\partial x$ will be smaller than $\partial\overline{u'v'}/\partial y$ by a factor of order δ/x. In the y-component equation, we can show that the terms on the left-hand side and the larger of the two viscous terms are again of order $u_e^2\delta/x^2$, and since $\overline{u'^2}$ and $\overline{u'v'}$ are of the same order, $\partial\overline{u'v'}/\partial x$ is of order $(\delta/x)(\partial\overline{v'^2}/\partial y)$. Since in the x-component equation we expect $\partial\overline{u'v'}/\partial y$ to be of the same order as the acceleration terms on the left-hand side (that is, u_e^2/x), it follows that the term $\partial\overline{v'^2}/\partial y$ in the y-component equation is of order u_e^2/x, so that the pressure difference across the layer must be of order $\rho u_e^2\delta/x$. [The other terms in the y-component equation are small so that $\partial\overline{v'^2}/\partial y$ nearly balances $(-1/\rho)\,\partial p/\partial y$; see Problem 3.2.[1]]

Therefore, in turbulent flow we can again neglect the x-wise stress-gradient term, in this case $\partial\overline{u'^2}/\partial x$, and also neglect the pressure difference across the shear layer. However, both of these neglected terms are of order δ/x times the retained terms, whereas in laminar flow the difference in order was $(\delta/x)^2$. This is not directly related to the fact that $d\delta/dx$ is larger in turbulent flow, but the two facts taken together imply that the thin-shear-layer approximation is not so accurate in turbulent flow as in laminar flow. Thus, up to order δ/x, the x-component momentum equation for uncoupled turbulent flow becomes

$$u\frac{\partial u}{\partial x} + v\frac{\partial u}{\partial y} = u_e\frac{du_e}{dx} + \nu\frac{\partial^2 u}{\partial y^2} - \frac{\partial}{\partial y}\overline{u'v'} + f_x, \qquad (3.12)$$

where we have again used the result that $\partial p/\partial y = 0$ and the differentiated form of Bernoulli's equation,

$$u_e\frac{du_e}{dx} = -\frac{1}{\rho}\frac{dp_e}{dx}. \qquad (3.13)$$

In analogy to the momentum equation for turbulent flow, we must take $\overline{T'u'}$ and $\overline{T'v'}$ in Eq. (3.8) to be of the same order, but we can neglect $\partial\overline{T'u'}/\partial x$ compared with $\partial\overline{T'v'}/\partial y$. The resulting enthalpy equation for uncoupled turbulent flow becomes, to order δ/x,

$$u\frac{\partial T}{\partial x} + v\frac{\partial T}{\partial y} = \frac{k}{\rho c_p}\frac{\partial^2 T}{\partial y^2} - \frac{\partial}{\partial y}\overline{T'v'}. \qquad (3.14)$$

---

[1] This implies that $p + \rho\overline{v'^2}$ is independent of y, and since v' is zero outside the shear layer or at a solid surface, it follows that the pressure difference across the layer is of *smaller* order than $\rho u_e^2\delta/x$: however, changes of this order occur between the layer edge and the region of maximum $\overline{v'^2}$.

Axisymmetric Flows

For simplicity all the previous equations have been derived only for two-dimensional flow, with w and all z-wise derivatives of the time-averaged quantities taken as zero. Fully three-dimensional flows with nonzero z-wise derivatives are governed by equations containing more terms (see Problem 2.4, for example). However, the special case of axisymmetric flow, such as flow in a circular pipe or circular jet or the boundary layer on a body of circular cross section ("body of revolution") can be treated by transforming the three-dimensional equations to cylindrical x, y, θ coordinates and then neglecting variations with respect to the circumferential coordinate θ. This is a suitable system for circular pipes and jets in which the flow extends all the way to the axis and is at most at a small angle to the axis—according to the above qualitative thin-shear-layer arguments, which apply without change if y is replaced by r. For boundary layers on bodies of revolution, however, we need the curvilinear (x, y) coordinate system shown in Fig. 3.2, in which the continuity, momentum, and enthalpy equations for an uncoupled thin shear layer can be written as

$$\frac{\partial r^K u}{\partial x} + \frac{\partial r^K v}{\partial y} = 0, \tag{3.15}$$

$$u\frac{\partial u}{\partial x} + v\frac{\partial u}{\partial y} = -\frac{1}{\rho}\frac{dp}{dx} + \frac{1}{\rho r^K}\frac{\partial}{\partial y}\left[r^K\left(\mu\frac{\partial u}{\partial y} - \rho\overline{u'v'}\right)\right] + f_x, \tag{3.16}$$

$$u\frac{\partial T}{\partial x} + v\frac{\partial T}{\partial y} = \frac{1}{\rho c_p}\frac{1}{r^K}\frac{\partial}{\partial y}\left[r^K\left(k\frac{\partial T}{\partial y} - \rho c_p\overline{T'v'}\right)\right]. \tag{3.17}$$

Here K is a "flow index," equal to unity in axisymmetric flow and zero in two-dimensional flow. Note that μ and k have been taken inside the derivatives merely for convenience; they are still constant.

In general r is related to r_0, the radius of the surface $y = 0$, by

$$r(x, y) = r_0(x) + y\cos\phi(x), \tag{3.18}$$

where

$$\phi = \tan^{-1}\frac{dr_0}{dx}. \tag{3.19}$$

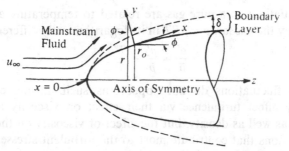

Figure 3.2. Notation and coordinate system for an axisymmetric body.

Defining

$$t \equiv \frac{y \cos \phi}{r_0},$$ (3.20)

we can write Eq. (3.18) as

$$\frac{r}{r_0} = 1 + t.$$ (3.21)

Here t represents the deviation of r from r_0 and is called the *transverse curvature* term. In many axisymmetric flows the body radius is quite large relative to the boundary-layer thickness, so the transverse curvature effect is negligible. In that case, Eqs. (3.15)–(3.17) become

$$\frac{1}{r_0^K} \frac{\partial}{\partial x} \left(r_0^K u \right) + \frac{\partial v}{\partial y} = 0,$$ (3.22)

$$u \frac{\partial u}{\partial x} + v \frac{\partial u}{\partial y} = -\frac{1}{\rho} \frac{dp}{dx} + \frac{1}{\rho} \frac{\partial}{\partial y} \left(\mu \frac{\partial u}{\partial y} - \rho \overline{u'v'} \right) + f_x,$$ (3.23)

$$u \frac{\partial T}{\partial x} + v \frac{\partial T}{\partial y} = \frac{1}{\rho c_p} \frac{\partial}{\partial y} \left(k \frac{\partial T}{\partial y} - \rho c_p \overline{T'v'} \right),$$ (3.24)

where r_0 depends only on x and not on y. We note that Eqs. (3.23) and (3.24) now have exactly the same form as Eqs. (3.12) and (3.14) for two-dimensional flow. In Problem 3.5, we discuss a transformation known as the *Mangler transformation*, which transforms the axisymmetric equations for small t into those of an equivalent two-dimensional flow.

In circular pipes and jets we use Eqs. (3.15)–(3.17) again, with $K = 1$ and with r_0 and ϕ in Eq. (3.18) equal to zero so that $y = r$. It is usual to leave Eqs. (3.16) and (3.17) otherwise unaltered for compactness, rather than expand the right-hand side by using $\partial r / \partial r \equiv 1$. Equation (3.15) can, of course, be simplified to

$$\frac{\partial u}{\partial x} + \frac{1}{r} \frac{\partial}{\partial r} (rv) = 0$$ (3.25)

since $\partial r / \partial x = 0$. The reader should note, and understand, the differences between Eqs. (3.22) and (3.25).

3.2 Estimates of Density Fluctuations in Coupled Turbulent Flows

Density fluctuations in a pure gas are related to temperature and pressure fluctuations by the gas law $p = \rho R T$ in its logarithmically differentiated form

$$\frac{\rho'}{\bar{\rho}} = \frac{p'}{\bar{p}} - \frac{T'}{\bar{T}}.$$ (3.26)

Temperature fluctuations do not appear, as such, in the equations of motion; they affect turbulence via their effect on viscosity and thermal conductivity as well as density, but the effect of viscosity on the large-scale turbulent motions that contribute most to the turbulent stresses is small in any case. Pressure fluctuations and of course mean-pressure gradients appear in the instantaneous equations, even in constant-density (uncoupled) flow, but provided that the ratio of the pressure fluctuation p' to the mean absolute pressure $\bar{p}$ is small, the effect of pressure fluctuations should be nearly the same as if that ratio were *negligible*. It will appear below that density fluctuations in high-speed shear layers are generated mainly by temperature fluctuations; $p'/\bar{p}$ *is* quite small compared with $\rho'/\bar{\rho}$. Therefore the discussion of compressibility effects can be based on $\rho'/\bar{\rho}$, or $T'/\bar{T}$, alone; below, in places where T'/T is used for convenience, the reader concerned with nearly isothermal mixtures of gases should read $-\rho'/\bar{\rho}$.

Morkovin [1], analyzing the measurements of Morkovin and Phinney [2] and Kistler [3], observed that total temperature fluctuations in supersonic boundary layers with small rates of heat transfer through the walls were much smaller than (static) temperature fluctuations. Figure 3.3 shows that the root-mean-square (rms) total temperature fluctuation, normalized by a factor that approximately collapses results at different Mach numbers, is no more than 0.05. The result derived as Eq. (3.28), below, implies that *static* temperature fluctuations normalized in the same way would be about $2\sqrt{\overline{u'^2}}/u_e$, which is several times larger. Now the instantaneous total temperature is defined by

$$T_0 + T_0' = T + T' + \frac{1}{2c_p}\left[(u+u')^2 + (v+v')^2 + (w+w')^2\right], \quad (3.27a)$$

but all velocity fluctuations are small compared with u, and in a two-dimensional thin shear layer with flow predominantly in the x direction, $v \ll u$ and $w = 0$. Thus expanding the squared terms and neglecting small quantities,

$$T_0 + T_0' \simeq T + T' + \frac{u^2}{2c_p} + \frac{uu'}{c_p}. \quad (3.27b)$$

Now, if the total temperature fluctuation T_0' is small compared with the

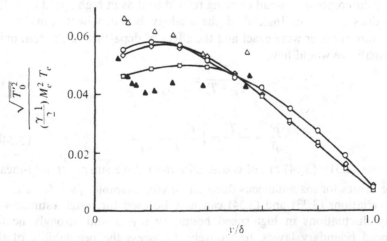

Figure 3.3. Root-mean-square fluctuations in *total* temperature in supersonic boundary layers (after Morkovin ([1]).

(static) temperature fluctuation T', then subtracting from Eq. (3.27b) its own time average $T_0 = T + u^2/2c_p$, we have the result

$$T' \simeq -\frac{uu'}{c_p}. \tag{3.28}$$

This result is qualitatively plausible for high-speed boundary layers because the mean temperature is high where the velocity is low (near the surface, where dissipation of kinetic energy into heat by viscous and turbulent stresses is highest), and we therefore expect *fluctuations* of velocity and temperature to tend to have opposite signs at any instant. Noting that the speed of sound a is given by

$$a^2 = (\gamma - 1)c_p T \tag{3.29}$$

and that the local Mach number M is defined by

$$M = \frac{u}{a}, \tag{3.30}$$

we can rewrite (3.28) as

$$\frac{T'}{T} \simeq -(\gamma - 1)M^2\frac{u'}{u}, \tag{3.31}$$

which shows at once that since u'/u is usually small, T'/T is usually small if $(\gamma - 1)M^2$ is not large compared with unity. Now if pressure fluctuations are small, the gas law, Eq. (3.26), reduces to

$$\frac{\rho'}{\bar{\rho}} \simeq \frac{-T'}{T}, \tag{3.32}$$

and therefore we have the key result that for high-speed flows

$$\frac{\rho'}{\bar{\rho}} \simeq (\gamma - 1)M^2\frac{u'}{u}. \tag{3.33}$$

In low-speed boundary layers on heated walls, the velocity and temperature fluctuations again tend to have opposite signs. However, since the driving temperature difference $T_w - T_e$ is imposed separately from the velocity difference u_e instead of being related to it as in high-speed flow, Eq. (3.33) does not hold. Instead, if the analogy between heat transfer and momentum transfer were exact and the effect of departure of Pr from unity were small, we would have

$$\frac{T'}{T_w - T_e} \simeq -\frac{u'}{u_e} \tag{3.34a}$$

or

$$\frac{\rho'}{\bar{\rho}} = \frac{-T'}{T} = \left(\frac{T_w - T_e}{T}\right)\frac{u'}{u_e}. \tag{3.34b}$$

Equations (3.31)–(3.34) are of course still valid if we substitute root-mean-square values for instantaneous fluctuations (for example, $\sqrt{\overline{u'^2}}$ for u').

The relations (3.33) and (3.34) can now be used for rough estimates of density fluctuations in high-speed boundary layers and strongly heated low-speed boundary layers, respectively, to assess the negligibility of the terms in the mass, momentum, and energy equations in which density fluctuations appear. The relations are likely to be less satisfactory in free shear layers, where the velocity and temperature fluctuations are less closely related, but they should still be useful for rough estimates.

The largest value of $\sqrt{u'^2}/u_e$ reached in a low-speed boundary layer in zero pressure gradient is about 0.1. Values in high-speed flows may be even lower, but if we use the low-speed figure in the root-mean-square version of the above formulas, we can make generous estimates of $\sqrt{T'^2}/T$ or $\sqrt{\rho'^2}/\bar{\rho}$ for high-speed boundary layers or for low-speed boundary layers on strongly-heated walls. Typical figures are given in Table 3.1; to express the temperature fluctuations in high-speed flow as fractions of T_e, we have assumed that the maximum temperature fluctuation occurs where $u/u_e = 0.5$. The figures for infinite temperature ratio or Mach number are not realistic, but they serve to show that the ratio of temperature fluctuation to local (or wall) temperature does not rise indefinitely. This is easy enough to see in the case of Eq. (3.34), which reduces, for $T_w/T_e \to \infty$, to $\sqrt{T'^2}/T_w = \sqrt{u'^2}/u_e$. In the case of Eq. (3.33), the explanation for the approach to an asymptotic value is that if the heat transfer from a high-speed flow to a surface is not too large, the surface temperature is not much less than the total temperature of the external stream and therefore rises rapidly with Mach number. Values given in Table 3.1(b) are for the "recovery" temperature, the temperature reached by an insulated (adiabatic) surface.

The recovery temperature is always *less* than the external-stream stagnation temperature H_e/c_p except at the stagnation point itself, whether the Prandtl number is greater or less than unity. The difference is usually expressed in terms of the *recovery factor*

$$r = \frac{h_{w,ad} - h_e}{H_e - h_e},$$ (3.35)

where subscript ad denotes adiabatic-wall condition. For a perfect gas, Eq. (3.35) becomes, writing T_r for $T_{w,ad}$,

$$r = \frac{T_r - T_e}{u_e^2/2c_p} = \frac{T_r/T_e - 1}{\frac{1}{2}(\gamma - 1)M_e^2},$$ (3.36a)

or

$$T_r = T_e\left[1 + \frac{r}{2}(\gamma - 1)M_e^2\right],$$ (3.36b)

using the definitions of total enthalpy and of Mach number M.

The above analysis is inexact and neglects the decrease in typical values of $\sqrt{u'^2}/u_e$ as M_e increases, but it leads to such small values for the ratio of root-mean-square temperature (or density) fluctuation to the mean temperature (or density) that even rather large errors in the analysis would still allow us to conclude that density fluctuations are small in practice, both in low-speed flows with high heat transfer and in high-speed adiabatic-wall flows.

Thus most terms containing ρ' can be dropped from the mass, momentum, and enthalpy equations for thin shear layers. As an example, let us consider the continuity equation

$$\frac{\partial}{\partial x}(\rho u + \overline{\rho' u'}) + \frac{\partial}{\partial y}(\rho v + \overline{\rho' v'}) = 0. \tag{2.39}$$

Using Eq. (3.33), we see that in a high-speed boundary layer

$$\overline{\rho' v'} \simeq (\gamma - 1) M^2 \rho \frac{\overline{u'v'}}{u} = \rho v \left[(\gamma - 1) M^2 \frac{\overline{u'v'}}{uv} \right]. \tag{3.37}$$

We cannot expect the quantity in square brackets in Eq. (3.37) to be of an order smaller than unity in any supersonic flow ($M > 1$). Therefore, we *cannot* neglect $\overline{\rho' v'}$ compared with ρv in the continuity or momentum equations. (Strictly speaking, we should have considered the gradients of ρv and $\overline{\rho' v'}$ with respect to y, but according to the arguments of Section 3.1, these are of order $1/\delta$ times the typical values of the quantities themselves.)

The term $\overline{\rho' u'}$ can be evaluated approximately as a fraction of ρu in the same way—in fact one merely changes v to u and v' to u' everywhere in Eq. (2.39) (see Problem 3.6). The result is that $\overline{\rho' u'}$ is negligible compared with ρu as long as $(\gamma - 1) M^2$ is not an order of magnitude greater than unity.

Discussion of terms such as $\partial(\overline{\rho' u'^2})/\partial x$ involves triple products of fluctuating velocity—in this case $\overline{u'^3}$. Here we merely remark that these

Table 3.1 Approximate estimates of temperature fluctuations assuming $\sqrt{\overline{u'^2}} / u_e = 0.1$.

(a) Low-speed flow over heated wall

$(T_w - T_e)/T_e$	0.25	0.5	1	2	4	∞
$T_w - T_e$ for $T_e = 300 K$	75	150	300	600	1200	∞
$\sqrt{\overline{T'^2}}/T_e$ from Eq. (3.34)	0.025	0.05	0.1	0.2	0.4	∞
$\sqrt{\overline{T'^2}}/T_w$	0.02	0.033	0.05	0.067	0.08	0.10

(b) High-speed flow over adiabatic wall (zero heat transfer to surface) (T_r is "recovery" temperature (Eq. (3.36).))

M_e	1	2	3	4	5	∞
$(T_r - T_e)/T_e$	0.178	0.712	1.6	2.85	4.45	∞
$\sqrt{\overline{T'^2}}/T_e$ from Eq. (3.33)	0.04	0.16	0.36	0.64	1.0	∞
$\sqrt{\overline{T'^2}}/T_r$	0.017	0.047	0.069	0.083	0.092	0.112

Note: The last line of each section of the table is the more meaningful because maximum temperature fluctuations occur near the wall.

terms are even smaller than terms such as $\partial(u\overline{\rho'u'})/\partial x$ and can therefore be neglected in nearly all circumstances (see Bradshaw [4]).

3.3 Equations for Coupled Turbulent Flows

The conservation equations for compressible flows given in Section 2.4 are somewhat lengthy and will not be repeated here.

The continuity equation for a compressible two-dimensional turbulent flow is given by Eq. (2.39), and the x-component momentum equation is given by Eq. (2.41) or (2.42). The form of the x-component momentum equation that is valid for compressible laminar or turbulent flow in a thin shear layer can again be obtained by order-of-magnitude arguments; the arguments for the main terms go through exactly as for constant-property flow, but the terms containing density fluctuations in turbulent flow need some care. Therefore, this section is mainly concerned with the turbulence terms, and the equations for coupled laminar flow are obviously special cases of the turbulent-flow equations.

As discussed in Section 3.2, the terms $\overline{\rho'u'}$ and $\overline{\rho'v'}$ are generally of the same order, and $\overline{\rho'v'}$ can be as large as ρv even at moderate supersonic Mach numbers or in low-speed flows with absolute temperature variations as small as 2:1. However, $\overline{\rho'u'} \ll \rho u$, and we can neglect it both in the first term and in the bracketed stress terms of Eq. (2.42). It is *usually* possible to neglect $\overline{\rho'u'^2}$ compared with $\overline{\rho u'^2}$ and $\overline{\rho'u'v'}$ compared with $\overline{\rho u'v'}$ because $\rho'/\bar\rho$ is fairly small; however, in shear layers with large mean density variations $\overline{\rho'u'v'}$ can be several percent of $\overline{\rho u'v'}$.

Momentum and Continuity Equations

We now apply the order-of-magnitude arguments of Section 3.1 to the remaining terms of Eq. (2.42). The viscous-stress terms, which in Eq. (2.42) were so complicated that we left them in σ notation, simplify greatly because the bulk-viscosity terms in Eq. (2.6) for the σ stresses, involving the dilatation $\partial u/\partial x + \partial v/\partial y$, are small. We finally obtain

$$\rho u \frac{\partial u}{\partial x} + \overline{\rho v}\frac{\partial u}{\partial y} = -\frac{dp}{dx} + \frac{\partial}{\partial y}\left(\mu \frac{\partial u}{\partial y}\right) - \frac{\partial}{\partial y}\left(\overline{\rho u'v'} + \overline{\rho'u'v'}\right) + \rho f_x + \overline{\rho'f_x'},$$

(3.38)

where $\overline{\rho v}$ is an abbreviation for $\rho v + \overline{\rho'v'}$, and we have retained the $\overline{\rho'u'v'}$ term as a reminder, although we shall neglect it in later discussions.

Using the above argument about $\overline{\rho'v'}$ and its companions, we see that in a thin shear layer the continuity equation (2.39) becomes

$$\frac{\partial \rho u}{\partial x} + \frac{\partial}{\partial y}\overline{\rho v} = 0.$$

(3.39)

The y-component momentum equation is usually taken as $\partial p/\partial y = 0$, so that, as in constant-density flow, $\partial p/\partial x$ can be replaced by $-\rho_e u_e\, du_e/dx$, where ρ_e is the external-stream density. At high Mach numbers the changes of pressure due to streamline curvature in the xy plane or to the Reynolds-stress gradient in the y direction, $\partial(\overline{\rho v'^2}+\overline{\rho'v'^2})/\partial y$, may be a significant fraction of the absolute pressure. The pressure difference across the shear layer due to streamline curvature with radius R is of order $\rho_e u_e^2 \delta/R$, where ρ_e and u_e are the density and velocity, respectively, at the edge of the shear layer (say). Since the edge Mach number M_e is given by $M_e^2 = \rho_e u_e^2/\gamma p_e$, the ratio of the pressure change induced by curvature to the absolute pressure at the edge, p_e, is of order $\gamma M_e^2 \delta/R$. The pressure change due to the Reynolds stress is approximately $\overline{\rho v'^2}$, and $\overline{\rho v'^2}$ is of the same order as $\overline{\rho u'v'}$ and therefore, in the case of a boundary layer, of the same order as the surface shear stress τ_w. Therefore the ratio of a typical pressure change due to Reynolds stress to the absolute pressure is of order τ_w/p_e; substituting for p_e from the above definition of M_e, we find

$$\frac{\tau_w}{p_e} = \frac{\tau_w}{\rho_e u_e^2/2}\left(\frac{\gamma M_e^2}{2}\right) = c_f \frac{1}{2}\gamma M_e^2.$$

Although c_f decreases as M_e increases, the ratio of stress-induced pressure difference to absolute pressure increases, particularly on cold walls where c_f is higher.

The Enthalpy Equation

The enthalpy equation for compressible flow remains slightly more complicated than the constant-property equation even after the thin-shear-layer simplifications are made, but the dissipation term Φ defined by Eq. (2.27) simplifies greatly, because only y-component derivatives need be included and even then $\partial v/\partial y$ is always small. For two-dimensional steady flow we obtain, from Eq. (2.49),

$$c_p\left(\overline{\rho u}\frac{\partial T}{\partial x} + \overline{\rho v}\frac{\partial T}{\partial y}\right) = \frac{\partial}{\partial y}\left(k\frac{\partial T}{\partial y}\right) - c_p\frac{\partial}{\partial y}\left(\overline{\rho T'v'}+\overline{\rho'T'v'}\right) + u\frac{dp}{dx}$$

$$+ \mu\left(\frac{\partial u}{\partial y}\right)^2 + \rho\varepsilon. \tag{3.40}$$

Note that according to the thin-shear-layer approximation, p is a function only of x in a two-dimensional flow.

The question of whether the pressure-gradient term and the viscous dissipation term in the enthalpy equation are negligible depends on the Mach number, and has nothing to do with the validity of the thin-shear-layer approximation. This is, however, a convenient place to discuss the question because the terms are algebraically so much simpler in a thin shear layer. The criterion for the neglect of the dissipation term $\mu(\partial u/\partial y)^2$ in Eq.

(3.40) compared with the molecular conduction term is

$$\nu\left(\frac{\partial u}{\partial y}\right)^2 \ll \frac{1}{\rho}\frac{\partial}{\partial y}\left(k\frac{\partial T}{\partial y}\right).$$

To an adequate order-of-magnitude approximation in a flow with a typical temperature change ΔT in the y direction and a typical velocity scale u_e (for example, the flow speed at the outer edge of a boundary layer), this becomes

$$\nu\left(\frac{u_e}{\delta}\right)^2 \ll \frac{k}{\rho}\frac{\Delta T}{\delta^2},$$

or equivalently, multiplying by $\delta^2/[(\gamma-1)c_p T_e \nu]$, where T_e is the temperature at the shear-layer edge, say,

$$\frac{u_e^2}{(\gamma-1)c_p T_e} \equiv M_e^2 \ll \frac{1}{(\gamma-1)\mathrm{Pr}}\frac{\Delta T}{T_e}.$$

Since $(\gamma-1)\mathrm{Pr}$ is of order unity in gas flows, the condition for neglect of the dissipation term in the enthalpy equation is simply that the Mach number shall be much less than the square root of the fractional change in absolute temperature imposed by heat transfer to or from the flow; if $\Delta T = 10°C$ and $T = 300$ K, then $M \ll 0.2$. In a laminar boundary layer the condition for the neglect of $u\,dp/dx$ is similar, relying on the thin-shear-layer result $\delta^2 \sim \nu x/u_e$. In a turbulent boundary layer it is even weaker (see Problem 3.7). In high-speed flows where dissipation provides the main heat source, the dissipation and conduction terms are necessarily of the same order, showing at once that $\Delta T/T$ is of order M^2 in this case.

If the pressure-gradient and viscous dissipation terms are negligible, in which case $\overline{\rho'T'v'}$ will be negligible as well, Eq. (3.40) reduces to

$$\rho c_p\left(u\frac{\partial T}{\partial x} + v\frac{\partial T}{\partial y}\right) = \frac{\partial}{\partial y}\left(k\frac{\partial T}{\partial y}\right) - c_p\frac{\partial}{\partial y}\left(\rho\overline{T'v'}\right). \tag{3.41}$$

The Total-Enthalpy Equation

To the thin-shear-layer approximation, the time-average total enthalpy for a two-dimensional turbulent flow

$$H \equiv h + \tfrac{1}{2}\left(u^2 + v^2 + \overline{u'^2} + \overline{v'^2} + \overline{w'^2}\right) \tag{3.42a}$$

reduces to

$$H = h + \tfrac{1}{2}u^2, \tag{3.42b}$$

and for a perfect gas it reduces to

$$H = c_p T + \tfrac{1}{2}u^2 \tag{3.42c}$$

The conservation equation for total enthalpy in a general flow, obtained from the conservation equation for thermal plus kinetic energy by adding a "conservation" equation for $h - e \equiv p/\rho$ (see Problem 2.7), becomes rather complicated when turbulent terms are added. To the thin-shear-layer approximation, however, we can use the definition Eq. (3.42c) and add a conservation equation for $u^2/2$ to the (static) enthalpy equation (3.40). The conservation equation for $u^2/2$ is just u times the x-component momentum equation, Eq. (3.38), which can be arranged as

$$\rho u \frac{\partial}{\partial x}\left(\frac{u^2}{2}\right) + \overline{\rho v}\frac{\partial}{\partial y}\left(\frac{u^2}{2}\right) = -u\frac{dp}{dx} + u\frac{\partial}{\partial y}\left(\mu\frac{\partial u}{\partial y}\right) - u\frac{\partial}{\partial y}\left(\rho\overline{u'v'} + \overline{\rho'u'v'}\right)$$

$$+ \rho u f_x + u\overline{\rho'f_x'}. \tag{3.43}$$

Adding this to Eq. (3.40), we get, substituting the definition Eq. (3.42c) for $c_p T + u^2/2$,

$$\rho u\frac{\partial H}{\partial x} + \overline{\rho v}\frac{\partial H}{\partial y} = \frac{\partial}{\partial y}\left(k\frac{\partial T}{\partial y} + \mu u\frac{\partial u}{\partial y}\right) - c_p\frac{\partial}{\partial y}\left(\rho\overline{T'v'} + \overline{\rho'T'v'}\right)$$

$$- u\frac{\partial}{\partial y}\left(\rho\overline{u'v'} + \overline{\rho'u'v'}\right) + \rho\varepsilon + \rho u f_x + u\overline{\rho'f_x'} \tag{3.44a}$$

or, using Eq. (3.42c) and the definition of Prandtl number $\text{Pr} = \mu c_p/k$,

$$\rho u\frac{\partial H}{\partial x} + \overline{\rho v}\frac{\partial H}{\partial y} = \frac{\partial}{\partial y}\left[\frac{\mu}{\text{Pr}}\frac{\partial H}{\partial y} + \mu\left(1 - \frac{1}{\text{Pr}}\right)u\frac{\partial u}{\partial y}\right] - c_p\frac{\partial}{\partial y}\left(\rho\overline{T'v'} + \overline{\rho'T'v'}\right)$$

$$- u\frac{\partial}{\partial y}\left(\rho\overline{u'v'} + \overline{\rho'u'v'}\right) + \rho\varepsilon + \rho u f_x + u\overline{\rho'f_x'}. \tag{3.44b}$$

Now ε, the rate of dissipation of turbulent kinetic energy into thermal internal energy, is approximately equal, over most parts of the flow, to the rate of production of turbulent energy, which in turn is equal to the product of the turbulent shear stress $\rho\overline{u'v'} + \overline{\rho'u'v'}$ and the mean velocity gradient $\partial u/\partial y$ (the product of a stress and a rate of strain is equal to a rate of doing work). The error incurred in assuming production equal to dissipation is of the same order as approximations already made, for instance, in going from Eq. (3.42a) to Eq. (3.42b). We therefore merge the shear-stress term and the dissipation ($\equiv$ production) term as $\partial[u(\rho\overline{u'v'} + \overline{\rho'u'v'})]/\partial y$. If we now write

$$\tau \equiv \mu\frac{\partial u}{\partial y} - \rho\overline{u'v'} - \overline{\rho'u'v'}, \tag{3.45a}$$

$$\dot{q} \equiv -k\frac{\partial T}{\partial y} + c_p\rho\overline{T'v'} + c_p\overline{\rho'T'v'} \tag{3.45b}$$

for compactness, Eq. (3.44a) becomes

$$\rho u\frac{\partial H}{\partial x} + \overline{\rho v}\frac{\partial H}{\partial y} = \frac{\partial}{\partial y}(-\dot{q} + u\tau) + \rho u f_x + u\overline{\rho'f_x'}. \tag{3.46}$$

For a laminar flow, Eq. (3.44b) reduces to

$$\rho u\frac{\partial H}{\partial x} + \rho v\frac{\partial H}{\partial y} = \frac{\partial}{\partial y}\left[\frac{\mu}{\text{Pr}}\frac{\partial H}{\partial y} + \mu\left(1 - \frac{1}{\text{Pr}}\right)u\frac{\partial u}{\partial y}\right]. \tag{3.47a}$$

If the Prandtl number is unity, then the total-enthalpy equation becomes

$$\rho u\frac{\partial H}{\partial x} + \rho v\frac{\partial H}{\partial y} = \frac{\partial}{\partial y}\left(\mu\frac{\partial H}{\partial y}\right). \tag{3.47b}$$

Axisymmetric Flows

The coupled versions of the axisymmetric flow equations given as Eqs. (3.15)–(3.17) can be written down without further comment as

$$\frac{\partial}{\partial x}\left(r^{K}\rho u\right)+\frac{\partial}{\partial y}\left(r^{K}\overline{\rho v}\right)=0, \tag{3.48}$$

$$\rho u\frac{\partial u}{\partial x}+\overline{\rho v}\frac{\partial u}{\partial y}=-\frac{dp}{dx}+\frac{1}{r^{K}}\frac{\partial}{\partial y}\left[r^{K}\left(\mu\frac{\partial u}{\partial y}-\overline{\rho\,u'v'}-\overline{\rho'u'v'}\right)\right]$$
$$+\rho f_{x}+\overline{\rho'f_{x}'}, \tag{3.49}$$

$$\rho u\frac{\partial T}{\partial x}+\overline{\rho v}\frac{\partial T}{\partial y}=\frac{c_{p}}{r^{K}}\frac{\partial}{\partial y}\left[r^{K}\left(\frac{k}{c_{p}}\frac{\partial T}{\partial y}-\overline{\rho T'v'}-\overline{\rho'T'v'}\right)\right], \tag{3.50}$$

for a perfect gas. The axisymmetric version of the total-enthalpy equation, Eq. (3.44a), is

$$\rho u\frac{\partial H}{\partial x}+\overline{\rho v}\frac{\partial H}{\partial y}=\frac{1}{r^{K}}\frac{\partial}{\partial y}\left\{r^{K}\left[k\frac{\partial T}{\partial y}-\rho c_{p}\overline{T'v'}-c_{p}\overline{\rho'T'v'}\right.\right.$$
$$\left.\left.+u\left(\mu\frac{\partial u}{\partial y}-\overline{\rho u'v'}-\overline{\rho'u'v'}\right)\right]\right\}+\rho\varepsilon+\rho u f_{x}+\overline{u\rho'f_{x}'}. \tag{3.51a}$$

Similarly the axisymmetric version of Eq. (3.44b) is

$$\rho u\frac{\partial H}{\partial x}+\overline{\rho v}\frac{\partial H}{\partial y}=\frac{1}{r^{K}}\frac{\partial}{\partial y}\left\{r^{K}\left[\frac{\mu}{\mathrm{Pr}}\frac{\partial H}{\partial y}+\mu\left(1-\frac{1}{\mathrm{Pr}}\right)u\frac{\partial u}{\partial y}\right]\right\}$$
$$-\frac{c_{p}}{r^{K}}\frac{\partial}{\partial y}\left[r^{K}\left(\rho\overline{T'v'}+\overline{\rho'T'v'}\right)\right]$$
$$-\frac{u}{r^{K}}\frac{\partial}{\partial y}\left[r^{K}\left(\rho\overline{u'v'}+\overline{\rho'u'v'}\right)\right]+\rho u f_{x}+u\overline{\rho'f_{x}'}. \tag{3.51b}$$

Using the compact notation of Eq. (3.45), Eqs. (3.49), (3.50), and (3.51a) can also be written as

$$\rho u\frac{\partial u}{\partial x}+\overline{\rho v}\frac{\partial u}{\partial y}=-\frac{dp}{dx}+\frac{1}{r^{K}}\frac{\partial}{\partial y}\left(r^{K}\tau\right)+\rho f_{x}+\overline{\rho'f_{x}'}, \tag{3.52}$$

$$\rho u\frac{\partial T}{\partial x}+\overline{\rho v}\frac{\partial T}{\partial y}=-\frac{1}{r^{K}}\frac{\partial}{\partial y}\left(-r^{K}\dot{q}\right), \tag{3.53}$$

$$\rho u\frac{\partial H}{\partial x}+\overline{\rho v}\frac{\partial H}{\partial y}=-\frac{1}{r^{K}}\frac{\partial}{\partial y}\left[r^{K}\left(-\dot{q}+u\tau\right)\right]+\rho u f_{x}+u\overline{\rho'f_{x}'}. \tag{3.54}$$

As before, in the above equations the flow index K is unity for an axisymmetric flow and zero for a two-dimensional flow.

3.4 Integral Equations

The equations that result from integrating the momentum and enthalpy equations across the width of the shear layer (i.e., from $y = 0$ to $y = \delta$) are useful in qualitative discussion and in simple types of calculation methods called *integral methods*. For this reason we discuss the derivation of these equations for two-dimensional flows below and quote their axisymmetric versions without proof later in this section.

Two-Dimensional Flows

We first consider the momentum equation, Eq. (3.52), neglect body forces, write the viscous shear stress plus the turbulent shear stress as τ, and replace $-dp/dx$ by $\rho_e u_e \, du_e/dx$ to get

$$\rho u \frac{\partial u}{\partial x} + \overline{\rho v} \frac{\partial u}{\partial y} = \rho_e u_e \frac{du_e}{dx} + \frac{\partial \tau}{\partial y}. \tag{3.55}$$

We then multiply the continuity equation, Eq. (3.48), by u and add the resulting expression to Eq. (3.55),

$$\frac{\partial}{\partial x}(\rho u^2) + \frac{\partial}{\partial y}(u\overline{\rho v}) = \rho_e u_e \frac{du_e}{dx} + \frac{\partial \tau}{\partial y}. \tag{3.56}$$

We now integrate the above equation with respect to y across the shear layer, choosing the upper limit of integration as $y = h$, where h is independent of x and sufficiently greater than the shear-layer thickness δ so that all quantities reach their external-flow values and $u \rightarrow u_e$, etc. Then Eq. (3.56) becomes

$$\int_0^h \frac{\partial}{\partial x}(\rho u^2) \, dy + u_e \rho_h v_h = \int_0^h \rho_e u_e \frac{du_e}{dx} \, dy - \tau_w, \tag{3.57}$$

where τ_w is the wall shear stress. The expression

$$\rho_h v_h = -\int_0^h \frac{\partial}{\partial x}(\rho u) \, dy$$

follows from the continuity equation, Eq. (3.39); thus

$$\int_0^h \left[\frac{\partial}{\partial x}(\rho u^2) - u_e \frac{\partial}{\partial x}(\rho u) - \rho_e u_e \frac{du_e}{dx} \right] dy = -\tau_w \tag{3.58}$$

or, after rearranging,

$$\int_0^h \left\{ -\frac{\partial}{\partial x}[\rho u(u_e - u)] - \frac{du_e}{dx}(\rho_e u_e - \rho u) \right\} dy = -\tau_w. \tag{3.59}$$

Now since $u_e - u = 0$ for $y \geq h$, both parts of the integrand contribute only when $y < h$ and so are independent of h. Therefore, the first part of the integral can be written as

$$-\frac{d}{dx} \int_0^h \rho u(u_e - u) \, dy,$$

the x derivative now being ordinary rather than partial because the definite integral is independent of y. With a little rearrangement and a change of sign, and replacing h by ∞ because the integrands are zero for large y, we get

$$\frac{d}{dx}\left[\rho_e u_e^2 \int_0^\infty \frac{\rho u}{\rho_e u_e}\left(1-\frac{u}{u_e}\right)dy\right] + \rho_e u_e \frac{du_e}{dx}\int_0^\infty \left(1-\frac{\rho u}{\rho_e u_e}\right)dy = \tau_w. \qquad (3.60)$$

The momentum thickness θ, defined by

$$\theta = \int_0^\infty \frac{\rho u}{\rho_e u_e}\left(\frac{u_e - u}{u_e}\right)dy, \qquad (3.61)$$

is a measure of the momentum-flux deficit caused by the presence of the boundary layer. Similarly, the displacement thickness δ^*, which is a measure of deficiency of mass flow rate, can be defined as

$$\delta^* = \int_0^\infty \left(1-\frac{\rho u}{\rho_e u_e}\right)dy. \qquad (3.62)$$

Introducing Eqs. (3.61) and (3.62) into Eq. (3.60), we obtain

$$\frac{d}{dx}\left(\rho_e u_e^2 \theta\right) + \rho_e u_e \frac{du_e}{dx}\delta^* = \tau_w, \qquad (3.63a)$$

or

$$\frac{d}{dx}\left(\rho_e u_e^2 \theta\right) = \tau_w + \delta^* \frac{dp}{dx}. \qquad (3.63b)$$

Noting the isentropic relation

$$\frac{1}{\rho_e}\frac{d\rho_e}{dx} = -\frac{M_e^2}{u_e}\frac{du_e}{dx}, \qquad (3.64)$$

and using the definitions of local skin-friction coefficient

$$c_f = \frac{\tau_w}{\frac{1}{2}\rho_e u_e^2} \qquad (3.65)$$

and shape factor

$$H = \frac{\delta^*}{\theta}, \qquad (3.66)$$

we can write Eq. (3.63) as

$$\frac{d\theta}{dx} + \frac{\theta}{u_e}\frac{du_e}{dx}\left(H+2-M_e^2\right) = \frac{c_f}{2}. \qquad (3.67)$$

This is the *momentum integral equation for a two-dimensional compressible laminar or turbulent flow*. For a constant-density flow, it becomes

$$\frac{d\theta}{dx} + \frac{\theta}{u_e}\frac{du_e}{dx}\left(H+2\right) = \frac{c_f}{2}, \qquad (3.68)$$

where now θ, δ^*, and c_f are given by

$$\theta = \int_0^\infty \frac{u}{u_e}\left(1 - \frac{u}{u_e}\right) dy, \tag{3.69a}$$

$$\delta^* = \int_0^\infty \left(1 - \frac{u}{u_e}\right) dy, \tag{3.69b}$$

$$c_f = \frac{\tau_w}{\frac{1}{2}\rho u_e^2}. \tag{3.69c}$$

Equation (3.67) also applies to free shear layers if we take $y = 0$ at or below the lower edge of the layer and put $c_f = 0$.

To derive the enthalpy integral equation for two-dimensional compressible flows without body force, we start with the total enthalpy equation, Eq. (3.46). Using the continuity equation, Eq. (3.39), we can write Eq. (3.46) as

$$\frac{\partial}{\partial x}(\rho u H) + \frac{\partial}{\partial y}(\overline{\rho v}H) = \frac{\partial}{\partial y}(-\dot{q} + u\tau). \tag{3.70}$$

We now integrate the above equation with respect to y from $y = 0$ to $y = h > \delta$ to get

$$\int_0^h \frac{\partial}{\partial x}(\rho u H)\, dy + \rho_h v_h H_e = \dot{q}_w. \tag{3.71}$$

As in the derivation of the momentum integral equation, we substitute for $\rho_h v_h$ from the continuity equation and write Eq. (3.71) as

$$\int_0^h \frac{\partial}{\partial x}[\rho u(H_e - H)]\, dy = -\dot{q}_w \tag{3.72}$$

and then as

$$\frac{d}{dx}\left[\rho_e u_e \int_0^\infty \frac{\rho u}{\rho_e u_e}\left(1 - \frac{H}{H_e}\right) dy\right] = -\frac{\dot{q}_w}{H_e}, \tag{3.73}$$

using reasoning similar to that which led to the writing of Eq. (3.59) as Eq. (3.60). Since $H_e - H = 0$ (to sufficient accuracy) for $y \geqslant h$, the integrand of Eq. (3.73) contributes only for $y < h$, and so the result is independent of h.

Equation (3.73) shows that the rate of increase of total-enthalpy deficit per unit span (in the z direction) is equal to the rate of heat transfer *from* the fluid to a unit area of the surface. Comparing this equation with Eq. (3.60), we see that the rate of increase of deficit in each case is affected by transfer into the surface (enthalpy transfer $-\dot{q}_w$, momentum transfer τ_w); the momentum integral equation contains an additional term depending on pressure gradient, but as we have already seen in Section 3.3, the *total* enthalpy is unaffected by pressure gradients as such.

If we introduce θ_H by

$$\theta_H = \int_0^h \frac{\rho u}{\rho_e u_e}\left(\frac{H - H_e}{H_w - H_e}\right) dy, \tag{3.74}$$

then we can write the total-enthalpy integral equation (3.72) for a two-dimensional compressible laminar or turbulent flow as

$$\frac{d}{dx}\left[\rho_e u_e (H_w - H_e)\theta_H\right] = \dot{q}_w.$$ (3.75)

In Eq. (3.74), θ_H is a measure of total-enthalpy-flux *surplus* caused by the presence of the thermal boundary layer. For an incompressible flow, where H is equal to the static enthalpy h, the total-enthalpy thickness θ_H is equal to the static-enthalpy thickness

$$\theta_h = \int_0^\infty \frac{\rho u}{\rho_e u_e}\left(\frac{h - h_e}{h_w - h_e}\right) dy,$$ (3.76)

where h can be replaced by $c_p T$ if c_p is constant. Noting this and taking ρ constant, we can write Eq. (3.75) for an incompressible flow as

$$\frac{d}{dx}\left[u_e (T_w - T_e)\theta_h\right] = \frac{\dot{q}_w}{\rho c_p}.$$ (3.77)

When the wall temperature is *uniform*, Eq. (3.77) can be written as

$$\frac{1}{u_e}\frac{d}{dx}(u_e\theta_T) = \text{St}$$ (3.78a)

or as

$$\frac{d\theta_T}{dx} + \frac{\theta_T}{u_e}\frac{du_e}{dx} = \text{St}.$$ (3.78b)

Here St denotes the Stanton number defined by

$$\text{St} = \frac{\dot{q}_w}{\rho c_p (T_w - T_e) u_e},$$ (3.79)

and

$$\theta_T = \int_0^h \frac{u}{u_e}\left(\frac{T - T_e}{T_w - T_e}\right) dy.$$ (3.80)

Axisymmetric Flows

For axisymmetric compressible flows, the momentum integral equation can be derived in a similar manner. With the definitions of the *areas* θ and δ^* now given by

$$\theta = \int_0^\infty \frac{r}{r_0}\frac{\rho u}{\rho_e u_e}\left(1 - \frac{u}{u_e}\right) dy,$$ (3.81a)

$$\delta^* = \int_0^\infty \frac{r}{r_0}\left(1 - \frac{\rho u}{\rho_e u_e}\right) dy,$$ (3.81b)

it can be written as

$$\frac{d\theta}{dx} + \frac{\theta}{u_e}\frac{du_e}{dx}(H+2-M_e^2) + \frac{\theta}{r_0}\frac{dr_0}{dx} = \frac{c_f}{2},$$ (3.82a)

or

$$\frac{1}{r_0}\frac{d}{dx}(r_0\theta) + \frac{\theta}{u_e}\frac{du_e}{dx}(H+2-M_e^2) = \frac{c_f}{2}.$$ (3.82b)

Similarly the total-enthalpy integral equation can be written as

$$\frac{1}{r_0}\frac{d}{dx}[r_0\rho_e u_e \theta_H (H_w - H_e)] = \dot{q}_w.$$ (3.83)

For an incompressible flow with uniform wall temperature, Eq. (3.83) reduces, using Eq. (3.79), to

$$\frac{d\theta_T}{dx} + \frac{\theta_T}{u_e}\frac{du_e}{dx} + \frac{\theta_T}{r_0}\frac{dr_0}{dx} = \text{St}.$$ (3.84)

3.5 Boundary Conditions

The full equations of motion and heat transfer discussed in Chapter 2 generally require boundary conditions on all sides of the domain in which the solution is to be obtained. (For a detailed discussion, see Chapter 4 of [5], and for a discussion on partial differential equations in general, see [6].) If we restrict ourselves to steady two-dimensional flows, this implies specifying all of the variables or their gradients on four sides of the two-dimensional computational domain, which we can assume for simplicity to be a rectangle. The thin-shear-layer equations are parabolic and require boundary conditions on only three sides of a rectangular domain; conditions at the downstream boundary do not have to be specified because the equations prohibit upstream influence of conditions at $x = x_1$ (see Fig. 3.4) on the solution at $x < x_1$, and this is one of the main simplifications achieved by using the thin-shear-layer equations, since it implies that solutions can be obtained by a single marching sweep from the upstream boundary to the downstream boundary.

Uncoupled Flows

We first consider uncoupled flows, for simplicity. The velocity-field boundary conditions on u and v are that the streamwise component of mean velocity u is defined on three sides of the computational domain whereas v is defined only on one of the streamwise sides, conventionally the lower side. The pressure is *not* an unknown in the thin-shear-layer equations since the

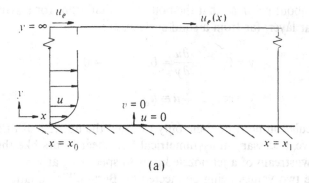

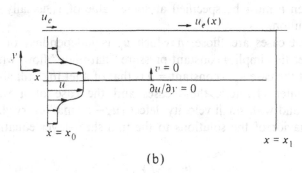

Figure 3.4. Boundary conditions for thin shear layers.

neglect of the normal-component momentum equation implies that the pressure changes through the layer do not affect the solution. If the flow adjacent to the shear layer is "inviscid" so that Bernoulli's equation applies, specifying the streamwise component of velocity at the inviscid edge is equivalent to specifying the pressure, because the normal component of velocity is small enough for its contribution to the total pressure to be neglected.

Thus the conventional initial and boundary conditions for a thin-shear-layer calculation of a boundary layer on the plane $y = 0$ [Fig. 3.4(a)] are

$$0 < y < \infty, \qquad u = u(y), \qquad x = x_0 \quad \text{(initial conditions)}. \qquad (3.85a)$$

$$\left. \begin{aligned} y = 0, \qquad & u = 0, \qquad & v = 0, \\ y = \infty, \qquad & u = u_e(x), \end{aligned} \right\} \qquad x \geq x_0. \qquad (3.85b)$$

Here ∞ represents a value of y large compared with the thickness of the shear layer; in practice, the outer boundary condition can be applied at a value of y (y_e, say) about 1.5 times larger than the conventional "total" thickness of the shear layer, i.e., the distance from the line $y = 0$ to the line on which $u = 0.995u_e$. By implication, the above boundary conditions apply to a boundary layer. Most free shear layers are symmetrical about the line $y = 0$ and therefore have $v = 0$ upon that line, but the specification of $u = 0$ on $y = 0$ must be replaced by the specification of $\partial u / \partial y = 0$ as a condition of symmetry about $y = 0$, so that the boundary conditions for a symmetrical free thin shear layer, far from a solid surface, are

$$y = 0, \qquad \frac{\partial u}{\partial y} = 0, \qquad v = 0,$$

$$y = \infty, \qquad u = u_e.$$

(3.86)

Thus the calculation need be done only for, say, the upper half of the flow. The alternative, necessary in asymmetrical free shear layers like the mixing layer just downstream of a jet nozzle lip, is to specify u_e at $y = -\infty$ as well as $y = \infty$, the two values being connected by Bernoulli's equation and the condition that the static pressure difference across the shear layer shall be negligible; then v must be specified at some value of y, usually from the outer flow solution.

The simplest cases are those in which u_e is independent of x; for a boundary layer this implies constant-pressure "flat-plate" flow, whereas for a turbulent jet the case $u_e = $ constant $= 0$ is that of a jet in "still air." As we shall see in later chapters, these cases and the case of a wake with $u_e = $ constant and with small velocity defect ($u_e - u$ small everywhere) lead to simple behavior of the solutions to the thin-shear-layer equations such that

$$\frac{u}{\Delta u} = f\left(\frac{y}{\delta}\right)$$

(3.87a)

or

$$\frac{u_e - u}{\Delta u} = f\left(\frac{y}{\delta}\right),$$

(3.87b)

where Δu is the velocity change across the shear layer (e.g., the centerline velocity for a jet in still air) and the functions are independent of x except insofar as δ depends on x. These are *similar solutions*, so called because if the equations of motion are reformulated as equations for the function f, they become ordinary differential equations for $f(y/\delta)$ and all velocity profiles $u(y)$ are geometrically similar if plotted as in Eqs. (3.87). We shall see below that departures from similar solutions in real-life shear layers are often small enough so that the similarity form of the equations achieves a useful reduction in the dependence of the solution on x; that is, $f(y/\delta)$ varies less rapidly with x than $u(y)$ and is therefore easier to compute.

Turbulent flows may need extra boundary conditions in practice, although in principle the above boundary conditions apply again if applied to both the mean and the fluctuating parts of the velocity. Details depend on the turbulence model employed, and one of the main difficulties is that the turbulent flow very near a solid surface [in the viscous sublayer (see Section 6.1)] is exceptionally complicated physically, although the number of variables on which the flow depends is small. Therefore it is fairly easy to correlate experimental data for the sublayer in such a way as to give the right mean velocity at the edge of the sublayer, but it is rather difficult to extend any turbulence model rationally into the sublayer. It follows that in some calculation methods the boundary condition actually employed is not the true boundary condition at the surface but a simulated boundary condition using empirical data at the edge of the sublayer. In Chapter 6 we discuss both methods of treating the sublayer.

Similar comments apply to the heat-field equations. If the thin-shear-layer approximation is not made, the temperature must be specified on all boundaries; the alternative is to specify the heat-flux rate, but this is realistic only for a solid surface heated electrically or by nuclear energy. In heat exchangers, the surface temperature and heat-flux rate are coupled by the properties of the solid surface, and solutions for the fluid flow and the solid heat conduction must be matched. The simplest initial and boundary conditions again apply to a thin shear layer, with

$$0 < y < \infty, \qquad T = T(y), \qquad x = x_0 \quad \text{(initial conditions),} \qquad (3.88a)$$

$$\left. \begin{array}{l} y = 0, \qquad T = T_w(x), \\[1ex] y = \infty, \qquad T = T_e, \end{array} \right\} \quad x \geq x_0, \qquad (3.88b)$$

where again we have taken the case of a boundary layer on the solid surface $y = 0$ for simplicity, assuming that the surface temperature T_w (subscript w denoting wall) is known; alternatively $\dot{q}_w = -k(\partial T/\partial y)_w$ can be assumed known, giving a "gradient" boundary condition rather like the symmetry boundary condition for the velocity field in Eq. (3.86). Symmetrical heated jets and wakes of course have $\dot{q} = -k(\partial T/\partial y) = 0$ on the line of symmetry.

As mentioned above, the conventional assumption for thin-shear-layer calculations is that the external-flow velocity is known before the start of the calculation, whereas in fact the effect of the shear layer on the external flow may be considerable so that the two must be matched iteratively; see [7]. In the entry region of a duct, boundary layers grow on the walls, and, before they meet, the flow between them is inviscid—in the sense that molecular and turbulent transfer of heat and momentum can be neglected—and also very nearly parallel to the duct axis. We can thus assume that the pressure and axial velocity are constant across the core region so that the core flow can be treated as inviscid and one-dimensional. This makes the calculation of the viscous-inviscid interaction particularly easy, though not trivial; we have to calculate the core flow velocity that gives a constant rate of mass flow down the duct (assuming the walls to be impermeable), but the core-flow acceleration and the pressure gradient applied to the boundary layers are connected by Bernoulli's equation. Once the boundary layers have met, Bernoulli's equation no longer applies to the core flow, and we have to calculate the interaction between the boundary layers on the duct walls. The general scheme is the same, but we have to use the thin-shear-layer equations everywhere rather than using Bernoulli's equation in the core. These difficulties with the momentum equation do not much affect the solution of the heat-transfer equations once the velocity field is known.

Coupled Flows

In uncoupled flows the velocity field can be calculated first and then used as an input to the heat-field equations. This is appropriate if the density and viscosity can be assumed independent of temperature and if buoyancy terms are negligible. However, if the temperature variation is large enough for viscosity or density changes to be significant (note that, in a gas, significant pressure changes usually *imply* significant temperature changes and therefore need not be considered separately), the velocity-field equations become coupled to the heat-field equations as described in Section 1.3. In these cases the velocity-field equations and the heat-field equations must be solved simultaneously. As we shall see in Chapter 13, this is not a major difficulty, because the velocity-field equations are nonlinear and must therefore be solved iteratively at each streamwise "step" of the marching process. Therefore, it is relatively simple to include, in this iteration loop, a solution of the heat-field equations to allow for the variation of fluid properties in the velocity-field equations.

In coupled flows the boundary conditions are the same as in uncoupled ones. The sublayer matching procedure for turbulent wall flows becomes more complicated because simultaneous equations for velocity and temperature must now be solved, and it therefore becomes preferable to extend the turbulence-model solution down to the wall.

Buoyant thin shear layers again present no special difficulties with boundary conditions, whether or not we neglect density differences outside the buoyancy term itself. Of course, buoyancy introduces more degrees of freedom into the problem and results in the appearance of flows that have no direct counterpart in the nonbuoyant case, such as plumes.

3.6 Thin-Shear-Layer Equations: Summary

The thin-shear-layer approximation discussed in this chapter is a formal approximation (unlike the empirical approximations used to deal with turbulent flow, for example) and becomes exact as the rate of growth of the shear layer (boundary layer, wake or jet) tends to zero. The main advantage of the thin-shear-layer or boundary-layer approximation is that the equation for the conservation of the component of momentum normal to the shear layer can be collapsed so that the change of pressure across the thickness of the shear layer is negligible. The thin-shear-layer approximation also leads to the neglect of streamwise diffusion of momentum and heat, so that the equations reduce to the statement that the rate of change of the conserved property along a mean streamline is equal to the sum of the effects of molecular or turbulent diffusion *normal* to the shear layer and of the effects of pressure gradients, body forces, or heat sources. Of course, a thin-shear-layer calculation must, in general, be matched to the solution for the flow outside the shear layer. In flows more complicated than those in simple ducts or over airfoils and blades, the thin-shear-layer approximation may be less accurate, but it is useful for engineering purposes in nearly all heat-transfer problems that do not involve separation of the flow from a solid surface and can even be used, with care, in some separated flows.

Although many flows of engineering interest are three-dimensional, only mildly three-dimensional flows like those over swept wings can be calculated accurately with a reasonable expenditure of computing time, and the choice of computational technique and turbulence model requires rather more knowledge of fluid mechanics than can be contained in this book. We have therefore restricted the discussion mainly to two-dimensional and axisymmetric flows. However, it should be emphasized that three-dimensional flows embody more degrees of freedom rather than more physical principles.

The conservation principles discussed in Chapters 2 and 3 hold for any fluid, but the equations are, of course, simpler if the fluid obeys the gas law with constant specific heat. The equations derived in Chapters 2 and 3 are valid for more general constitutive relations, except where the gas law has been explicitly used in the derivation: in practice it is only in a gas that the kinetic energy becomes a significant fraction of the thermal energy, because liquid flows are usually slow. The approximation of constant density is acceptable in most liquid-flow problems—and indeed in many gas-flow problems—except for the buoyancy terms in the equations, which may be significant even if the percentage change in density is small.

Problems

3.1 Show that in a two-dimensional boundary layer whose thickness δ is growing at the rate $d\delta/dx$, the ratio of $\partial^2 u/\partial x^2$ to $\partial^2 u/\partial y^2$, assumed small in the derivation of Eq. (3.9), is actually of order $(d\delta/dx)^2$.

3.2 Show that the thin-shear-layer (TSL) equation for x-component momentum can be rewritten, correct to the TSL approximation, as

$$\left(\frac{\partial P}{\partial s}\right)_{\psi} = \frac{\partial \sigma_{xy}}{\partial y}$$

where P is the total pressure $P = p + 1/2\rho(u^2 + v^2)$ and $(\partial/\partial s)_\psi$ is the gradient along a mean streamline (arc distance s). Discuss the order of magnitude of neglected terms.

3.3 Consider a two-dimensional flow in the xy-plane and by using the balance of mass, momentum and energy into a control volume of dimensions dx, dy within the boundary layer, where

$$\frac{\partial^2 u}{\partial y^2} \gg \frac{\partial^2 u}{\partial x^2}, \qquad \frac{\partial^2 T}{\partial y^2} \gg \frac{\partial^2 T}{\partial x^2},$$

derive Eqs. (3.5), (3.10b) and (3.11). Show all your assumptions.

3.4 Show that in a laminar flow with heat transfer the "compression work" term

$u(dp/dx)$ in the enthalpy equation, Eq. (3.40), is small compared to the heat-flux term $(1/c_p)(\partial/\partial y)(k\,\partial T/\partial y)$ if the temperature difference across the shear layer, ΔT, is *large* compared to $\mathrm{Pr}\,u_e^2/c_p$.

3.5 Since the governing equations for two-dimensional and axisymmetric flows differ from each other only by the radial distance $r(x, y)$, the axisymmetric flow equations, Eqs. (3.15) to (3.17) can be placed in a nearly two-dimensional form by using a transformation known as the *Mangler transformation*. In the case of flow over a body of revolution of radius r_0 (a function of x), we find that if the boundary-layer thickness is small compared with r_0, so that $r(x, y) = r_0(x)$, this transformation puts them exactly in two-dimensional form. We define the Mangler transformation by

$$d\bar{x} = \left(\frac{r_0}{L}\right)^{2K} dx, \qquad d\bar{y} = \left(\frac{r}{L}\right)^{K} dy \qquad (P3.1)$$

to transform an axisymmetric flow with coordinates (x, y), into a two-dimensional flow with coordinates $(\bar{x}, \bar{y})$. In Eq. (P3.1) L is an arbitrary reference length. If a stream function in Mangler variables $(\bar{x}, \bar{y})$ is related to a stream function ψ in (x, y) variables by

$$\bar{\psi}(\bar{x}, \bar{y}) = \left(\frac{1}{L}\right)^{K} \psi(x, y),$$

then

(a) show that the relation between the Mangler transformed velocity components $\bar{u}$ and $\bar{v}$ in $(\bar{x}, \bar{y})$ variables and the velocity components u and v in (x, y) variables is:

$$u = \bar{u}$$

$$v = \left(\frac{L}{r}\right)^{K}\left[\left(\frac{r_0}{L}\right)^{2K}\bar{v} - \frac{\partial\bar{y}}{\partial x}\bar{u}\right]. \qquad (P3.2)$$

(b) By substituting from Eqs. (P3.2) into Eqs. (3.15)–(3.17), show that for laminar flows the Mangler-transformed continuity momentum and energy equations are:

$$\frac{\partial\bar{u}}{\partial\bar{x}} + \frac{\partial\bar{v}}{\partial\bar{y}} = 0$$

$$\bar{u}\frac{\partial\bar{u}}{\partial\bar{x}} + \bar{v}\frac{\partial\bar{u}}{\partial\bar{y}} = -\frac{1}{\rho}\frac{dp}{d\bar{x}} + v\frac{\partial}{\partial\bar{y}}\left[(1+t)^{2K}\frac{\partial\bar{u}}{\partial\bar{y}}\right] \qquad (P3.3)$$

$$\bar{u}\frac{\partial T}{\partial\bar{x}} + \bar{v}\frac{\partial T}{\partial\bar{y}} = \frac{k}{\rho c_p}\frac{\partial}{\partial\bar{y}}\left[(1+t)^{2K}\frac{\partial T}{\partial\bar{y}}\right] \qquad (P3.4)$$

where

$$t = -1 + \left(1 + \frac{2L\cos\phi}{r_0^2}\bar{y}\right)^{1/2}. \qquad (P3.5)$$

Note that for $t = 0$, Eqs. (P3.3) and (P3.4) in the $(\bar{x}, \bar{y})$ plane are exactly in the same form as those for two-dimensional flows in the (x, y) plane.

3.6 By repeating the arguments used to derive a typical value for $\overline{\rho'v'}$ in Eq. (3.37), show that a typical value for $\overline{\rho'u'}$ in a nearly-adiabatic high-speed shear layer is

$$\rho(\gamma-1)M^2\frac{\overline{u'^2}}{u}$$

where M is the Mach number.

3.7 Show that in a coupled turbulent flow the condition for the neglect of the "compression work" term $u\,dp/dx$ in the enthalpy equation, Eq. (3.40)—compared to the "heat transfer" term $c_p\,\partial(\rho\overline{T'v'})/\partial y$, say—is

$$M_e^2 \ll \frac{1}{10}\frac{1}{\gamma-1}\frac{\Delta T}{T_e}.$$

[Hint: The factor 10 is an approximation to δ/θ in a turbulent boundary layer.]

3.8 Show that in a boundary layer of thickness δ on a surface of longitudinal curvature radius R below a stream of Mach number M_e, the ratio of the pressure difference across the layer to the absolute pressure at the edge is of order $\gamma M_e^2\delta/R$.

3.9 Show that in a turbulent boundary layer the ratio of the pressure change induced by the Reynolds normal stress to the absolute pressure at the edge is of order $\gamma M_e^2 c_f/2$, where c_f is the skin-friction coefficient. [Hint: assume that $\overline{\rho v'^2}$ is of the same order as the shear stress $-\rho\overline{u'v'}$].

3.10 Show that if in the decelerating turbulent boundary layer in an expanding passage (a diffuser) the skin-friction term in the momentum-integral equation, Eq. (3.68), is negligible and H can be taken as constant, Eq. (3.68) gives

$$\frac{\theta}{\theta_0} = \left(\frac{u_e}{u_{e,0}}\right)^{-(H+2)}$$

where subscript 0 denotes initial conditions.

3.11 Show that for an incompressible zero-pressure gradient flow over a wall at uniform temperature, Eq. (3.78) can be written as

$$\frac{d\theta_T}{dx} = \text{St}. \tag{P3.6}$$

3.12 While Eq. (3.10b) expresses conservation of momentum at each point within the boundary layer, Eq. (3.67) expresses conservation of momentum for the boundary layer as a whole. Another useful equation is the so-called kinetic-energy integral equation which expresses the physical fact that the rate of change of the flux of kinetic-energy defect within the boundary layer is equal to the rate at which kinetic energy is dissipated by viscosity. It is given by

$$\frac{d}{dx}\left(\frac{1}{2}\rho u_e^3\delta^{**}\right) = \mu\int_0^\infty\left(\frac{\partial u}{\partial y}\right)^2 dy \tag{P3.7}$$

Here δ^{**} denotes the kinetic-energy thickness which measures the flux of kinetic energy defect within the boundary layer as compared with an inviscid flow. It is defined for constant-density flow by

$$\delta^{**} = \int_0^\infty \frac{u}{u_e}\left(1-\frac{u^2}{u_e^2}\right) dy. \tag{P3.8}$$

Suggestion: Using the procedure outlined below, derive Eq. (P3.7).

a. First multiply Eq. (3.10b) by u and integrate it across the layer.
b. Integrate of the term involving v by "parts."
c. Note that

$$\int_0^\infty u \frac{\partial^2 u}{\partial y^2} dy = - \int_0^\infty \left(\frac{\partial u}{\partial y} \right)^2 dy .$$

d. Using the steps b and c in step a, after multiplication by minus 2, show that the resulting expression can be put in the same form as Eq. (P3.7).

References

[1] Morkovin, M. V.: Effects of compressibility on turbulent flows, in *The Mechanics of Turbulence*, 367 Gordon and Breach, New York, 1961.

[2] Morkovin, M. V. and Phinney, R. E.: Extended applications of hot-wire anemometry to high speed turbulent boundary layers, Dept. of Aeronautics, John Hopkins Univ. Rep. AFOSR TN-58-469, AD-158279, 1958.

[3] Kistler, A. L.: Fluctuation measurements in a supersonic turbulent boundary layer. *Phys. Fluids* **2**:290, 1959.

[4] Bradshaw, P.: Compressible turbulent shear layers. *Ann. Rev. Fluid Mech.* **9**: 33, 1977.

[5] Cebeci, T. and Bradshaw, P.: *Momentum Transfer in Boundary Layers*. Hemisphere, Washington, DC, 1977.

[6] Hildebrand, F. B.: *Advanced Calculus for Applications*. Prentice-Hall, Englewood Cliffs, NJ, 1962.

[7] Cebeci, T. (ed.): *Numerical and Physical Aspects of Aerodynamic Flows* **II**, Springer-Verlag, New York, 1984.

CHAPTER 4

Uncoupled Laminar Boundary Layers

In this and the following chapter, we begin our application of the concepts and equations introduced in the previous chapters with the simplest possible case: laminar flow, with temperature or concentration differences so small that the density, viscosity, and conductivity appearing in the conservation equations for mass, momentum, and enthalpy can be taken as constant, so that the equations are uncoupled. Later, in Chapters 6, 7, and 11, we discuss the more advanced problems of turbulence and of variable density, and in Chapter 9 we discuss the effect of buoyancy on momentum and heat transfer. Even in uncoupled flows, we need to define a "reference" temperature at which to evaluate the fluid properties; in weakly coupled flows it may be sufficient to evaluate the properties at some average temperature of the fluid and then use the *uncoupled* equations. In boundary layers an obvious choice is the arithmetic mean of the surface and external-stream temperatures, called *mean film temperature* and defined by $(T_w + T_e)/2 \equiv T_m$; in the duct flows discussed in Chapter 5 we use a slightly more sophisticated *bulk fluid temperature*, defined in Eq. (5.4).

Although turbulent flows are in general more important, they are of course usually preceded by a region of laminar flow ending in a "transition" region as, for example, on the surface of a turbomachine blade. In addition, substantial regions of laminar flow are found in the smaller veins of the human body, in small-scale fluidic devices, and in journal bearings. In some situations of engineering importance, efforts are made to increase the proportion of boundary-layer flow that is laminar, to decrease the overall drag or heat transfer. Conversely, in some low-Reynolds-number flows undesirable laminar regions occur, leading to flow separation or loss in heat-transfer rates.

Despite the fact that the boundary-layer equations presented in Chapter 3 are nonlinear partial differential equations, there are certain cases when transformations allow them to be reduced to a system of ordinary differential equations, which are relatively easy to solve numerically. This process, which makes use of the similarity of velocity (and temperature) profiles in transformed coordinates, is very useful in momentum and heat-transfer problems and is discussed in Section 4.1. The *physics* of the problems is just the same as in nonsimilar cases; so similarity solutions are useful as simple but nontrivial examples of the physical processes. Furthermore, the similarity transformation is a help in numerical solutions, even in *nonsimilar* flows.

In the remaining sections we make use of this similarity concept to obtain solutions of the equations for external flow, i.e., shear layers with an effectively infinite unsheared stream on one (or both) sides, for both two-dimensional and axisymmetric flows. In Section 4.2 we discuss the similarity solutions of momentum and energy equations for two-dimensional boundary layers, and in Section 4.3 we extend the discussion to nonsimilar flows. Section 4.4 deals with axisymmetric flows.

Section 4.5 is devoted to wall jets, i.e., jets blowing along walls as in Fig. 1.2(c), which appear frequently in film-cooling applications.

4.1 Similarity Analysis

Let us consider a two-dimensional uncoupled laminar flow with negligible body force. The governing equations are given by Eqs. (3.5), (3.10b), and (3.11); that is,

$$\frac{\partial u}{\partial x} + \frac{\partial v}{\partial y} = 0, \tag{3.5}$$

$$u\frac{\partial u}{\partial x} + v\frac{\partial u}{\partial y} = -\frac{1}{\rho}\frac{dp}{dx} + v\frac{\partial^2 u}{\partial y^2}, \tag{3.10b}$$

$$u\frac{\partial T}{\partial x} + v\frac{\partial T}{\partial y} = \frac{v}{\mathrm{Pr}}\frac{\partial^2 T}{\partial y^2}. \tag{3.11}$$

For an external flow we can replace $-1/\rho\, dp/dx$ in Eq. (3.10b) by $u_e\, du_e/dx$, and for given values of v, Pr, $u_e(x)$, T_e, and $T_w(x)$, we can write the solution of these equations in the form

$$\frac{u}{u_e} = \phi_1(x, y), \tag{4.1a}$$

$$\frac{T_w - T}{T_w - T_e} = \phi_2(x, y), \tag{4.1b}$$

subject to the boundary conditions given by Eqs. (3.85b) and (3.88b);

that is,

$$y = 0, \qquad u = v = 0, \qquad T = T_w, \tag{4.2a}$$

$$y = \delta, \qquad u = u_e, \qquad T = T_e. \tag{4.2b}$$

There are special cases in which Eq. (4.1) can be written, with $\Delta T(x)$ denoting $T_w - T_e$, as

$$\frac{u}{u_e} = \phi_1(\eta), \tag{4.3a}$$

$$\frac{T_w - T}{\Delta T} = \phi_2(\eta), \tag{4.3b}$$

where η, called a *similarity variable*, is a special function of x and y; in practice, as we shall see later, it is proportional to $y/\delta(x)$. In such cases the number of independent variables is reduced from two (x and y) to one (η), so that Eqs. (3.5), (3.10b), and (3.11) become ordinary differential equations for u, v, and T. In jet flows in still air, we replace u_e by the centerline velocity u_c and T_w by the centerline temperature T_c, and we define ΔT to be equal to $T_c - T_e$. In wakes in a uniform stream we write Eq. (4.3) in the following form:

$$\frac{u_c - u}{\Delta u} = \phi_1(\eta), \tag{4.4a}$$

$$\frac{T_c - T}{\Delta T} = \phi_2(\eta), \tag{4.4b}$$

where in this case the velocity and temperature scales, Δu and ΔT, respectively, are $u_c - u_e$ and $T_c - T_e$.

A formal method of deriving similarity variables is the so-called group theoretic method discussed by Hansen [1] and Na [2]. It is convenient to apply it to the single equation obtained by defining a stream function $\psi(x, y)$ to satisfy Eq. (3.5), giving

$$u = \frac{\partial \psi}{\partial y}, \qquad v = -\frac{\partial \psi}{\partial x}, \tag{4.5}$$

and by substituting u and v from Eq. (4.5) into Eq. (3.10b), again neglecting the body force f_x in the latter. We get

$$\frac{\partial \psi}{\partial y} \frac{\partial^2 \psi}{\partial x \, \partial y} - \frac{\partial \psi}{\partial x} \frac{\partial^2 \psi}{\partial y^2} = u_e \frac{du_e}{dx} + v \frac{\partial^3 \psi}{\partial y^3}. \tag{4.6}$$

The idea of this approach is to introduce the "linear transformation" defined by

$$x = A^{\alpha_1} \bar{x}, \qquad y = A^{\alpha_2} \bar{y}, \qquad \psi = A^{\alpha_3} \bar{\psi}, \qquad u_e = A^{\alpha_4} \bar{u}_e. \tag{4.7}$$

Here α_k ($k = 1, \ldots, 4$) are constants and A is called the *parameter of transformation*.

Under this transformation, Eq. (4.6) can be written as

$$A^{2\alpha_3-\alpha_1-2\alpha_2}\left(\frac{\partial\bar{\psi}}{\partial\bar{y}}\frac{\partial^2\bar{\psi}}{\partial\bar{x}\,\partial\bar{y}}-\frac{\partial\bar{\psi}}{\partial\bar{x}}\frac{\partial^2\bar{\psi}}{\partial\bar{y}^2}\right)=A^{2\alpha_4-\alpha_1}\bar{u}_e\frac{d\bar{u}_e}{d\bar{x}}+\nu A^{\alpha_3-3\alpha_2}\frac{\partial^3\bar{\psi}}{\partial\bar{y}^3}.$$

$$(4.8)$$

Comparing this equation with Eq. (4.6), we see that these two equations will be invariant (unaltered by the transformation) if the powers of A in each term are the same. If this is true, then Eq. (4.8) gives

$$2\alpha_3-\alpha_1-2\alpha_2=2\alpha_4-\alpha_1=\alpha_3-3\alpha_2. \qquad (4.9)$$

Partial solution of Eq. (4.9) gives

$$\alpha_3=\alpha_1-\alpha_2, \qquad \alpha_4=2\alpha_3-\alpha_1. \qquad (4.10)$$

We define a new variable $\alpha=\alpha_2/\alpha_1$ and write Eq. (4.10) as

$$\frac{\alpha_3}{\alpha_1}=1-\alpha, \qquad \frac{\alpha_4}{\alpha_1}=2\frac{\alpha_3}{\alpha_1}-1=1-2\alpha. \qquad (4.11)$$

Knowing the relations among the α, we can find a relationship between the barred and unbarred quantities. This can be done by eliminating the parameter of transformation A. For example, from Eq. (4.7), we can write

$$A=\left(\frac{x}{\bar{x}}\right)^{1/\alpha_1}=\left(\frac{y}{\bar{y}}\right)^{1/\alpha_2}=\left(\frac{\psi}{\bar{\psi}}\right)^{1/\alpha_3}=\left(\frac{u_e}{\bar{u}_e}\right)^{1/\alpha_4}$$

or

$$\frac{y}{x^\alpha}=\frac{\bar{y}}{\bar{x}^\alpha}, \qquad \frac{\psi}{x^{1-\alpha}}=\frac{\bar{\psi}}{\bar{x}^{1-\alpha}}, \qquad \frac{u_e}{x^{1-2\alpha}}=\frac{\bar{u}_e}{\bar{x}^{1-2\alpha}}. \qquad (4.12)$$

These combinations of variables in Eq. (4.12) are seen to be invariant under these linear transformations, Eq. (4.7), and are called *absolute variables*. According to Morgan's theorem [1], they are the similarity variables provided that the *boundary conditions* for the velocity field in Eq. (4.2) can also be transformed and expressed independent of x. We now put

$$\eta=\frac{y}{x^\alpha}, \qquad f(\eta)=\frac{\psi}{x^{1-\alpha}}, \qquad (4.13)$$

and

$$h(\eta)=\frac{u_e}{x^{1-2\alpha}}=\text{const}=C, \qquad (4.14)$$

where the function $h(\eta)$ is a constant, since the mainstream velocity u_e is a function of x only and thus cannot be a nonconstant function of η because this introduces dependence on y.

Sometimes it is more convenient to write the second expression in Eq. (4.13) as

$$\psi=x^{1-\alpha}f(\eta)\frac{x^\alpha}{x^\alpha}=x^{1-2\alpha}x^\alpha f(\eta)$$

or

$$\psi = \frac{u_e}{C} x^\alpha f(\eta) \tag{4.15}$$

by making use of Eq. (4.14). Using the definition of the first expression in Eq. (4.5) and the similarity variables defined by the first expression of Eq. (4.13) and by Eq. (4.15), we can write

$$u = \frac{\partial \psi}{\partial y} = \frac{\partial \psi}{\partial \eta} \frac{\partial \eta}{\partial y} = \frac{u_e}{C} \frac{\partial f}{\partial \eta},$$

or with a prime denoting differentiation with respect to η,

$$\frac{u}{u_e} = \frac{f'(\eta)}{C}. \tag{4.16}$$

It is now a simple matter to transform Eq. (4.6) and its boundary conditions. To conform with the usual notation, we let $m = 1 - 2\alpha$, so that

$$u_e = Cx^m \tag{4.17}$$

and

$$\eta = \frac{y}{x^{(1-m)/2}} = \frac{y}{x^{1/2}} x^{m/2} = \left(\frac{u_e}{Cx}\right)^{1/2} y, \tag{4.18a}$$

$$\psi = \frac{u_e x^{(1-m)/2}}{C} f(\eta) = x^{(m+1)/2} f(\eta) = \left(\frac{u_e x}{C}\right)^{1/2} f(\eta). \tag{4.18b}$$

Now ψ should have the dimension (velocity × length), and η [and $f(\eta)$] should be dimensionless. So far we have ignored the undoubted presence of ν among the governing variables, and if we introduce ν into Eq. (4.18), we can correct the dimensions, uniquely, to obtain

$$\psi = (u_e \nu x)^{1/2} f(\eta), \tag{4.19a}$$

$$\eta = \left(\frac{u_e}{\nu x}\right)^{1/2} y. \tag{4.19b}$$

This transformation is called the *Falkner–Skan transformation* and will be used extensively for external boundary-layer flows. We see that Eq. (4.19b) implies that the thickness of a "similar" laminar boundary layer, corresponding to a given value of η, is proportional to $\sqrt{\nu x / u_e}$, as already shown in Section 3.1. Equation (4.19a) shows that ψ is proportional to $u_e \sqrt{\nu x / u_e}$, that is, to freestream velocity × width of shear layer. These physical interpretations of the Falkner–Skan nondimensionalization carry over to duct flows (Chapter 5) and free shear layers (Chapter 8), although the algebra may differ.

The Falkner–Skan transformation given by Eq. (4.19) can be used to reduce the boundary-layer equations (3.5), (3.10b), and (3.11) into ordinary differential equations for similar flows (see Problem 4.1). It can also be used for nonsimilar flows for convenience in numerical work, as we shall discuss later—briefly, it reduces, even if it does not eliminate, dependence on x. For two-dimensional uncoupled flows we treat the nonsimilar case by allowing the dimensionless stream function f to vary with x also,

$$\psi = (u_e \nu x)^{1/2} f(x, \eta). \qquad (4.19c)$$

Then by using the chain-rule relations that connect the parameters in the xy plane to those in the $x\eta$ plane, that is,

$$\left(\frac{\partial}{\partial x}\right)_y = \left(\frac{\partial}{\partial x}\right)_\eta + \left(\frac{\partial}{\partial \eta}\right)_x \frac{\partial \eta}{\partial x}, \qquad (4.20a)$$

$$\left(\frac{\partial}{\partial y}\right)_x = \left(\frac{\partial}{\partial \eta}\right)_x \frac{\partial \eta}{\partial y}, \qquad (4.20b)$$

and the definition of stream function given by Eq. (4.5) together with the transformation defined by Eqs. (4.19b) and (4.19c), we get

$$\left(\frac{\partial \psi}{\partial x}\right)_y = f \frac{d}{dx}(u_e \nu x)^{1/2} + (u_e \nu x)^{1/2} \frac{\partial f}{\partial x} + (u_e \nu x)^{1/2} f' \frac{\partial \eta}{\partial x},$$

$$\left(\frac{\partial \psi}{\partial y}\right)_x = (u_e \nu x)^{1/2} f' \left(\frac{u_e}{\nu x}\right)^{1/2} = u_e f', \qquad \frac{\partial^2 \psi}{\partial y^2} = u_e f'' \left(\frac{u_e}{\nu x}\right)^{1/2},$$

$$\left(\frac{\partial T}{\partial x}\right)_y = \left(\frac{\partial T}{\partial x}\right)_\eta + T' \frac{\partial \eta}{\partial x}, \qquad \left(\frac{\partial T}{\partial y}\right)_x = T' \left(\frac{u_e}{\nu x}\right)^{1/2},$$

where the primes denote differentiation with respect to η. With this procedure and with the replacement of the $-1/\rho \, dp/dx$ term in Eq. (3.10b) by $u_e \, du_e/dx$, we can write the transformed momentum and energy equations for two-dimensional uncoupled laminar flows as

$$f''' + \frac{m+1}{2} ff'' + m\left[1 - (f')^2\right] = x\left(f' \frac{\partial f'}{\partial x} - f'' \frac{\partial f}{\partial x}\right), \qquad (4.21)$$

$$\frac{1}{\text{Pr}} T'' + \frac{m+1}{2} fT' = x\left(f' \frac{\partial T}{\partial x} - T' \frac{\partial f}{\partial x}\right). \qquad (4.22)$$

Here m, the index in Eq. (4.17), is a dimensionless pressure-gradient parameter defined by

$$m = \frac{x}{u_e} \frac{du_e}{dx}. \qquad (4.23)$$

Just as we express u in terms of dimensionless quantities (for example, $u/u_e = f'$), we also express T in terms of dimensionless quantities. For flows with *specified wall temperature* we find it convenient to use a quantity defined by

$$g = \frac{T_w - T}{T_w - T_e}, \qquad (4.24)$$

which, like u/u_e, is zero at the wall and unity in the external stream. When the *wall heat flux* is specified, this definition is not appropriate since the wall temperature T_w is not known. For this reason we define

$$T = T_e + T_e(1 - g)\phi(x) \tag{4.25}$$

with $g = g(x, \eta)$ and with ϕ denoting a function to be specified.

In the case of specified wall temperature, we choose

$$\phi(x) = \frac{T_w - T_e}{T_e} \tag{4.26a}$$

and recover the definition of g given by Eq. (4.24).

In the case of specified wall heat flux, that is,

$$\left(\frac{\partial T}{\partial y}\right)_w = -\frac{\dot{q}_w(x)}{k} \quad \text{at} \quad y = 0,$$

we can write

$$\left(\frac{\partial T}{\partial \eta}\right)_w = -\frac{\dot{q}_w(x)}{k}\sqrt{\frac{\nu x}{u_e}}.$$

Differentiating Eq. (4.25) with respect to η and equating it to the above expression, taking $g'_w = 1$ for simplicity, the definition of $\phi(x)$ becomes

$$\phi(x) = \frac{\dot{q}_w(x)x}{k\sqrt{R_x}\,T_e}, \tag{4.26b}$$

where R_x denotes the Reynolds number,

$$R_x = \frac{u_e x}{\nu}. \tag{4.27}$$

Introducing Eq. (4.25) into Eq. (4.22), we can rewrite the dimensionless transformed energy equation as

$$\frac{1}{\mathrm{Pr}}g'' + \frac{m+1}{2}fg' + n(1 - g)f' = x\left(f'\frac{\partial g}{\partial x} - g'\frac{\partial f}{\partial x}\right). \tag{4.28}$$

Here, like m, n is a dimensionless parameter. For the case of specified wall temperature it is defined by

$$n = \frac{x}{(T_w - T_e)}\frac{d}{dx}(T_w - T_e), \tag{4.29a}$$

and for the case of specified wall heat flux it is defined by

$$n = \frac{x}{\phi}\frac{d\phi}{dx}. \tag{4.29b}$$

In terms of similarity variables and dimensionless quantities, the boundary conditions for the momentum and energy equations follow from Eqs. (4.2). For the case of specified wall temperature, including mass transfer at the wall, they are

$$\eta = 0, \quad f' = 0, \quad f = -\frac{1}{(u_e \nu x)^{1/2}}\int_0^x v_w \, dx, \quad g = 0, \tag{4.30a}$$

$$\eta = \eta_e, \quad f' = 1, \quad g = 1. \tag{4.30b}$$

We see from the above equations that the dimensionless temperature is zero at the wall and unity at the edge, just like the dimensionless streamwise velocity u/u_e. For the case of specified wall heat flux, the boundary condition on g in Eq. (4.30a) is replaced by

$$\eta = 0, \qquad g' = 1. \tag{4.30c}$$

The complicated formula for the dimensionless stream function at the surface f_w in Eq. (4.30a) reflects the fact that the surface is not a streamline if there is transpiration through the surface, with normal component velocity v_w. For flows with suction, f_w is positive, and for flows with blowing, f_w is negative; on solid surfaces it is zero.

In Eq. (4.30b) η_e is the (transformed) distance from the surface at which u equals u_e to adequate numerical accuracy, say $u/u_e = 0.9999$; η_e seldom exceeds 8 in laminar boundary layers, and this value is therefore suitable as an outer boundary for numerical calculations. It is analogous to the outer limit h used in Section 3.4. Note that η_e is somewhat greater than the value of η corresponding to $y = \delta$, where $u/u_e = 0.995$ according to the definition of δ used in this book. In physical coordinates the velocity and thermal boundary-layer thicknesses, $\delta(x)$ and $\delta_t(x)$, respectively, usually increase with increasing downstream distance for both laminar and turbulent flows. In transformed coordinates η_e is approximately constant for most laminar flows (a big advantage from a computational point of view). For $\text{Pr} < 1$ or $\text{Pr} > 1$, the transformed thermal boundary-layer thickness δ_t becomes, respectively, larger or smaller than the transformed velocity boundary-layer thickness δ, but δ_t/δ, of course, remains constant down the length of a *similar* flow.

4.2 Two-Dimensional Similar Flows

For two-dimensional uncoupled similar laminar flows, the dimensionless stream function f and temperature g are functions of η only. As a result, in Eqs. (4.21) and (4.28), the derivatives of f, f', and g with respect to x disappear, and the equations become

$$f''' + \frac{m+1}{2} ff'' + m\left[1 - (f')^2\right] = 0, \tag{4.31}$$

$$\frac{g''}{\text{Pr}} + \frac{m+1}{2} fg' + n(1-g)f' = 0. \tag{4.32}$$

The solutions of Eq. (4.31) and (4.32) are independent of x provided that the boundary conditions [Eq. (4.30)] and the pressure- and temperature-gradient parameters m and n are independent of x. These requirements are satisfied if f_w, m, and n are constants.

With the requirement that f_w is constant, Eq. (4.30) leads to the following boundary conditions for Eqs. (4.31) and (4.32):

$$\eta = 0, \quad f = \text{const}, \quad f' = 0, \quad g = 0 \quad \text{or} \quad g' = 1, \quad (4.33a)$$

$$\eta = \eta_e, \quad f' = 1, \quad g = 1. \qquad (4.33b)$$

The requirement that m is constant leads to Eq. (4.17) and that n is constant leads to

$$T_w - T_e = C_1 x^n \qquad (4.34a)$$

for the case of specified wall temperature and to

$$\phi = \frac{\dot{q}_w(x) x}{k \sqrt{R_x} T_e} = C_2 x^n \qquad (4.34b)$$

for the case of specified wall heat flux, with both C_1 and C_2 being constant. Thus, as discussed in Section 4.1, similar boundary-layer flows for momentum and energy equations are obtained when the external velocity varies with the surface distance x as prescribed by Eq. (4.17) and when either the wall temperature or heat flux vary as prescribed by Eq. (4.34) and the boundary conditions are independent of x. We note that when flow conditions satisfy Eq. (4.17) and the boundary conditions in Eq. (4.33) but not Eq. (4.34a) or Eq. (4.34b), then although the momentum equation (4.21) will have similarity solutions, the energy equation (4.28) will not.

Solutions for the Velocity Field

The solutions of the Falkner–Skan equation, Eq. (4.31), are sometimes referred to as *wedge-flow* solutions. The flow is deflected through an angle of magnitude $\beta\pi/2$ (see Fig. 4.1), where

$$\beta = \frac{2m}{m+1}$$

and can have either sign; negative values can be realized, in principle, near the rear of a body. The solution of the Falkner–Skan equation for attached boundary layers is limited to values of m in the range $-0.0904 \le m \le \infty$.

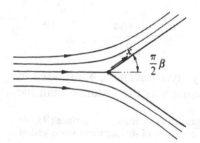

$\frac{\pi}{2}\beta$

Figure 4.1. Flow past a wedge. In the neighborhood of the leading edge external velocity distribution is $u_e(x) = Cx^{\beta/2}$ $^{\beta)}$.

Table 4.1 Solutions of the Falkner–Skan equation for zero mass transfer

m	f_w''	δ_1^*	θ_1	H
1	1.23259	0.64791	0.29234	2.216
$\frac{1}{3}$	0.75745	0.98536	0.42900	2.297
0	0.33206	1.72074	0.66412	2.591
-0.05	0.21351	2.1174	0.75148	2.818
-0.0904	0.0	3.427	0.86880	3.949

For $m = -0.0904$, it is found that $f_w'' \equiv 0$ (that is, the wall shear stress τ_w is zero) and the boundary layer is on the point of separation[1] at all x.

The main use of the Falkner–Skan transformation is in numerical work in nonsimilar boundary layers, and only two of the wedge flows are common in practice. The flow over a flat plate at zero incidence, with constant external velocity u_e, is known as *Blasius flow* and corresponds to $m = 0$. The second common case is $m = 1$ (wedge half-angle 90°), which is a two-dimensional stagnation flow, known as *Hiemenz flow*.

Solutions for a range of m are, however, useful since they illustrate the main features of the response of boundary layers to pressure gradients. In particular, we can demonstrate the variation of velocity profile shapes, skin-friction coefficient c_f, and the displacement and momentum thickness δ^* and θ. In terms of transformed variables the definitions of c_f, δ^*, and θ follow from Eq. (3.69) and are given by

$$c_f = \frac{2f_w''}{\sqrt{R_x}}, \tag{4.35}$$

$$\frac{\delta^*}{x} = \frac{1}{\sqrt{R_x}} \int_0^{\eta_e} (1 - f') \, d\eta = \frac{\eta_e - f_e + f_w}{\sqrt{R_x}} \equiv \frac{\delta_1^*}{\sqrt{R_x}}, \tag{4.36}$$

$$\frac{\theta}{x} = \frac{1}{\sqrt{R_x}} \int_0^{\eta_e} f'(1 - f') \, d\eta \equiv \frac{\theta_1}{\sqrt{R_x}}. \tag{4.37}$$

Values of these quantities and of the profile shape parameter $H \equiv \delta^*/\theta$ are tabulated for a range of m in Table 4.1, and the velocity profiles are plotted in similarity variables in Fig. 4.2.

For the example of a flow with zero pressure gradient, the above equations and the values of Table 4.1 lead to

$$c_f = \frac{0.664}{\sqrt{R_x}}, \qquad \frac{\delta^*}{x} = \frac{1.72}{\sqrt{R_x}}, \qquad \frac{\theta}{x} = \frac{0.664}{\sqrt{R_x}}.$$

Several arbitrary definitions of the boundary-layer thickness, δ, are used in practice and in text books. Here we shall define it as the distance from the

---

[1] Strictly speaking, "separation" implies that the boundary layer leaves the surface ($d\delta/dx$ no longer $\ll 1$). In practice this is equivalent to a reversal of sign of surface shear stress and of flow direction near the surface, i.e., recirculation.

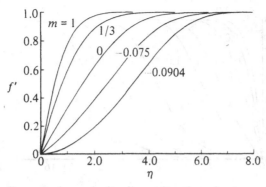

Figure 4.2. The dimensionless velocity f' as a function of η for various values of m.

surface at which $f' = 0.995$, Fig. 4.2 ($m = 0$), leads to

$$\frac{\delta}{x} = \frac{5.3}{\sqrt{R_x}}.$$

Table 4.2 presents some properties of the solutions of the Blasius equation [Eq. (4.31) with $m = 0$] with suction and/or blowing at the wall that are important in mass-transport problems in which the contaminant is injected or absorbed at the surface. Figure 4.3 shows the corresponding velocity profiles for several values of the mass-transfer parameter $f_w = -2v_w/u_e\sqrt{R_x}$. In this case the wall mass-transfer velocity v_w must vary as $1/\sqrt{x}$ for similarity. The effect of the mass-transfer parameter is approximately analogous to that of the pressure-gradient parameter; increasing the blowing rate, like increasing a positive ("adverse") pressure gradient, decreases the slope of the velocity profile near the wall, and at $f_w = -1.238$, it becomes zero. With increasing suction rate, as in increasing "favorable" (negative) pres-

Table 4.2 Solutions of the Blasius equation with suction (positive f_w) or blowing (negative f_w) at the wall

$f_w = -2\dfrac{v_w}{u_e}\sqrt{R_x}$	f_w''	δ_1^*	θ_1	H
2.0	1.16812	0.73561	0.33624	2.188
1.50	0.94449	0.86761	0.38898	2.230
1.0	0.72828	1.04911	0.45646	2.30
0.5	0.52254	1.31187	0.54508	2.407
0	0.33206	1.72074	0.66412	2.591
−0.250	0.24492	2.03821	0.73984	2.755
−0.500	0.16507	2.46571	0.83014	2.970
−0.750	0.09433	3.1243	0.93866	3.328
−1.000	0.03618	4.36353	1.07236	4.069
−1.238	0.00017	12.8814	1.23834	10.402

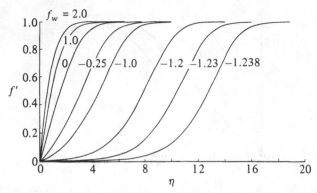

Figure 4.3. The variation of dimensionless velocity f' across the boundary layer on a flat plate with mass transfer.

sure gradient, the slope of the velocity profile increases. The reason for the (approximate) analogy is that blowing (say) injects new fluid with zero x-component momentum into the boundary layer, whereas adverse pressure gradient reduces the x-component momentum of the existing boundary-layer fluid directly.

Solutions for the Temperature Field

The Falkner–Skan equation is a nonlinear differential equation and requires special care to solve because boundary conditions are applied at two points ($\eta = 0$ and $\eta = \eta_e$). However, it is an easy matter to solve the similarity form of the energy equation, Eq. (4.32), with its two-point boundary conditions, once the solution $f(\eta)$ of the Falkner–Skan equation has been determined. This is because Eq. (4.32) is a *linear* equation for g (with variable coefficients). The linearity implies that if $\hat{g}(\eta)$ is a solution of Eq. (4.32), then so is any constant multiple of $\hat{g}$, say, $\lambda \hat{g}$. Thus from any solution $\hat{g}(\eta)$, say, satisfying

$$\hat{g}(0) = 0, \qquad \hat{g}'(0) = 1, \tag{4.38}$$

we use $\lambda = 1/\hat{g}(\eta_e)$ to get a solution satisfying the boundary conditions of the energy equation in Eq. (4.33). To carry this out numerically, we use a standard method, say the Runge-Kutta method, which actually solves the second-order equation (4.32) in terms of two first-order equations, using an auxiliary variable p defined by

$$\hat{g}' = p, \tag{4.39a}$$

$$p' = -\Pr\frac{m+1}{2}fp - \Pr nf'(1-\hat{g}), \tag{4.39b}$$

with *one-point* boundary conditions (initial values) at $\eta = 0$,

$$\hat{g}(0) = 0, \qquad p(0) \equiv \hat{g}'(0) = 1. \tag{4.39c}$$

Then the solution of Eq. (4.32) satisfying its boundary conditions $g = 0$ at $\eta = 0$ and $g = 1$ at $\eta = \eta_e$ is given by

$$g(\eta) = \frac{\hat{g}(\eta)}{\hat{g}(\eta_e)}, \tag{4.40a}$$

$$g'(\eta) = \frac{p(\eta)}{\hat{g}(\eta_e)}. \tag{4.40b}$$

When the wall temperature is uniform ($n = 0$), the solution of the energy equation

$$g'' + \Pr \frac{m+1}{2} fg' = 0 \tag{4.41}$$

can be reduced to a quadrature, specifically, a weighted integral of $f(\eta)$. We simply write Eq. (4.41) as

$$\frac{g''}{g'} \equiv (\ln g')' = -\frac{\Pr}{2}(m+1)f. \tag{4.42}$$

Integrating Eq. (4.42) once, we get

$$g' = C \exp\left[\int_0^\eta -\frac{\Pr}{2}(m+1)f(\eta)\, d\eta\right], \tag{4.43}$$

where C is a constant of integration. Integrating once more and noting that $g(0) = 0$, we can write the solution of Eq. (4.41) as

$$g(\eta) = g(0) + C\int_0^\eta \exp\left[\int_0^\zeta -\Pr\frac{m+1}{2}f(z)\, dz\right] d\zeta, \tag{4.44}$$

where z and ζ are dummy variables of integration. To find C, we use the "edge" boundary condition, $g(\eta_e) = 1$; we can then write Eq. (4.44) as

$$g = \frac{\displaystyle\int_0^\eta \exp\left[-\Pr\frac{m+1}{2}\int_0^\zeta f(z)\, dz\right] d\zeta}{\displaystyle\int_0^{\eta_e} \exp\left[-\Pr\frac{m+1}{2}\int_0^\zeta f(z)\, dz\right] d\zeta}. \tag{4.45}$$

From Eq. (4.45) we see that the calculation of the temperature profile requires the values of the stream function f as a function of η for the specified value of m. From Eq. (4.45) we also see that the value of the Prandtl number $\Pr$ has a considerable influence on the behavior of the solution. These conclusions hold for nonuniform wall temperature also.

For a flat-plate boundary layer, $m = 0$, and Eq. (4.45) simplifies further. Noting that integrating Eq. (4.31) once with respect to η, with $m = 0$, gives

$$-\frac{1}{2}\int f\, d\eta = \ln f'',$$

we can write Eq. (4.45) as

$$g = \frac{\int_0^\eta (f'')^{Pr} \, d\eta}{\int_0^{\eta_e} (f'')^{Pr} \, d\eta}.$$ (4.46)

When $Pr = 1$, Eq. (4.46) yields the relation given in Eq. (1.10); that is,

$$g = \frac{f' - f'(0)}{f'(\eta_e) - f'(0)} = f' = \frac{u}{u_e}$$ (4.47)

Solutions for Very Small or Very Large Prandtl Number

It is interesting, and simple, to examine the solutions of the energy equation (4.41) for various values of m for fluids with very small or very large Prandtl number when Eq. (4.45) becomes even simpler. To show this, let us first consider the case when $Pr \ll 1$, found in liquid metals, which have high thermal conductivity but low viscosity, so that the velocity boundary-layer thickness is very small compared with the thermal boundary-layer thickness. A good approximation can be made by assuming f' to be unity throughout the thermal boundary layer, the error being significant only within the very thin velocity boundary layer. With this assumption, Eq. (4.45) can be written as

$$g = \frac{\int_0^\eta e^{-(Pr/4)(m+1)\eta^2} \, d\eta}{\int_0^{\eta_e} e^{-(Pr/4)(m+1)\eta^2} \, d\eta}.$$ (4.48)

Using the definition of the error function,

$$\mathrm{erf}(u) \equiv \frac{2}{\sqrt{\pi}} \int_0^u e^{-\zeta^2} \, d\zeta,$$

we can write Eq. (4.48) as

$$g = \mathrm{erf}\left(\frac{\sqrt{(m+1)Pr}}{2} \eta \right),$$ (4.49)

and from the known properties of the error function it follows that

$$g'_w = \sqrt{\frac{(m+1)Pr}{\pi}} = 0.564\sqrt{(m+1)Pr}.$$ (4.50)

Therefore, using Eq. (4.50) and the definition of local Nusselt number,

$$\mathrm{Nu}_x = -\frac{x}{T_w - T_e}\left(\frac{\partial T}{\partial y}\right)_w,$$ (4.51a)

which in terms of transformed variables can also be written as

$$\text{Nu}_x = g'_w \sqrt{R_x}, \tag{4.51b}$$

and also using the definition of Stanton number,

$$\text{St} = \frac{\dot{q}_w}{\rho c_p (T_w - T_e) u_e} = \frac{-k(\partial T / \partial y)_w}{\rho c_p (T_w - T_e) u_e} = \frac{\text{Nu}_x}{\text{Pr} R_x}, \tag{4.51c}$$

we get

$$\text{Nu}_x = 0.564 \sqrt{(m+1)} \, \text{Pr}^{1/2} R_x^{1/2}, \tag{4.52a}$$

$$\text{St} = 0.564 \sqrt{(m+1)} \, \text{Pr}^{-1/2} R_x^{-1/2}. \tag{4.52b}$$

Since c_f also varies as $R_x^{-1/2}$, we see that the *Reynolds analogy factor* $\text{St}/\frac{1}{2} c_f$ is independent of x (see Table 4.3).

When $\text{Pr} \gg 1$ (heavy, high-viscosity oils), the velocity boundary layer is very much thicker than the thermal boundary layer. A good approximation in this case is to assume that the velocity profile is linear throughout the thermal boundary layer with its slope given by the value at the wall, f''_w. Thus, writing

$$f' = f''_w \eta$$

so that

$$f = \frac{f''_w \eta^2}{2},$$

we can integrate Eq. (4.45) to get g'_w so that (see Problem 4.12)

$$\text{Nu}_x = 1.12 \left(\frac{m+1}{12} f''_w \right)^{1/3} \text{Pr}^{1/3} R_x^{1/2}. \tag{4.53}$$

Thus the Stanton number now varies as $\text{Pr}^{-2/3} R_x^{-1/2}$, and the analogy

Table 4.3 Similarity solutions of the energy equation for a uniformly heated surface

m	\multicolumn{4}{c}{$g'_w = \dfrac{\text{Nu}_x}{\sqrt{R_x}}$}	$\dfrac{\text{St}}{\frac{1}{2} c_f}$ for $\text{Pr}=1$
	0.1	0.72	1	10	
1	0.2191	0.5017	0.5708	1.3433	0.463
$\frac{1}{3}$	0.1723	0.3884	0.4402	1.0135	0.581
0.1	0.1497	0.3315	0.3741	0.8414	0.753
0	0.1389	0.2957	0.3321	0.7289	1.000
−0.01	0.1373	0.2911	0.3267	0.7141	1.045
−0.05	0.1294	0.2680	0.2995	0.6386	1.40
−0.0904	0.1085	0.2031	0.2224	0.4071	∞

factor $St/\frac{1}{2}c_f$ varies as $Pr^{-2/3}$. Note that since for a flat-plate flow $f_w'' = 0.332$, Eq. (4.53) becomes

$$Nu_x = 0.339 Pr^{1/3} R_x^{1/2} \qquad (4.54a)$$

Similarly for a stagnation-point flow with $f_w'' = 1.23259$ (see Table 4.1), Eq. (4.53) reduces to

$$Nu_x = 0.661 Pr^{1/3} R_x^{1/2} \qquad (4.54b)$$

Sample Numerical Solutions

The energy equation in similarity form, Eq. (4.32), can be solved numerically for different boundary conditions and for various values of m, n, and Pr. Here we present the results for several values of m and Pr for a uniformly heated surface ($n = 0$). Table 4.3 presents the dimensionless heat-flux parameter g_w' for various values of m and Pr, and Figs. 4.4 and 4.5 show the corresponding temperature profiles. The profiles in Fig. 4.4, which are for a fluid with $Pr = 1$, show that pressure gradient has some effect on the temperature profiles but much less than on the velocity profiles (Fig. 4.2). Of course, this effect is exerted via changes in the velocity profile; pressure gradient as such does not appear in the energy equation in constant-property flow (Eq. 3.11). The temperature profiles in Fig. 4.5 are for a flat-plate flow with fluids of different Prandtl number. The profile thickness is greatly affected by the Prandtl number, but the effect on profile *shape* is rather smaller.

Figure 4.6 shows the variation of $Nu_x/\sqrt{R_x}$ ($\equiv g_w'$) with Prandtl number, for $m = 1$ and 0. From the numerical results in Fig. 4.6 we see that the analytic solution for low Prandtl number, Eq. (4.52), is a good approximation when $Pr \leq 0.01$ and the high-Prandtl-number solution, Eq. (4.54), is a good approximation for $m = 0$ when $Pr \geq 1$ and for $m = 1$ when $Pr > 2$. The reason is that the low-Pr approximation requires $\delta_t \gg \delta$, whereas the high-Pr

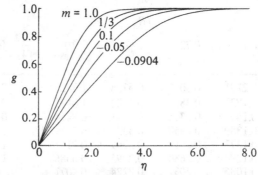

Figure 4.4. The effect of pressure gradient on temperature profiles for $Pr = 1$.

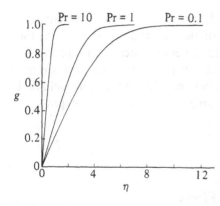

Figure 4.5. The effect of Prandtl number on temperature profiles for zero-pressure gradient flow ($m = 0$). The velocity profile shape is the same as that of $g(\eta)$ at $Pr = 1$.

approximation merely requires $\partial u / \partial y \simeq$ constant for $y < \delta_t$, which (Fig. 4.2) is a good approximation for $m = 0$ if $\delta_t < 0.7\delta$ approximately. From Fig. 4.6 we also observe that when the Prandtl number is near unity, the variation of $Nu_x / \sqrt{R_x}$ with Prandtl number can be approximated by

$$Nu_x = 0.332 Pr^{0.33} R_x^{1/2} \qquad (4.55a)$$

for $m = 0$ and by

$$Nu_x = 0.57 Pr^{0.40} R_x^{1/2} \qquad (4.55b)$$

for $m = 1$. Except for the one-half power of Reynolds number, the numerical values in these formulas are *empirical*, chosen to give the best fit to the numerical solutions. In particular the Prandtl number exponent of 0.33 in Eq. (4.55a) is not an asymptotically exact result, as is the $Pr^{1/3}$ factor in Eq.

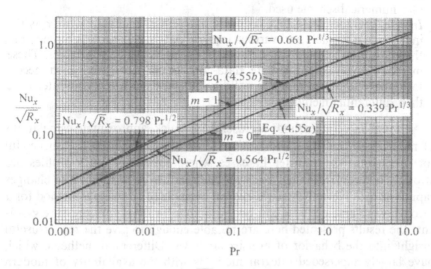

Figure 4.6. Variation of the ratio $Nu_x / \sqrt{R_x}$ with Prandtl number for two values of m. The solid lines denote the numerical solutions of the energy equation (4.32) for $n = 0$.

(4.54). Note that the last line in Table 4.3, for $m = -0.0904$, represents the separation profile (Table 4.1) where the surface shear stress is zero but the heat transfer rate is not; Reynolds' analogy between heat and momentum transfer is, at best, useful only when the effects of pressure gradient are not too large, because the essential difference between the temperature-field and velocity-field equations is that in uncoupled flow the pressure gradient appears only in the latter.

4.3 Two-Dimensional Nonsimilar Flows

When the external velocity distribution of the flow does not vary according to Eq. (4.17), that is, when m is not a constant, or when for a flow with mass transfer m is constant but the dimensionless stream function at the wall f_w (Eq. (4.30a)) is *not* constant, then the velocity field becomes nonsimilar and the similarity equation for the velocity field given in Section 4.2 does not apply. Since a necessary (but not sufficient) condition for similar solutions of the energy equation is the similarity of the velocity field, the solution of the energy equation also becomes nonsimilar and cannot be obtained from the similarity equation discussed in the previous section.

The solution for nonsimilar flows can be obtained by either of the following:

Differential methods: solving the partial differential equations for conservation of mass, momentum, and energy derived in Chapter 3 so that the only inaccuracies are those introduced by the thin-shear-layer approximation itself or by finite-difference errors associated with the particular numerical scheme used.

Integral methods: based on ordinary differential equations with x as the independent variable and various integral parameters such as momentum thickness θ and profile shape parameter H as dependent variables. These methods involve approximate data-correlation formulas, which necessarily have a limited range of validity but take much less computer time than differential methods.

Although integral methods are useful for quick, approximate calculations of momentum transfer in nonsimilar flows, they have been less successful for predicting heat transfer in cases where the temperature profiles are strongly nonsimilar because the wall temperature or heat-flux rate changes rapidly with x. Nevertheless, we give an example of an integral method for a zero-pressure-gradient flow and two examples for flows with pressure gradient; the results presented here are reliable enough to give the reader useful insight into the behavior of nonsimilar flows. Differential methods, which have largely superseded integral methods with the availability of modern computers, are discussed in Chapters 13 and 14, but we give one sample solution here, following the presentation of integral methods.

Pohlhausen's Method

A simple integral method for calculating laminar boundary layers is Pohlhausen's method in which velocity and temperature profiles are assumed that satisfy the momentum integral equation (3.68) and the energy integral equation (3.78b) together with a set of boundary conditions given by Eqs. (4.2). Additional "boundary conditions" are also obtained by evaluating the momentum equation (3.10b) at the wall with $v_w = 0$, that is,

$$\nu \frac{\partial^2 u}{\partial y^2} = \frac{1}{\rho} \frac{dp}{dx} = -u_e \frac{du_e}{dx}, \qquad (4.56a)$$

and also some additional boundary conditions from differentiating the edge boundary conditions with respect to y, namely,

$$y \to \delta \qquad \frac{\partial u}{\partial y}, \quad \frac{\partial T}{\partial y}, \quad \frac{\partial^2 u}{\partial y^2}, \quad \frac{\partial^2 T}{\partial y^2}, \quad \frac{\partial^3 u}{\partial y^3}, \quad \frac{\partial^3 T}{\partial y^3}, \dots \to 0.$$

$$(4.56b)$$

Note that Eqs. (4.56) are *properties of the solution* of the PDEs and *not* boundary conditions for the PDEs.

To illustrate this method we consider a flat-plate flow on which the wall temperature T_w is equal to the freestream temperature T_e from the leading edge to the point x_0 (Fig. 4.7). For values of x greater than x_0 we assume that the wall temperature is constant. A hydrodynamic boundary layer is formed starting at the leading edge $x = 0$, whereas a thermal boundary layer originates at $x = x_0$.

To obtain the solutions of the momentum integral equation (3.68), which for this case becomes

$$\frac{d\theta}{dx} = \frac{c_f}{2}, \qquad (4.57)$$

we shall assume a third-order polynomial for u,

$$u = a + by + cy^2 + dy^3, \qquad (4.58)$$

and evaluate the constants a, b, c, and d from the velocity boundary conditions given in Eqs. (4.2), together with Eq. (4.56a) and the requirement that $\partial u/\partial y = 0$ at $y = \delta$ from Eq. (4.56b); we do not have enough free

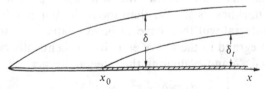

Figure 4.7. Flat plate with an unheated starting length.

coefficients to specify $\partial^2 u / \partial y^2 = 0$ at $y = \delta$. This procedure yields

$$\frac{u}{u_e} = \frac{3}{2}\frac{y}{\delta} - \frac{1}{2}\left(\frac{y}{\delta}\right)^3.\tag{4.59}$$

Substitution of Eq. (4.59) into the definitions of θ given by Eq. (3.69a) and c_f given by (3.69c) and of the resulting expressions into Eq. (4.57) yields, after some rearrangement,

$$\delta\frac{d\delta}{dx} = \frac{140}{13}\frac{\nu}{u_e}.\tag{4.60}$$

The integration of Eq. (4.60) for the boundary condition $\delta = 0$ at $x = 0$ yields the boundary-layer thickness

$$\delta = 4.64\sqrt{\frac{\nu x}{u_e}}.\tag{4.61}$$

The local skin-friction coefficient can next be evaluated by substituting Eq. (4.61) into Eq. (4.59), which gives the complete velocity profile, and then evaluating the gradient at the wall and using Eq. (3.69c). The resulting expression can be expressed in the form

$$c_f = \frac{0.646}{\sqrt{R_x}},\tag{4.62}$$

a result that is within 3 percent of the exact Blasius solution [Eq. (4.35, with $f_w'' = 0.332$, so that the coefficient is 0.664].

To obtain the solutions of the energy integral equation (3.78b), which for this case becomes

$$\frac{d\theta_T}{dx} = \text{St},\tag{4.63}$$

we again assume a third-order polynomial for T and evaluate its constants from the temperature boundary conditions. The resulting temperature profile can be written in the form

$$\frac{T - T_w}{T_e - T_w} = \frac{3}{2}\frac{y}{\delta_t} - \frac{1}{2}\left(\frac{y}{\delta_t}\right)^3.\tag{4.64}$$

Equations (4.59) and (4.64) can now be used to obtain a solution of Eq. (4.63). Noting the definition of θ_T, we can write

$$\theta_T = \int_0^h \left[\frac{3}{2}\frac{y}{\delta} - \frac{1}{2}\left(\frac{y}{\delta}\right)^3\right]\left[1 - \frac{3}{2}\frac{y}{\delta_t} + \frac{1}{2}\left(\frac{y}{\delta_t}\right)^3\right]dy.$$

For the calculations that follow we will assume that the thermal boundary-layer thickness is smaller than the hydrodynamic boundary-layer thickness. The integral will then have to be evaluated only to $y = \delta_t$, since for $y > \delta_t$, the integrand in the above equation is identically zero. Introducing the ratio $s = \delta_t/\delta$ and evaluating the integral, we get

$$\theta_T = \delta\left(\tfrac{3}{20}s^2 - \tfrac{3}{280}s^4\right)$$

The second term in the parentheses is small compared with the first term because of the assumption that $s < 1$. It will, therefore, be neglected. Substitution of the integral into Eq. (4.63) and using Eq. (4.64) together with the definition of Stanton number, we can write

$$\frac{3}{20}\frac{d}{dx}(s^2\delta) = \frac{3}{2}\frac{k}{\rho c_p u_e \delta_t} = \frac{3}{2}\frac{k}{\rho c_p u_e \delta s}$$

or

$$\delta s^3 \frac{d\delta}{dx} + 2s^2\delta^2\frac{ds}{dx} = \frac{10k}{\rho c_p u_e}.$$

Combining this relation with Eqs. (4.60) and (4.61) gives

$$s^3 + 4xs^2\frac{ds}{dx} = \frac{13}{14}\frac{1}{Pr} \qquad (4.65)$$

This differential equation can be solved (see problem 4.14) to yield

$$s^3 = \frac{13}{14Pr} + Cx^{-3/4} \qquad (4.66)$$

The constant C is evaluated from the boundary condition $s = 0$ at $x = x_0$. This gives

$$s = \frac{1}{1.026\,Pr^{1/3}}\left[1 - \left(\frac{x_0}{x}\right)^{3/4}\right]^{1/3} \qquad (4.67a)$$

When the flat plate is heated along its entire length ($x_0 = 0$), the ratio of boundary-layer thicknesses become

$$s = \frac{0.977}{Pr^{1/3}} \qquad (4.67b)$$

Note that for highly-viscous oils, which have a Prandtl number of the order of 1000, the thermal boundary-layer thickness is of the order of one-tenth that of the hydrodynamic boundary layer. In general the Prandtl numbers of gases are smaller than 1. In this case, $s > 1$, and we cannot simplify the expression for θ_T as we did in the preceding calculation. However, since the smallest value for the Prandtl number in a gas is about 0.6, the largest value of s is 1.18. Thus the error introduced in Eq. (4.67) is small. The only materials that have very low Prandtl numbers are liquid metals. The solution given by Eq. (4.67) will not be applicable for them.

In heat-transfer problems it is convenient and useful to define a heat-transfer coefficient $\hat{h}$ by

$$\hat{h} = \frac{\dot{q}_w}{T_w - T_e} = -\frac{k}{T_w - T_e}\left(\frac{\partial T}{\partial y}\right)_w, \qquad (4.68a)$$

which can also be related to the local Nusselt number, Eq. (4.51a), by

$$\hat{h} = Nu_x\frac{k}{x}. \qquad (4.68b)$$

For the temperature profile given by Eq. (4.64), we can write Eq. (4.68) as

$$\hat{h} = \frac{3}{2}\frac{k}{\delta_t},$$

(4.69)

which shows that the heat-transfer coefficient is inversely proportional to the thermal boundary-layer thickness.

Introduction of Eq. (4.61) and the relation (4.67a) for the ratio of the boundary-layer thicknesses into Eq. (4.69) yields

$$\hat{h} = \frac{0.332 k \, \mathrm{Pr}^{1/3}}{\sqrt{\nu x / u_e}\left[1 - (x_0/x)^{3/4}\right]^{1/3}}.$$

(4.70)

Substitution of this equation into Eq. (4.68b) yields an expression for local Nusselt number,

$$\mathrm{Nu}_x = \frac{0.332 \mathrm{Pr}^{1/3} R_x^{1/2}}{\left[1 - (x_0/x)^{3/4}\right]^{1/3}},$$

(4.71)

which for the case of a plate heated along its entire length ($x_0 = 0$) reduces to

$$\mathrm{Nu}_x = 0.332 \mathrm{Pr}^{1/3} R_x^{1/2}.$$

(4.72)

This result is seen to be identical to the exact solution fitted by Eq. (4.55).

Equation (4.72) was derived with the assumption that the hydrodynamic boundary-layer thickness was greater than the thermal boundary-layer thickness. When this is not the case, that is, the thermal boundary layer is thicker, Eckert and Gross [3] recommend the use of the following equation for a flat plate heated along its entire length with Prandtl numbers ranging from 0.005 to 0.05:

$$\mathrm{Nu}_x = \frac{\sqrt{\mathrm{Pr}\, R_x}}{1.55\sqrt{\mathrm{Pr}} + 3.09\sqrt{0.372 - 0.15\,\mathrm{Pr}}}$$

(4.73)

Thwaites' Integral Method for the Velocity Field

A useful integral method for calculating momentum transfer in uncoupled laminar flows with pressure gradient but with no mass transfer is Thwaites' method [4]. It is based on the observation that most experimental data and accurate numerical solutions for laminar boundary layers lie close to a unique linear function,

$$\frac{u_e}{\nu}\frac{d\theta^2}{dx} = f\left(\frac{\theta^2}{\nu}\frac{du_e}{dx}\right) = A - B\frac{\theta^2}{\nu}\frac{du_e}{dx}.$$

(4.74)

Here A and B are constants that need to be specified. Equation (4.74) is a first-order ordinary differential equation that can be integrated to give

$$\theta^2 = \frac{\nu A \int_0^x u_e^{B-1}\, dx}{u_e^B} + \theta_i^2 \frac{u_{e_i}^B}{u_e^B}, \tag{4.75}$$

where the subscript i denotes initial conditions.

Thwaites chooses $A = 0.45$, $B = 6$ and writes Eq. (4.75) as

$$\theta^2 = \frac{0.45\nu}{u_e^6} \int_0^x u_e^5\, dx + \theta_i^2 \left(\frac{u_{e_i}}{u_e}\right)^6, \tag{4.76a}$$

or in dimensionless form,

$$\left(\frac{\theta}{L}\right)^2 = \frac{0.45}{(u_e^*)^6 R_L} \int_0^{x^*} u_e^{*5}\, dx^* + \left(\frac{\theta}{L}\right)_i^2 \left(\frac{u_{e_i}^*}{u_e^*}\right)^6. \tag{4.76b}$$

Here with L and u_{ref} denoting reference length and velocity, respectively, x^*, u_e^*, and R_L denote dimensionless quantities defined by

$$x^* = \frac{x}{L}, \qquad u_e^* \equiv \frac{u_e}{u_{\text{ref}}}, \qquad R_L \equiv \frac{u_{\text{ref}} L}{\nu}. \tag{4.77}$$

The subscript i in Eq. (4.76) denotes the initial conditions at $x^* = 0$. For example, the term $(\theta/L)_i^2(u_{e_i}^*/u_e^*)^6$ is equal to zero in calculations starting from a stagnation point, because $u_{e_i}^* = 0$ (θ_i is *not* zero, as can be seen from the similar-flow solution for $m = 1$ in Table 4.1).

Once θ is calculated, the other boundary-layer parameters H and c_f can be determined, as functions of the pressure-gradient parameter $\lambda \equiv (\theta^2/\nu)\, du_e/dx$, from the appropriate data-fit relations given below. For $0 \le \lambda \le 0.1$, a suitable data fit is

$$R_\theta \frac{c_f}{2} = 0.225 + 1.61\lambda - 3.75\lambda^2 + 5.24\lambda^3,$$

$$H = 2.61 - 3.75\lambda + 5.24\lambda^2. \tag{4.78a}$$

For $-0.1 \le \lambda \le 0$, a suitable data fit is

$$R_\theta \frac{c_f}{2} = 0.225 + 1.472\lambda + \frac{0.0147\lambda}{0.107 + \lambda},$$

$$H = 2.472 + \frac{0.0147}{0.107 + \lambda}. \tag{4.78b}$$

Here

$$R_\theta = \frac{u_e \theta}{\nu}, \qquad c_f = \frac{\tau_w}{\frac{1}{2}\rho u_e^2}. \tag{4.78c}$$

Comparisons of accurate numerical solutions for laminar boundary layers with Eq. (4.74), carried out by the authors using a program similar to that described in Chapter 13, show that the choices $A = 0.45$, $B = 6$, made by Thwaites in 1950 on the basis of the few calculations available at that time, still cannot be bettered.

Smith and Spalding's Integral Method for the Temperature Field

For heat transfer in constant-property laminar boundary-layer flows with variable u_e but *uniform* surface temperature, one may use a method similar to Thwaites' method to compute heat transfer [5]. The *conduction thickness*,

$$\delta_c \equiv \frac{k(T_w - T_e)}{\dot{q}_w} \equiv -\frac{T_w - T_e}{(\partial T/\partial y)_w} \tag{4.79}$$

can be expressed in nondimensional form, with an accuracy comparable to that of Thwaites' correlation Eq. (4.74), as

$$\frac{u_e}{v}\frac{d\delta_c^2}{dx} = F\left(\frac{\delta_c^2}{v}\frac{du_e}{dx}, \text{Pr}\right). \tag{4.80}$$

For similar laminar flows, the heat-transfer relationship is given by

$$\text{Nu}_x = \frac{\dot{q}_w x}{k(T_w - T_e)} = \frac{-x(\partial T/\partial y)_w}{T_w - T_e} = CR_x^{1/2}, \tag{4.81}$$

where C is a function of m and Pr. Combining the definition of δ_c with Eq. (4.81) yields

$$\delta_c^2 = \frac{x v}{C^2 u_e} \tag{4.82}$$

Thus the parameters appearing in Eq. (4.80) can be expressed as

$$\frac{u_e}{v}\frac{d\delta_c^2}{dx} = \frac{1-m}{C^2}, \qquad \frac{\delta_c^2}{v}\frac{du_e}{dx} = \frac{m}{C^2}. \tag{4.83}$$

Computations for fixed Prandtl numbers and a variety of pressure gradients have shown that the relationship given by Eq. (4.80) is, like Thwaites' correlation, almost a linear one but with the constants A and B depending on the Prandtl number. Thus we can integrate the linear version of Eq. (4.80) in the same way as Eq. (4.74) to get

$$\delta_c^2 = \frac{vA \int_0^x u_e^{B-1}\, dx}{u_e^B} + \delta_{c_i}^2 \frac{u_{e_i}^B}{u_e^B} \tag{4.84}$$

analogous to Eq. (4.75) and with subscript i again denoting initial conditions. The heat-transfer coefficient, expressed as a local Stanton number, can be obtained from

$$\text{St} = \frac{\dot{q}_w}{\rho c_p u_e (T_w - T_e)} = \frac{k}{\rho c_p u_e \delta_c} = \frac{\text{Nu}_x}{R_x \text{Pr}}, \tag{4.85}$$

using the definition of δ_c in Eq. (4.79). Substituting from Eq. (4.84) and

Table 4.4 Constants in Eq. (4.86) for various
Prandtl numbers

Pr	c_1	c_2	c_3
0.7	0.418	0.435	1.87
0.8	0.384	0.450	1.90
1.0	0.332	0.475	1.95
5.0	0.117	0.595	2.19
10.0	0.073	0.685	2.37

writing in nondimensional form, the local Stanton number can be written as

$$\text{St} = \frac{c_1(u_e^*)^{c_2}}{\left[\int_0^{x^*} (u_e^*)^{c_3} dx^*\right]^{1/2}} \frac{1}{\sqrt{R_L}}. \tag{4.86}$$

Here $c_1 = \text{Pr}^{-1}A^{-1/2}$, $c_2 = B/2 - 1$, $c_3 = B - 1$ (see Table 4.4). The Reynolds analogy factor $\text{St}/\frac{1}{2}c_f$ in nonsimilar flows usually depends on x; it could be obtained (in a complicated algebraic form that will not be given here) by combining the above expression for St with Thwaites' result for c_f obtained from Eq. (4.78) by substituting for R_θ from Eq. (4.76).

Integral Solutions for Nonsimilar Flows

As an example of a two-dimensional nonsimilar laminar flow with no mass transfer, we consider flow past a uniformly heated circular cylinder of radius r_0, normal to the axis. The external velocity distribution given by the inviscid flow theory is

$$u_e = 2u_\infty \sin\theta, \tag{4.87}$$

where $\theta = x/r_0$ is now an *angle*; x is measured around the circumference.

Figure 4.8 shows the variation of the dimensionless wall shear stress $(\tau_w/\rho u_\infty^2)(u_\infty r_0/\nu)^{1/2}$. The results indicated by the solid line are obtained by Thwaites' method, and those indicated by the dashed line are obtained by the similarity solution for a stagnation-point (Hiemenz) flow with $m = 1$ (see Section 4.2). In the vicinity of the stagnation point, i.e., for small values of the angle θ, Eq. (4.87) can be approximated as

$$u_e = 2u_\infty \theta = \frac{2u_\infty x}{r_0}. \tag{4.88}$$

Using the definition of wall shear in the Falkner–Skan variables, Eq. (4.19), we can write

$$\tau_w = \mu\left(\frac{\partial u}{\partial y}\right)_w = \mu u_e f_w'' \sqrt{\frac{u_e}{\nu x}}.$$

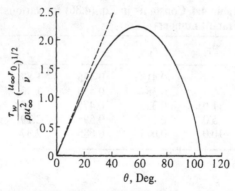

Figure 4.8. The variation of dimensionless wall shear distribution around the circumference of a circular cylinder. The result indicated by the solid line was obtained by Thwaites' method and that indicated by the dashed line by the similarity solution, Eq. (4.89).

Substituting Eq. (4.88) into the above equation and recalling that $f_w'' = 1.23259$ for $m = 1$ (see Table 4.1), we get

$$\frac{\tau_w}{\rho u_\infty^2}\left(\frac{u_\infty r_0}{\nu}\right)^{1/2} = 2\sqrt{2}\,\theta \times 1.23259 = 3.486\theta. \qquad (4.89)$$

We see from Fig. 4.8 that the similarity solution agrees well with the solutions from Thwaites' method up to $\theta = \pi/6$ rad $= 30°$. This is to be expected since, in that range, the flow is nearly a stagnation-point flow, with $u_e \approx 2(u_\infty/r_0)x$. Figure 4.8 also shows that flow separation takes place at $\theta = 1.83$ rad $= 105°$. Separation, of course, seriously invalidates Eq. (4.87) at all θ; a realistic calculation would involve iterating an inviscid external-flow solution with a solution for the boundary layer and the separated flow region (see [6]).

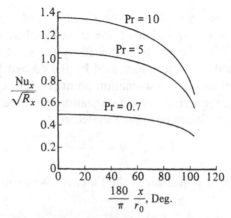

Figure 4.9. The variation of dimensionless heat-flux parameter, $\mathrm{Nu}_x/\sqrt{R_x}$ around the circumference of a circular cylinder at three Prandtl numbers.

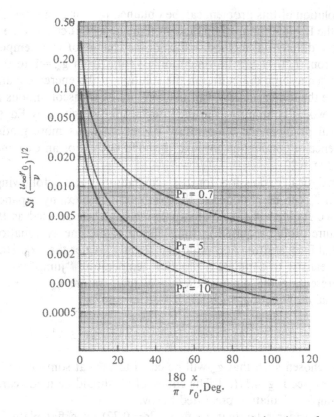

Figure 4.10. The variation of local Stanton number around the circumference of a circular cylinder at three Prandtl numbers.

Figure 4.9 shows the variation of $\mathrm{Nu}_x/\sqrt{R_x}$, and Fig. 4.10 shows that of local Stanton number St, for three values of Prandtl number. As in the similarity solutions of Table 4.3, the value of $\mathrm{Nu}_x/\sqrt{R_x}$ ($\equiv g'_w$) or St at separation is not zero; separated-flow calculations would yield nonzero values even for $1.83 < \theta < \pi$. At separation, $\mathrm{St}/\frac{1}{2}c_f = \infty$ as in the similarity solutions with $c_f = 0$ everywhere ($m = -0.0904$).

Differential Solution for a Flat Plate with Unheated Starting Length

As another example, we consider a laminar incompressible flow on a flat plate on which heating starts at $x = x_0$ rather than at the origin of the velocity boundary layer, $x = 0$ (see Fig. 4.7). (We recall that if the heating started at $x = 0$ and was uniform along the plate, then the energy equation would have a similarity solution.)

The solution of this problem can be obtained by using an integral method such as the Pohlhausen method discussed previously. It can also be obtained by using a differential method, although the change of wall temperature at $x = x_0$, from T_e to T_w or in dimensionless form from $g_w = 1$ to 0 (see Fig. 4.11), presents some difficulties in the solution of the energy equation. This is because the dimensionless wall temperature g_w is discontinuous at $x = x_0$ and Nu_x would be infinite as in the integral solution given by Eq. (4.71). In a practical problem the temperature rise at $x = x_0$ is more gradual, as a consequence of conduction through the solid body. For an extreme case of this, see Fig. 6.20.

An analytical expression for the temperature profile following a step change in T_w can be obtained in the neighborhood of x_0 by expanding u, v, and T (see Problem 4.13). That expression can then be used as the initial temperature profile required in the solution of the energy equation in full differential form starting at a value of x slightly greater than x_0. However, it is more realistic to simulate a practical temperature "jump" as a gradual rise, representing the dimensionless wall temperature g_w for $x > x_0$ by a simple analytic function, such as

$$g_w = \frac{1}{2}\left(1 + \cos \pi \frac{x - x_0}{a}\right) \qquad x_0 \leq x \leq x^*. \qquad (4.90)$$

Here a is chosen such that g_w will be equal to zero at some $x = x^*$; that is, $a = x^* - x_0$ [see Fig. 4.11(a)]. The choice of x^* should be made carefully, as in the sample calculation presented below.

Let us consider a laminar air flow ($\text{Pr} = 0.72$) on a flat plate 5 m long, heated for $x > 1$ m only, and calculate the variation of dimensionless heat-flux parameter g_w' along the plate by using the computer program

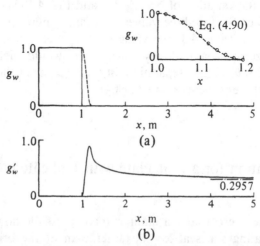

(a)

(b)

Figure 4.11. Flat-plate flow with an unheated starting length. The dashed line in (a) corresponds to the fairing shown in the insert: (a) Distribution of specified wall temperature, g_w, and; (b) Computed wall temperature gradient g_w'.

presented in Section 13.5—what follows can be understood without reference to that section.

Since the velocity field is similar, we can start the velocity and temperature calculations at $x = x_0 = 1$ m. We solve the momentum and energy
equations, Eqs. (4.21) and (4.28) by setting $m = 0$, $n = 0$ subject to the
boundary conditions

$$\eta = 0, \qquad f = f' = 0, \qquad g = 0,$$
$$\eta = \eta_e, \qquad f' = 1, \qquad g = 1, \qquad \qquad (4.91)$$

for a Prandtl number of 0.72. Since the boundary condition for the energy
equation (i.e., the temperature) is the same at the wall as at the edge, the
initial computer solution will consist of a Blasius solution for the velocity
field and unity for the temperature field. For $x > x_0 = 1$, we let the wall
boundary condition of the energy equation take on the values of g_w
according to Eq. (4.90) and solve the momentum and energy equations at
each specified value of x.

Figure 4.11 shows the distribution of x values used to perform the
heat-transfer calculations. Note that at $x^* = 1.2$ m, $g_w = 0$, making $a = 0.2$.
Even with the "step" spread over a distance $0.2 x_0$, the peak dimensionless
heat-transfer rate g_w' is as much as twice the asymptotic value. Note also
that a relatively large number of x values were used in $x_0 \leq x \leq x^*$ due to
the rapid variation of g_w as given by Eq. (4.90). For $x > x^*$, a few x values
are sufficient. We also note from Fig. 4.11(b) that the dimensionless heat-flux
parameter g_w' tends to approach its similarity value $g_w' = 0.2957$ at large x.

4.4 Axisymmetric Flows

We see from Eqs. (3.15) to (3.17) that the boundary-layer equations for
two-dimensional and axisymmetric flows differ from each other only by the
appearance of the radial distance $r(x, y)$. For this reason, the axisymmetric
flow equations can be placed in a nearly two-dimensional form by using the
Mangler transformation (Problem 3.5). When the boundary-layer thickness is
small compared with the body radius r_0, so that $r(x, y) = r_0(x)$, this
transformation puts the equations exactly in two-dimensional form. With
(x, y) denoting the axisymmetric flow coordinates, $(\bar{x}, \bar{y})$ the two-dimensional coordinates, and L a reference length, the Mangler transformation is
defined by Eq. (P3.1),

$$d\bar{x} = \left(\frac{r_0}{L}\right)^{2K} dx, \qquad d\bar{y} = \left(\frac{r}{L}\right)^{K} dy. \qquad (P3.1)$$

As in two-dimensional flows, it is advantageous to use transformed
variables for axisymmetric flows. This can easily be done by combining the
Falkner–Skan transformation with the Mangler transformation. The resulting transformation, known as the Falkner–Skan–Mangler transformation,

can be written as

$$d\bar{x} = \left(\frac{r_0}{L}\right)^{2K} dx, \tag{4.92a}$$

$$d\eta = \left(\frac{u_e}{\nu\bar{x}}\right)^{1/2}\left(\frac{r}{L}\right)^K dy, \tag{4.92b}$$

$$\psi(x, y) = L^K (u_e \nu\bar{x})^{1/2} f(\bar{x}, \eta). \tag{4.92c}$$

We note that for a two-dimensional flow, since the flow index K is zero and $\bar{x} = x$, the above transformation reduces to the Falkner–Skan transformation given by Eqs. (4.19b) and (4.19c). The appearance of L in Eq. (4.92c) is due to the definition of stream function, which for an axisymmetric flow is

$$r^K u = \frac{\partial \psi}{\partial y}, \qquad r^K v = -\frac{\partial \psi}{\partial x}, \tag{4.93}$$

but use of Eq. (4.92a) shows that η as defined by Eq. (4.92b) is still dimensionless.

By using the chain rule and the relations given by Eqs. (4.92a) and (4.92b), we can write the following relations connecting the parameters in the xy plane to those in the $\bar{x}\eta$ plane:

$$\left(\frac{\partial}{\partial x}\right)_y = \left(\frac{r_0}{L}\right)^{2K}\left[\left(\frac{\partial}{\partial \bar{x}}\right)_\eta + \frac{\partial \eta}{\partial \bar{x}}\left(\frac{\partial}{\partial \eta}\right)_{\bar{x}}\right], \tag{4.94a}$$

$$\left(\frac{\partial}{\partial y}\right)_x = \left(\frac{u_e}{\nu\bar{x}}\right)^{1/2}\left(\frac{r}{L}\right)^K\left(\frac{\partial}{\partial \eta}\right)_{\bar{x}}. \tag{4.94b}$$

With the definition of stream function, Eq. (4.93), and with the relations given by Eqs. (4.92) and (4.94), we can write the transformed momentum and energy equations for two-dimensional or axisymmetric uncoupled laminar flows as

$$(bf'')' + \frac{m+1}{2}ff'' + m\left[1 - (f')^2\right] = \bar{x}\left(f'\frac{\partial f'}{\partial \bar{x}} - f''\frac{\partial f}{\partial \bar{x}}\right), \tag{4.95}$$

$$(eT')' + \frac{m+1}{2}fT' = \bar{x}\left(f'\frac{\partial T}{\partial \bar{x}} - T'\frac{\partial f}{\partial \bar{x}}\right). \tag{4.96}$$

Here

$$b = (1+t)^{2K}, \qquad e = \frac{b}{\text{Pr}}, \tag{4.97}$$

and t is a transverse curvature parameter that denotes the deviation of the local radius $r(x, y)$ from the body radius $r_0(x)$ (see Problem 2.3). In terms of transformed variables it is given by

$$t = -1 + \left[1 + \left(\frac{L}{r_0}\right)^2\frac{2\cos\phi}{L}\left(\frac{\nu\bar{x}}{u_e}\right)^{1/2}\eta\right]^{1/2}, \tag{4.98}$$

where ϕ is the surface angle defined by $\tan\phi = dr_0/dx$, so that $r = r_0 + y\cos\phi$. Note from the definition of $\bar{x}$ that L cancels out of the definition of

t: it can be seen that if $(\nu\bar{x}/u_e)^{1/2}$ is small compared with r_0 (implying that δ/r_0 is small), t becomes small also. The quantity m is again a dimensionless pressure-gradient parameter identical to the one defined by Eq. (4.23) except that the x are replaced by $\bar{x}$; that is,

$$m = \frac{\bar{x}}{u_e}\frac{du_e}{d\bar{x}}. \tag{4.99}$$

As in two-dimensional flows, we use the definition of dimensionless temperature g given by Eq. (4.25) and rewrite Eq. (4.96) as

$$(eg')' + \frac{m+1}{2}fg' + n(1-g)f' = \bar{x}\left(f'\frac{\partial g}{\partial \bar{x}} - g'\frac{\partial f}{\partial \bar{x}}\right). \tag{4.100}$$

Here, for a specified wall temperature, n again denotes a dimensionless temperature parameter or heat-flux parameter identical to those defined by Eq. (4.29) except that the x are replaced by $\bar{x}$. The definition of ϕ in Eq. (4.26b) now becomes

$$\phi(x) = \left(\frac{L}{r_0}\right)^K\frac{\dot{q}_w(\bar{x})\bar{x}}{kT_e R_{\bar{x}}^{1/2}}, \tag{4.101}$$

where

$$R_{\bar{x}} = \frac{u_e\bar{x}}{\nu}. \tag{4.102}$$

The boundary conditions for the transformed momentum and energy equations given by Eqs. (4.95) and (4.100) are

$$\eta = 0, \quad f' = 0, \quad f = -\frac{1}{(u_e\nu\bar{x})^{1/2}}\int_0^{\bar{x}}\bar{v}_w d\bar{x}, \quad g = 0, \quad \text{or} \quad g' = 1,$$

$$\tag{4.103a}$$

$$\eta = \eta_e, \quad f' = 1, \quad g = 1, \tag{4.103b}$$

where $\bar{v}_w \equiv v_w(L/r_0)^K$. In the case of no mass transfer, $f_w \equiv 0$.

When the transverse curvature effect is negligible, the momentum and energy equations for uncoupled axisymmetric laminar flows also admit similarity solutions. As in two-dimensional flows, the stream function $f(\bar{x}, \eta)$ and the dimensionless temperature $g(\bar{x}, \eta)$ in Eqs. (4.95) and (4.100) now become functions of η, which with negligible transverse curvature effect is

$$\eta = \left(\frac{u_e}{\nu\bar{x}}\right)^{1/2}\left(\frac{r_0}{L}\right)y. \tag{4.104}$$

The resulting momentum and energy equations and their boundary conditions are identical in form to those for two-dimensional flows given by Eqs. (4.31) to (4.33). The main difference comes in the definitions of m, n, f_w, and g'_w. Although for similarity they are constants, the axisymmetric definitions contain $\bar{x}$, whereas the two-dimensional ones contain x. The relation between x and $\bar{x}$ is given by Eq. (4.92a). Thus the body radius $r_0(x)$ appears in the definitions and, for example, makes the value of m for axisymmetric stagnation-point flow different from the value for a two-dimensional flow. To illustrate the relations between two-dimensional and three-dimensional parameters further, let us consider flow near the stagnation point of an axisymmetric body with a blunt nose, for which

$$u_e = Cx. \tag{4.105}$$

Since for similarity $u_e \sim (\bar{x})^m$ [see Eq. (4.99)] and since near the stagnation point $r_0 \sim x$, we find from Eq. (4.92a) that

$$x \sim \bar{x}^{1/3}. \tag{4.106}$$

Thus the value of m for a stagnation-point flow on a blunt nose is $\frac{1}{3}$, in contrast to a two-dimensional one for which the value of m is 1.

Axisymmetric bodies with sharp conical noses also admit similarity solutions near the nose. The external velocity distributions vary with the distance x from the vertex as

$$u_e = Cx^s. \tag{4.107}$$

Here C is simply a scaling constant and the exponent s is a function of the vertex semi-angle of the cone as shown in Fig. 4.12. A vertex semi-angle θ of 90° corresponds to a blunt-nosed body, so that $s=1$ as shown by Eq. (4.105). To find the value of m for those flows for which u_e is given by Eq. (4.107), we again assume $r_0 \sim x$ and find that

$$u_e \sim \bar{x}^{s/3}, \tag{4.108}$$

a result that indicates that for each specified vertex semi-angle of the cone, the value of m is $s/3$.

We note from this discussion that we can still use the results given in Tables 4.1 and 4.2 for the similarity solutions of the momentum and energy equations of an axisymmetric flow provided that we find the "equivalent" m for the body under consideration.

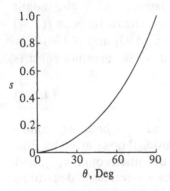

Figure 4.12. Values of s in Eq. (4.107) governing potential flow about semi-infinite cones (after Hess and Faulkner [7]).

Nonsimilar Flows

The calculation of nonsimilar axisymmetric flows is in principle similar to that of two-dimensional flows. We can either solve Eqs. (4.95) and (4.100) subject to Eq. (4.103) by a differential method or solve simplified ordinary differential equations for profile parameters (integral methods). In the latter case, for example, we can extend the two-dimensional integral methods of Thwaites and Smith–Spalding (Section 4.3) to axisymmetric flows by using the Mangler transformation. If we denote the two-dimensional variables by the subscript 2 and the axisymmetric ones by the subscript 3 [these correspond, respectively, to barred and unbarred variables in Eq. (4.92a)], remembering that surface distances transform according to Eq. (4.92a) and distances normal to the surface transform according to Eq. (4.92b) (no transverse curvature effect), then

$$\theta_2 = \left(\frac{r_0}{L}\right)^K \theta_3, \qquad dx_2 = \left(\frac{r_0}{L}\right)^{2K} dx_3. \qquad (4.109)$$

Using Eq. (4.109), we can write Thwaites' formula, Eq. (4.76b), as

$$\left(\frac{\theta_3}{L}\right)^2 = \frac{0.45 R_L^{-1}}{(u_e^*)^6 (r_0^*)^{2K}} \int_0^{x_3^*} (u_e^*)^5 (r_0^*)^{2K} dx_3^* + \left(\frac{\theta_3}{L}\right)_i^2 \left(\frac{u_{e_i}^*}{u_e^*}\right)^6, \qquad (4.110)$$

where

$$r_0^* = \frac{r_0}{L}, \qquad x_3^* = \frac{x_3}{L}.$$

For the stagnation region of an axisymmetric body with a blunt nose,

$$\left(\frac{\theta_3}{L}\right)^2 = R_L^{-1} \frac{0.056}{(du_e^*/dx_3^*)_i}. \qquad (4.111)$$

Once θ_3 is calculated from Eq. (4.110), then the variables δ^*, H and c_f can be calculated from Eqs. (4.78).

Similarly, Eq. (4.84) can be extended to axisymmetric flows by the Mangler transformation and can be written as

$$r_0^2 (\delta_c)_3^2 = \frac{\nu A \int_0^{x_3} r_0^2 u_e^{B-1} dx_3}{u_e^B} \qquad (4.112a)$$

or, in terms of dimensionless quantities, as

$$St = \frac{c_1 (r_0^*)^K (u_e^*)^{c_2}}{\left[\int_0^{x_3^*} (u_e^*)^{c_3} (r_0^*)^{2K} dx_3^*\right]^{1/2}} R_L^{-1/2}. \qquad (4.112b)$$

Here the constants c_1, c_2, and c_3 are the same as those given in Table 4.4.

   As an example of an axisymmetric flow, let us consider a laminar flow past a uniformly heated sphere of radius R_0 in a cross flow with a velocity

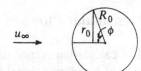

Figure 4.13. Notation for sphere.

u_∞, for comparison with the flow past a cylinder in a cross flow that was discussed in Section 4.3. The velocity distribution for inviscid flow is

$$u_e = \tfrac{3}{2} u_\infty \sin \phi, \tag{4.113}$$

where $\phi(\equiv x/R_0)$ is measured from the stagnation point (see Fig. 4.13).

Figure 4.14 shows the variation of the dimensionless wall shear stress $(\tau_w/\rho u_\infty^2)(u_\infty R_0/\nu)^{1/2}$ around the circumference of the sphere with reference length L taken equal to R_0. The results indicated by the solid line are obtained by Thwaites' method using Eq. (4.110) for the external velocity distribution given by Eq. (4.113) with x/R_0 taken equal to $\sin \phi$, and those indicated by the dashed line by the similarity solution for axisymmetric stagnation flow. The latter results were obtained by following a procedure similar to that for a two-dimensional stagnation point on the circular cylinder. Instead of Eq. (4.88) and the equations that follow it, we have

$$u_e = \frac{3}{2} \frac{u_\infty x}{R_0}. \tag{4.114}$$

Using the definition of wall shear and the similarity variable given by Eq. (4.104), we can write

$$\tau_w = \mu \left(\frac{\partial u}{\partial y} \right)_w = \mu u_e f_w'' \sqrt{\frac{u_e}{\nu \bar{x}}} \frac{r_0}{R_0}. \tag{4.115}$$

Noting that

$$\bar{x} = \frac{1}{R_0^2} \frac{x^3}{3}$$

and that for $m = \tfrac{1}{3}$, $f_w'' = 0.75745$ (see Table 4.1), we can write Eq. (4.115) in

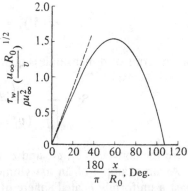

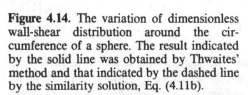

Figure 4.14. The variation of dimensionless wall-shear distribution around the circumference of a sphere. The result indicated by the solid line was obtained by Thwaites' method and that indicated by the dashed line by the similarity solution, Eq. (4.11b).

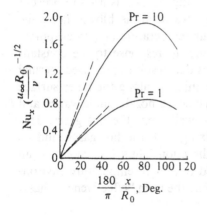

Figure 4.15. The variation of dimensionless heat transfer distribution around the circumference of a sphere. The results indicated by the solid lines were obtained from Eq. (4.112) and those indicated by the dashed lines of the similarity solution, Eq. (4.117).

dimensionless quantities as

$$\frac{\tau_w}{\rho u_\infty^2}\left(\frac{u_\infty R_0}{\nu}\right)^{1/2} = \frac{9}{2\sqrt{2}}\left(\frac{x}{R_0}\right)(0.75745) = 2.41\left(\frac{x}{R_0}\right). \quad (4.116)$$

We see from Fig. 4.14 that, as in flow past a circular cylinder, the similarity solution agrees well with the solution obtained from Thwaites' method up to $x/R_0 < \pi/9$ rad $= 20°$. Also the flow separation takes place at $x/R_0 = 1.885$ rad, which corresponds to $108°$.

Figure 4.15 shows the variation of $\mathrm{Nu}_x\left(\dfrac{u_\infty R_0}{\nu}\right)^{-1/2}$ on the sphere for two values of Prandtl number. The results indicated by the solid line are obtained from Eq. (4.112) and from the definition of the Stanton number $\mathrm{St} = \mathrm{Nu}_x/\mathrm{Pr}\,R_x$, and those indicated by the dashed line are obtained from the similarity solution

$$\mathrm{Nu}_x\left(\frac{u_\infty R_0}{\nu}\right)^{-1/2} = 2.12 g_w'\left(\frac{x}{R_0}\right), \quad (4.117)$$

where g_w' is obtained from the solution of the energy equation, with $m = \frac{1}{3}$. Again, we see that the similarity solutions are in good agreement with nonsimilar solutions for $x/R_0 < \pi/9$ rad.

4.5 Wall Jets and Film Cooling

Wall jets (e.g., Fig. 4.16) are used in a wide range of engineering situations. They form the basis for boundary-layer control on airfoils and wings where, for example, strong tangential blowing over the trailing-edge prevents flow separation and increases lift for low-speed flight. Injection of warm air through slots, providing film heating of the wing surface, is used to prevent icing. In the simple example of demisting, warm air is blown along the surface of a windshield to raise the local air temperature and to evaporate water droplets and films. In the gas turbine, in response to the constant demand for the increased cycle temperatures that result in increased thermal efficiency, a stream of relatively cool gas is injected along the inner surface of combustors and along the external surface of nozzle guide vanes and blades to create a layer that insulates the wall from the hot gases. This coolant film is gradually destroyed by mixing with the hot gases, and its effectiveness decreases in the downstream direction. Thus, film cooling is an effective way of protecting surfaces exposed to a high-temperature environment. In the literature relating to film cooling, the term "effectiveness" has a specific meaning and is defined by

$$\eta^* = \frac{h_w - h_e}{h_c - h_e},$$ (4.118a)

where c, e, and w refer to the enthalpy h of the coolant fluid in the exit flow of the slot, in the external stream, and at the wall, respectively. Frequently, results are referred to an adiabatic wall, and the effectiveness becomes

$$\eta^* = \frac{h_{aw} - h_e}{h_c - h_e},$$ (4.118b)

so that for a flow with constant specific heat,

$$\eta^* = \frac{T_{aw} - T_e}{T_c - T_e}.$$ (4.118c)

Here T_{aw} is called the *adiabatic wall temperature*; it corresponds to the limiting case of a perfectly insulated (i.e., adiabatic) surface on which heat flux would be zero. The heat flux with film cooling is then calculated from

$$\dot{q}_w = \hat{h}(T_w - T_{aw}).$$ (4.119)

Note that in the case of no blowing, T_{aw} would be equivalent to the freestream temperature T_e, or to the recovery temperature T_r in the case of compressible flow.

Another way to protect a surface from a high-temperature environment is to use transpiration cooling. In this case the surface is usually a porous

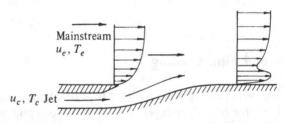

Figure 4.16. A typical film cooling geometry and velocity profiles for tangential injection.

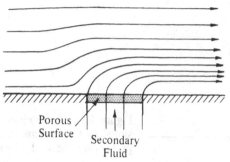

Figure 4.17. Sketch of transpiration cooling.

material, and the secondary (injected) fluid enters the boundary layer through the permeable surface (see Fig. 4.17). This procedure is used primarily to protect the region where the secondary fluid enters the boundary layer. Thus it differs from the film cooling that is used to protect the surface just at the location of coolant addition as well as the protection of the region downstream of the injection location. Contrary to film cooling, the use of transpiration cooling, however, is somewhat limited since the small pore size of the material leads to clogging and subsequent maldistribution of coolant flow. Ablation cooling is a related technique but relies more on cooling by latent heat of vaporization than on transpiration of precooled material.

In practice, the films are almost always turbulent. Depending on the slot arrangement and on the nature of the downstream body, they may lead to two- or three-dimensional flow with or without local regions of separation. Usually, however, the flow downstream of the immediate vicinity of the slot is two-dimensional. Laminar wall jets are to be found mainly in small-scale fluidic devices or high-velocity situations where low densities ensure low values of Reynolds number. In general, the latter cases are complicated by shock-wave patterns that will not be considered here. As usual, solutions for uncoupled laminar-flow problems are simple and educative and allow comparison with the corresponding turbulent flows to be considered in Section 6.7. The configuration shown in Fig. 4.16, and the results presented below and in Section 6.7, relate to two-dimensional flows and may be regarded, from the points of view of boundary-layer control and film cooling, as idealizations that designers of necessarily three-dimensional configurations strive to achieve.

Figure 4.18 shows the diagram of a two-dimensional slot arrangement ahead of an adiabatic wall, the configuration for all calculations presented in this section. This arrangement includes the slot and freestream flows found in most applications; if required, the arrangement can easily be modified to allow the representation of wall jets with zero freestream velocity, and the adiabatic wall boundary condition ($\dot{q}_w = 0$) can be replaced by specified wall heat-flux rate or by specified wall temperature.

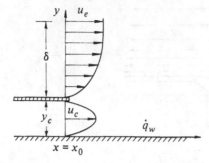

Figure 4.18. A sketch of simple two-dimensional slot arrangement and initial velocity profiles for the wall jet problem.

Other configurations are possible and the reader is referred to the review by Goldstein [8] on film cooling in turbulent flows, which, although written some years ago, is an excellent introduction for the student.

In order to solve the boundary-layer equations for the configuration shown above, we again find it convenient to express Eqs. (3.5), (3.10b), and (3.11) in dimensionless quantities. For this purpose, we use the Falkner–Skan transformation defined by Eqs. (4.19b) and (4.19c),

$$\eta = \left(\frac{u_e}{\nu x}\right)^{1/2} y, \qquad \psi = (u_e \nu x)^{1/2} f(x, \eta). \tag{4.120}$$

We also define a dimensionless temperature g,

$$g = \frac{T_c - T}{T_c - T_e}, \tag{4.121}$$

which is related to the film-cooling effectiveness, Eq. (4.118c), by

$$\eta^* = 1 - g_w. \tag{4.122}$$

In terms of the transformed variables of Eq. (4.120), the momentum and energy equations for a zero-pressure-gradient flow can then be written in the form

$$f''' + \frac{1}{2} ff'' = \xi \left(f' \frac{\partial f'}{\partial \xi} - f'' \frac{\partial f}{\partial \xi} \right), \tag{4.123}$$

$$\frac{g''}{\text{Pr}} + \frac{1}{2} fg' = \xi \left(f' \frac{\partial g}{\partial \xi} - g' \frac{\partial f}{\partial \xi} \right). \tag{4.124}$$

Here with y_c denoting the slot height, ξ is a dimensionless x distance, x/y_c.

Equations (4.123) and (4.124) require initial and boundary conditions. At some $\xi = \xi_0$, the *initial* conditions correspond to

$$y < y_c, \qquad u = u_1(y), \qquad T = T_1(y), \tag{4.125a}$$

$$y > y_c, \qquad u = u_2(y), \qquad T = T_2(y). \tag{4.125b}$$

The boundary conditions are

$$y = 0, \qquad u = v = 0, \qquad \frac{\partial T}{\partial y} \doteq 0, \tag{4.126a}$$

$$y = \delta, \qquad u = u_e, \qquad T = T_e. \tag{4.126b}$$

In terms of dimensionless quantities, the initial conditions become

$$f' = \phi_1(\eta), \qquad g = \phi_3(\eta) \qquad \eta < \eta_c, \qquad (4.127a)$$

$$f' = \phi_2(\eta), \qquad g = \phi_4(\eta) \qquad \eta > \eta_c, \qquad (4.127b)$$

and the boundary conditions become

$$\eta = 0, \qquad f = f' = 0, \qquad g' = 0, \qquad (4.128a)$$

$$\eta = \eta_e, \qquad f' = 1, \qquad g = 1. \qquad (4.128b)$$

These initial and boundary conditions imply that the flow properties depend on the ratio of slot to freestream velocity u_c/u_e, the ratio of slot to freestream temperature T_c/T_e, the Reynolds number based on slot height $u_e y_c/\nu$, the adiabatic wall boundary condition, and the initial profile shapes.

The numerical solution of Eqs. (4.123) and (4.124) subject to the initial and boundary conditions given by Eqs. (4.127) and (4.128) requires the initial profiles to be smooth and free of discontinuities: computation of the flow immediately downstream of a slot lip of finite thickness is a considerable task. A typical velocity profile, like that shown in Fig. 4.18, must be smoothed in the region separating the slot flow from the boundary layer as shown in Fig. 4.19. For the demonstration calculations below, we represent the initial boundary-layer profile $u_2(y)$ by a sine function and the slot profile $u_1(y)$ by a parabola, with a fairing between them (see Section 14.2). The parabola of course corresponds to fully developed duct flow in the slot.

The solution of the energy equation also requires initial conditions, and similar difficulties near the slot lip must be overcome. If we assume that the boundary-layer flow has a uniform temperature T_e, so that $T_2(y) = T_e$, and the slot flow has a uniform temperature T_c, so that $T_1(y) \equiv T_c$, then the initial dimensionless temperature profile g, which is 1 for $y > y_c$ and 0 for $y < y_c$, can be approximated by the expression

$$g = \frac{1}{2}\left[1 + \tanh\beta\left(\frac{y}{\delta} - \frac{y_c}{\delta}\right)\right]. \qquad (4.129)$$

Here β and y_c/δ are specified constants. Figure 4.20 shows a comparison between the exact and approximate expressions. We see that for $y_c/\delta = 0.25$, by choosing β equal to either 10 or 20, we can approximate the initial temperature profile satisfactorily. Approximations for initial conditions like

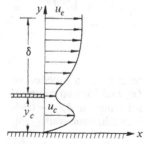

Figure 4.19. Smoothed velocity profiles of Fig. 4.18.

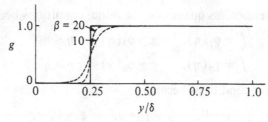

Figure 4.20. Smooth dimensionless temperature profiles (dashed lines) according to Eq. (4.129). Solid line denotes the uniform temperature profile.

the above are frequently needed to allow engineering problems to be solved by relatively simple methods. They facilitate the process of obtaining a solution without significantly affecting the numerical results. Of course a tabulation of experimental data for initial conditions could be used instead.

Figures 4.21 and 4.22 show the results obtained by solving Eqs. (4.123), and (4.124) with $Pr = 1$, subject to the initial velocity and temperature profiles and boundary conditions discussed above. The calculations used the numerical method and the computer program discussed in Chapters 13 and 14.

Figure 4.21 presents the dimensionless wall shear parameter f''_w $= \sqrt{R_L} \sqrt{\xi} c_f / 2$ as a function of dimensionless distance $\xi - \xi_0$ for five different values of the velocity ratio u_c / u_e and for two values of δ / y_c. Here R_L denotes a Reynolds number defined by $u_e y_c / \nu$, $\xi - \xi_0$ is the distance from the slot exit and ξ_0 is the distance from the effective origin of the

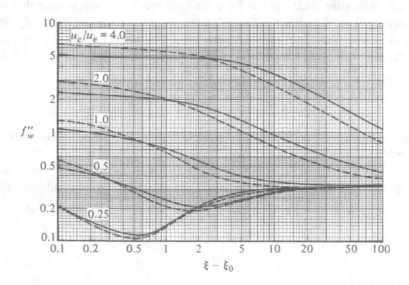

Figure 4.21. Variation of dimensionless wall shear parameter f''_w as a function of distance $\xi - \xi_0$ [$\equiv (x - x_0)/y_c$] for five velocity ratios u_c / u_e and for two values of δ / y_c. The dashed lines correspond to $\delta / y_c = 1.95$ and the solid lines to $\delta / y_c = 0.95$. u_c denotes the maximum slot velocity. Calculations are for an adiabatic wall.

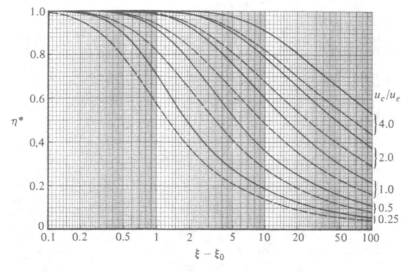

Figure 4.22. Adiabatic wall effectiveness η^* as a function of $\xi - \xi_0$ for five velocity ratios u_c/u_e and for values of δ/y_c. The dashed lines correspond to $\delta/y_c = 1.95$ and the solid lines to $\delta/y_c = 0.95$.

boundary layer on the upper lip of the slot to the plane of the slot exit. For the higher ratios of slot to freestream velocity, the wall shear parameter is nearly constant with distance and with velocity ratio. At the lowest velocity ratio, $u_c/u_e = 0.25$, the freestream flow begins to influence the slot flow at a value of $\xi - \xi_0$ of around 0.5 with a consequent increase in the wall shear parameter for a short downstream distance. The influence of δ/y_c is clearly complicated but of secondary importance. At large distances from the slot, we expect the wall shear parameter f''_w to approach its flat-plate boundary-layer value of 0.332. Figure 4.21 shows that at small velocity ratios, say $u_c/u_e = 0.25$, for $\delta/y_c = 0.95$ and 1.95 at $\xi - \xi_0 = 10$, f''_w is 0.30, close to its asymptotic value of 0.332. At a velocity ratio of 1.0 and at the same location, f''_w is 0.35 for $\delta/y_c = 0.95$. This indicates that at smaller velocity ratios, the asymptotic wall shear is approached more rapidly and the influence of the ratio of boundary-layer thickness to slot height is smaller.

Figure 4.22 presents the calculated values of the adiabatic-wall effectiveness η^* as a function of $\xi - \xi_0$ for five different values of u_c/u_e and for two values of δ/y_c. They indicate that, for example, for $\delta/y_c = 0.95$, with a velocity ratio of 4, an effectiveness of unity is sustained to a downstream distance of $\xi - \xi_0 = 3$ and with a velocity ratio of 0.25 to $\xi - \xi_0 = 0.2$. As might be expected, the higher values of velocity ratio with corresponding increase of slot fluid result in higher values of effectiveness except in the near-slot region where the freestream fluid has not penetrated to the wall. The region of unity effectiveness is longer for the larger velocity ratios. The thickness of the external boundary layer has more influence on the effectiveness than on the wall shear parameter downstream of the slot, consistent

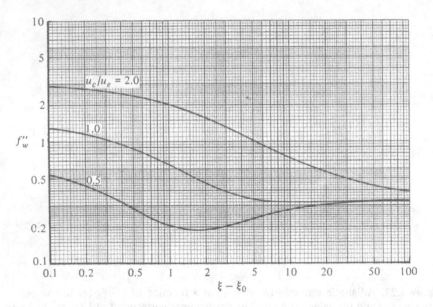

Figure 4.23. Variation of dimensionless wall shear parameter f_w'' as a function of distance $\xi - \xi_0$ on an isothermal surface for $\delta/y_c = 0.95$ and $Pr = 0.72$.

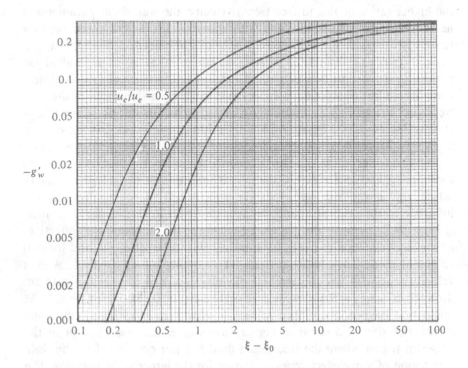

Figure 4.24. Variation of dimensionless wall heat flux parameter g_w' as a function of distance $\xi - \xi_0$ on an isothermal surface for $y/\delta_c = 0.95$ and $Pr = 0.72$.

with increased mixing of slot and freestream fluid occurring with increased boundary-layer thickness. Large values of δ/y_c can be encountered in practice, with consequent further reductions in effectiveness.

Care must be taken in the use of Figs. 4.21 and 4.22 and especially in making comparisons with the corresponding results for turbulent flow shown in Figs. 6.27 and 6.28. The results of Figs. 4.21 and 4.22 (and also Figs. 4.23 and 4.24) are, in their nondimensional form, independent of Reynolds number and a particular choice of fluid, geometry and u_e implies a value of boundary-layer thickness, and therefore δ/y_c, which may not accord with the results shown in the figure. It can be expected that the relationship between η^* and δ/y_c, for example, will be nonlinear with η^* increasing substantially as δ/y_c tends to zero. Problems 4.33 and 4.34 require the use of Figs. 4.21 and 4.24 and will help in their interpretation.

Figures 4.23 and 4.24 show the results for an isothermal flat plate. The calculations were made for the three velocity ratios with $Pr = 0.72$ and $\delta/y_c = 0.95$ by using the same initial velocity and temperature profiles as in previous calculations. The main difference between these calculations and those shown in Figures 4.21 and 4.22 lies in the wall boundary condition used for the energy equations; whereas the adiabatic-wall case used the condition $g'_w = 0$ [see Eq. (4.128a)], the isothermal-wall case used $g_w = 0$. Note that the dimensionless wall temperature gradient g'_w is

$$g'_w = \frac{\dot{q}_w \xi^{1/2}}{(T_c - T_e) R_L^{1/2}} \frac{L}{k}. \tag{4.130}$$

Problems

4.1 Using the Falkner–Skan transformation given by Eqs. (4.19), show that the momentum and energy equations, Eqs. (3.10b) and (3.11) and their boundary conditions, Eqs. (4.2), can be expressed in the forms given by Eqs. (4.31) to (4.33). Assume that wall temperature is specified and use the definition of dimensionless temperature given by Eq. (4.24).

4.2 In the derivation of the Falkner–Skan transformation discussed in Sec. 4.1, we have assumed that there is no mass transfer at the wall, that is: the normal velocity at the wall, v_w, is zero.

(a) By using the procedure of Sec. 4.1, find $v_w(x)$ for similarity when there is mass transfer.
    Hint: Define

$$v_w = A^{as} \bar{v}_w$$

and recall that

$$v_w = -\frac{\partial \psi}{\partial x}.$$

(b) For similarity Eq. (4.31) and its boundary conditions must be independent of x. This means that for mass transfer, $v_w(x)$ and $u_e(x)$ must be such that the dimensionless stream function at the wall, f_w, is constant. Using this, find the variation of v_w with x from Eq. (4.30a).

(c) Show that for a laminar flat-plate flow with mass transfer we must have

$$f_w = -2\frac{v_w}{u_e}\sqrt{R_x}$$

for similarity.

4.3 The equation governing the vertical velocity v in small perturbations of long wave-lengths in a uniformly-stratified shear flow—for example, the lee waves found downwind of mountain ranges—may be written in dimensionless form as

$$\left(\frac{\partial}{\partial t} - iy\right)^2 \frac{\partial^2 v}{\partial y^2} - N^2 v = 0$$

where $i = \sqrt{-1}$ and N is the Brunt–Väisälä frequency. Using the procedure of Sec. 4.1, find the similarity form of this equation.

Hint: Let $t = A^{\alpha_1}\bar{t}$, $y = A^{\alpha_2}\bar{y}$.

4.4 The Von Mises transformation is sometimes used to place the boundary-layer equations into a more amenable form for solution. In this transformation new independent variables (x, ψ) are taken with ψ denoting the stream function

$$\psi = \int_0^y \rho u\, dy$$

Show that with this transformation the momentum and energy equations given by Eqs. (3.10b) and (3.11), subject to the boundary conditions given by Eqs. (4.2) can be written as

$$u\frac{\partial u}{\partial x} - u_e\frac{du_e}{dx} = vu\frac{\partial}{\partial\psi}\left(u\frac{\partial u}{\partial\psi}\right) \tag{P4.1}$$

$$\frac{\partial T}{\partial x} = \frac{v}{\text{Pr}}\frac{\partial}{\partial\psi}\left(u\frac{\partial T}{\partial\psi}\right) \tag{P4.2}$$

$$\psi = 0; \qquad u = 0, \qquad T = T_w \tag{P4.3a}$$

$$\psi \to \infty; \qquad u \to u_e, \qquad T \to T_e. \tag{P4.3b}$$

Note that when u is determined from Eqs. (P4.1) and (P4.3), then v follows from the continuity equation (3.5). Note also that with this transformation, momentum and energy equations are placed in the form of the nonlinear unsteady heat conduction equation,

$$\frac{\partial T}{\partial t} = \frac{\partial}{\partial x}\left[a(t, x)\frac{\partial T}{\partial x}\right].$$

The most useful applications of this transformation have been in laminar heat transfer problems.

4.5 (a) Show that if we integrate Eq. (4.31) across the boundary layer, use the definitions of dimensionless displacement and momentum thickness, δ_1^* and θ_1 (see Eqs. (4.36) and (4.37) respectively), and note that f'' is zero at the edge of the boundary layer because $\partial u/\partial y$ is zero there, we get, for no mass transfer,

$$-f_w'' + \frac{m+1}{2}\int_0^{\eta_e} ff''\, d\eta + m(\delta_1^* + \theta_1) = 0. \tag{P4.4}$$

(b) Show that Eq. (P4.4) can be written as

$$-f_w'' + m\delta_1^* + \left(\frac{3m+1}{2}\right)\theta_1 = 0 \qquad (P4.5)$$

by noting that the integral in Eq. (P4.4) is the dimensionless momentum thickness θ_1 defined by Eq. (4.37). Note that Eq. (P4.5) provides a relation between the three boundary-layer parameters, f_w'', θ_1, and δ_1^* for a given m. In this way θ_1 can be obtained exactly without evaluating Eq. (4.37) numerically.

4.6 As in similar velocity layers, we can also obtain useful integral relations for similar thermal layers for uniform wall temperature. Setting $n = 0$ and assuming no mass transfer at the surface, show that Eq. (4.32) can be integrated with respect to η from $\eta = 0$ to $\eta = \eta_e$ for $g'(\eta_e) = 0$, $g(\eta_e) = 1$, to get

$$-g_w' + \Pr\frac{m+1}{2}\left[f(\eta_e) - \int_0^{\eta_e} f'g\,d\eta\right] = 0. \qquad (P4.6)$$

Show that Eq. (P4.6) can be written as

$$-g_w' + \Pr\frac{m+1}{2}\theta_{T_1} = 0 \qquad (P4.7)$$

where

$$\theta_{T_1} = \int_0^{\eta_e} f'(1-g)\,d\eta.$$

For similar thermal boundary layers with uniform wall temperature, this equation can be used to compute θ_T without evaluating Eq. (3.80) numerically.

4.7 Show that for a similar velocity and thermal boundary-layer flow with porous wall and nonuniform wall temperature, Eqs. (4.31) and (4.32) can be written as

$$-f_w'' + \frac{m+1}{2}f_w + m\delta_1^* + \frac{3m+1}{2}\theta_1 = 0 \qquad (P4.8)$$

$$-g_w' + \frac{1}{2}\Pr(m+1)f_w + \Pr\left(\frac{m+1}{2} + n\right)\theta_{T_1} = 0. \qquad (P4.9)$$

4.8 Air at 20°C and 1 atm. flows at a velocity of 10 m s^{-1} past a 1.5 m-long flat plate whose surface temperature is maintained at 120°C. Use arithmetic film temperature T_f to evaluate the fluid properties.

(a) Calculate the local Nusselt number, skin-friction coefficient and Stanton number at the end of the plate.

(b) If the width of the plate is 0.5 m, calculate the total heat transfer from the plate.

4.9 Glycerine at 40°C flows at a velocity of 3 m s^{-1} past a 4 m long flat plate whose surface is maintained at 20°C. Calculate per unit width, the heat transferred to the plate.

4.10 Air at standard temperature and pressure flows normal to a uniformly heated 2 cm diameter circular cylinder at a velocity of 1 m s^{-1}.

(a) Show that the Nusselt number based on cylinder radius is

$$\mathrm{Nu} = \sqrt{2}\,g_w'\sqrt{R}$$

where

$$R = \frac{u_\infty r_0}{\nu}$$

(b) Taking $Pr = 0.72$, $\nu = 1.5 \times 10^{-5}$ m^{2}s^{-1}, calculate the displacement thickness δ^* and the enthalpy thickness θ_T at the stagnation point.

4.11 Calculate the nondimensional displacement thickness, momentum thickness and skin-friction coefficient on a semi-infinite circular cone whose half angle is 120 degrees. Note that this flow is "under" the cone as shown in the sketch. The external velocity varies as:

$$u_e \sim s^{2.2}$$

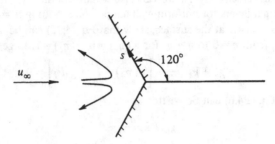

Problem 4.11.

4.12 Derive Eq. (4.53). (See Appendix C.)

4.13 When there is a jump in the wall temperature distribution so that $\partial T/\partial x$ is discontinuous, say at $x = x_0$, a solution of the energy equation in the neighborhood of x_0 can be obtained analytically by using expansions of u, v and T. Let us assume that the velocity field is known and that we are interested in the solution of the energy equation with g given by Eq. (4.24). The energy equation

$$u\frac{\partial g}{\partial x} + v\frac{\partial g}{\partial y} = \frac{\nu}{Pr}\frac{\partial^2 g}{\partial y^2} \tag{P4.10}$$

is subject to

$$y = 0, \quad g = 0 \text{ for } x < x_0, \quad g = 1 \text{ for } x > x_0 \tag{P4.11a}$$

$$y = \delta_t, \quad g = 0 \quad x > x_0. \tag{P4.11b}$$

For $x > x_0$, we assume that u and v do not change much in the x-direction in the neighborhood of the discontinuity and we expand them by

$$u = \sum_{k=1}^{\infty} \lambda_k y^k, \quad v = \sum_{k=2}^{\infty} \mu_k y^k. \tag{P4.12}$$

We expand the dimensionless g by

$$g = \sum_{n=0}^{\infty} z^{n/3}G_n(\zeta) \tag{P4.13}$$

where

$$z = x - x_0, \quad \zeta = y/z^{1/3}. \tag{P4.14}$$

(a) Show that for $n = 0$ with primes denoting differentiation with respect to ζ and $\lambda_1 = (\partial u/\partial y)_w$, Eq. (P4.10) can be written as

$$G_0'' + \frac{1}{3}\lambda_1\frac{Pr}{\nu}\zeta^2 G_0' = 0. \tag{P4.15}$$

(b) Show that the solution of Eq. (P4.15) is

$$G_0(\zeta) = 1 + A \int_0^\zeta \exp\left(-\frac{1}{9}\lambda_1 \frac{\text{Pr}}{\nu} \zeta^3\right) d\zeta \qquad (P4.16)$$

where

$$A = -\frac{3}{\Gamma(1/3)}\left(\frac{\text{Pr}\lambda_1}{9\nu}\right)^{1/3} \qquad (P4.17)$$

and the Gamma function (see Appendix C) $\Gamma(1/3)$ is equal to 2.679 approximately.

(c) Show that for a flat-plate laminar flow

$$\lambda_1 = 0.332\frac{u_e}{x_0}\sqrt{R_{x_0}} \qquad (P4.18)$$

where

$$R_{x_0} = \frac{u_e x_0}{\nu}.$$

(d) Show that

$$\text{Nu}_{x_0} = CR_{x_0}^{1/2}\text{Pr}^{1/3}\frac{1}{[(x/x_0)-1]^{1/3}} \qquad (P4.19)$$

where

$$\text{Nu}_{x_0} = \frac{\dot{q}_w}{(T_w - T_e)}\frac{x_0}{k}$$

$$C = \frac{3}{\Gamma(1/3)}\left(\frac{0.332}{9}\right)^{1/3}.$$

4.14 Derive Eq. (4.66).

Hint: Let $s^3 = y$.

4.15 Consider an incompressible laminar flow over a flat plate which is unheated for $0 \le x \le x_0$ but for $x > x_0$, the plate is subjected to a constant heat flux. Using Eq. (3.77) and with velocity profile approximated by Eq. (4.59) and with the temperature profile approximated by

$$\frac{T - T_e}{T_w - T_e} = 1 - \frac{3}{2}\frac{y}{\delta_t} + \frac{1}{2}\left(\frac{y}{\delta_t}\right)^3$$

show that

(a) $$\text{Nu}_x = \frac{\dot{q}_w x}{(T_w - T_e)k} = 0.417R_x^{1/2}\text{Pr}^{1/3}\left(1 - \frac{x_0}{x}\right)^{-1/3}. \qquad (P4.20)$$

(b) $$T_w - T_e = 2.40\dot{q}_w\frac{x}{k}R_x^{-1/2}\text{Pr}^{-1/3}\left(1 - \frac{x_0}{x}\right)^{1/3}. \qquad (P4.21)$$

Hint: Note that the integration of Eq. (3.77) yields

$$u_e(T_w - T_e)\Theta_T = \frac{\dot{q}_w}{\rho c_p}(x - x_0).$$

4.16 Air at 20°C and 1 atm flows at a velocity of 10 m s^{-1} past a 1.5 m long flat plate. Assuming that the heating starts at 1 m from the leading edge and the plate is 0.4 m wide, compute the power required to maintain the surface temperature at 80°C.

4.17 When the surface temperature of a flat plate is an arbitrary function of x, the energy equation can be solved accurately and easily for specified values of dimensionless temperature g_w by using the Fortran program given in Sec. 13.6. Approximate solutions for constant-property flow over a flat plate with non-uniform wall temperature can also be obtained by taking into account the fact that the energy equation for constant-property flow is a linear one so that we may use the superposition law for linear differential equations. We represent the wall temperature variation as one consisting of a sum of step functions ΔT_{wi} shown below and compute the total heat flux $\dot{q}_w$ from

$$\dot{q}_w = \sum_i \hat{h}_i \Delta T_{wi}. \tag{P4.22}$$

Here $\hat{h}_i$ denotes the heat transfer coefficient due to a step function in wall temperature at x_i [see Eq. (4.70)] and can be written as

$$\hat{h}_i = \frac{0.332 k \,\mathrm{Pr}^{1/3}\sqrt{R_x}}{x\left[1-(x_i/x)^{3/4}\right]^{1/3}} \qquad x > x_i. \tag{P4.23}$$

Using this property of the energy equation, consider the problem 4.16. Assume that the plate has the following temperature distribution: 0 to 0.5 m, 50°C, 0.5 to 1 m, 60°C, 1.0 to 1.5 m, 80°C. Find the local heat flux at $x = 1.5$ m. Note that in this problem $x_1 = 0$, $x_2 = 0.5$ m and $x_3 = 1.0$ and the first step in temperature is $T_{w1} - T_e$.

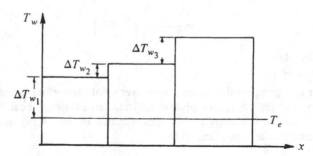

Problem 4.17. Stepwise variation of wall temperature.

4.18 Show that Thwaites' formula, in its general form Eq. (4.74), follows if the velocity profile shape in a laminar boundary layer is uniquely related to the pressure gradient parameter $(\theta^2/\nu)\, du_e/dx$. Hint: This implies that the velocity gradient at the wall, suitably nondimensionalized, is a function of the pressure-gradient parameter.

4.19 Equation (4.86) is based on the assumption given by Eq. (4.80). For a Prandtl number equal to 10, determine the relationship

$$\frac{u_e}{\nu}\frac{d\delta_c^2}{dx} = F\left(\frac{\delta_c^2}{\nu}\frac{du_e}{dx}\right)$$

for wedge flows, $u_e = Cx^m$. That is, derive an expression for each of the terms in the above equation and indicate how you could make a plot which would describe the relationship between them.

Hint: For wedge flows, evaluate $u_e/\nu \, (d\delta_c^2/dx)$ and $\delta_c^2/\nu \, (du_e/dx)$ in terms of m and $g_w' (\equiv \mathrm{Nu}_x/\sqrt{R_x})$. Note that $\delta_c/x = \mathrm{Nu}_x$.

4.20 Heat transfer from an isothermal flat plate can be obtained by an approximate procedure due to Leveque. This procedure considers the solution of the energy equation (3.11) in the region near the wall with Eq. (3.11) approximated to

$$u\frac{\partial T}{\partial x} = \frac{\nu}{\mathrm{Pr}} \frac{\partial^2 T}{\partial y^2}.$$ (P4.24)

It also assumes that close to the wall u varies linearly with y, that is,

$$u = \lambda y.$$ (P4.25)

Eq. (P4.24) with u given by Eq. (P4.25) can be transformed to an ordinary differential equation by introducing a new variable ξ, where

$$\xi = y\left(\frac{\lambda \mathrm{Pr}}{9\nu x}\right)^{1/3}.$$ (P4.26)

Then the resulting equation becomes

$$\frac{d^2 T}{d\xi^2} + 3\xi^2 \frac{dT}{d\xi} = 0.$$ (P4.27)

For the boundary conditions given by

$$\xi = 0, \qquad T = T_w : \xi \to \infty; \qquad T \to T_e,$$ (P4.28)

Eq. (P4.27) has the following solution

$$\frac{T - T_w}{T_e - T_w} = \frac{1}{0.893} \int_0^\xi e^{-\xi^3} \, d\xi.$$ (P4.29)

(a) Show that according to the Leveque solution the local Nusselt number Nu_x defined by Eq. (4.51a) is

$$\mathrm{Nu}_x = \frac{x}{0.893}\left(\frac{\lambda \mathrm{Pr}}{9\nu x}\right)^{1/3}.$$ (P4.30)

(b) Using the Blasius solution to evaluate λ, obtain an expression for Nu_x and compare its accuracy with the exact solutions for small, moderate and high Prandtl number fluids. Discuss the results.

4.21 For a two-dimensional flow near a stagnation point ($u_e = Cx$), calculate δ^*, θ, τ_w by Thwaites' method and $\dot{q}_w$ by the Smith-Spalding method. Compare the results with the exact solutions.

4.22 Repeat problem 4.21 for a flat plate.

4.23 Repeat problem 4.21 for an axisymmetric stagnation-point flow ($u_e = Cx^{1/3}$).

4.24 Using Thwaites' method, compute θ/L, δ^*/L and c_f for a flow in which the external velocity varies linearly with distance as

$$\frac{u_e}{u_\infty} = 1 - a\frac{x}{L}$$ (P4.31)

as a function of x/L for $a = 1/8$ with $R_L = 10^6$. Find also the location of flow separation. Note that this flow is usually referred to as Howarth's flow.
Hint: A listing for integrating Eq. (4.76) by the trapezoidal rule and computing $R_\theta c_f/2$ according to Eq. (4.78b) is given below. Note that for convenience uniform Δx-spacing (0.02) is taken.

```
         REAL LAMDA(51)
         DIMENSION X(51),UE(51),DUEDX(51),THETA(51),FX(51),EL(51)
C
         NP      = 51
         RL      = 1000000.0
         DO 10 I= 1,NP
         X(I)    = 0.02*FLOAT(I-1)
         UE(I)   = 1.0-X(I)/8.0
         DUEDX(I)= -0.125
   10 CONTINUE
C
         SUM     = 0.0
         FX(1)   = UE(1)**5
         DO 20 I= 2,NP
         C       = 0.45/UE(I)**6
         FX(I)   =UE(I)**5
         SUM     = SUM + 0.5*(FX(I)+FX(I-1))*(X(I)-X(I-1))
         THETA(I)= SQRT(C*SUM/RL)
         LAMDA(I)= THETA(I)**2*DUEDX(I)*RL
         IF (LAMDA(I) .LT. 0.0)   GO TO 15
         EL(I)      = 0.225 + LAMDA(I)*(1.61+LAMDA(I)*(-3.75+5.24*LAMDA(I)))
         GO TO 20
   15 EL(I)   = 0.225 + 1.472*LAMDA(I)+0.0147*LAMDA(I)/(0.107+LAMDA(I))
   20 CONTINUE
C
         WRITE (6,100)  (I,X(I),UE(I),THETA(I),LAMDA(I),EL(I),  I=2,NP)
  100 FORMAT (1H0,3X,1HJ,8X,1HX,14X,2HUE,11X,5HTHETA,10X,5HLAMDA,12X,1HL/
     1         (1H ,2X,I2,5E15.6) )
C
         STOP
         END
```

4.25 For two-dimensional steady flows, the separation point is defined as the point where the wall shear stress τ_w is equal to zero, that is,

$$\left(\frac{\partial u}{\partial y}\right)_w = 0.$$

A simple empirical technique due to Stratford[9] can be used to give a rough estimate of separation position method without solving the laminar boundary-layer equations. For a given pressure distribution, for example, $c_p(x)$, the expression

$$c_p^{1/2} \times \left(\frac{dc_p}{dx}\right) \tag{P4.32}$$

is calculated around the body. Separation is predicted when it reaches a value of 0.102. Here c_p is defined as

$$c_p = 1 - \left(\frac{u_e}{u_0}\right)^2 \tag{P4.33}$$

where u_0 is the velocity at the beginning of the adverse pressure gradient. This method can also be used to predict the wall shear (τ_w) distribution in two-dimensional incompressible laminar flows by using the following equation

$$c_p\left(x\frac{dc_p}{dx}\right)^2 = 0.0104\left(1 - \frac{\tau_w}{\tau_B}\right)^3\left(1 + 2.02\frac{\tau_w}{\tau_B}\right). \tag{P4.34}$$

Here τ_B denotes the wall shear for Blasius flow ($m = 0$):
Calculate the local skin-friction coefficient distribution and the location of flow separation for Howarth's flow (Eq. P4.31) by using this method. Compare your results with those of Thwaites, taking $\lambda = -0.09$ at separation.

4.26 Calculate the Stanton number distribution for the external velocity distribution given by Eq. (P4.31) for Pr $=1$ and 10 and plot the ratio of St/($c_f/2$).
Hint: See the listing in Problem 4.28.

4.27 The equation of an ellipse whose center is at $(a,0)$ can be written as

$$\frac{(x-a)^2}{a^2} + \frac{y^2}{b^2} = 1. \tag{P4.35}$$

According to inviscid flow theory, the external velocity distribution around the ellipse is given, for zero angle of attack, by

$$u_e(s) = u_\infty(1+t)\cos\beta. \tag{P4.36}$$

Here s denotes the surface distance, t is the thickness ratio of the ellipse ($\equiv b/a$), and β is the angle between the line tangent to the body and the positive x-axis, that is,

$$\beta = \tan^{-1}\frac{dy}{dx}. \tag{P4.37}$$

Show that for a circular cylinder ($t=1$), Eq. (P4.36) reduces to Eq. (4.87).

4.28 Air at 15°C and at a pressure of 1 atm is flowing over a uniformly heated ellipse of thickness ratio 1/4 and wall temperature of $T_w = 30°C$ at zero angle of attack at a velocity of 10 m s^{-1}. Assuming the flow to be laminar up to separation, compute

(a) Local skin-friction coefficient
(b) Location of flow separation
(c) Wall heat flux rate, in W m^{-2}.

Take Pr = 0.7, $a = 1$.
Hint: A listing for integrating Eq. (4.86) by the trapezoidal rule is given below.

```
      DIMENSION X(101),Y(101),S(101),UE(101),ST(101)
C
      C1       = 0.418
      C2       = 0.435
      C3       = 1.870
      T        = 0.25
      NP       = 100
      RL       = 10000.0
      S(1)     = 0.0
      UE(1)    = 0.0
      ST(1)    = 0.0
      DO 10 I=1,NP
      X(I)     = 0.02*FLOAT(I-1)
      Y(I)     = T*SQRT(1.0 - (X(I)-1.0)**2)
   10 CONTINUE
C
      DO 20 I=2,NP
      S(I)     = S(I-1)+SQRT((X(I)-X(I-1))**2+(Y(I)-Y(I-1))**2)
      BETA     = ATAN((1.0 - X(I))/Y(I)*T*T)
      UE(I)    = (1.0 + T)*COS(BETA)
   20 CONTINUE
C
      SUM      = 0.0
      FX1      = UE(1)**C3
      DO 30 I=2,NP
      C        = C1*UE(I)**C2
      FX2      = UE(I)**C3
      SUM      = SUM + 0.5 * (FX1+FX2)*(S(I)-S(I-1))
      ST(I)    = C / SQRT(SUM*RL)
      FX1      = FX2
   30 CONTINUE
C
      WRITE(6,100) (I,X(I),Y(I),UE(I),ST(I),I=1,NP)
  100 FORMAT(1H0,4X,1HI,7X,1HX,13X,1HY,13X,2HUE,12X,2HST/(1H ,2X,I3,1X,
     1        4E14.6))
C
      STOP
      END
```

4.29 Repeat parts (a) and (b) of problem 4.28 by using Stratford's method.

4.30 Air at 15°C and at a pressure of 1 atm is flowing over a uniformly heated NACA 0012 airfoil at a wall temperature of 25°C and at a velocity of 20 m s^{-1}. Assuming the flow to be laminar up to separation, with $c = 1$ m, compute

(a) Local skin-friction coefficient
(b) Location of flow separation
(c) Wall heat flux rate, in W m^{-2}.

The coordinates of the airfoil and its external (inviscid) velocity distribution are given below.

$\dfrac{x}{c} \times 10^2$	$\dfrac{y}{c} \times 10^2$	$\dfrac{u_e}{u_\infty}$	$\dfrac{x}{c} \times 10^2$	$\dfrac{y}{c} \times 10^2$	$\dfrac{u_e}{u_\infty}$
0	0	0	25	5.941	1.174
1.25	1.894	1.005	40	5.803	1.135
2.5	2.614	1.114	50	5.294	1.108
5.0	3.555	1.174	60	4.563	1.080
7.5	4.200	1.184	70	3.664	1.053
10	4.683	1.188	80	2.623	1.022
15	5.345	1.188	90	1.448	0.978
20	5.737	1.183	95	0.807	0.952
			100	0.126	0.915

4.31 A prolate spheroid is an ellipsoid of revolution whose length along its symmetry axis is greater than the diameter of its largest circular cross section. The equation of a prolate spheroid whose center is at $(a, 0)$ can be written, analogous to Eq. (P4.35) for an ellipse, as

$$\frac{(x-a)^2}{a^2} + \frac{r_0^2}{b^2} = 1. \tag{P4.38}$$

According to inviscid flow theory, for zero angle of attack, the external velocity distribution around the prolate spheroid is given by

$$u_e(s) = u_\infty A \cos \beta. \tag{P4.39}$$

Here s represents the surface distance and β is given by Eq. (P4.37). The parameter A is a function of the thickness ratio $t(\equiv b/a)$ of the elliptic profile. It is given by

$$A = \frac{(1-t^2)^{3/2}}{\sqrt{1-t^2} - \frac{1}{2}t^2 \ln\left[(1+\sqrt{1-t^2})/(1-\sqrt{1-t^2})\right]}. \tag{P4.40}$$

Show that for a sphere ($t = 1$), Eq. (P4.39) reduces to Eq. (4.113).

4.32 Water at 10°C and at a pressure of 1 atm is flowing over a uniformly heated prolate spheroid of thickness ratio 4 and wall temperature of $T_w = 25$°C at a velocity of 5 m s^{-1}. Assuming the flow to be laminar, with $a = 0.1$ m, compute

(a) Local skin-friction coefficient
(b) Location of flow separation
(c) Wall heat flux rate, in W m^{-2}.

4.33 In a low velocity application of film cooling, where laminar flow can be assumed and the free-stream velocity of air is 1 m s^{-1},

    (a) Determine the distance from the slot at which the wall temperature rises to a value corresponding to

$$T_w - T_e = 0.5(T_c - T_e)$$

    for a velocity ratio of 1.0 and $\delta/y_c = 1.95$.

    (b) Obtain the corresponding value of $c_f/2$ for a value of y_c of 3 mm.

    (c) With the assistance of the equations (see page 80)

$$\frac{\delta^*}{x} = 1.72\, R_x^{-0.5} \quad \text{and} \quad \frac{\delta}{\delta^*} = 3.08$$

    calculate the length Reynolds number of the boundary layer on the upper surface of the slot and consider its achievement in practice.

    (d) If the value of u_e is increased to 100 m s^{-1} with other parameters unchanged, what difficulties would you expect?

4.34 Consider the flow situation of problem 4.33 with the adiabatic-wall replaced by a wall with a heat-flux distribution which leads to a uniform wall temperature and the value of δ/y_c changed to 0.95. If the free-stream temperature is 350°K and the wall temperature 300°K, calculate the local heat-transfer rate at $(x - x_0)/y_c = 16$ and state, with reasons, whether this value would increase or decrease if the value of δ/y_c was increased to 1.95.

References

[1] Hansen, A. G.: *Similarity Analysis of Boundary-Value Problems in Engineering.* Prentice-Hall, Englewood Cliffs, NJ, 1964.

[2] Na, T. Y.: *Computational Methods in Engineering Boundary Value Problems.* Academic, New York, 1979.

[3] Eckert, E. R. G. and Gross, J. F.: *Introduction to Heat and Mass Transfer.* McGraw-Hill, New York, 1968.

[4] Thwaites, B. (ed.): *Incompressible Aerodynamics.* Clarendon, Oxford, 1960.

[5] Kays, W. M. and Crawford, M. E.: *Convective Heat and Mass Transfer.* McGraw-Hill, New York, 1980. See also Smith, A. G. and Spalding, D. B.: Heat transfer in a laminar boundary layer with constant fluid properties and constant wall temperature. *J. Roy. Aero. Soc.* **62**:60, 1958.

[6] Cebeci, T. (ed.): *Numerical and Physical Aspects of Aerodynamic Flows II.* Springer-Verlag, New York, 1984.

[7] Hess, J. L. and Faulkner, S. M.: Accurate values of the exponent governing potential flow about semi-infinite cones. *AIAA J.* **3**:767, 1965.

[8] Goldstein, R. J.: Film cooling, in *Advances in Heat Transfer*, **7**. Academic, New York, 1971, 321.

[9] Rosenhead, L. (ed): *Laminar Boundary Layers.* Oxford University Press, London, 1963.

CHAPTER 5

Uncoupled Laminar Duct Flows

In duct flows, e.g., Fig. 1.2(a), the velocity field is said to be *fully developed* when u/u_0 is a function of y only, independent of x. In this case, it follows from the continuity equation that the normal velocity component v is zero everywhere, and the solution of the momentum equation for two-dimensional or axisymmetric uncoupled laminar flows can be obtained exactly and very easily. Clearly this is a particularly simple case of profile similarity, in which the y scaling factor is independent of x. Equally clearly, it applies only to flows in ducts of constant cross section, specifically circular ducts (pipes) and rectangular ducts whose cross section is so wide compared with its height that the flow at some distance from the side walls is near enough two-dimensional.

The *thermal* layer is said to be *fully developed* when the dimensionless temperature profile G, defined by

$$G = \frac{T_w - T}{T_w - T_m},\tag{5.1}$$

is independent of x and depends only on y, that is,

$$\frac{dG}{dx} = 0.\tag{5.2}$$

Note that the temperature itself continues to change with x in general, since the rate of heat transfer through the duct wall is in general nonzero. A fully developed velocity field is a prerequisite for a fully developed thermal field. In coupled flows, where temperature changes are large enough to affect the density, complete similarity is not possible, as we shall see in Chapter 12.

124

In this chapter we shall discuss three different types of surface conditions for heat transfer in duct flows. Section 5.1 deals with the case in which *both* velocity and temperature fields are fully developed, in constant-area circular and noncircular ducts. Section 5.2 deals with the case in which the duct contains an unheated entrance section followed by a heated section. We shall assume that the unheated entrance section is sufficiently long that the velocity field is fully developed when it enters the heated part of the duct. In this case we shall discuss the solutions of the energy (enthalpy) equation with $v = 0$,

$$u \frac{\partial T}{\partial x} = \frac{v}{\Pr} \frac{1}{r^K} \frac{\partial}{\partial y} \left(r^K \frac{\partial T}{\partial y} \right), \tag{5.3}$$

as the temperature field is developing. Finally, Section 5.3 deals with the case in which a constant-area duct is heated over its entire length, so that the thermal boundary layer starts at the entrance of the duct and the velocity field and the temperature field develop simultaneously.

Often, heat-transfer calculations for flows in ducts are made on the basis of the temperature difference between the average temperature of the fluid and the wall temperature of the duct. That is, the average temperature plays the role of a baseline temperature, the same role that is played by the free stream temperature T_e in external flows. This average temperature, which is sometimes called the *mixed mean fluid temperature* or *bulk fluid temperature*, is defined by

$$T_m = \frac{\int \rho u T \, dA}{A \rho_m u_m}. \tag{5.4}$$

Here dA is an element of cross-sectional area, and $A \rho_m u_m$ is the mass flow rate which is, of course, independent of x even if A varies. The quantity $\rho_m u_m$ defines the bulk-average mass velocity, with the mean (average) velocity u_m (equal to the uniform velocity u_0 at inlet) defined by

$$u_m = \frac{\int u \, dA}{A}. \tag{5.5}$$

For present purposes, ρ is taken as constant and cancels from Eq. (5.4). The reason why T_m is the right average to use in one-dimensional analysis is that $c_p \int \rho u T \, dA$ is the rate of axial flow of enthalpy.

5.1 Fully Developed Duct Flow

Constant-Area Ducts

When the velocity profile is fully developed, v and $\partial u / \partial x$ are zero, and the momentum equation for two-dimensional or axisymmetric flow in a constant-area duct can be written as

$$\frac{dp}{dx} = \frac{\mu}{r^K} \frac{d}{dy} \left(r^K \frac{du}{dy} \right). \tag{5.6}$$

Since the pressure gradient is independent of y, this equation can be integrated twice with respect to y to obtain the desired velocity profile. In the case of a circular duct of radius r_0 with y measured radially inward from the wall, we have $r = r_0 - y$, and $dr = -dy$, and since $K = 1$, we can write Eq. (5.6) as

$$\frac{dp}{dx} = \frac{\mu}{r} \frac{d}{dr} \left(r \frac{du}{dr} \right). \tag{5.7}$$

Applying the boundary conditions

$$x = x_i, \qquad p = p_i, \qquad x = x_0, \quad p = p_0, \tag{5.8a}$$

$$r = r_0, \qquad u = 0, \qquad r = 0, \quad \frac{du}{dr} = 0, \tag{5.8b}$$

we obtain the velocity profile for axisymmetric Poiseuille flow as

$$u = \frac{(p_i - p_0) r_0^2}{4 \mu L} \left[1 - \left(\frac{r}{r_0} \right)^2 \right], \tag{5.9}$$

where $L = x_0 - x_i$ so that $(p_i - p_0)/L = -dp/dx$. Equation (5.9) shows that u is a parabolic function of r, zero at the wall, and a maximum on the axis. Problem 5.1 provides further details of the parabolic profile; for example, the bulk-average velocity u_m defined by Eq. (5.4) is just half the maximum. Here we note only that if we define the *friction factor* f relating the pressure drop, diameter $d \equiv 2r_0$, and representative dynamic pressure $\frac{1}{2} \rho u_m^2$ as

$$f = \frac{-d}{\frac{1}{2} \rho u_m^2} \left(\frac{dp}{dx} \right) \equiv \frac{-2 r_0}{\frac{1}{2} \rho u_m^2} \left(\frac{dp}{dx} \right) = \frac{\tau_w}{\frac{1}{8} \rho u_m^2}, \tag{5.10}$$

the parabolic solution leads to

$$f = \frac{64}{R_d}, \tag{5.11}$$

where R_d is defined by

$$R_d = \frac{u_m d}{\nu}. \tag{5.12}$$

If the velocity field is fully developed, the velocity u in the energy equation (5.3) is simply given by Eq. (5.9), and v is zero. If the temperature field is also fully developed, as defined by Eqs. (5.1) and (5.2), the energy equation (5.3) can be simplified even further. Differentiating Eq. (5.1) with respect to x, adding and subtracting dT_m/dx, and using Eq. (5.2), we get

$$\frac{\partial(T-T_m)}{\partial x} = (1-G)\frac{d}{dx}(T_w - T_m), \qquad (5.13)$$

or

$$\frac{\partial T}{\partial x} = (1-G)\frac{dT_w}{dx} + G\frac{dT_m}{dx}. \qquad (5.14)$$

This expression can now be substituted into the left-hand side of Eq. (5.3), and if we can deduce dT_w/dx and dT_m/dx from the boundary conditions, the resulting expression can be treated as an ordinary differential equation. This equation can be solved rather easily if we note that the temperature difference $T_w - T_m$ is related to the wall heat-flux rate $\dot{q}_w$ by

$$\dot{q}_w = -k\left(\frac{\partial T}{\partial y}\right)_w = k(T_w - T_m)\left(\frac{\partial G}{\partial y}\right)_w. \qquad (5.15)$$

Let us consider a duct with a *uniformly heated wall* ($\dot{q}_w = $ constant), which arises in a number of applications such as nuclear reactor fuel elements, electric resistance heating, etc. Since the temperature profile is assumed fully developed, G [and therefore $(\partial G/\partial y)_w$] is independent of x, and Eq. (5.15) therefore shows that

$$T_w - T_m = \text{const},$$

from which

$$\frac{dT_w}{dx} = \frac{dT_m}{dx}. \qquad (5.16)$$

Substituting Eq. (5.16) into (5.14),

$$\frac{\partial T}{\partial x} = \frac{dT_w}{dx} = \frac{dT_m}{dx},$$

and the resulting expression into Eq. (5.3), we get

$$u\frac{dT_m}{dx} = \frac{\nu}{\text{Pr}}\frac{1}{r^K}\frac{\partial}{\partial y}\left(r^K\frac{\partial T}{\partial y}\right). \qquad (5.17)$$

For a circular pipe, following the procedure that led to Eq. (5.7), we can write Eq. (5.17), as

$$u\frac{dT_m}{dx} = \frac{\nu}{\text{Pr}}\frac{1}{r}\frac{\partial}{\partial r}\left(r\frac{\partial T}{\partial r}\right) \qquad (5.18)$$

with the boundary conditions

$$r = 0, \quad \frac{\partial T}{\partial r} = 0; \quad r = r_0, \quad T = T_w, \qquad (5.19)$$

and with u given by Eq. (5.9) in the form

$$u = 2u_m\left[1 - \left(\frac{r}{r_0}\right)^2\right], \qquad (5.20)$$

we can integrate Eq. (5.18) twice with respect to r and evaluate the two constants of integration from Eq. (5.19). This gives the temperature profile

$$T = T_w - \frac{1}{2} \frac{\text{Pr}}{\nu} u_m r_0^2 \frac{dT_m}{dx} \left[\frac{3}{4} + \frac{1}{4} \left(\frac{r}{r_0} \right)^4 - \left(\frac{r}{r_0} \right)^2 \right]. \tag{5.21}$$

To obtain the mixed mean temperature, we substitute Eqs. (5.20) and (5.21) into Eq. (5.4) and get

$$T_w - T_m = \frac{11}{48} \frac{\text{Pr}}{\nu} u_m r_0^2 \frac{dT_m}{dx}. \tag{5.22}$$

We now define the Nusselt number by

$$\text{Nu} = \frac{\dot{q}_w}{(T_w - T_m)} \frac{d}{k}, \tag{5.23a}$$

or in terms of heat-transfer coefficient $\hat{h}$,

$$\text{Nu} = \frac{\hat{h} d}{k}. \tag{5.23b}$$

Here $d(\equiv 2r_0)$ is the pipe diameter. Since

$$\dot{q}_w = -k \left(\frac{\partial T}{\partial y} \right)_w = k \left(\frac{\partial T}{\partial r} \right)_{r=r_0} = \hat{h}(T_w - T_m),$$

we can write the Nusselt number as

$$\text{Nu} = \left(\frac{\partial T}{\partial r} \right)_{r=r_0} \frac{d}{T_w - T_m}. \tag{5.24}$$

Differentiating Eq. (5.21) with respect to r and substituting the resulting expression and Eq. (5.22) into Eq. (5.24) we get

$$\text{Nu} = \tfrac{48}{11} = 4.364. \tag{5.25}$$

When the circular duct has a *uniform wall temperature* rather than a uniform heat flux (as in the case of heat transfer in a water-tube boiler), then since

$$\frac{dT_w}{dx} = 0, \tag{5.26}$$

Eq. (5.14) becomes

$$\frac{\partial T}{\partial x} = G \frac{dT_m}{dx}. \tag{5.27}$$

A similar approach for this case yields (see Problem 5.6)

$$\text{Nu} = 3.658, \tag{5.28}$$

a result that is 16 percent less than the solution for constant heat flux. Note that $\dot{q}_w$ and $T_w - T_m$ both tend to zero far downstream as the whole heats up to the wall temperature, whereas if $\dot{q}_w$ is held constant, T_m and T_w continue to increase indefinitely.

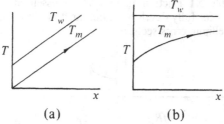

(a)   (b)

Figure 5.1. Variation of mixed mean fluid temperature in a circular tube with fully developed velocity and temperature profiles, (a) constant wall heat flux, (b) constant surface temperature.

The corresponding temperature variations with the duct length for these two cases are shown in Fig. 5.1.

For two-dimensional flows the flow index $K = 0$, and Eq. (5.3) becomes

$$u\frac{\partial T}{\partial x} = \frac{\nu}{\text{Pr}}\frac{\partial^2 T}{\partial y^2}.\tag{5.29}$$

As in the axisymmetric case, the solutions of Eq. (5.29) can be obtained for the cases corresponding to uniform wall temperature and constant heat-flux rate as discussed in Problem 5.7.

Ducts of Noncircular Cross Section

For fully developed laminar flows in ducts of any cross section, the momentum and energy equations can be written as

$$0 = -\frac{1}{\rho}\frac{dp}{dx} + \nu\left(\frac{\partial^2 u}{\partial y^2} + \frac{\partial^2 u}{\partial z^2}\right),\tag{5.30}$$

$$u\frac{\partial T}{\partial x} = \frac{k}{\rho c_p}\left(\frac{\partial^2 T}{\partial y^2} + \frac{\partial^2 T}{\partial z^2}\right).\tag{5.31}$$

In laminar flow we can assume that the pressure is constant across a given section and the secondary flow is negligible (that is, $v = w = 0$). In turbulent flow, the Reynolds stresses generate secondary flow, which greatly complicates the problem. The solutions of Eqs. (5.30) and (5.31) are discussed in detail by Shah and London [1]. Some of the results corresponding to the boundary conditions of constant surface temperature and heat-flux rate in tubes of elliptic cross section and rectangular cross section as well as in some miscellaneous cross sections are given in Tables 5.1 to 5.3. The Nusselt number, with subscript T denoting the uniform-surface-temperature case

Table 5.1 Friction factors and Nusselt numbers for fully developed laminar flows in ducts of elliptic cross section

$\dfrac{b}{a}$	fR_{d_e}	Nu_T	Nu_H
1.00	16.000	3.658	4.364
0.80	16.098	3.669	4.387
0.50	16.823	3.742	4.558
0.25	18.240	3.792	4.880
0.125	19.146	3.725	5.085
0.0625	19.536	3.647	5.176
0	19.739	3.488	5.225

Table 5.2 Friction factors and Nusselt numbers for fully developed laminar flows in ducts of rectangular cross section

$\dfrac{b}{a}$	fR_{d_e}	Nu_T	Nu_H
1.00	14.227	2.976	3.608
0.714	14.565	3.077	3.734
0.50	15.548	3.391	4.123
0.25	18.233	4.439	5.331
0.125	20.584	5.597	6.490
0.05	22.477	—	7.451
0	24.000	7.541	8.235

Table 5.3 Friction factors and Nusselt numbers for fully developed laminar flows in ducts of miscellaneous cross sections

$\dfrac{b}{a}$	fR_{d_e}	Nu_T	Nu_H
Equilateral triangle, △	13.337	2.47.	3.111
Hexagon, ⬡	15.054	3.34	4.002
Semicircle, ◠	15.767	—	4.089

Table 5.4 Friction factors and Nusselt numbers for fully developed laminar flow parallel to a bundle of circular tubes with constant heat-flux rate

$\dfrac{\lambda}{d}$	fR_{d_e}	Nu
1.0	6.503	—
1.001	—	1.26
1.01	7.634	1.52
1.03	12.441	2.14
1.05	15.478	2.82
1.10	20.377	4.62
1.20	24.950	7.48
1.30	27.417	9.19
1.50	31.035	11.26
2.00	39.384	15.27
3.011	58.46	—
4.00	79.003	—

and H the uniform-heat-flux-rate case, is defined by

$$\mathrm{Nu} = \frac{\dot{q}_w}{(T_w - T_m)}\frac{d_e}{k}. \tag{5.32}$$

Here d_e denotes the equivalent or "hydraulic" diameter $4A/s$, where A is the cross-sectional area of the duct and s is its perimeter; for a two-dimensional duct, d_e is twice the duct height, whereas for a circular pipe, it equals the actual diameter. Equation (5.23) is a special case of Eq. (5.32) for a circular pipe.

Similarly, the Reynolds number R_{d_e} is defined by

$$R_{d_e} = \frac{u_m d_e}{\nu}. \tag{5.33}$$

In Tables 5.1 to 5.3, the Nusselt number results for constant heat-flux rate are obtained with the restriction that the temperature across the periphery of the duct be constant. The definition of friction factor is still given by Eq. (5.10) provided that we replace d by d_e.

Another geometry that is of interest occurs in nuclear power reactors where a bundle of circular-cross-section fuel rods is located inside a tube

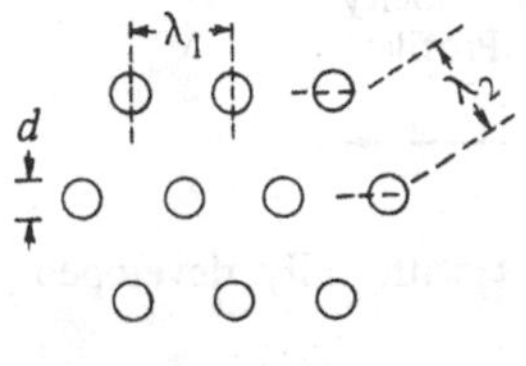

Figure 5.2. Arrangement of circular rods. Note that in this case the spacings λ_1 and λ_2 are equal and denoted by λ.

and generates constant heat-flux rate (see Fig. 5.2). A good approximation to the problem, if there are a large number of rods, is to consider the rod bundle as infinite in lateral extent. Solutions for this case have been obtained by Sparrow, et al. [2] for triangular and square arrays in longitudinal flows (see Table 5.4). The governing parameter is the ratio of spacing to the tube diameter λ/d. For a triangular array the hydraulic diameter is given by

$$d_e = d \left[\frac{2\sqrt{3}}{\pi} \left(\frac{\lambda}{d} \right)^2 - 1 \right].$$

5.2 Thermal Entry Length for a Fully Developed Velocity Field

In this section we shall discuss the prediction of laminar heat transfer in symmetrical two-dimensional or axisymmetric (circular) ducts for the case when the velocity field is fully developed at the start of heat transfer, $x = x_0$, say; the fully developed temperature field discussed in the previous subsection is attained asymptotically at large x. Figure 5.3 shows a sketch of the flow under consideration. The incoming flow has uniform temperature T_e, reference velocity u_0 (say, the average duct velocity), and a fully developed velocity profile. Heat transfer begins at $x = x_0$, so that for $x < x_0$, $T_w = T_e$, and for $x > x_0$ the boundary conditions are

$$y = 0, \quad T = T_w(x) \quad \text{or} \quad \frac{\partial T}{\partial y} = - \frac{\dot{q}_w(x)}{k}, \tag{5.34a}$$

$$y = L, \quad \frac{\partial T}{\partial y} = 0. \tag{5.34b}$$

Here, L corresponds to the half-width (strictly, half-height) of the duct for a two-dimensional flow and to the radius r_0 for an axisymmetric flow. The

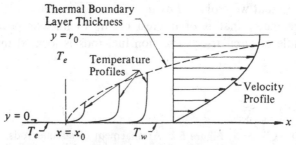

Figure 5.3. Thermal entry problem in a symmetrical duct with fully developed velocity field.

temperature is given by the solution of the energy equation

$$u\frac{\partial T}{\partial x} = \frac{\nu}{\Pr}\frac{1}{r^K}\frac{\partial}{\partial y}\left(r^K\frac{\partial T}{\partial y}\right),$$ (5.3)

subject to Eqs. (5.34) with u assumed known everywhere.

Before we discuss the solution of Eqs. (5.3) and (5.34), we express them in terms of dimensionless quantities defined by

$$\hat{u} = \frac{u}{u_0}, \qquad \hat{x} = \frac{x - x_0}{L}\frac{1}{\Pr R_L}, \qquad \hat{y} = \frac{y}{L}, \qquad \hat{r} = \frac{r}{L} = \frac{r_0}{L} - \hat{y},$$ (5.35)

where $R_L = u_0 L/\nu$. Note that $\Pr R_L = u_0 L/(k/\rho c_p)$, so that $\hat{x}$ does not depend on ν. We now write Eq. (5.3) as

$$\hat{u}(\hat{r})^K\frac{\partial T}{\partial \hat{x}} = \frac{\partial}{\partial \hat{y}}\left[(\hat{r})^K\frac{\partial T}{\partial \hat{y}}\right].$$ (5.36)

Note that $\Pr$ no longer appears explicitly, that is, our scaling has eliminated the fluid properties completely.

Transformed Variables

Near the origin of the thermal boundary layer, the thermal boundary-layer thickness δ_t is very small compared with the pipe radius r_0 or with the half-width of the duct. [This implies that the half-width or radius L used in Eq. (5.35) is not directly relevant, and indeed it will drop out in the analysis.] To maintain computational accuracy, we use a similarity variable defined by

$$\eta = \frac{\hat{y}}{\sqrt{\hat{x}}} \equiv y\sqrt{\frac{u_0 \Pr}{\nu(x - x_0)}},$$ (5.37)

which is simply the Falkner–Skan variable η with ν replaced by $\nu/\Pr \equiv k/\rho c_p$. We also use the dimensionless temperature g defined in Eq. (4.25), that is,

$$T = T_e + T_e(1 - g)\phi(\hat{x}),$$ (4.25)

and with primes now denoting differentiation with respect to η, we write the energy equation (5.36) and its boundary conditions as

$$(a_1 g')' + a_2\frac{\eta}{2}g' - a_3 g = a_2\hat{x}\frac{\partial g}{\partial \hat{x}} - a_3,$$ (5.38)

$$\eta = 0, \qquad g = 0 \quad \text{or} \quad g' = 1; \qquad \eta = \eta_e, \qquad g = 1.$$ (5.39)

Here, for convenience[1], we have defined parameters a_1, a_2, and a_3 by

$$a_1 = (\hat{r})^K, \qquad a_2 = (\hat{r})^K\hat{u}, \qquad a_3 = \hat{x}a_2\hat{n}.$$ (5.40a)

---

[1] In Chapter 7 we shall show that by using the concept of eddy conductivity, ε_h, we can write the energy equation for turbulent flow in the same form as in laminar flow just by redefining a_1 as

$$a_1 = (\hat{r})^K\left(1 + \frac{\Pr\varepsilon_h}{\nu}\right)$$

where ε_h is a function of y.

As before, for a specified wall temperature we define

$$\hat{n} = \frac{1}{T_w - T_e} \frac{d}{d\hat{x}} (T_w - T_e), \tag{5.40b}$$

and for a specified wall heat flux, with $\phi(\hat{x}) = \dot{q}_w(\hat{x}) L \sqrt{\hat{x}} / k T_e$,

$$\hat{n} = \frac{1}{\phi(\hat{x})} \frac{d\phi(\hat{x})}{d\hat{x}}. \tag{5.40c}$$

Note that $L\sqrt{\hat{x}}$ is actually independent of the half-width or radius L; in a two-dimensional duct the solution depends on L only via the velocity u, provided $\delta_t < L$, but in a circular pipe the radius enters via the quantities defined in Eq. (5.40a).

Primitive Variables

The similarity parameter η provides a very useful computational advantage in the early stages of the flow because the transformed thermal boundary-layer thickness is nearly constant ($\eta_e \approx 8$) and temperature profiles are nearly similar in (η, g) coordinates, but we abandon it in favor of the primitive variable $\hat{y} \equiv y/L$ at some $\hat{x} = \hat{x}_s$ before the calculated thermal boundary-layer thickness exceeds the duct half-width or pipe radius L. In this case, with primes now denoting differentiation with respect to $\hat{y}$ and the dimensionless temperature again defined by Eq. (4.25), we write Eq. (5.36) and its boundary conditions as

$$(a_1 g')' - a_3 g = a_2 \frac{\partial g}{\partial \hat{x}} - a_3, \tag{5.41}$$

$$\hat{y} = 0, \quad g = 0 \quad \text{or} \quad g' = 1; \quad \hat{y} = 1, \quad g' = 0. \tag{5.42}$$

Here the definitions of a_1, a_2, and $\hat{n}$ are identical to those given by Eq. (5.40) except that now

$$a_3 = a_2 \hat{n}, \quad \phi(\hat{x}) = \frac{\dot{q}_w(\hat{x}) L}{k T_e}. \tag{5.43}$$

Sample Solutions

Although for some simple boundary conditions the energy equation can be solved by analytical methods, as discussed in Section 13.3, numerical methods must be used to solve this equation in general cases. The results shown here were obtained by using a modified version of the computer program discussed in Section 13.4.

Figures 5.4 and 5.5 show the distribution of the local Nusselt number Nu for a fully developed velocity field with heat transfer for $x > x_0$ only, in a

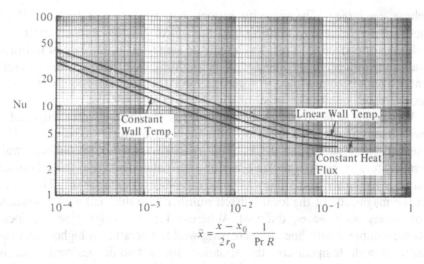

Figure 5.4. Variation of local Nusselt number in the thermal entry region of a circular pipe with different wall boundary conditions. Here $R \equiv u_m r_0 / \nu$.

circular pipe and in a two-dimensional duct, for three cases: constant wall temperature, constant wall heat flux, and a wall temperature that varies linearly with $x - x_0$. Here the local Nusselt number is defined by Eq. (5.32).

We see from Fig. 5.4 that in a circular pipe with uniform wall temperature, local Nusselt number attains its fully developed value of 3.658, Eq. (5.28), at $\hat{x}$ equal to about 0.1. Thus the thermal entry length l_t for this case is

$$\frac{l_t}{r_0} = 0.1 \, \mathrm{Pr} \, R_d = 0.2 \, \mathrm{Pr} \, u_m \frac{L}{\nu} = 0.2 u_m \frac{L}{\kappa}, \tag{5.44}$$

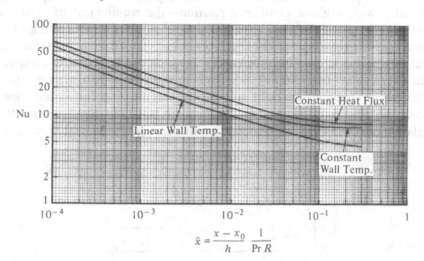

Figure 5.5. Variation of local Nusselt number in the thermal entry region of a two-dimensional duct with different wall boundary conditions, $R = u_m h / \nu$.

where $\kappa = k/\rho c_p$. The viscosity ν does not appear; $u_m L/\kappa$ is a Péclet number. In the case of uniform heat flux, local Nusselt number attains its fully developed value of 4.364 at about the same $\hat{x}$ as in the case of uniform wall temperature. In the linear wall temperature case, however, the thermal entry length becomes slightly bigger than in either of the previous cases. The local Nusselt number becomes nearly constant at $\hat{x} \simeq 0.3$.

We see the same trend in the magnitude of the thermal entry length of a two-dimensional duct subject to different wall boundary conditions. The thermal entry length for the case of constant heat flux or uniform wall temperature is nearly the same as for a circular duct. The local Nusselt number for either case attains its constant value around $\hat{x} = 0.1$. The trend in the magnitude of the local Nusselt number with different wall boundary conditions is, however, different. Whereas for a circular pipe the local Nusselt number with linearly increasing wall temperature is higher than for uniform wall temperature, the situation for a two-dimensional duct is reversed; i.e., the values of local Nusselt number with linearly increasing wall temperature are lower than those for a uniform wall temperature.

5.3 Hydrodynamic and Thermal Entry Lengths

In this section we discuss the prediction of laminar heat transfer in symmetrical two-dimensional or axisymmetric ducts for the case when both the velocity and the temperature fields are developing. Figure 5.6 shows a sketch of the flow under consideration. Again we denote the incoming temperature and velocity of the flow by T_e and u_0, respectively.

The prediction of the nonsimilar velocity field in the entrance region of the duct requires the solution of the continuity and momentum equations, together with a global continuity relation—the requirement of constant mass flow rate, which should be regarded as a boundary condition (implying zero normal velocity at the duct surface) rather than an equation. Once the velocity field is known, then the energy equation can be solved to compute the (nonsimilar) temperature field. Even in the case when the momentum and energy equations are being solved together, the procedure is the same: solve the continuity, momentum, and energy equations together with the relation obtained from the conservation of mass. As in external flows or in

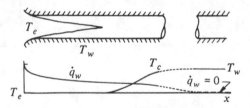

Figure 5.6. Entry flow in a uniformly heated duct: T_e is the fluid temperature at entry to the duct and T_c is the centerline temperature.

the thermal-entry-length problem, the governing equations can be solved either in primitive coordinates (that is, x, y or x, r) or in transformed coordinates. Here we shall follow the procedure used by Cebeci and Chang [3] and use both coordinate systems in the solution of the continuity, momentum, and energy equations.

Transformed Variables

In the early developing stages of the flow, we solve the governing equations in transformed variables. For this purpose we employ a modified version of the Falkner–Skan–Mangler transformation discussed in Section 4.4. Here we restrict our attention to ducts of constant cross section; therefore, the transformation of the x distance according to the first relation given in Eq. (P3.1), has no effect, but we do transform the y coordinate according to the second relation in Eq. (P3.1). That is, for uncoupled duct flows with constant cross section we define the modified Falkner–Skan–Mangler transformation by

$$d\eta = \left(\frac{u_0}{\nu x}\right)^{1/2}\left(\frac{r}{L}\right)^{K} dy, \qquad \psi(x, y) = L^{K}(u_0\nu x)^{1/2}f(x, \eta). \quad (5.45)$$

Here L denotes a reference length, which again, for convenience, is chosen as the half-width of a two-dimensional duct or the radius of a circular pipe. If $y \ll L$ in a pipe, $r/L \equiv (L - y)/L$ is nearly unity and has no effect in Eq. (5.45).

Using this transformation and following a procedure similar to that discussed in Section 4.4, we can now express the momentum and energy equations, Eqs. (3.15) and (3.17) for laminar flow, and their boundary conditions, in transformed variables. Again we use the definition of the stream function that satisfies the continuity equation, Eq. (4.93), and by using relations similar to those defined by Eqs. (4.94), we write the transformed momentum and energy equations in the following form:

$$(bf'')' + \tfrac{1}{2}ff'' = x\frac{dp^*}{dx} + x\left(f'\frac{\partial f'}{\partial x} - f''\frac{\partial f}{\partial x}\right), \quad (5.46)$$

$$(eg')' + \tfrac{1}{2}fg' + nf'(1 - g) = x\left(f'\frac{\partial g}{\partial x} - g'\frac{\partial f}{\partial x}\right). \quad (5.47)$$

Here a prime denotes differentiation with respect to η, and, for the case of specified wall temperature, the parameter n again denotes a dimensionless temperature gradient in the streamwise direction,

$$n = \frac{x}{T_w - T_e}\frac{d}{dx}(T_w - T_e), \quad (5.48a)$$

where T_e is the entry temperature and thus the temperature at the edge of the thermal layer. For specified wall heat flux, with $\phi(x) = \dot{q}_w(x)x/kT_e R_x^{1/2}$

$$n = \frac{x}{\phi}\frac{d\phi}{dx}, \quad (5.48b)$$

where

$$R_x = \frac{u_0 x}{\nu}. \tag{5.49}$$

The dimensionless parameters b, e, and p^* are defined by

$$b = (1-t)^{2K}, \qquad e = \frac{b}{\mathrm{Pr}}, \qquad p^* = \frac{p}{\rho u_0^2}, \tag{5.50a}$$

where

$$t = 1 - \left[1 - 2\eta\left(\frac{x}{L}\frac{1}{R_L}\right)^{1/2}\right], \qquad R_L = \frac{u_0 L}{\nu}. \tag{5.50b}$$

In terms of the transformed variables given by Eq. (5.45), the boundary conditions for the momentum and energy equations follow from Eq. (4.2), and can be expressed by equations very similar to those defined by Eq. (4.30); that is,

$$\eta = 0, \qquad f' = 0, \qquad f = -\frac{1}{(u_0\nu x)^{1/2}}\int_0^x v_w\,dx, \qquad g = 0 \quad \text{or} \quad g' = 1, \tag{5.51a}$$

$$\eta = \eta_e, \qquad f' = \frac{u_e}{u_0}, \qquad g = 1. \tag{5.51b}$$

Primitive Variables

As in Section 5.2 we change to the "primitive" variables x and y (or r) instead of x and η at some point before either the velocity boundary layer or the thermal boundary layer has reached the centerline; recall that if $\mathrm{Pr} < 1$ the thermal boundary layer thickness δ_t is larger than δ. If the Prandtl number is sufficiently *large*, so that the temperature profile grows very slowly, one may even use fully developed velocity profiles in the solution of the energy equation for all x, thus avoiding the solution of the momentum equation. As we shall see later, this introduces little error in the calculation of thermal entry length for flows with high Prandtl number.

When we use primitive variables, we define a dimensionless distance Y by

$$dY = \left(\frac{u_0}{\nu L}\right)^{1/2}\left(\frac{r}{L}\right)^K dy, \qquad \hat{x} = \frac{x}{L} \tag{5.52a}$$

and a dimensionless stream function $F(\hat{x}, Y)$ by

$$\psi = (u_0\nu L)^{1/2}L^K F(\hat{x}, Y), \tag{5.52b}$$

in effect, replacing the x of Eq. (5.45) by the half-width L. With $p^* = p/\rho u_0^2$

as before, we can write the momentum and energy equations as

$$(bF'')' = \frac{dp^*}{d\hat{x}} + F'\frac{\partial F'}{\partial \hat{x}} - F''\frac{\partial F}{\partial \hat{x}}, \tag{5.53}$$

$$(eg')' + n(1-g)F' = F'\frac{\partial g}{\partial \hat{x}} - g'\frac{\partial F}{\partial \hat{x}}. \tag{5.54}$$

Here the definitions of b and e are identical to those given in Eq. (5.50a). The primes denote differentiation with respect to Y; n and t, with $\phi(\hat{x}) = \dot{q}_w(\hat{x})L/kT_e\sqrt{R_L}$, are given by

$$n = \begin{cases} \dfrac{d}{(T_w - T_e)d\hat{x}}(T_w - T_e) & \text{(specified wall temperature)}, \\[2mm] \dfrac{1}{\phi}\dfrac{d\phi}{d\hat{x}} & \text{(specified wall heat flux),} \end{cases} \tag{5.55a}$$

$$t = 1 - \left(1 - \frac{2Y}{\sqrt{R_L}}\right)^{1/2}. \tag{5.55b}$$

The wall boundary conditions for Eqs. (5.53) and (5.54) with specified wall temperature or heat flux are

$$Y = 0, \quad F' = 0, \quad F = -\frac{1}{(u_0\nu L)^{1/2}}\int_0^x v_w\,dx, \quad g = 0 \quad \text{or} \quad g' = 1. \tag{5.56a}$$

At the centerline, symmetry requires $\partial u/\partial y$ and $\partial T/\partial y$ to be zero in untransformed coordinates. In terms of dimensionless variables Y and g, the parameters $\partial u/\partial Y$ and $\partial g/\partial Y$ at the centerline, where Y_c is equal to $R_L/2$ for a circular pipe and $\sqrt{R_L}$ for a plane duct, are indeterminate (of the form $0/0$) for a circular pipe. To determine the centerline boundary condition, we consider Eqs. (5.53) and (5.54) and find that at $Y = Y_c$,

$$F_c'' = -\frac{1}{2}\sqrt{R_L}\left[\frac{dp^*}{d\hat{x}} + \frac{1}{2}\frac{d}{d\hat{x}}(F_c')^2\right],$$

$$g_c' = -\frac{1}{2}\sqrt{R_L}\,\mathrm{Pr}\,F_c'\left[n(g_c - 1) + \frac{\partial g_c}{\partial \hat{x}}\right]. \tag{5.56b}$$

As mentioned at the beginning of this section, the presence of the pressure-gradient term in the momentum equation introduces an additional unknown to the system given by Eqs. (5.46), (5.47), and (5.51) or by Eqs. (5.53), (5.54), and (5.56). Thus another equation is needed. This additional equation can be obtained from the conservation of mass and should be regarded as a boundary condition rather than an equation. For example, in the case of a flow in a plane duct of half-height L, mass balance gives

$$u_0 L = \int_0^L u\,dy. \tag{5.57}$$

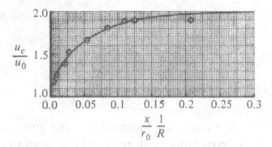

Figure 5.7. Variation of the centerline velocity ratio in the entrance region of a pipe for laminar flow. Here $R = u_m r_0 / \nu$. The circles denote the experimental data of Pfenninger [5].

In terms of dimensionless variables Y and stream function $F(x, Y)$ defined by Eqs. (5.52), we can write the above equation as

$$\sqrt{R_L} = F(\hat{x}, \sqrt{R_L}). \tag{5.58}$$

When transformed variables are used, Eq. (5.58) can also be expressed in the form

$$f(x, \eta_{sp}) = \eta_{sp}, \tag{5.59}$$

where $\eta_{sp} = \sqrt{R_L(L/x)}$. Similar equations can also be written for a circular duct (see Problem 5.22). The solution of the system of equations in transformed variables or in primitive variables consisting of Eqs. (5.46),

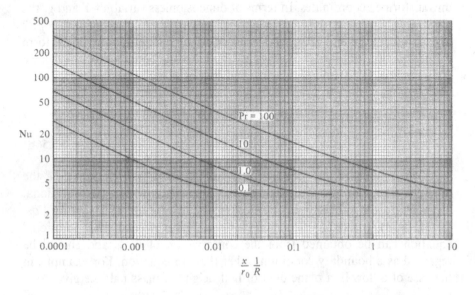

Figure 5.8. Local Nusselt number distribution for laminar flow in a circular duct with uniform wall temperature. Here Nu is defined by Eq. (5.24).

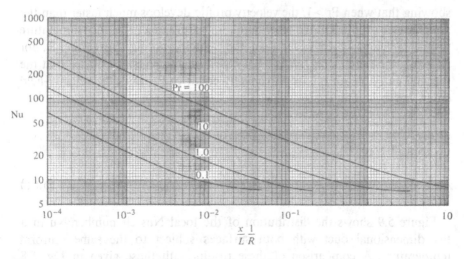

Figure 5.9. Nusselt number distribution for laminar flow in a plane channel with uniform wall temperature. Here Nu is defined by Eq. (5.23b), with diameter d replaced by the channel half-width L and R is defined by $u_m 2h/2$.

(5.47), and (5.51) for transformed variables and Eqs. (5.53), (5.54), and (5.56) for primitive variables, can be obtained by treating the pressure difference between successive x stations as a nonlinear eigenvalue (as described in Section 14.1) or as an unknown (as described by Bradshaw et al. [4]).

Sample Solutions

Figures 5.7 through 5.9 show the results obtained by using the numerical method described in Section 14.1. Figures 5.7 and 5.8 show the variations of dimensionless centerline velocity and local Nusselt number Nu for laminar flow in a circular duct with uniform wall temperature. The calculations, and the measurements of Pfenninger [5], in Fig. 5.7 show that the dimensionless centerline velocity has reached its asymptotic value of 2 at $(x/r_0)/R = 0.20$, where $R = u_0 r_0 / \nu \equiv R_d/2$. Thus the entry length l_v for a laminar flow in a circular duct is

$$\frac{l_v}{r_0} = 0.20R. \tag{5.60}$$

The results in Fig. 5.8 show that for $Pr = 1$, the Nusselt number defined by Eq. (5.23) becomes nearly constant (equal to 3.658) for $(x/r_0)/R \geq 0.2$: compare Eq. (5.44). This means that for $(x/r_0)/R > 0.2$, the temperature profile is fully developed; thus the entry length for the temperature field is the same as the length for the velocity field, as we should expect for $Pr = 1$. For $Pr = 10$, the Nusselt number becomes nearly constant at $(x/r_0)/R = 2$, showing that when $\mathrm{Pr} > 1$, the velocity profile develops much faster than the temperature profile. On the other hand, for values of $\mathrm{Pr} < 1$, the temperature profile develops much faster than the velocity profile. For example, for $\mathrm{Pr} = 0.1$, the ratio $(x/r_0)/R$ is 0.03. Based on these results, we see that the thermal entry length l_t is approximately equal to Pr times the velocity-field entry length l_v; that is,

$$l_t = \mathrm{Pr}\, l_v$$

or, using Eq. (5.60),

$$\frac{l_t}{r_0} = 0.20 R\,\mathrm{Pr}. \tag{5.61}$$

Figure 5.9 shows the distribution of the local Nusselt number Nu in a two-dimensional duct with both surfaces subject to the same uniform temperature. A comparison of these results with those given in Fig. 5.8 shows that the thermal entry length in a two-dimensional duct is nearly the same as the thermal entry length in a circular duct. The same is true of the hydrodynamic entry lengths in the two cases.

Problems

The solution to the following problems can be achieved without using the computer program of Section 13.3. For those wishing to become familiar with the use of computer programs for solving problems of fluid mechanics and heat transfer, we recommend that Problems 5.17 to 5.20 be solved with this program.

5.1 Show that for a fully developed laminar flow in a circular pipe,

(a) The average velocity u_m is related to the maximum velocity u_{max} by

$$u_m = \frac{u_{max}}{2}. \tag{P5.1}$$

(b) The volume flow rate, $Q(\equiv u_0 \pi r_0^2)$ can be written in the form

$$Q = \frac{\pi r_0^4}{8\mu L}(p_i - p_0) \tag{P5.2}$$

which is known as the Hagen–Poiseuille equation.

5.2 Show that for a fully developed laminar flow between two parallel plates, the solution of the momentum equation (5.6), subject to the boundary conditions

$$y = 0, \qquad u = 0; \qquad y = 2h, \qquad u = u_0 = \text{const}$$

is

$$u = u_0 \frac{y}{2h} - \frac{h^2}{2\mu}\frac{dp}{dx}\frac{y}{h}\left(2 - \frac{y}{h}\right). \tag{P5.3}$$

5.3 The type of flow represented by Eq. (P5.3) is known as Couette flow. When both plates are stationary ($u_0 \equiv 0$), Eq. (P5.3) becomes

$$u = -\frac{h^2}{2\mu}\frac{dp}{dx}\frac{y}{h}\left(2-\frac{y}{h}\right),\qquad (P5.4)$$

the case of plane Poiseuille flow. Thus Eq. (P5.3) comprises the linear velocity distribution of Eq. (P5.4), due to the shear flow, between the two plates with no imposed pressure gradient, together with the quadratic velocity distribution caused by the pressure parameter, $P[\equiv -(h^2/2\mu u_0)(p_0-p_i)/L]$.

The flow represented by Eq. (P5.4) is similar to that existing in the narrow clearance between a journal bearing and a stationary shaft, where inertial effects are negligible and the oil is forced along the annulus. Show that

(a) the average velocity u_m is related to the maximum velocity u_{max} by

$$\frac{u_{max}}{u_m}=\frac{3}{2}\qquad (P5.5)$$

(b) and the dimensionless pressure drop can be written as

$$\frac{p_i-p_0}{\rho u_m^2}=\frac{3}{R_h}\frac{L}{h}\quad \text{where}\quad R_h=\frac{u_m h}{\nu}.\qquad (P5.6)$$

(c) Plot u/u_0 against $y/2h$ from Eq. (P5.3) for values of $P=-2,\ -1,\ 0,\ 1,\ 2$ to illustrate the effect of pressure gradient on the velocity profile.

5.4 Derive Eq. (5.28) by following a procedure similar to that used for the constant heat flux rate case. Note that since G is not known prior to the integration, an initial temperature profile must be assumed. Use the temperature profile, Eq. (5.21), corresponding to the constant heat flux rate case for this purpose and integrate Eq. (5.3) with respect to r. Note that $\partial T/\partial x$ now is given by Eq. (5.27). Once a temperature profile is obtained in this way, integrate the energy equation to obtain a new profile. For each temperature profile obtain a mixed mean temperature and a Nusselt number. Repeat this procedure until the successive values of Nusselt number are such that

$$|Nu^{(i+1)}-Nu^{(i)}|<\varepsilon$$

where ε is a tolerance parameter, say 10^{-4}.

5.5 Using the trapezoidal rule, integrate Eq. (5.18) numerically for a fully developed thermal boundary layer with constant heat flux rate in the range $0\le \bar{r}\le 1$. Take $\Delta\bar{r}=0.1,\ 0.05,\ 0.025$ and 0.01 and discuss the effect $\Delta\bar{r}$ has on the computed results.

Hint: Express T, u and r in terms of dimensionless quantities. For example, let

$$\bar{u}=\frac{u}{u_m},\qquad \bar{r}=\frac{r}{r_0},\qquad p=\frac{T}{r_0^2(Pr/\nu)u_m(dT_m/dx)}$$

and write Eq. (5.18) as

$$\frac{d}{d\bar{r}}\left(\bar{r}\frac{dp}{d\bar{r}}\right)=\bar{u}\bar{r}.$$

Integrating from the centerline and setting $dp/d\bar{r} = 0$ at $\bar{r} = 0$, we get

$$\bar{r}\frac{dp}{d\bar{r}} = g.$$

Integrating one more time and taking the mean of the integral, namely, $g_m/\bar{r}_m$, we obtain, with $\bar{p} = p(\bar{r}) - p(0)$,

$$\bar{p} = \int_0^{\bar{r}} \left(\frac{g}{\bar{r}}\right) d\bar{r}.$$

The dimensionless bulk temperature p_m is obtained from

$$p_m \equiv 2\int_0^1 \bar{r}\bar{u}p \, d\bar{r}$$

and Nusselt number, Nu, from

$$\text{Nu} = \left(\frac{\partial T}{\partial r}\right)_{r=r_0}\frac{d}{T_w - T_m} = \left(\frac{\partial T}{\partial \bar{r}}\right)_{\bar{r}=1}\frac{2}{T_w - T_m} = \frac{2}{\bar{p}_w - \bar{p}_m}\left(\frac{\partial p}{\partial \bar{r}}\right)_{\bar{r}=1}.$$

A listing using the above notation is given below. Note that $\bar{p}_m = \bar{p}_m - p(0)$ and $\bar{p}_w = \bar{p}(1)$.

```
      DIMENSION R(51),U(51),PB(51),G(51)
C
      NP      = 51
      DO 10 J= 1,NP
      R(J)    = 0.02 * FLOAT(J-1)
      U(J)    = 2.0 * (1.0 - R(J)**2)
   10 CONTINUE
C
      G(1)    = 0.0
      PB(1)   = 0.0
      RU1     = R(1) * U(1)
      DO 20 J= 2,NP
      RU2     = R(J) * U(J)
      G(J)    = G(J-1) + 0.5 * (RU1 + RU2) * (R(J) - R(J-1))
      GM      = 0.5 * (G(J) + G(J-1))
      RM      = 0.5 * (R(J) + R(J-1))
      PB(J)   = PB(J-1) + GM /RM * (R(J) - R(J-1))
      RU1     = RU2
   20 CONTINUE
      T1      = R(1) * U(1) * PB(1)
      PBM     = 0.0
      DO 30 J= 2,NP
      T2      = R(J) * U(J) * PB(J)
      PBM     = PBM + (T1 +T2) * (R(J) -R(J-1))
      T1      = T2
   30 CONTINUE
      CNU     = 2.0 /(PB(NP) - PBM) * G(NP)
      WRITE( 6, 100) CNU
      WRITE( 6, 110) (J, R(J), PB(J), J=1,NP)
      STOP
C
  100 FORMAT(1H0,3X,'NUSSELT NO =',2X,E14.6,/)
  110 FORMAT(1H0,3X,1HJ,8X,1HR,12X,2HPB/(1H ,2X,I2,2E14.6))
      END
```

5.6 Repeat problem 5.5 for the case of uniform wall temperature.

5.7 Show that for a fully developed laminar thermal boundary layer in a two-dimensional channel, the solution of the energy equation is given by

$$\text{Nu} = 7.60 \tag{P5.7a}$$

for the constant surface temperature case and by

$$Nu = 8.23 \tag{P5.7b}$$

for the constant heat flux rate case.

5.8 For liquid metals (very low Prandtl number) the thermal boundary layers fill the tube near the entrance while the velocity distribution is still very nearly uniform (slug flow). An approximation for this case can be obtained by solving Eq. (5.3) with u as a constant. Show that under these conditions for a laminar flow in a circular pipe with constant heat flux rate,

$$Nu = 8.00.$$

Note that thermal diffusion along the tube can be important with liquid metals but is ignored in this question.

5.9 Air at atmospheric pressure and initially at 120°C is heated as it flows through a 1 cm diameter tube at a velocity of 3 m s^{-1}. Calculate the heat transfer per unit length of tube if a constant-heat-flux condition is maintained at the wall and the wall temperature is 5°C above the air temperature. Assume the flow is fully developed.

5.10 Oil flows at a rate corresponding to a bulk velocity of 1 m s^{-1}, and is cooled from 80°C to 60°C, in an 0.02 m-diameter tube whose surface is maintained at 15°C. How long must the tube be to accomplish this cooling for a fully developed flow? Assume the following properties for oil: $c_p = 2.1$ kJ kg^{-1} K^{-1}, $k = 0.1$ W m^{-1} K^{-1}, $\nu = 6 \times 10^{-6}$ m^2 s^{-1}, $\kappa = 6 \times 10^{-5}$ m^2 s^{-1}.

5.11 Aviation fuel flows at a rate of 2 kg h^{-1} in a 2 m long, 0.05 m-diameter tube and is heated from 10°C to 50°C. Using Fig. 5.4 and assuming that the wall heat flux is constant, calculate and plot the tube surface temperature and fluid bulk temperature as a function of tube length. Assume that the fluid properties are constant and are given by

$$\rho = 700 \text{ kg m}^{-3}, \qquad k = 0.3 \times 10^{-3} \text{ kW m}^{-1} \text{ K}^{-1}$$
$$c_p = 2.3 \text{ kJ kg}^{-1} \text{ K}^{-1}, \qquad \nu = 3 \times 10^{-3} \text{ m}^2 \text{ s}^{-1}$$

5.12 Glycerin at a temperature of 20°C flows in a pipe of 0.005 m in diameter and 0.6 m in length. If the pressure drop is 2 bar and the pipe surface temperature is maintained at 30°C, and $T_m = 25$°C find:
 (a) hydrodynamic entrance length
 (b) thermal entrance length
 (c) the local heat transfer rate at $x = 0.2, 0.4, 0.6$ m.
 The properties of glycerin are as follows:

$$\rho = 1300 \text{ kg m}^{-3} \qquad \nu = 3 \times 10^{-3} \text{ m}^2 \text{ s}^{-1}$$
$$c_p = 2.3 \text{ kJ kg}^{-1} \text{ K}^{-1} \qquad k = 0.3 \times 10^{-3} \text{ kW m}^{-1} \text{ K}^{-1}.$$

5.13 Air at 1 bar and at 260°K flows in an 2 cm-diameter, 3 m long pipe at a velocity of 1 m s^{-1}. If the last 2 m portion of the pipe is heated by keeping the wall temperature at 310°K, find the local heat transfer rates at $x = 1.5$ and 3 m. Assume that the air temperature is constant over the first 1 m of the pipe and $T_m = 285$°C.

5.14 Consider the same flow as in problem 5.13 except that the heat transfer to the 2 m section is applied at a uniform rate of 25 W m^{-2}. Find the surface temperature distribution along the heated section.

5.15 Apply the Leveque solution (see problem 4.20) to a laminar flow in a circular pipe with uniform wall temperature and develop an expression for local and average Nusselt number for the thermal entry region. Compare your results with more accurate solutions and discuss any discrepancies.

5.16 (a) Consider a laminar flow in a circular duct in which the velocity field is developed and the temperature field is developing, but in which the thermal shear layers have not yet merged. Using the definition of dimensionless temperature given by Eq. (4.25) and the definition of bulk temperature given by Eq. (5.4), show that the dimensionless bulk temperature g_m is

$$g_m = 1 + 2\int_0^{\hat{y}_e}(1 - \hat{y})\,\hat{u}(g - 1)\,d\hat{y}. \qquad (P5.8)$$

Here $\hat{y}_e$ is the dimensionless thermal boundary-layer thickness.

(b) Show that when the thermal shear layers have merged, the dimensionless bulk temperature is

$$g_m = 2\int_0^1(1 - \hat{y})\,\hat{u}g\,d\hat{y}. \qquad (P5.9)$$

(c) Show that when the thermal shear layers are merged, the local Nusselt number defined by Eq. (5.23) can be written as

$$\text{Nu} = \frac{2}{g_m}\,g_w'. \qquad (P5.10)$$

where $g_w' = (\partial g/\partial\hat{y})_w$. When the thermal shear layers are not merged

$$g_w' = \left(\frac{\partial g}{\partial\eta}\right)_w\frac{1}{\sqrt{\hat{x}}}. \qquad (P5.11)$$

5.17 Among the cases for which analytical methods can be used to solve Eq. (5.3) is that of laminar flow in a circular pipe with uniform wall temperature and fully-developed velocity profiles. This problem, which is known as the Graetz problem, provides a closed-form solution to Eq. (5.3) and can be written in the following form:

$$\frac{T - T_e}{T_w - T_e} = \sum_{n=0}^{\infty} C_n\phi_n\left(\frac{r}{r_0}\right)\exp\left[-\frac{\lambda_n^2(x^+)}{\text{Pe}}\right]$$

where C_n are coefficients, $\phi_n(r/r_0)$ are functions of r/r_0 determined by the boundary conditions $x^+ = x/r_0$, Pe denotes the Peclet number ($\equiv R_d\text{Pr}$), and λ_n^2 are exponents also determined by the boundary conditions. Sellars, Tribus and Klein [6] show that for a circular pipe with uniform wall temperature and fully-developed velocity profile, the Nusselt number defined by Eq. (5.23) is given by Eq. (13.53), that is

$$\text{Nu} = \frac{\displaystyle\sum_{n=0}^{\infty} G_n\exp\{-\lambda_n^2(x^+/\text{Pe})\}}{2\displaystyle\sum_{n=0}^{\infty}(G_n/\lambda_n^2)\exp\{-\lambda_n^2(x^+/\text{Pe})\}}.$$

The constants G_n and eigenvalues λ_n in Eq. (13.53) are given in Table 13.1.

(a) Evaluate and plot the variation of Nu with x^+ and, using the computer program given in Sec. 13.3, reproduce the solutions of Table 13.2.

(b) For $x^+ < 0.001$ Pe, the solutions of Eq. (13.53) can be approximated by

$$Nu = \frac{1.357}{(x^+/Pe)^{1/3}}.$$

Compare your numerical results with this equation.

5.18 Sellars, Tribus and Klein show that for a circular pipe with constant heat flux and fully-developed velocity profiles, the Nusselt number defined by Eq. (5.23) is given by the following formula

$$Nu = \frac{1}{\dfrac{11}{48} + \dfrac{1}{2} \displaystyle\sum_{m=0}^{\infty} \dfrac{\exp\left[-\beta_m^2 x^+/Pe\right]}{\beta_m^4 \phi_m'(-\beta_m^2)}}. \qquad (P5.12)$$

The eigenfunctions ϕ_m' and eigenvalues β_m^2 appearing in this equation are given in Table P5.1.

(a) Modify the computer program given in Sec. 13.3 to obtain solutions of this problem numerically and compare the numerical results with those given by Eq. (P5.12) for 10^{-4} Pe $< x^+ < \infty$. Note that for $x^+ > 0.25$ Pe, the solutions of Eq. (P5.12) tend to a constant value given by Eq. (5.25).

(b) For $x^+ < 0.001$ Pe, the solutions of Eq. (P5.12) can be approximated by

$$Nu = \frac{1.639}{(x^+/Pe)^{1/3}}.$$

Compare the numerical results with this equation.

Table P5.1. Infinite-series solution-functions for a circular pipe with constant heat flux thermal-entry length

m	β_m^2	$-\phi_m'(-\beta_m^2)$
0	25.64	8.854×10^{-3}
1	84.62	2.062×10^{-3}
2	176.40	9.435×10^{-4}

5.19 Sellars et al. have also obtained analytical solutions for an incompressible laminar flow in a circular pipe with linear wall temperature. They showed that for this case, the local Nusselt number is given by the expression

$$Nu = \frac{\dfrac{1}{2} - 4 \displaystyle\sum_{n=0}^{\infty} \left(G_n/\lambda_n^2\right) \exp\left(\dfrac{-\lambda_n^2 x^+}{Pe}\right)}{\dfrac{88}{768} - 8 \displaystyle\sum_{n=0}^{\infty} \left(G_n/\lambda_n^4\right) \exp\left(-\dfrac{\lambda_n^2 x^+}{Pe}\right)} \qquad (P5.13)$$

where λ_n, G_n are the same functions tabulated in Table 13.1. The solution of this equation for values of $x^+ > 0.5$ Pe also tends to a constant value given by Eq. (5.25), see Fig. 5.3.

(a) Modify the computer program given in Sec. 13.3 to obtain solutions of this problem numerically and compare the numerical results with those given by Eq. (P5.13) for 10^{-4} Pe $< x^+ < \infty$.

(b) For $x^+ \le 0.001$ Pe, the solutions of Eq. (P5.13) can be approximated by

$$Nu = \frac{2.035}{(x^+/Pe)^{1/3}}.$$

Compare your numerical results with this equation.

5.20 For certain boundary conditions the solutions of Eq. (5.3) can be obtained by analytical methods without resorting to numerical methods.

(a) Show that for a two-dimensional slug flow, the solution of Eq. (5.3) subject to the boundary conditions

$$T(0, y) = T_e; \; T(x, \pm h) = T_w; \text{ and in the line of symmetry } \left(\frac{\partial T}{\partial y}\right)_{y=0} = 0$$

can be written as

$$g(x^*, y^*) = \sum_{n=0}^{\infty} \frac{4}{\lambda_n} \sin \frac{\lambda_n}{2} \exp\left[-\frac{\lambda_n^2}{4} \frac{x^*}{G_z}\right] \cos \frac{\lambda_n}{2} y^*. \quad \text{(P5.14)}$$

Here $x^* = x/h$, $y^* = y/h$ and

$$g = \frac{T - T_w}{T_e - T_w}, \quad G_z = R_h Pr, \quad \lambda_n = (2n + 1)\pi.$$

(b) Using the definition of local Nusselt number defined by Eq. (5.32) with d_e replaced by x, show that

$$Nu_x = \frac{2x^* \sum\limits_{n=0}^{\infty} \exp\left[-\frac{\lambda_n^2}{4} \frac{x^*}{G_z}\right]}{\int_0^1 g \, dy^*}. \quad \text{(P5.15)}$$

(c) Modify the computer program of Sec. 13.3 to obtain solutions to this problem numerically and compare them with Eq. (P5.15).

5.21 In the case of a flow in a circular duct, mass balance gives

$$u_0 \pi r_0^2 = \int_0^{r_0} 2\pi r u \, dr. \quad \text{(P5.16)}$$

Show that Eq. (P5.16) in transformed and primitive variables defined by Eqs. (5.45) and (5.52) can be written as

$$f(x, \eta_{sp}) = \eta_{sp} \quad \text{(P5.17)}$$

$$F(x, \sqrt{R_L}) = \sqrt{R_L}/2 \quad \text{(P5.18)}$$

respectively.

5.22 Show that the centerline distance Y_c is equal to $1/2 \sqrt{R_L}$ for a circular duct and to $\sqrt{R_L}$ for a plane duct.

5.23 Derive Eqs. (5.56b).

References

[1] Shah, R. K. and London, A. L.: Laminar flow forced convection in ducts, in *Advances in Heat Transfer*, Supplement 1. Academic, New York, 1978.

[2] Sparrow, E. M., Loeffler, A. L., and Hubbard, H. A.: Heat transfer to longitudinal laminar flow between cylinders. *J. Heat Transfer* **83**:415, 1961.

[3] Cebeci, T. and Chang, K. C.: A general method for calculating momentum and heat transfer in laminar and turbulent duct flows. *Numerical Heat Transfer*, **1**: 39, 1978.

[4] Bradshaw, P., Cebeci, T., and Whitelaw, J. H.: *Engineering Calculation Methods for Turbulent Flows*. Academic, London, 1981.

[5] Pfenninger, W.: Further laminar flow experiments in a 40-foot long two-inch diameter tube. Northrop Aircraft, Hawthorne, CA, Rept. AM-133, 1951.

[6] Sellars, J. R., Tribus, M., and Klein, J. S.: Heat transfer to laminar flow in a round tube or flat conduit—the Graetz problem extended. *Trans. ASME* **78**: 441, 1956.

CHAPTER 6
Uncoupled Turbulent Boundary Layers

The main distinction between the treatment of turbulent flow in this chapter and Chapter 7 and the treatment of laminar flows in Chapters 4 and 5 is that whereas the diffusivities of momentum and heat are known transport properties in laminar flow, the effective diffusivities in turbulent flow are not. Our knowledge of these turbulent diffusivities depends on measurements of flow characteristics such as mean velocity and temperature gradients, together with corresponding turbulent fluxes of momentum (the Reynolds stresses) and of heat, and is usually expressed in terms of empirically based models.

The vast majority of engineering and environmental flows are turbulent, and those where density differences are sufficiently small can be regarded as uncoupled. Thus, the velocity and heat-transfer properties of the water around the body of a swimmer, of the flow of air over the radiator or oil cooler of a moving automobile, or of warm gas flowing in a domestic heating duct can be determined by solving the equations of momentum and energy consecutively. Chapters 10 to 12 deal with situations where the equations are coupled by significant density variations, and simultaneous solution of the equations is required. In Chapter 9 the density variations are presumed small enough for the coupling between the energy equation and the momentum equation to be confined to the buoyancy term of the momentum equation; in Chapters 10 to 12 the density variations are not presumed small, and the coupling may involve all terms containing thermodynamic properties. It is useful to note, however, that present turbulence models for problems requiring the solution of coupled equations are mainly based on experiments with *small* temperature and density variations and on solutions of uncoupled equations.

The analogy between heat and momentum transport apparent from the steady laminar-flow equations is maintained, perhaps less obviously, in the time-averaged equations for turbulent flow. Therefore, one is immediately attracted to the idea of solving turbulent heat-transfer problems by assuming plausible values of Reynolds analogy factors or of the apparent "turbulent" Prandtl number based on the ratio of the turbulent diffusivities of momentum and heat. Until quite recently, indeed, such assumptions were the only ones available, but more advanced models of turbulent heat transfer—making more demands on experimental data—are now coming into engineering use, following advances in turbulence models for momentum transfer.

The dependence of turbulent heat-(and momentum-) transfer calculations on empirical data implies that their accuracy cannot exceed that of the data except by good luck. In general, heat-transfer measurements are less accurate than velocity-field measurements. Even the existing measurements of the rate of heat transfer through a solid surface below a low-speed constant-pressure boundary layer, perhaps the most basic quantity of all, are somewhat scattered, with a standard deviation of order 10 percent. In contrast, the rate of momentum transfer from a constant-pressure boundary layer (specifically, the skin-friction coefficient c_f at a given Reynolds number) is known to an accuracy of about 2 percent, perhaps better. Measurements of heat-transfer rates within the turbulent flow, that is, $\rho c_p \overline{T'u'}$ and $\rho c_p \overline{T'v'}$ in the case of two-dimensional flow, involve simultaneous measurements of velocity and temperature fluctuations and are inevitably less accurate than measurements of momentum-transfer rates, which depend on velocity fluctuations alone. Even mean-temperature measurements are difficult compared to the ease, and relative confidence, with which mean velocity can be measured with a pitot tube. All these difficulties mean that the empirical data needed for heat-transfer calculations are rarely certain to better than 10 percent. Furthermore, since an essential input to any fluid-flow heat-transfer problem is the behavior of the mean velocity field, turbulent heat-transfer calculations for a given flow can never be more accurate than turbulent momentum-transfer calculations for the same flow and are likely to be significantly less accurate.

Eddy Viscosity and Turbulent Prandtl Number

The simplest way of using empirical data for momentum transfer in shear layers is to define an "eddy viscosity" analogous to the molecular viscosity $\mu = $ stress/rate of strain, and then to use experimental data to produce a correlation formula for the eddy viscosity in terms of the shear-layer thickness, freestream velocity, and other quantities (see Section 6.6). For the shear stress $-\rho\overline{u'v'}$ in a two-dimensional thin shear layer, the eddy kinematic viscosity is defined by

$$\varepsilon_m = \frac{-\overline{u'v'}}{\partial u / \partial y}. \tag{6.1}$$

Analogous quantities can be defined for turbulent heat-transfer rates. Just as $-\rho\overline{u'v'}$ is the most important of the six[1] independent Reynolds (turbulent) stresses in a thin shear layer, being the rate of turbulent transfer of x-component momentum in the y direction (normal to the surface in the case of a boundary layer), $\rho c_p \overline{T'v'}$ is the most important of the three turbulent heat-flux rates, being the rate of turbulent transport of enthalpy in the y direction. We define the eddy diffusivity of heat for this quantity by

$$\varepsilon_h = \frac{-\overline{T'v'}}{\partial T / \partial y}. \tag{6.2}$$

Note that a minus sign still appears; in heat transfer, as in momentum transfer, we expect transport *down* the gradient of the quantity in question. The eddy diffusivity of heat has the same dimensions as the eddy kinematic viscosity, namely, velocity $\times$ length. It is easy enough to see that if the mean and fluctuating temperatures were everywhere proportional to the mean and fluctuating x-component velocities, the eddy diffusivity of heat, defined by Eq. (6.2), would equal the eddy kinematic viscosity defined by Eq. (6.1). As mentioned in Chapter 1, we can never expect an exact analogy between heat and momentum transfer in turbulent flow even for exactly analogous boundary conditions, because pressure fluctuations appear in the velocity-field equations but not in the temperature-field equations. However, we may hope that the ratio $\varepsilon_m / \varepsilon_h$, which we call the *turbulent Prandtl number* by analogy with the molecular Prandtl number $\mathrm{Pr} \equiv \nu/\kappa \equiv \mu c_p / k$, will usually be of order unity and sufficiently well behaved to form the basis of data correlations. We therefore define

$$\mathrm{Pr}_t \equiv \frac{\varepsilon_m}{\varepsilon_h} = \frac{\overline{u'v'}/(\partial u/\partial y)}{\overline{T'v'}/(\partial T/\partial y)}. \tag{6.3}$$

We shall see in Section 6.1 that the use of similarity arguments for the turbulent flow near a solid surface (say, the inner 10 or 20 percent of the thickness of a thermal boundary layer) predicts that Pr_t should be a constant *in that region*. The strong dependence of turbulent wall flows on inner-layer behavior implies that reasonably good predictions can be obtained by assuming Pr_t to be constant across the layer. This is not a proof

---

[1] $-\rho\overline{u'^2}, -\rho\overline{u'v'}, -\rho\overline{v'^2}, -\rho\overline{w'^2}, -\rho\overline{u'w'}, -\rho\overline{v'w'}$. The last two are zero by symmetry in two-dimensional flow, i.e., flow in which the mean z-component velocity w is equal to zero, but $-\rho\overline{w'^2}$ is not. Recall from Section 2.4 that $-\rho\overline{u_i'u_j'}$ is a stress acting in the x_i direction on a surface normal to the x_j direction; for example, $-\rho\overline{u'v'}$ is a shear stress acting in the x direction on a surface normal to the y direction.

that Pr_t really is a universal constant or even that it has any physical significance except as a symbol to represent the right-hand side of Eq. (6.3); its value in the inner layer is about 0.9, but in the outer layer of a wall flow or in a free shear layer it is as small as 0.5, and depends significantly on flow geometry. However, empirical correlations for Pr_t are very frequently used in heat-transfer calculations, as we shall see in Section 6.2.

More Advanced Turbulence Models

More advanced methods of predicting turbulent momentum and heat transfer rely on exact "transport" equations for $\overline{u'v'}$ and $\overline{T'v'}$ that can be obtained from the momentum and enthalpy equations. The left-hand sides of these equations give the rate of change of $\overline{u'v'}$ and $\overline{T'v'}$ along a mean streamline, that is, $d\overline{u'v'}/dt$ and $d\overline{T'v'}/dt$ using the substantial derivative, or "transport operator," defined by Eq. (2.10) without its $\partial/\partial t$ term. The right-hand sides contain complicated time averages of products of fluctuating quantities, as well as $\partial u/\partial y$ or $\partial T/\partial y$, so that the equations are not directly solvable. However, the use of empirical data correlations for the quantities on the right-hand sides of these transport equations yields closed systems of equations that, at least qualitatively, represent the fact that it is the *rate of change* of $\overline{u'v'}$ or $\overline{T'v'}$ that depends on the local conditions. The formulas (6.1) and (6.2) nominally imply that the *value* of $\overline{u'v'}$ or $\overline{T'v'}$ depends on local conditions and therefore apply in principle only to flows whose local conditions are changing very slowly with distance downstream. The range of application of Eqs. (6.1) and (6.2) is really limited only by the empirical data correlations used for ε_m and ε_h, but transport–equation methods are gaining favor for the more complicated kinds of flow. For reviews of engineering calculation methods for turbulent flow, see [1] and [2].

6.1 Composite Nature of a Turbulent Boundary Layer

As was discussed in Sections 4.1 and 4.2, the laminar boundary-layer equations admit similarity solutions for some flows with special boundary conditions. These include a variety of two-dimensional and axisymmetric external and free shear flows. For example, in the case of a laminar flat-plate flow, we have seen that the dimensionless velocity and temperature profiles, u/u_e and $(T_w - T)/(T_w - T_e)$, respectively, reduce to a single curve when plotted against a dimensionless y coordinate called the *similarity variable*, $\eta \equiv (u_e/vx)^{1/2}y$. Regardless of the Reynolds number of the flow or of the local skin friction, the geometrical similarity of the velocity profiles is maintained. Except for some free shear flows, there is no choice of one dimensionless y coordinate that leads to the collapse of the complete velocity profile for *turbulent* shear layers into a single curve, because the viscous-dependent part of the profile (very near a solid surface) and the Reynolds-stress-dependent part of the profile require *different* length-scaling parameters. Recall that Reynolds stresses, and turbulent mixing rates in general, are usually much larger than the corresponding viscous stresses or diffusion rates: at a solid surface, however, velocity fluctuations, like mean velocities, are zero: therefore, the Reynolds shear stress $-\rho\overline{u'v'}$ is zero, and the whole of the shear stress on the wall is applied as a viscous stress. As the distance from the wall increases, the Reynolds shear stress increases very rapidly, and the viscous stress decreases to maintain near-constant total shear stress. The very thin viscous-dependent region is called the *viscous sublayer* (for details see Section 6.2).

When an obstacle is placed in a laminar flat-plate boundary-layer flow, the velocity profiles downstream from the obstacle do not at first resemble the Blasius profile shown in Fig. 4.2. However, at low Reynolds numbers, if the layer is allowed to develop far enough downstream, the velocity profiles slowly return to the Blasius profile. In turbulent boundary layers, the effect of disturbances near the surface (even well outside the very thin viscous sublayer) disappears quite soon, because of the greater diffusivity, and the velocity profiles quickly return to "normal" boundary-layer profiles. The phenomenon was experimentally investigated by Klebanoff and Diehl [3]. Analysis of the data of Fig. 6.1 shows that the inner part of the turbulent layer returns more quickly to normal than the outer part of the layer, which suggests that the flow close to the wall is relatively insensitive to the flow conditions away from the wall and to the upstream conditions. Figures 6.2 and 6.3 further illustrate that effect. Here, turbulent flow in a rectangular channel passes from a rough surface to a smooth one and vice versa. The figures show that in both cases the shear stress near the wall very rapidly assumes the new value corresponding to the local surface conditions, whereas further from the wall the shear stress, which nearly equals the

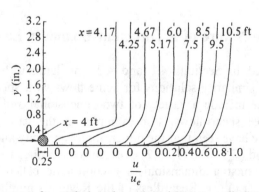

Figure 6.1. Response of a turbulent boundary layer to wall disturbances. Mean-velocity distribution of a turbulent boundary layer on a flat plate behind a cylindrical rod in contact with the surface at $x = 4$ ft from the leading edge. After Klebanoff and Diehl [3].

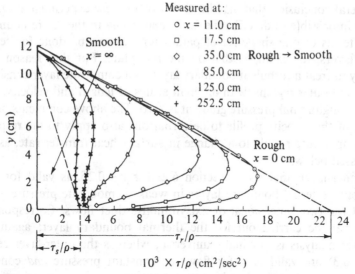

Figure 6.2. Variation of the shear-stress distribution in turbulent flow through a rectangular channel. Flow passes from a rough surface to a smooth one at $x = 0$. After Jacobs [4].

Reynolds shear stress $-\rho\overline{u'v'}$ here because the viscous shear stress $\mu\,\partial u/\partial y$ is small, changes very slowly. In fact, a new state of equilibrium is established only at rather long distances from the start of the rough surface. Although the experiment is for channel flow, the basic phenomenon applies also to boundary layers; it also applies to perturbations caused by changes in surface mass-transfer rate rather than roughness.

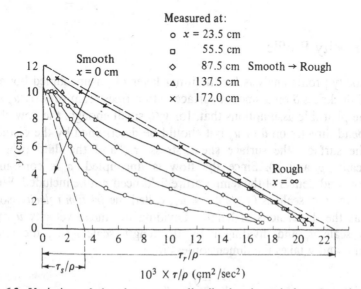

Figure 6.3. Variation of the shear-stress distribution in turbulent flow through a rectangular channel. Flow passes from a smooth surface to a rough one. After Jacobs [4].

A general conclusion that may be drawn from these experimental facts is that it is impossible to describe the flow phenomena in the entire boundary layer in terms of one single set of parameters, as can be done for certain laminar layers, for example, that on the flat plate. For that reason, it is necessary to treat a turbulent boundary layer as a composite layer consisting of inner and outer regions, with different scales of length and velocity, even when the longitudinal pressure gradient is zero. The above comments on the response of the velocity profile to perturbations also apply to the response of the temperature profile to a change in surface heat-transfer rate, as will be discussed below.

The inner-layer analysis of Section 6.2, for $y < 0.2\delta$, is valid for two-dimensional external boundary layers in weak or moderate pressure gradients and for two-dimensional or axisymmetric duct flows. Corresponding analyses can be carried out for the thermal boundary layer; again, the inner-layer analysis is nominally universal, whereas the outer-layer results (Section 6.3) are valid only for flows at constant pressure *and* constant surface temperature or surface heat-transfer rate.

In the following Sections 6.2 to 6.4, we present the velocity-profile analysis, step by step, followed by the corresponding analysis for the temperature profile. Note that we assume constant-property flow so that the effects of density and viscosity variations on the velocity field are neglected. For the extension to variable-property flow, see Chapter 11.

6.2 The Inner Layer

The Velocity Profile

The velocity-profile analysis for the inner layer of an uncoupled boundary layer of thickness δ on a smooth surface with a freestream velocity u_e starts from the plausible assumptions that, for y/δ small enough, the flow should not depend directly on δ or u_e but should be determined by the distance y from the surface, the surface shear stress, τ_w, and the fluid properties (specifically, ρ and μ). Since the flow is uncoupled, ρ is constant by definition, and details of the temperature field need not be included. We can define a velocity scale as $(\tau_w/\rho)^{1/2} \equiv u_\tau$, called the *friction velocity*, so that ν/u_τ has the dimensions of length. Dividing the mean velocity u by the velocity scale u_τ and dividing y by the length scale ν/u_τ, the relation between u and y takes the dimensionless form

$$\frac{u}{u_\tau} = f_1\left(\frac{u_\tau y}{\nu}\right), \tag{6.4}$$

called the *law of the wall*. As yet, f_1 is unknown.

The above process of dimensional analysis can be performed more rigorously, with checks to see that all the dimensionless groups have been identified. This is scarcely needed for the present simple case, but it is essential in more complicated cases like the law of the wall for temperature discussed below. A convenient process is the "matrix-elimination" method presented by E. S. Taylor [5]; its advantage over other methods is that dimensionless groups that are already known can be inserted easily, making the analysis much shorter than in methods that start from a position of total ignorance. For an example of its use, let us apply it to the inner-layer velocity profile. We first identify all the relevant variables, namely u, y, τ_w, ρ, and μ, and construct a matrix whose columns give the mass, length, and time dimensions of the variables in each row:

$$
\begin{array}{c|ccc}
 & M & L & T \\
\hline
u & 0 & 1 & -1 \\
y & 0 & 1 & 0 \\
\tau_w & 1 & -1 & -2 \\
\mu & 1 & -1 & -1 \\
\rho & 1 & -3 & 0
\end{array}
$$

The dimensions of τ_w, for example, are those of pressure or stress and can be constructed by noting that ρu^2 has the same dimensions as pressure, which are therefore $ML^{-3}(L^2T^{-2})$. We now eliminate the mass dimension by dividing all but one of the variables containing the mass dimension by the remaining variable containing mass; our knowledge of fluid dynamics prompts us to choose the density ρ as the dividing variable:

$$
\begin{array}{c|ccc}
 & M & L & T \\
\hline
u & 0 & 1 & -1 \\
y & 0 & 1 & 0 \\
\dfrac{\tau_w}{\rho} & 0 & 2 & -2 \\
\dfrac{\mu}{\rho} & 0 & 2 & -1 \\
\rho & 1 & -3 & 0
\end{array}
$$

Obviously the density cannot appear in any dimensionless group in this problem except as τ_w/ρ or μ/ρ—there is no other way of canceling its mass dimension, and we can therefore drop it from the matrix.

Next we eliminate the time dimension, simply because in this case the T (time) column contains more zeros than the L (length) column. Again using our knowledge of fluid flow to choose physically useful combinations of variables, we get

$$
\begin{array}{c|cc}
 & L & T \\
\hline
\dfrac{u}{\sqrt{\tau_w/\rho}} & 0 & 0 \\
y & 1 & 0 \\
\dfrac{\mu/\rho}{\sqrt{\tau_w/\rho}} & 1 & 0
\end{array}
$$

By inspection, the length dimension can be eliminated by forming $y\sqrt{\tau_w/\rho}/(\mu/\rho)$. No other independent dimensionless groups can be constructed; so we have

$$f\left[\frac{u}{\sqrt{\tau_w/\rho}}, \frac{y\sqrt{\tau_w/\rho}}{\mu/\rho}\right] = 0,$$

which, with our usual notation, is equivalent to Eq. (6.4).

The Temperature Profile

In the case of the temperature profile we expect that the list of variables will again include y, τ_w, ρ, and μ since the thermal field depends on the velocity field; u need not be included explicitly since it is given by Eq. (6.4). Since the velocity field depends on the rate of transfer of momentum to the surface, τ_w, we may expect the temperature field to depend, analogously, on the rate of heat transfer from the surface to the flow, $\dot{q}_w$. We also expect the fluid properties k and c_p to appear. It should be noted that $\dot{q}_w/\rho c_p$ has the dimensions of velocity × temperature and that the modern convention is to measure $\dot{q}_w$ in mechanical units, for example, kWm^{-2}, so that the "mechanical equivalent of heat" does not appear. Also note that the dimensions of energy per unit mass are the same as those of (velocity)^2 and need not be specified separately. Further, if the velocity-field analysis is restricted to $y/\delta < 0.2$, we expect that by analogy the temperature-field analysis will be restricted to $y/\delta_t < 0.2$ approximately, where δ_t is the thickness of the temperature profile. If the surface is heated all the way from the origin of the boundary layer, all the turbulent fluid will be heated and—as long as the molecular Prandtl number is not orders of magnitude different from unity—δ_t would be almost equal to δ if analogous definitions of δ and δ_t were used. However, we shall see below that the inner-layer analysis for the temperature field can also be applied with good engineering accuracy at a sufficient distance downstream of step changes in $\dot{q}_w$ (so that $\delta_t < \delta$), for $y/\delta_t < 0.2$; obviously the restriction to $y/\delta_t < 0.2$ is stronger than the restriction $y/\delta < 0.2$ in this case. Finally we note that the temperature, like the velocity, must be measured as a difference from the surface value; we write it as $T_w - T$ so that it will usually be positive for positive $\dot{q}_w$.

We now have eight variables, y, τ_w, ρ, μ, k, c_p, $\dot{q}_w$, and $T_w - T$, with *four* dimensions, mass, length, time, and temperature. The theory of dimensional analysis—or Taylor's elimination procedure—shows that we should therefore have $8 - 4 = 4$ dimensionless groups, and the most convenient arrangement of the relation between $T_w - T$ and y (the temperature profile) is obtained by Taylor's elimination procedure (see Problem 6.1):

$$\frac{T_w - T}{\dot{q}_w/\rho c_p u_\tau} = f_2\left(\frac{u_\tau y}{\nu}, \frac{\mu c_p}{k}, \frac{\rho u_\tau^3}{\dot{q}_w}\right), \tag{6.5a}$$

where $\dot{q}_w/\rho c_p u_\tau$ is called the *friction temperature* and is given symbol T_τ by analogy with the friction velocity u_τ. The second group on the right is the molecular Prandtl number, a property of the fluid. The final group, which would be difficult to derive by intuition and might be forgotten altogether if the above check on the total number of dimensionless groups were omitted, can be thought of as the ratio of the rate at which the moving fluid does work against the surface shear stress $\tau_w \equiv \rho u_\tau^2$ (proportional to τ_w times the velocity scale u_τ) to the rate of supply of energy by heat transfer through the surface, $\dot{q}_w$. In low-speed (constant-property) flow this ratio is very small, so that "kinetic heating" of the flow by dissipation of mechanical energy into thermal energy, via work done against τ_w, is negligible. We ignore the final group in Eq. (6.5a) in what follows, remarking only that dimensional analysis, being a branch of mathematics rather than of physics, frequently yields correct but unimportant dimensionless parameters. Therefore, the version of the law of the wall for temperature that is used in practice in uncoupled turbulent flows is

$$\frac{T_w - T}{T_\tau} = f_2\left(\frac{u_\tau y}{\nu}, \mathrm{Pr}\right). \tag{6.5b}$$

The Logarithmic Profiles

The next step in the velocity-profile analysis after the derivation of Eq. (6.4) is to differentiate it; we obtain, after rewriting,

$$\frac{\partial u}{\partial y} = \frac{u_\tau}{y} f_3\left(\frac{u_\tau y}{\nu}\right), \tag{6.6}$$

where $f_3(Z) = Z(df_1/dZ)$ for any variable Z. Now $u_\tau y/\nu$ is a form of Reynolds number, and we may expect that if $u_\tau y/\nu$ is large enough, the effects of viscosity on the local behavior of the velocity profile should be small, leading to

$$\frac{\partial u}{\partial y} = \frac{u_\tau}{\kappa y}, \tag{6.7}$$

where κ is an absolute constant, *not* the thermal diffusivity. It is found by experiment that Eq. (6.7) is valid for $u_\tau y/\nu > 50$ approximately, and that κ is about 0.40–0.41. Analogously, we expect the effect of viscosity on the temperature profile to be small for $u_\tau y/\nu > 50$ approximately; also, the effect of thermal conductivity on the temperature profile should be small if the group

$$\frac{u_\tau y}{\nu} \mathrm{Pr} \equiv \frac{u_\tau y}{k/\rho c_p} \tag{6.8}$$

is large (presumably greater than about 50). This group is an example of a Péclét number and represents the ratio of convective transport of heat to molecular diffusion; the denominator $k/\rho c_p$ is the "thermal diffusivity" or "thermometric conductivity", and we shall see below that the *turbulent* diffusivity of heat in the inner layer is proportional to the numerator $u_\tau y$. Note that if Pr is greater than unity (fluid with high viscosity and/or low conductivity), the condition for viscous effects to be negligible, $u_\tau y/\nu > 50$, is stronger (larger lower limit on y) than the condition for conductive effects to be negligible, $(u_\tau y/\nu)\text{Pr} > 50$, and if Pr < 1, the condition for viscous effects is weaker than the condition for conductive effects. If both conditions are satisfied, then we expect μ and k to be unimportant in determining the local behavior of $\partial T/\partial y$, and the formal derivative of Eq. (6.5), without the kinetic heating parameter $\rho u_\tau^3/\dot{q}_w$, namely,

$$\frac{\partial(T_w - T)}{\partial y} = \frac{T_\tau}{y} f_4\left(\frac{u_\tau y}{\nu}, \frac{\mu c_p}{k}\right), \tag{6.9}$$

(analogous to Eq. (6.6)) will reduce to

$$\frac{\partial(T_w - T)}{\partial y} = \frac{T_\tau}{\kappa_h y}, \tag{6.10}$$

where κ_h is an absolute constant, expected to be roughly equal to κ. It is found by experiment that the lower limit on the validity of Eq. (6.10) is indeed the smaller of $u_\tau y/\nu \approx 50$ and $(u_\tau y/\nu)\text{Pr} \approx 50$—the limits being ill-defined because Eq. (6.10), like Eq. (6.7), becomes valid asymptotically—and that κ_h is about 0.44. The difference between κ and κ_h is just about significant in view of the experimental difficulties of measuring $\dot{q}_w$ and $T(y)$ and then differentiating the latter, and an overlapping range of values for κ and κ_h is quoted in the literature.

Integrating Eq. (6.7) with respect to y and requiring compatibility with Eq. (6.4) gives

$$\frac{u}{u_\tau} = \frac{1}{\kappa} \ln \frac{u_\tau y}{\nu} + c, \tag{6.11a}$$

or

$$u^+ = \frac{1}{\kappa} \ln y^+ + c, \tag{6.11b}$$

where c is a constant found experimentally to be about 5.0–5.2. Equation (6.11) is the logarithmic law for velocity. Analogously, integrating Eq. (6.10) and requiring compatibility with Eq. (6.5) gives

$$\frac{T_w - T}{T_\tau} = \frac{1}{\kappa_h} \ln \frac{u_\tau y}{\nu} + c_h, \tag{6.12a}$$

or

$$T^+ = \frac{1}{\kappa_h} \ln y^+ + c_h, \tag{6.12b}$$

where c_h is a function of Pr. It might be more logical to write the argument of the logarithm as the Peclet number $y^+ \text{Pr}$, that is, incorporating the thermal conductivity rather than the viscosity, but this would merely change the dependence of c_h on Pr without eliminating it altogether, and the form shown is easier to compare with Eq. (6.11).

Recall the limits of validity:

The logarithmic velocity profile, Eq. (6.11), is valid for $50\nu/u_\tau < y < 0.2\delta$ approximately.
The logarithmic temperature profile, Eq. (6.12), is valid for $\max[50\nu/u_\tau, 50\nu/(u_\tau \mathrm{Pr})] < y < 0.2\delta_t$ approximately.

In flows with transpiration through the surface, or significant streamwise pressure gradient, the shear stress τ will vary with y even in the inner layer, and if $\dot{q}_w$ varies with x or if there are heat sources within the fluid, the heat-transfer rate $\dot{q}$ will also vary with y. In these cases it is believed that local values rather than surface values should be used in the right-hand sides of Eqs. (6.7) and (6.10), leading to the equations

$$\frac{\partial u}{\partial y} = \frac{(\tau/\rho)^{1/2}}{\kappa y} = \frac{\left(\overline{-u'v'}\right)^{1/2}}{\kappa y}, \qquad (6.13)$$

$$\frac{\partial (T_w - T)}{\partial y} = \frac{(\dot{q}/\rho c_p)}{(\tau/\rho)^{1/2}\kappa_h y} = \frac{\overline{T'v'}}{\left(\overline{-u'v'}\right)^{1/2}\kappa_h y} \qquad (6.14)$$

with approximately the same ranges of validity as the logarithmic laws that they replace. Obviously, within this range of validity the molecular contributions to τ and $\dot{q}$ are negligible.

Equation (6.13) seems to be valid in flows with transpiration (Problem 6.5), even for large values of $\partial\tau/\partial y$, but may not be a significant improvement over Eq. (6.7) in boundary layers that are changing rapidly in the streamwise direction, because the basic analysis on which both equations depend assumes that the effect of streamwise rates of change is small. Equation (6.14) has more limited support, deriving mainly from high-speed flows in which $\dot{q}$ varies with y because of kinetic heating; data in high-speed flows also support the use of Eq. (6.13) in cases where ρ, rather than τ, varies with y.

The Viscous and Conductive Sublayers

Equations (6.11) and (6.12) apply far enough from the surface for viscous and conductive effects, including viscous shear stress and conductive heat transfer, to be negligible. Close enough to the surface for *turbulent* shear stress to be negligible (in practice $y^+ < 3$, approximately) the viscous stress law gives simply

$$\tau_w = \mu \frac{du}{dy}, \qquad (6.15a)$$

$$u = \frac{\tau_w y}{\mu}, \qquad (6.15b)$$

or

$$u^+ = y^+, \tag{6.15c}$$

which is, as it should be, a special case of Eq. (6.4). In the case of the temperature profile, the heat-conduction law gives, for $y^+ \mathrm{Pr} < 3$ and $y^+ < 3$ (say),

$$\dot{q}_w = -k\frac{dT}{dy}, \tag{6.16a}$$

or

$$T_w - T = \frac{\dot{q}_w y}{k}, \tag{6.16b}$$

or

$$T^+ = y^+ \mathrm{Pr}, \tag{6.16c}$$

which is a special case of Eq. (6.5b).

In the viscous sublayer, the law of the wall, Eq. (6.4), can be represented by using one of several empirical approaches. Van Driest [6] showed that a good approximation to the sublayer velocity profile could be obtained by replacing Eq. (6.13) by the *empirical fit*

$$\frac{\partial u}{\partial y} = \frac{\left(-u'v'\right)^{1/2}}{\kappa y [1 - \exp(-y/A)]}. \tag{6.17}$$

Here A is a "damping-length" constant for which the best dimensionally correct empirical choice is about $26\nu/u_\tau$. It can be seen that Eq. (6.17) tends to Eq. (6.13) for $y \gg A$ (actually for $u_\tau y/\nu > 50$, as implied above). Now $-\overline{u'v'}$ can be eliminated by noting that the total shear stress $-\rho\overline{u'v'} + \mu(du/dy)$ is very nearly independent of y and equal to $\tau_w \equiv \rho u_\tau^2$. This gives

$$\nu\frac{du}{dy} + (\kappa y)^2 \left[1 - \exp\left(\frac{-y}{A}\right)\right]^2 \left(\frac{du}{dy}\right)^2 = u_\tau^2. \tag{6.18}$$

Solving this quadratic equation for du/dy and rewriting in dimensionless variables gives

$$\frac{du^+}{dy^+} = \frac{-1 + \sqrt{1 + 4a}}{2a}, \tag{6.19}$$

where $a = (\kappa y^+)^2 [1 - \exp(-y^+/A^+)]^2$ and $A^+ = 26$. This equation can be integrated formally to obtain

$$u^+ = \int_0^{y^+} \frac{2}{1 + \sqrt{1 + 4a}}\, dy^+, \tag{6.20}$$

but the actual evaluation of the integral must be done numerically. The resulting velocity profile is shown in Fig. 6.4.

Equation (6.5), like Eq. (6.4), can also be extended to include the (conductive) sublayer by using a procedure similar to that used for the

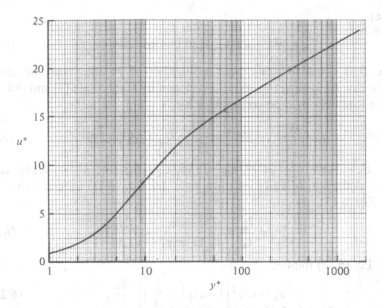

Figure 6.4. Dimensionless velocity distribution in the inner layer.

velocity profile. The requirement that the heat-transfer rate is independent of y and equal to $\dot{q}_w$ gives

$$\frac{k}{\rho c_p}\frac{dT}{dy} - \overline{T'v'} = -\frac{\dot{q}_w}{\rho c_p}. \tag{6.21}$$

Using the definition of eddy conductivity given by Eq. (6.2), we can write Eq. (6.21) as

$$\left(\frac{1}{\mathrm{Pr}} + \varepsilon_h^+\right)\frac{dT}{dy} = -\frac{\dot{q}_w}{\rho c_p v}, \tag{6.22a}$$

where $\varepsilon_h^+ \equiv \varepsilon_h/v$, or in dimensionless quantities as

$$\frac{dT^+}{dy^+} = \frac{1}{1/\mathrm{Pr} + \varepsilon_h^+}. \tag{6.22b}$$

To integrate this equation we need an expression for ε_h^+. We see from Eq. (6.22) that in the conductive sublayer where ε_h^+ is quite small, the term $1/\mathrm{Pr}$ dominates the temperature distribution if $\mathrm{Pr} < 1$. This means that in low-Prandtl-number fluids, the behavior of ε_h^+ very close to the wall is not crucial to the prediction of temperature distribution. On the other hand, if Pr is not small compared with unity, the dimensionless eddy-conductivity term plays an important role even very close to the surface. Van Driest's formula, Eq. (6.17), suggests replacing Eq. (6.14) by

$$\frac{\partial(T_w - T)}{\partial y} = \frac{\overline{T'v'}}{(-\overline{u'v'})^{1/2}\kappa_h y[1 - \exp(-y/B)]} \tag{6.23}$$

so that

$$\varepsilon_h = \left(\overline{-u'v'}\right)^{1/2} \kappa_h y \left[1 - \exp\left(-\frac{y}{B}\right)\right], \qquad (6.24)$$

where B is equal to ν/u_τ times a function of Pr that must be chosen empirically. According to a model developed by Cebeci (see [7] and [8]), it is given adequately by a power series in $\log$ Pr,

$$B = \frac{B^+ \nu}{u_\tau}, \qquad B^+ = \frac{1}{\text{Pr}^{1/2}} \sum_{i=1}^{5} C_i (\log_{10} \text{Pr})^{i-1}, \qquad (6.25)$$

where $C_1 = 34.96$, $C_2 = 28.79$, $C_3 = 33.95$, $C_4 = 6.3$, and $C_5 = -1.186$. If we use the definition of eddy viscosity given by Eq. (6.1), we can write this, using Eq. (6.17), as

$$\text{Pr}_t = \frac{\varepsilon_m}{\varepsilon_h} = \frac{\kappa}{\kappa_h} \frac{1 - \exp(-y/A)}{1 - \exp(-y/B)}, \qquad (6.26)$$

where Eq. (6.17) implies

$$\varepsilon_m = \kappa^2 y^2 \left[1 - \exp\left(-\frac{y}{A}\right)\right]^2 \frac{\partial u}{\partial y}, \qquad (6.27a)$$

or in dimensionless quantities

$$\varepsilon_m^+ = \kappa^2 y^{+2} \left[1 - \exp\left(-\frac{y^+}{A^+}\right)\right]^2 \frac{\partial u^+}{\partial y^+}. \qquad (6.27b)$$

Now $\varepsilon_m/\varepsilon_h$ is the ratio of the turbulent diffusivity of momentum to the turbulent diffusivity of heat, as displayed in Eq. (6.3); that is, it is equal to the turbulent Prandtl number Pr_t. We can therefore write Eq. (6.22b) as

$$\frac{dT^+}{dy^+} = \frac{1}{1/\text{Pr} + \varepsilon_m^+/\text{Pr}_t}, \qquad (6.28)$$

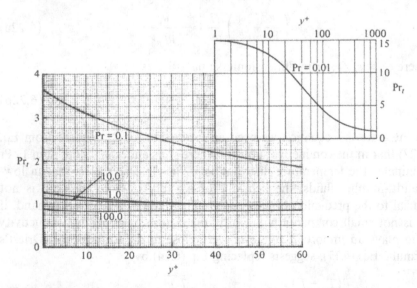

Figure 6.5. Variation of turbulent Prandtl number with y^+ at different molecular Prandtl numbers, Pr.

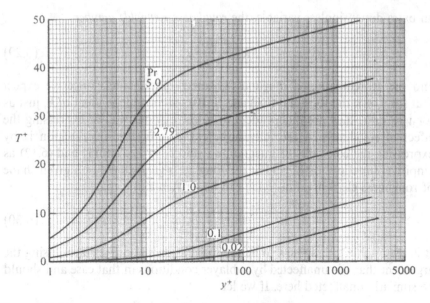

Figure 6.6. Mean-temperature distribution across the layer as a function of molecular Prandtl number.

which can be integrated numerically to give T^+ as a function of y^+ if we substitute for $Pr_t \equiv \varepsilon_m/\varepsilon_h$ and ε_m^+ from Eqs. (6.26) and (6.27b), obtaining $\partial u^+/\partial y^+$ from Eq. (6.19). The variation of Pr_t with y^+ for a range of values of Pr is shown in Fig. 6.5, and the resulting T^+ profiles are given in Fig. 6.6.

The Inner Layer on a Rough Surface

The discussion so far has dealt with the inner-layer similarity analysis of turbulent velocity and temperature profiles on a smooth surface with uniform temperature. We now extend the same discussion to rough surfaces; surface roughness has a significant effect on a turbulent boundary layer if the height of the roughness elements k corresponds to $u_\tau k/\nu > 5$ approximately.[1] Since in most cases the viscous sublayer is extremely thin, less than 1 percent of the shear-layer thickness in a boundary layer with $u_e\delta/\nu > 10^5$, the roughness elements must be very small if the surface is to be aerodynamically smooth.

The above dimensional analysis for a smooth surface can be extended to include rough surfaces by adding the roughness height k and the roughness geometry and packing density to the list of the variables used to derive Eqs. (6.4) and (6.5). The result is that (Problem 6.2) Eqs. (6.4) and (6.5) contain

---

[1]Like the use of κ for the constant in Eq. (6.7), the use of k for roughness height is standard notation, however confusing in a heat-transfer context.

an extra dimensionless variable, the *roughness Reynolds number*

$$k^+ = \frac{u_\tau k}{\nu},$$ (6.29)

and the aforementioned roughness-geometry parameters. Since we expect that for $y \gg k$ the formulas (6.6) and (6.10) will be independent of k, just as for $y \gg \nu/u_\tau$ they are independent of ν, the obvious way of expressing the effect of roughness in the fully turbulent part of the inner region is by expressing the constants of integration c and c_h in Eqs. (6.11) and (6.12) as empirical functions of k^+ and of the roughness geometry. For a given shape of roughness elements, Eq. (6.11) can be written as

$$u^+ = \frac{1}{\kappa}\ln y^+ + B_1(k^+).$$ (6.30)

It is reasonable to assume κ to be the same as on smooth surfaces, using the argument that κ is unaffected by sublayer conditions in that case and should be similarly unaffected here. If we let

$$B_2 = \frac{1}{\kappa}\ln k^+ + B_1(k^+),$$ (6.31)

we can write Eq. (6.30) as

$$u^+ = \frac{1}{\kappa}\ln y^+ - \frac{1}{\kappa}\ln k^+ + B_2 = \frac{1}{\kappa}\ln\frac{y}{k} + B_2.$$ (6.32)

If we let

$$\Delta u^+ = \frac{\Delta u}{u_\tau} = \frac{1}{\kappa}\ln k^+ + c - B_2,$$ (6.33)

where c has its original smooth-surface value, then we can write Eq. (6.32) as

$$u^+ = \frac{1}{\kappa}\ln y^+ + c - \Delta u^+.$$ (6.34)

Here Δu^+, like B_2, is a function of roughness geometry and density. Its relation with k^+ has been determined empirically for various types of roughness geometry as shown in Fig. 6.7.

We see from Eq. (6.34) that since for a given roughness geometry Δu^+ is a known function of k^+, the sole effect of the roughness is to shift the intercept, $c - \Delta u^+$, as a function of k^+ as shown in Fig. 6.8. For values of k^+ below approximately 5, the vertical shift Δu^+ approaches zero except for those roughnesses having such a wide distribution of particle sizes that there are some particles large enough to protrude through the viscous sublayer even though the average size is less than the thickness of the sublayer. For large values of k^+ the vertical shift is proportional to $\ln k^+$, with the constant of proportionality equal to $1/\kappa$, and by comparing Eqs. (6.32) and (6.34) we can see that this implies that B_2 is independent of k^+. This means

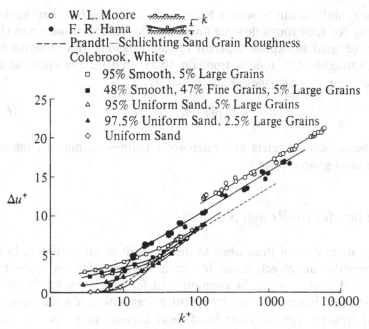

○ W. L. Moore ⌇⌇⌇⌇ ⌐ k
● F. R. Hama ⌇⌇⌇⌇ ⌐ ₋
---- Prandtl–Schlichting Sand Grain Roughness
       Colebrook, White
   □ 95% Smooth, 5% Large Grains
   ■ 48% Smooth, 47% Fine Grains, 5% Large Grains
   △ 95% Uniform Sand, 5% Large Grains
   ▲ 97.5% Uniform Sand, 2.5% Large Grains
   ◇ Uniform Sand

Figure 6.7. Effect of wall roughness on universal velocity profiles. After Clauser, ref. 9.

that the drag of the roughness elements is independent of viscosity, a reasonable result for nonstreamlined obstacles at large Reynolds number.

For surfaces covered with uniform roughness, three distinct flow regions in Fig. 6.7 can be identified. For close-packed sand-grain roughness, the approximate boundaries of these regions are as follows:

Hydraulically smooth: $k_s^+ \leq 5$.
Transitional: $5 \leq k_s^+ \leq 70$.
Fully rough: $k_s^+ \geq 70$.

Because molecular viscosity still plays some part in the transitional region, the geometry of roughness elements has a relatively large effect on

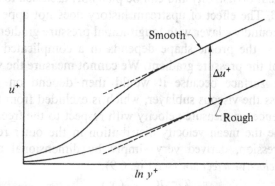

Figure 6.8. Mean velocity distribution on smooth and rough surfaces on a y^+, u^+ plot.

the velocity shift, as can be seen in Fig. 6.7. However, the fact that the shifts in velocity for fully rough flow are linear on the semilogarithmic plot (Fig. 6.7) can be used to express different roughness geometries in terms of a reference roughness. It follows from Eq. (6.33) that for the same velocity shift, Δu^+, any two roughnesses are related by

$$\frac{k}{k_s} = \exp\left[\kappa\left(B_2 - B_{2_s}\right)\right], \tag{6.35}$$

where the subscript s refers to a reference roughness, commonly taken as uniform sand-grain roughness.

Heat Transfer on Rough Surfaces

Now the adaptation of these ideas to the constant of integration c_h in Eq. (6.12) provides an object lesson in the differences between momentum transfer and heat transfer. Momentum transfer to a very rough surface ($k^+ > 200$, say) is accomplished by pressure drag with negligible molecular (viscous) transfer. On the other hand, heat transfer to or from a rough surface is always accomplished by molecular (conductive) transfer and so the constant of integration depends on k^+—strictly, on $k^+ \mathrm{Pr}$— *even for very large roughness Reynolds number*. The limiting form of c_h for large k^+, discussed in different notation by Jayatilleke, is therefore of the form [10]

$$c_h = f(k^+ \mathrm{Pr}). \tag{6.36}$$

6.3 The Outer Layer

The Velocity Profile

Outside the viscous sublayer, the velocity profile in a constant-property, steady two-dimensional boundary layer at constant pressure, i.e., constant u_e, is independent of viscosity and can be plausibly assumed to depend only on u_τ, y, and δ. The effect of upstream history does not appear explicitly, although in a boundary layer with longitudinal pressure gradient, that is, u_e depending on x, the profile shape depends in a complicated way on the whole history of the pressure gradient. We cannot measure the velocity with respect to the surface because it would then depend on the velocity difference across the viscous sublayer, which is excluded from the analysis; so for convenience we measure velocity with respect to the freestream value u_e and describe the mean velocity distribution in the outer region by the following expression, derived very simply by dimensional analysis and known as the *velocity-defect law* (see Fig. 6.9):

$$\frac{u_e - u}{u_\tau} = F_1\left(\frac{y}{\delta}\right). \tag{6.37}$$

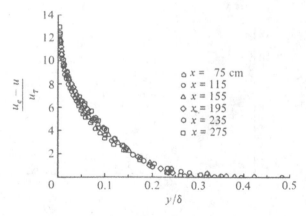

Figure 6.9. Universal plot of turbulent velocity profiles in zero pressure gradient. After Hoffmann and Perry [11].

For boundary layers in zero pressure gradient or flow in constant-area ducts with duct radius or half-height replacing δ, the function F_1 has been found empirically to be independent of Reynolds number (except in boundary layers at rather low Reynolds numbers, $u_e \theta / \nu < 5000$ approximately) and, most significantly, independent of surface roughness. For boundary layers on flat plates, the function F_1 is numerically different from that for duct flow, owing mainly to the presence of the free outer boundary.

The Temperature Profile

Analogously, if $\dot{q}_w$ is independent of x, we expect the temperature anywhere outside the viscous and conductive sublayers, measured with respect to the freestream value T_e, to depend only on T_τ, y, and δ; separate consideration of ρ, τ_w, q_w, etc., would lead to T_τ as a temperature scale, plus the appearance of the irrelevant parameter $\rho u_\tau^3 / \dot{q}_w$. Thus we expect

$$\frac{T_e - T}{T_\tau} = F_2\left(\frac{y}{\delta}\right) \tag{6.38}$$

independent of κ and ν (and therefore of Pr) outside the viscous and conductive sublayers (see Fig. 6.10). Note that by omitting δ_t we have restricted ourselves to the case of a boundary layer uniformly heated from the leading edge; in the more general case of an unheated starting length, δ_t / δ would appear as another argument of F_2 in Eq. (6.38), but as the function F_2 would depend inseparably on its variables, the simplicity of the analysis would be lost. If T_w, rather than $\dot{q}_w$, is constant, the above analysis applies, but the function F_2 is slightly different.

Equation (6.26) is an empirical formula for the turbulent Prandtl number defined by Eq. (6.3) and is nominally valid in the inner region of the thermal layer, say, $0 < y < 0.2\delta_t$. Outside the viscous and conductive sublayers, where the exponential damping factors in Eq. (6.26) become equal to unity,

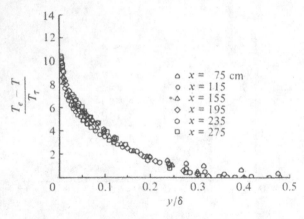

Figure 6.10. Universal plot of turbulent temperature profiles in zero pressure gradient. After Hoffmann and Perry [11].

Pr_t becomes equal to κ/κ_h, which is 0.91 if we take $\kappa = 0.40$, $\kappa_h = 0.44$ as elsewhere in this book, although values between 0.86 and 1.0 are quoted in the literature. According to experiments, this absolute constant value of Pr_t does not apply to the outer part of the flow, $y > 0.2\delta_t$, although a constant value is often used in engineering calculations; in free shear layers and near the outer edge of a boundary layer, Pr_t can be as low as 0.5. This suggests that Reynolds' analogy between heat and momentum transfer, which strictly implies $Pr_t = 1.0$, is adequate only in the inner layer of a turbulent wall flow.

6.4 The Whole Layer

The logarithmic velocity profile, Eq. (6.11), applies only over the first 10–20 percent of the boundary layer (or duct flow) outside the viscous sublayer. However, we can formally write the velocity profile for the outer layer in terms of the deviation from the logarithmic law:

$$u^+ = \frac{u}{u_\tau} = \frac{1}{\kappa}\ln\frac{u_\tau y}{\nu} + c + \frac{\Pi}{\kappa}w\left(\frac{y}{\delta}\right), \tag{6.39}$$

where w is a function of y/δ, Π is independent of y, and the notation is chosen for later convenience. Now Eq. (6.39) implies that the velocity-defect law for boundary layers in zero pressure gradient, Eq. (6.37), can be written as

$$\frac{u_e - u}{u_\tau} = F_1\left(\frac{y}{\delta}\right) = -\frac{1}{\kappa}\ln\frac{y}{\delta} - \frac{\Pi}{\kappa}\left[w\left(\frac{y}{\delta}\right) - w(1)\right], \tag{6.40}$$

which shows that Π, like F_1, is independent of x and must therefore be a constant. The function $w(y/\delta)$ must be derived from a fit to experimental data; the wholly empirical form,

$$w\left(\frac{y}{\delta}\right) = 1 - \cos\frac{\pi y}{\delta}, \tag{6.41}$$

is a popular and adequate data fit except that Eq. (6.39) then gives $\partial u / \partial y = u_\tau / \kappa \delta$ at $y = \delta$, instead of the correct value of zero. Note that for convenience we define $w(1) = 2$; Π is found by experiment to be about 0.55 in boundary layers for $u_e \theta / \nu > 5000$. The function $w(y/\delta)$ is called the *wake function* [12], and Π is the *wake parameter*; the names should not be taken to imply quantitative resemblance between boundary layers and wakes. Now by replacing the logarithmic function in Eq. (6.39) by the general law-of-the-wall function $f_1(u_\tau y / \nu)$ from Eq. (6.4), which includes the logarithmic law and the viscous sublayer profile, we get a velocity-profile formula for the whole layer,

$$u^+ \equiv \frac{u}{u_\tau} = f_1\left(\frac{u_\tau y}{\nu}\right) + \frac{\Pi}{\kappa}\left(1 - \cos \pi \frac{y}{\delta}\right). \tag{6.42}$$

A form analogous to Eq. (6.42) can be derived for the temperature profile in a boundary layer in zero pressure gradient with uniform wall temperature or wall heat-flux rate. Again Eq. (6.41) provides an adequate fit to the wake function; so combining Eqs. (6.5b) and (6.41), we can write

$$T^+ \equiv \frac{T_w - T}{T_\tau} = f_2\left(\frac{u_\tau y}{\nu}, \text{Pr}\right) + \frac{\Pi_h}{\kappa_h} w\left(\frac{y}{\delta}\right), \tag{6.43}$$

where Π_h is a constant differing slightly between the cases of uniform wall temperature and uniform wall heat-flux rate but in either case different from Π because $\text{Pr}_t \neq \kappa / \kappa_h$ in the outer layer. Π_h is independent of Pr and Reynolds number if the Reynolds number is high. Since Π depends on Reynolds number for $u_e \theta / \nu < 5000$, we must expect Π_h to do so as well, because if viscous effects on the turbulence change the fluctuating velocity field, they will affect heat transfer as well as momentum transfer. In principle, Π_h may depend on the thermal conductivity if the Peclet number $(u_e \theta / \nu)\text{Pr}$ is less than roughly 5000, according to our usual argument about the analogy between heat transfer and momentum transfer, but current experimental data are not sufficient to define the low-Reynolds-number and low-Peclet-number behaviors, and indeed the high-Reynolds-number value of Π_h is not known very accurately; it is about 0.3. Outside the viscous and conductive sublayers, Eq. (6.43) becomes

$$T^+ \equiv \frac{T_w - T}{T_\tau} = \frac{1}{\kappa_h} \ln \frac{u_\tau y}{\nu} + c_h + \frac{\Pi_h}{\kappa_h} w\left(\frac{y}{\delta}\right) \tag{6.44}$$

in analogy with Eq. (6.39); recall that c_h depends on Pr. Eqs. (6.39) and (6.44) can be used on rough surfaces if c is replaced by $c - \Delta u^+$, as in Eq. (6.34), and c_h is replaced by an expression such as Eq. (6.36).

The above formulas for velocity and temperature profiles can all be used in fully developed flow in circular pipes or two-dimensional ducts, again with uniform wall heat-flux rate or uniform differences between the wall temperature and the bulk-average temperature of the stream defined in Eq. (5.1). The boundary-layer thickness is replaced by the radius of the pipe or the half-height of the duct. Values of Π and Π_h are smaller than in the boundary layer—indeed Π for a pipe flow is so small that it is often neglected. Low-Reynolds-number effects on Π and Π_h in pipe or duct flow seem to be negligible, implying that the effects in boundary layers are associated with the irregular interface between the turbulent and nonturbulent flow, the "viscous superlayer".

In arbitrary pressure gradients, Eqs. (6.39) and (6.43) are usually still good fits to experimental data but Π and Π_h depend on the pressure distribution for all positions upstream. Eq. (6.43) is not necessarily a good fit to temperature profiles with arbitrary wall temperature or heat-flux distributions. Consider the case where a region of uniform wall heat-flux rate, in which Eq. (6.43) holds, is followed by a region of lower wall heat-flux rate, so that T_τ changes discontinuously while the temperature profile does *not*. Even the inner-layer formula for $T_w - T$, Eq. (6.5), breaks down at a step change in $\dot{q}_w$; it recovers fairly quickly, but the outer layer takes much longer.

6.5 Two-Dimensional Boundary Layers with Zero Pressure Gradient

As was discussed in the previous sections of this chapter, turbulent boundary-layer flows, unlike some laminar flows, do not admit similarity solutions, and as a result, solutions must always be obtained, as in nonsimilar laminar flows, by using either an integral method or a differential method (see Section 4.3). When there is no pressure gradient and the surface temperature is uniform, it is convenient to use an integral method and derive formulas for c_f, Nu_x, δ, δ_t, δ^*, θ, and H, as we shall discuss in this section. On the other hand, when the thermal layer is subject to a wide range of boundary conditions, then even if the pressure gradient is zero, it is more appropriate to use a differential method, as we shall see in Section 6.6 for two-dimensional flows.

Let us consider an uncoupled (constant-property) flow over a smooth flat plate (zero-pressure-gradient or constant-pressure flow) with uniform wall temperature. First we discuss the calculation of the hydrodynamic boundary-layer parameters. From the momentum integral equation, Eq. (3.68), we can write

$$\frac{dR_\theta}{dR_x} = \frac{c_f}{2}. \tag{6.45}$$

Here $R_\theta = u_e\theta/\nu$ and $R_x = u_e x/\nu$. Before we integrate this equation, we need to recall that if the Reynolds number is sufficiently large, we can identify three different flow regimes on such a surface (see Fig. 6.11). Starting from the leading edge, there is first a region $(0 < R_x < R_{x_{tr}})$ in which the flow is laminar. After a certain distance, there is a region $(R_{x_{tr}} < R_x < R_{x_t})$ in which transition from laminar to turbulent flow takes

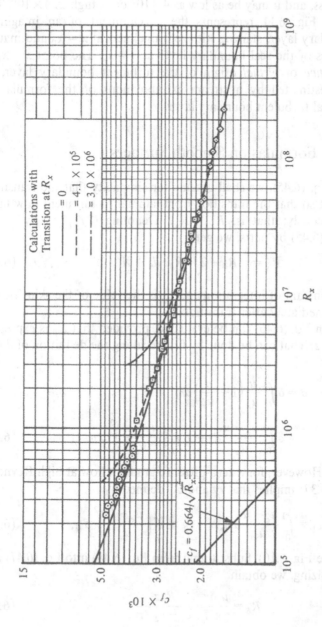

Figure 6.11. Local skin-friction coefficient on a smooth flat plate with three transition Reynolds numbers according to Eq. (6.51). The variation of laminar c_f with R_x is shown by Eq. (4.35) with $f''_w = 0.332$.

place, i.e., in which Reynolds stresses gradually grow. In the third region $(R_x \geq R_{x_t})$ the flow is fully turbulent. The Reynolds number $R_{x_{tr}}$ at the start of transition depends partly upon the turbulence in the freestream and greatly upon the surface conditions such as heating or cooling and smoothness or roughness, and it may be as low as 4×10^5 or as high as 4×10^6. The dashed line in Fig. 6.11 represents the development of an imaginary turbulent boundary layer, starting with zero thickness at $x = x_0$ and matching the thickness of the real boundary layer at $x = x_t$. The point $x = x_0$ is called the *effective* or *virtual* origin of the turbulent boundary layer. In practice, it is estimated by upstream extrapolations of the formula for growth of the real turbulent boundary layer.

Skin-Friction Formulas on Smooth Surfaces

In integrating Eq. (6.45), we shall assume for simplicity that the transition region is a point so that the transition from laminar to turbulent flow takes place instantaneously; that is, $R_{x_{tr}} = R_{x_t}$. Denoting $(2/c_f)^{1/2}$ by z and integrating Eq. (6.45) by parts, we get

$$R_x = z^2 R_\theta - 2 \int_{z_{tr}}^{z} R_\theta z \, dz + A_1, \tag{6.46}$$

where A_1 is a constant of integration and z_{tr} is the value of the skin-friction parameter z immediately after transition.

The integral in Eq. (6.46) can be evaluated, provided that R_θ is expressed as a function of z. That can be done by first recalling the definition of θ and writing it as

$$\theta = \delta \int_0^1 \frac{u}{u_e} \left(1 - \frac{u}{u_e} \right) d\eta$$

$$= \delta \int_0^1 \frac{u_e - u}{u_e} d\eta - \delta \int_0^1 \left(\frac{u_e - u}{u_e} \right)^2 d\eta, \tag{6.47}$$

where $\eta = y/\delta$. However, for a zero-pressure-gradient flow at high Reynolds numbers, Eq. (6.37) implies that c_1 and c_2, defined by

$$c_1 \equiv \int_0^1 \frac{u_e - u}{u_\tau} d\eta, \qquad c_2 \equiv \int_0^1 \left(\frac{u_e - u}{u_\tau} \right)^2 d\eta, \tag{6.48}$$

are constant (see Fig. 6.10). Substituting from Eq. (6.48) into Eq. (6.47) and nondimensionalizing, we obtain

$$R_\delta \equiv \frac{u_e \delta}{\nu} = \frac{R_\theta z}{c_1 - c_2/z}. \tag{6.49}$$

A relation between the boundary-layer thickness δ and local skin friction c_f can be obtained by considering one of the several empirical formulas developed for a turbulent velocity profile. Here we use the "wall-plus-wake" formula, Eq. (6.42).

Substituting Eq. (6.11b) into Eq. (6.42) and evaluating the resulting expression at the edge of the boundary layer, $y = \delta$, we can write

$$z = \frac{1}{\kappa}\ln\frac{R_\delta}{z} + c + \frac{2\Pi}{\kappa}. \tag{6.50}$$

With the values of c and Π taken as 5.0 and 0.55, respectively, we can now integrate Eq. (6.46) with the relations given by Eqs. (6.49) and (6.50). This integration allows the resulting expression to be written as [9]

$$(R_x - A_2)c_f = 0.324\exp\left[\frac{0.58}{\sqrt{c_f}}\left(1 - 8.125\sqrt{c_f} + 22.08c_f\right)\right]. \tag{6.51}$$

Here A_2 is an integration constant that depends on the values of c_f and R_x at transition. Figure 6.11 presents the results for three transition Reynolds numbers, $R_{x_{tr}} = 4\times10^5$ and 3×10^6, the first being the case when transition takes place at the leading edge. The value of $R_x = 4\times10^5$ corresponds to the approximate minimum value of R_x for which the flow can be turbulent. The highest value of R_x is a typical natural transition Reynolds number on a smooth flat plate in low-turbulence test rigs with no heat transfer. If the plate is heated, the location of natural transition in a gas flow moves upstream, decreasing the value of the transitional Reynolds number, whereas if the plate is cooled, the location of transition moves downstream. The reason is that since μ rises with gas temperature, the velocity gradient near the wall is reduced by heating, distorting the profile to a more unstable shape, and for cooling, the converse holds. In liquid flows μ falls with increasing fluid temperature, and the effect is reversed.

Putting $A_2 = 0$ in Eq. (6.51) (i.e., assuming that the turbulent boundary layer starts at $x = 0$ with negligible thickness), taking logarithms, and making further approximations lead to formulas like

$$\frac{1}{\sqrt{c_f}} = a + b\log c_f R_x,$$

where a and b are constants chosen to get the best agreement with experiment. Such less-rigorous formulas have been derived by many previous workers. Von Karman [13] took $a = 1.7$ and $b = 4.15$; i.e.,

$$\frac{1}{\sqrt{c_f}} = 1.7 + 4.15\log c_f R_x. \tag{6.52}$$

A formula for the average skin friction $\bar{c}_f$ (averaged over the distance x) that makes use of the above equation was obtained by Schoenherr [13]:

$$\frac{1}{\sqrt{\bar{c}_f}} = 4.13\log \bar{c}_f R_x. \tag{6.53}$$

Power-Law Velocity Profiles

By relating the profile parameter Π to the displacement thickness δ^* and to the momentum thickness θ (see Problem 6.7) as well as to the local skin-friction coefficient c_f, as is done in the analysis leading to Eq. (6.52), we can obtain relations between δ, c_f, δ^*, θ, and H. Much simpler but less accurate relations can be obtained by assuming that the velocity profile can be represented by the "power law"

$$\frac{u}{u_e} = \left(\frac{y}{\delta}\right)^{1/n}. \tag{6.54}$$

Here n is about 7 in zero-pressure-gradient flow, increasing slowly with Reynolds number. Using Eq. (6.54) and the definitions of δ^*, θ, and H, we can show that

$$\frac{\delta^*}{\delta} = \frac{1}{1+n}, \tag{6.55a}$$

$$\frac{\theta}{\delta} = \frac{n}{(1+n)(2+n)}, \tag{6.55b}$$

$$H = \frac{2+n}{n}. \tag{6.55c}$$

Other formulas obtained from power-law assumptions with $n = 7$, given by Schlichting [13], are the following equations valid only for Reynolds numbers R_x between 5×10^5 and 10^7:

$$c_f = 0.059 R_x^{-0.20}, \tag{6.56}$$

$$\bar{c}_f = 0.074 R_x^{-0.20}, \tag{6.57}$$

$$\frac{\delta}{x} = 0.37 R_x^{-0.20}, \tag{6.58}$$

$$\frac{\theta}{x} = 0.036 R_x^{-0.20}. \tag{6.59}$$

Equations (6.53) and (6.57) assume that the boundary layer is turbulent from the leading edge onward, that is, the effective origin is at $x = 0$. If the flow is turbulent but the Reynolds number is moderate, we should consider the portion of the laminar flow that precedes the turbulent flow. There are several empirical formulas for $\bar{c}_f$ that account for this effect. One is the formula quoted by Schlichting [13]. It is given by

$$\bar{c}_f = \frac{0.455}{(\log R_x)^{2.58}} - \frac{A}{R_x}, \tag{6.60}$$

and another is

$$\bar{c}_f = 0.074 R_x^{-0.20} - \frac{A}{R_x}, \qquad 5 \times 10^5 < R_x < 10^7. \tag{6.61}$$

Here A is a constant that depends on the transition Reynolds number $R_{x_{tr}}$. It is given by

$$A = R_{x_{tr}}(\bar{c}_{f_t} - \bar{c}_{f_l}),\qquad(6.62)$$

where $\bar{c}_{f_t}$ and $\bar{c}_{f_l}$ correspond to the values of average skin-friction coefficient for turbulent and laminar flow at $R_{x_{tr}}$. We note that although Eq. (6.61) is restricted to the indicated R_x range, Eq. (6.60) is valid for a wide range of R_x and has given good results up to $R_x = 10^9$.

Heat-Transfer Formulas on Smooth Surfaces with Specified Temperature

For a zero-pressure-gradient flow, the energy integral equation (3.78b) can be integrated to obtain the Stanton number St as a function of Reynolds number. This can be done by inserting the velocity profile expression given by Eq. (6.39) and the similar expression for temperature profile given by Eq. (6.44) into the definition of θ_T. Since we already have an expression for c_f, however, a simpler procedure would be to evaluate Eq. (6.44) at $y = \delta$ and make use of the relation given by Eq. (6.50). For example, at $y = \delta$, Eq. (6.44) becomes

$$\frac{T_w - T_e}{T_\tau} = \frac{1}{\kappa_h}\ln\frac{R_\delta}{z} + c_h + \frac{2\Pi_h}{\kappa_h}.\qquad(6.63)$$

Using the definition of T_τ and the local Stanton number St, the left-hand side of Eq. (6.63) becomes

$$\frac{T_w - T_e}{T_\tau} = \frac{1}{\mathrm{St}}\sqrt{\frac{c_f}{2}}.\qquad(6.64)$$

If we equate the two expressions for $\ln(R_\delta/z)$ obtained from Eqs. (6.50) and (6.63) and substitute Eq. (6.64) in the left-hand side of Eq. (6.63), we get, after rearranging,

$$\frac{\mathrm{St}}{c_f/2} = \frac{\kappa_h/\kappa}{1 - \{[c\kappa - c_h\kappa_h + 2(\Pi - \Pi_h)]/\kappa\}\sqrt{c_f/2}}.\qquad(6.65)$$

Because of the scatter in temperature-profile data, it is simplest to choose the value of the quantity in braces as one that gives the best agreement with St data. For air, this is

$$\frac{\mathrm{St}}{c_f/2} = \frac{1.11}{1 - 1.20\sqrt{c_f/2}}\qquad(6.66)$$

for $\kappa = 0.40$ and $\kappa_h = 0.44$. This equation for the Reynolds analogy factor $\mathrm{St}/(c_f/2)$ is quoted in the literature with a wide range of values for the empirical constants. For $c_f \approx 3\times10^{-3}$, a typical value for a laboratory boundary layer, the constants quoted here give $\mathrm{St}/\frac{1}{2}c_f \approx 1.16$, whereas at very high Reynolds number, where c_f is small, $St/\frac{1}{2}c_f$ asymptotes to 1.11 (see Simonich and Bradshaw [14]).

Substituting Eq. (6.56) into Eq. (6.66) yields

$$St = \frac{0.0327R_x^{-0.20}}{1-0.206R_x^{-0.10}}. \tag{6.67}$$

According to an extensive study conducted by Kader and Yaglom [15], the empirical formula

$$St = \frac{1}{Pr}\frac{\sqrt{c_f}}{4.3\ln R_x c_f + 3.8} \tag{6.68}$$

fits the existing data on air ($Pr = 0.7$) well. For fluids with $Pr \geq 0.7$, they recommend

$$St = \frac{\sqrt{c_f/2}}{2.12\ln R_x c_f + 12.5 Pr^{2/3} + 2.12\ln Pr - 7.2}. \tag{6.69}$$

Equations (6.68) and (6.69) utilize Von Karman's equation (6.52) for c_f with a slightly different constant ahead of the $\log c_f R_x$ term:

$$\frac{1}{\sqrt{c_f}} = 1.7 + 4.07\log c_f R_x.$$

For isothermal flat plates, Reynolds et al. [16] recommend the following empirical formula for Stanton number:

$$St Pr^{0.4}\left(\frac{T_w}{T_e}\right)^{0.4} = 0.0296R_x^{-0.20} \tag{6.70}$$

for $5\times10^5 < R_x < 5\times10^6$ and $0.5 \leq Pr \leq 1.0$.

The approximate procedure of Section 4.3 used to obtain an expression for Stanton number on an isothermal flat plate with unheated starting length can also be extended to turbulent flows by making suitable assumptions for velocity, temperature, and shear-stress profiles and by using eddy viscosity and turbulent Prandtl number concepts [17]. For example, from the definition of Stanton number with power-law profiles for velocity and temperature,

$$\frac{u}{u_e} = \left(\frac{y}{\delta}\right)^{1/n}, \qquad \frac{T_w - T}{T_w - T_e} = \left(\frac{y}{\delta_t}\right)^{1/n}, \tag{6.71a}$$

and with a shear-stress profile in the form

$$\frac{\tau}{\tau_w} = 1 - \left(\frac{y}{\delta}\right)^{(n+2)/n} \tag{6.71b}$$

and with $\tau/\rho = \varepsilon_m(\partial u/\partial y)$, $Pr_t = 1$, we can write

$$St = \frac{c_f}{2}\left(\frac{\delta_t}{\delta}\right)^{-1/n}. \tag{6.72}$$

Substituting this equation into the energy integral equation (3.78b) and using the momentum integral equation, which in this case is given by Eq. (4.57), the resulting expression can be written in the form

$$\int_{\delta_{x_0}}^{\delta} \frac{d\delta}{\delta} = \int_0^{\delta_t/\delta} \frac{n+1}{n+2} \frac{(\delta_t/\delta)^{2/n}}{1-(\delta_t/\delta)^{(2+n)/n}} d\left(\frac{\delta_t}{\delta}\right), \tag{6.73}$$

where δ_{x_0} denotes the hydrodynamic boundary-layer thickness at $x = x_0$ (see Fig. 4.7).

Integrating Eq. (6.73), we obtain

$$\frac{\delta}{\delta_{x_0}} = \left[1-\left(\frac{\delta_t}{\delta}\right)^{(2+n)/n}\right]^{-(n+1)/(n+2)}$$

or

$$\frac{\delta_t}{\delta} = \left[1-\left(\frac{\delta_{x_0}}{\delta}\right)^{(n+2)/(n+1)}\right]^{n/(2+n)}. \tag{6.74}$$

In the range $5 \times 10^5 \le R_x \le 10^7$, δ varies as $x^{4/5}$ [see Eq. (6.58)]. Thus Eq. (6.74) may be written as

$$\frac{\delta_t}{\delta} = \left[1-\left(\frac{x_0}{x}\right)^{4(n+2)/5(n+1)}\right]^{n/(2+n)}. \tag{6.75}$$

Substituting Eq. (6.75) into Eq. (6.72) and taking $n = 7$, we get

$$St = \frac{c_f}{2}\left[1-\left(\frac{x_0}{x}\right)^{9/10}\right]^{-1/9}. \tag{6.76}$$

For a plate heated from the leading edge ($x_0 = 0$), Eq. (6.76) becomes

$$St = St_T = \frac{c_f}{2},$$

where St_T denotes the Stanton number of the isothermal flat plate. With this notation, Eq. (6.76) can be written as

$$\frac{St}{St_T} = \left[1-\left(\frac{x_0}{x}\right)^{9/10}\right]^{-1/9}, \qquad x > x_0. \tag{6.77}$$

From the definition of heat-transfer coefficient $\hat{h}$ and Stanton number St, for an isothermal flat plate with unheated starting length,

$$\hat{h} = \rho u_e c_p St_T \left[1-\left(\frac{x_0}{x}\right)^{9/10}\right]^{-1/9}. \tag{6.78}$$

Similarly, with St_T given by Eq. (6.70),

$$St Pr^{0.4}\left(\frac{T_w}{T_e}\right)^{0.4} = 0.0296 R_x^{-0.20}\left[1-\left(\frac{x_0}{x}\right)^{9/10}\right]^{-1/9}. \tag{6.79}$$

For nonisothermal surfaces with arbitrary surface temperature, the heat flux at some distance x from the leading edge is, by superposition arguments [18],

$$\dot{q}_w = \int_{\xi=0}^{\xi=x} \hat{h}(\xi, x)\, dT_w(\xi), \tag{6.80}$$

where, with $\mathrm{St}_T(x)$ being evaluated at x,

$$\hat{h}(\xi, x) = \rho u_e c_p \mathrm{St}_T(x)\left[1-\left(\frac{\xi}{x}\right)^{9/10}\right]^{-1/9}. \tag{6.81}$$

The integration of Eq. (6.80) is performed in the "Stieltjes" sense rather than in the ordinary "Riemann" or "area" sense (see [19]). This must be done because the specified surface temperature may have a finite discontinuity, so that dT_w is undefined at some point. The Stieltjes integral may, however, be expressed as the sum of an ordinary or Riemann integral and a term that accounts for the effect of the finite discontinuities [18]. The integral of Eq. (6.80) may be written as

$$\dot{q}_w(x) = \int_{\xi=0}^{\xi=x} \hat{h}(\xi, x)\frac{dT_w(\xi)}{d\xi}\, d\xi + \sum_{n=1}^{N} \hat{h}(x_{0n}, x)[T_w(x_{0n}^+)-T_w(x_{0n}^-)]. $$

$$\tag{6.82}$$

Here N denotes the number of discontinuities and $T_w(x_{0n}^+)-T_w(x_{0n}^-)$ denotes the temperature jump across the nth discontinuity.

As an example, let us consider a plate whose temperature is equal to T_{w_1} from the leading edge to $x = x_0$ and equal to T_{w_2} for $x > x_0$. To find the heat flux for $x > x_0$, we note that since $dT_w/d\xi$ is zero except at $x = x_0$, the first term on the right-hand side is zero. Therefore, we concentrate our attention on the second term. Since $N = 2$, we can write

$$n=1, \quad \hat{h}(0, x) = \rho u_e c_p \mathrm{St}_T(x)[1-0]^{-1/9},$$

$$T_w(0^+)-T_w(0^-)=T_{w_1}-T_e,$$

$$n=2, \quad \hat{h}(x_0, x) = \rho u_e c_p \mathrm{St}_T(x)\left[1-\left(\frac{x_0}{x}\right)^{9/10}\right]^{-1/9},$$

$$T_w(x_0^+)-T_w(x_0^-)=T_{w_2}-T_{w_1}.$$

Thus the heat transfer for $x > x_0$ is given by

$$\dot{q}_w(x) = \rho u_e c_p \mathrm{St}_T(x)\left\{(T_{w_1}-T_e)+(T_{w_2}-T_{w_1})\left[1-\left(\frac{x_0}{x}\right)^{9/10}\right]^{-1/9}\right\}.$$

$$\tag{6.83}$$

Note that $\mathrm{St}_T(x)$ is computed with $T_w = T_{w_2}$.

Heat-Transfer Formulas on Smooth Surfaces with Specified Heat Flux

The analysis of thermal boundary layers on smooth surfaces with specified heat flux is similar to those with specified temperature. Based on the experiments of Reynolds, Kays, and Kline, the following empirical formula is recommended in Kays and Crawford [20]:

$$\text{St}\,\text{Pr}^{0.4} = 0.030 R_x^{-0.2}, \tag{6.84}$$

which is nearly identical to the one for specified temperature, Eq. (6.70). Note that the difference in Stanton-number formulas between the constant wall heat-flux case and the constant wall temperature case is considerably more in laminar flows, where the difference is 36 percent.

When the plate has an arbitrary heat-flux distribution on the surface and also includes an unheated section, the difference between the surface temperature and edge temperature can be calculated from the following formula given by Reynolds et al. [18]:

$$T_w(x) - T_e = \int_{\xi=0}^{\xi=x} g(\xi, x) \dot{q}_w(\xi)\, d\xi, \tag{6.85}$$

where, with Γ denoting the gamma function (see Appendix C),

$$g(\xi, x) = \frac{\frac{9}{10}\text{Pr}^{-0.6}R_x^{-0.8}}{\Gamma(\frac{1}{9})\Gamma(\frac{8}{9})(0.0287k)}\left[1 - \left(\frac{\xi}{x}\right)^{9/10}\right]^{-8/9}.$$

The nature of the integrand in Eq. (6.85) is such that integration is always performed in the usual Riemann sense, including integration across discontinuities. To illustrate this point further, let us consider a plate that is unheated for a distance x_0 from the leading edge and is heated at a uniform rate $\dot{q}_w$ for $x > x_0$. To find the wall temperature for $x > x_0$, we write Eq. (6.85) as

$$T_w(x) - T_e = \frac{3.42 \dot{q}_w}{\text{Pr}^{0.6}R_x^{0.8}k}\int_{x_0}^{x}\left[1 - \left(\frac{\xi}{x}\right)^{9/10}\right]^{-8/9} d\xi. \tag{6.86}$$

The integral can be evaluated in terms of beta functions (see Appendix C), and the resulting expression can be written as

$$T_w(x) - T_e = \frac{33.61 \dot{q}_w \text{Pr}^{0.4}R_x^{0.2}}{\rho c_p u_e}\frac{\beta_r(1/9, 10/9)}{\beta_1(1/9, 10/9)} \tag{6.87}$$

or, using the definition of Stanton number, as

$$\text{St} = \frac{0.030\,\text{Pr}^{-0.4}R_x^{-0.2}}{[\beta_r(\frac{1}{9}, \frac{10}{9})]/[\beta_1(\frac{1}{9}, \frac{10}{9})]}, \tag{6.88}$$

where $r = 1 - (x_0/x)^{9/10}$.

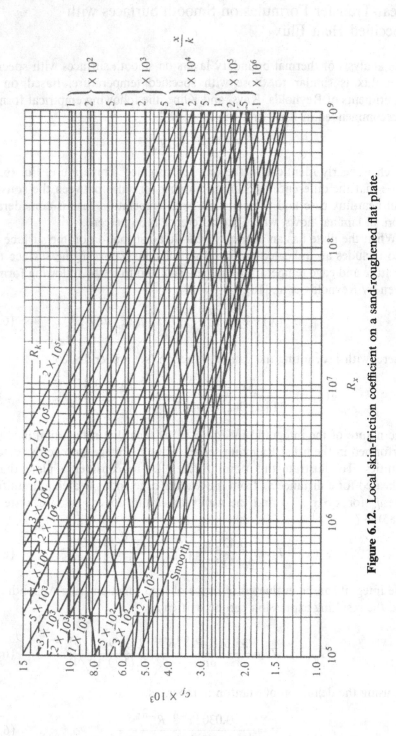

Figure 6.12. Local skin-friction coefficient on a sand-roughened flat plate.

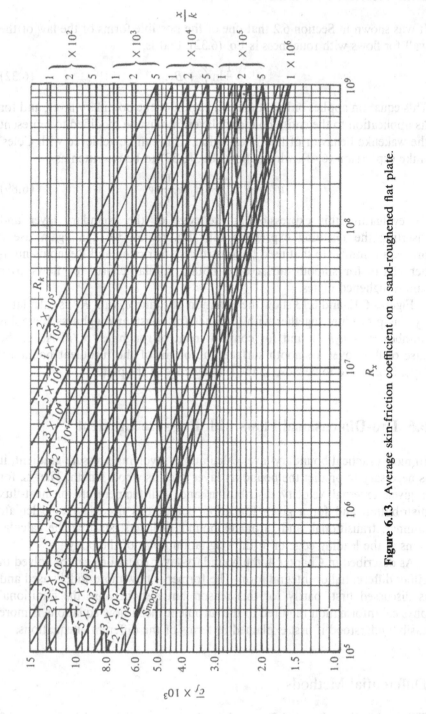

Figure 6.13. Average skin-friction coefficient on a sand-roughened flat plate.

Skin-Friction Formulas on Rough Surfaces

It was shown in Section 6.2 that one of the possible forms of the law of the wall for flows with roughness is Eq. (6.32), that is,

$$u^+ = \frac{1}{\kappa}\ln\frac{y}{k} + B_2. \tag{6.32}$$

This equation applies only in the inner region of the boundary layer, and for its application to the entire boundary layer, it must be modified to represent the wakelike behavior of the outer region. This can be achieved with Coles' wake expression (6.41), which allows Eq. (6.32) to be rewritten as

$$u^+ = \frac{1}{\kappa}\ln\frac{y}{k} + B_2 + \frac{\Pi}{\kappa}w. \tag{6.89}$$

By evaluating this expression at the edge of the boundary layer and inserting the resulting expression into Eq. (6.45), we can again use a procedure similar to that used to construct skin-friction and Reynolds-number charts for smooth surfaces to obtain similar charts for flows over sand-roughened plates.

Figures 6.12 and 6.13 show the variation of c_f and $\bar{c}_f$ with R_x as calculated by this procedure together with the lines for constant roughness Reynolds number $R_k = u_e k/\nu$ and for constant relative roughness x/k. As in the case of flow over a smooth surface, the origin of the turbulent boundary layer is assumed to be close to the leading edge of the plate.

6.6 Two-Dimensional Flows with Pressure Gradient

In most practical boundary-layer calculations involving pressure gradient, it is necessary to predict the boundary layer over its whole length. That is, for a given external velocity distribution and wall-temperature or heat-flux distribution and for a given transition point, it is necessary to calculate laminar, transitional, and turbulent boundary layers, starting the calculations at the leading edge or at the stagnation point of the body.

As described in Chapter 4, the boundary-layer equations can be solved in either differential or integral form. The former is now more widely used and is discussed first, partly for this reason but also because the additional physical information required in the turbulent mean-flow equations is more easily understood if first explained in terms of the differential equations.

Differential Methods

The basic principles of differential methods were discussed in Section 4.3 and are extended here to turbulent flows. The continuity, momentum, and energy equations for two-dimensional uncoupled turbulent flows are solved in the form

$$\frac{\partial u}{\partial x} + \frac{\partial v}{\partial y} = 0, \qquad (3.5)$$

$$u\frac{\partial u}{\partial x} + v\frac{\partial u}{\partial y} = u_e\frac{du_e}{dx} + v\frac{\partial^2 u}{\partial y^2} - \frac{\partial}{\partial y}\overline{u'v'}, \qquad (3.12)$$

$$u\frac{\partial T}{\partial x} + v\frac{\partial T}{\partial y} = \frac{k}{\rho c_p}\frac{\partial^2 T}{\partial y^2} - \frac{\partial}{\partial y}\overline{T'v'}, \qquad (3.14)$$

subject to the initial and boundary conditions given by Eqs. (3.85) and (3.88). The turbulent momentum and heat fluxes can be represented in various ways, and partly to make the explanations simple, we use the eddy-viscosity and turbulent-Prandtl-number concepts discussed in the previous sections. To do this, let us write the total (laminar plus turbulent) shear stress, using the eddy-viscosity concept, in the form

$$\tau = \mu\frac{\partial u}{\partial y} - \rho\overline{u'v'} = \rho v\left(1 + \varepsilon_m^+\right)\frac{\partial u}{\partial y}, \qquad (6.90)$$

where $\varepsilon_m^+ = \varepsilon_m/v$ and ε_m is the eddy viscosity defined by Eq. (6.1) and the total heat flux using the turbulent-Prandtl-number concept Eq. (6.3),

$$\dot{q} = -k\frac{\partial T}{\partial y} + \rho c_p\overline{T'v'} = -\mu c_p\left(\frac{1}{\mathrm{Pr}} + \frac{\varepsilon_m^+}{\mathrm{Pr}_t}\right)\frac{\partial T}{\partial y}. \qquad (6.91)$$

Inserting these relations into the above conservation equations and using the Falkner–Skan transformation defined by Eqs. (4.19b), and (4.19c) and the dimensionless temperature defined by Eq. (4.25), the equations can be expressed as

$$(bf'')' + \frac{m+1}{2}ff'' + m\left[1 - (f')^2\right] = x\left(f'\frac{\partial f'}{\partial x} - f''\frac{\partial f}{\partial x}\right), \qquad (6.92)$$

$$(eg')' + \frac{m+1}{2}fg' + n(1-g)f' = x\left(f'\frac{\partial g}{\partial x} - g'\frac{\partial f}{\partial x}\right), \qquad (6.93)$$

which are the same as Eqs. (4.21) and (4.28) with the viscous and conductive terms factored by

$$b = 1 + \varepsilon_m^+, \qquad e = \frac{1}{\mathrm{Pr}} + \frac{\varepsilon_m^+}{\mathrm{Pr}_t}, \qquad (6.94)$$

the ratios of the total (molecular plus turbulent) diffusivities to the molecular diffusivities. By setting $\varepsilon_m^+ = 0$ (laminar flows) we recover the laminar-flow equations. For turbulent flows, we need suitable empirical formulas for ε_m^+ and Pr_t so that we can use the "same" equations to solve laminar, transitional, and turbulent flows subject to Eq. (4.30), the usual boundary conditions for either laminar or turbulent flows.

Sometimes for turbulent flows it is more convenient to replace the true boundary conditions at $y = 0$ by new "boundary conditions" defined at some distance y_0 outside the viscous and conductive sublayers, to avoid integrating the equations through the region of large y gradients near the surface. Usually this y_0 is taken to be the distance given by

$$y_0 = \left(\frac{\nu}{u_\tau}\right) y_0^+, \qquad (6.95)$$

y_0^+ being a constant taken as about 50 for smooth surfaces. In that case, the "wall" boundary conditions for u, v, and T at $y = y_0$ for a specified wall temperature can be represented by

$$u_0 = u_\tau \left(\frac{1}{\kappa} \ln \frac{y_0 u_\tau}{\nu} + c\right), \qquad (6.96)$$

$$v_0 = -\frac{u_0 y_0}{u_\tau} \frac{du_\tau}{dx}, \qquad (6.97)$$

$$T_0 = T_w - T_\tau \left(\frac{1}{\kappa_h} \ln \frac{y_0 u_\tau}{\nu} + c_h\right). \qquad (6.98)$$

Here c and c_h are constants defined in Section 6.2, or the corresponding rough-surface constants, and Eq. (6.97) results from integrating the continuity equation with u given by Eq. (6.4). We also use relations for the changes in shear stress and heat-flux rate between $y = 0$ and $y = y_0$, as

$$u_\tau^2 = \frac{\tau_0}{\rho} - \alpha y_0,$$

where

$$\alpha = 0.3 \frac{du_0^2}{dx} - u_e \frac{du_e}{dx},$$

and

$$T_\tau = \frac{\dot{q}_w}{\rho c_p u_\tau} = \frac{\dot{q}_0 - \tau_0 u_0}{\rho c_p u_\tau}, \qquad \dot{q}_0 = -\mu c_p \left(\frac{1}{Pr} + \frac{\varepsilon_m^+}{Pr_t}\right)_0 \left(\frac{\partial T}{\partial y}\right)_0. \qquad (6.99)$$

Note that α is an approximation to the average $\partial \tau / \partial y$ in $0 < y \leq y_0$, where $0 < u \leq u_0$; the first term in its definition is an approximation to $u \, \partial u / \partial x + v \, \partial u / \partial y$. The corresponding term in the heat-transfer relations is the kinetic heating term $\tau_0 u_0$, which is small in low-speed flow. In terms of transformed coordinates, Eqs. (6.96)–(6.99) can be written as

$$\eta = \eta_0, \qquad (f_0')^2 = \alpha_2 f_0'' - \gamma_1,$$

$$x \frac{\partial f_0}{\partial x} + \frac{m+1}{2} f_0 = \frac{\eta_0 f_0'}{2} \left(1 + m + \frac{x}{u_\tau^2} \frac{du_\tau^2}{dx}\right), \qquad (6.100a)$$

$$g_0 = B^* \left(\frac{1}{Pr} + \frac{\varepsilon_m^+}{Pr_t}\right)_0 \left(\frac{u_e}{u_\tau}\right) \frac{g_0'}{\sqrt{R_x}}, \qquad (6.100b)$$

where b_0 denotes $b(\equiv 1 + \varepsilon_m^+)$ evaluated at $y = y_0$ and

$$\alpha_2 = \frac{B^2}{R_x^{1/2}} b_0, \qquad \gamma_1 = \frac{\alpha y_0}{u_\tau^2} B^2, \qquad B = \frac{1}{\kappa} \ln \frac{y_0 u_\tau}{\nu} + c,$$

$$B^* = \frac{1}{\kappa_h} \ln \frac{y_0 u_\tau}{\nu} + c_h.$$

$$(6.101)$$

Formulas for Turbulent Diffusivities

There are several eddy-viscosity and turbulent-Prandtl-number formulations that can be used to represent ε_m^+ and Pr_t, and here we use the one developed by Cebeci and Smith [12]. The accuracy of this formulation has been explored for a wide range of flows for which there are experimental data and has been found to give results sufficiently accurate for most engineering problems. According to this formulation, the turbulent boundary layer is treated as a composite layer consisting of inner and outer regions with separate analytic expressions for eddy viscosity in each region. The functions are empirical and based on limited ranges of experimental data. In the inner region of a smooth surface with or without mass transfer, the eddy-viscosity formula may be defined by

$$(\varepsilon_m)_i = l^2 \left| \frac{\partial u}{\partial y} \right| \gamma_{\mathrm{tr}} \gamma, \qquad 0 \le y \le y_c. \qquad (6.102)$$

Here the mixing length l is given by

$$l = \kappa y \left[1 - \exp\left(-\frac{y}{A} \right) \right], \qquad (6.103)$$

where $\kappa = 0.40$ and A is a damping-length constant, which may be represented by

$$A = 26 \frac{\nu}{N} u_\tau^{-1}, \qquad N = \left\{ \frac{p^+}{v_w^+} \left[1 - \exp(11.8 v_w^+) \right] + \exp(11.8 v_w^+) \right\}^{1/2},$$

$$(6.104a)$$

$$p^+ = \frac{\nu u_e}{u_\tau^3} \frac{du_e}{dx}, \qquad v_w^+ = \frac{v_w}{u_\tau}. \qquad (6.104b)$$

For flows with no mass transfer $(v_w^+ = 0)$,

$$N = (1 - 11.8 p^+)^{1/2}, \qquad (6.104c)$$

and obviously for a flow with no mass transfer or pressure gradient $N = 1$, and the definition of A reduces to that given in Eq. (6.17).

In Eq. (6.102) γ_{tr} is an intermittency factor which is required to represent the region of transition from laminar to turbulent flow. It is defined by an empirical correlation,

$$\gamma_{tr} = 1 - \exp\left[-G(x - x_{tr})\int_{x_{tr}}^{x}\frac{dx}{u_e}\right].$$ (6.105)

Here x_{tr} is the location of the start of transition and the factor G, which has the dimensions of velocity/(length)^2 and is evaluated at the transition location, is given by

$$G = 8.33 \times 10^{-4}\frac{u_e^3}{\nu^2}R_x^{-1.34}.$$ (6.106)

Equation (6.105) was obtained initially for uncoupled flows and, as we shall see in Chapter 11, is also useful for coupled adiabatic flows with Mach numbers less than 5: it is not applicable to nonadiabatic coupled flows since appreciable differences between the wall and freestream temperatures strongly influence the length of the transition region. According to the correlation given by Chen and Thyson [21], the extent of the transition region $R_{\Delta x}(\equiv R_{x_t} - R_{x_{tr}})$ for uncoupled flows is given by

$$R_{\Delta x} = 60R_{x_{tr}}^{2/3},$$ (6.107)

which shows that transitional boundary-layer flows are more extensive at lower Reynolds numbers and less important at higher Reynolds numbers.

In Eq. (6.102) γ is another intermittency factor that accounts for the fact that as the freestream is approached, the turbulence becomes intermittent; that is, the flow is turbulent for only a fraction γ of the time. It is given by the empirical expression

$$\gamma = \left[1 + 5.5\left(\frac{y}{y_0}\right)^6\right]^{-1},$$ (6.108)

where y_0 is defined as the y location where $u/u_e = 0.995$.

In the outer region the eddy-viscosity formula is defined by

$$(\varepsilon_m)_0 = \alpha\left|\int_0^\infty (u_e - u)\,dy\right|\gamma_{tr}\gamma, \qquad y_c \le y \le \delta.$$ (6.109)

In conventional boundary layers with $u < u_e$, this is simply $(\varepsilon_m)_0 = \alpha u_e\delta^*\gamma_{tr}\gamma$. Here α is a constant equal to 0.0168 when $R_\theta \ge 5000$. For lower values of R_θ, α varies with R_θ according to the empirical formula given by Cebeci and Smith [12]:

$$\alpha = 0.0168\frac{1.55}{1 + \Pi},$$ (6.110a)

where the variation of Π with R_θ can be correlated empirically by

$$\Pi = 0.55\left[1 - \exp\left(-0.243z_1^{1/2} - 0.298z_1\right)\right],$$ (6.110b)

where $z_1 = R_\theta/425 - 1$, with $R_\theta > 425$.

The condition used to define the inner and outer regions is the continuity of the eddy viscosity; from the wall outward the expression for the inner eddy viscosity is applied until $(\varepsilon_m)_i = (\varepsilon_m)_0$, which defines y_c.

Different expressions have been proposed for the turbulent Prandtl number, including those that express it as a function of distance y from the wall and of molecular Prandtl number Pr and those that treat it as a constant. For air, whose molecular Prandtl number is close to unity, the inclusion of Pr makes little difference, and in any case Pr affects Pr_t only for $u_\tau y/\nu Pr < 50$. According to the studies reported by Cebeci and Smith [12], a turbulent-Prandtl-number expression such as that given by Eq. (6.26) for the viscous and conductive sublayers produces nearly the same results for air as those obtained by a constant turbulent Prandtl number. The situation, on the other hand, is different when the molecular Prandtl number of the fluid differs greatly from unity. As we shall see later in Chapter 7, the choice of a variable turbulent Prandtl number in the sublayer yields results that agree much better with experiment than those obtained with a constant turbulent Prandtl number.

Sample Calculations

As an example of a two-dimensional turbulent flow with heat transfer (but no mass transfer), we consider an ellipse with a smooth surface as shown in Fig. 6.14. The external velocity distribution over the front part can be calculated to adequate accuracy from inviscid flow theory, and for zero angle of attack is given by Eqs. (P4.36) and (P4.37).

Figures 6.15 to 6.18 show the results obtained by solving Eqs. (6.92) and (6.93), subject to the boundary conditions given by Eqs. (4.30) for no mass transfer, with the Fortran program described in Section 13.5. The calculations were done for air, Pr = 0.72, and used the turbulence model given by Eqs. (6.102)–(6.110) for uniform wall temperature, with Pr_t constant and equal to 0.90. Two different Reynolds numbers, $R_{2a} = 10^6$ and 10^7, were used in the calculations. The location of transition was calculated from the empirical formula of Michel [22]:

$$R_{\theta_{tr}} = 1.174\left(1 + \frac{22400}{R_{x_{tr}}}\right)R_{x_{tr}}^{0.46}.\tag{6.111}$$

This equation applies only to uncoupled laminar flows and does not account for heat-transfer effects, which may influence transition. To use this equa-

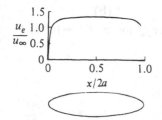

Figure 6.14. Velocity distribution in inviscid flow over an ellipse of thickness ratio 1 to 4.

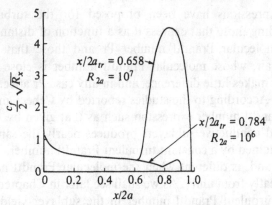

Figure 6.15. Effect of Reynolds number on wall shear parameter f_w'' ($\equiv c_f/2\sqrt{R_x}$) with axial distance $x/2a$.

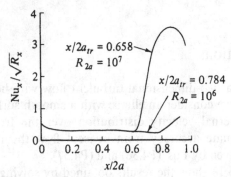

Figure 6.16. Effect of Reynolds number on wall heat-flux parameter g_w' ($\equiv$ $\mathrm{Nu}_x/\sqrt{R_x}$) with axial distance $x/2a$.

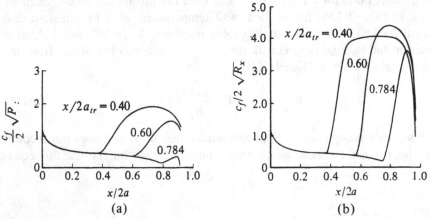

(a)   (b)

Figure 6.17. Effect of transition on wall shear parameter f_w'' for (a) $R_{2a} = 10^6$ and (b) $R_{2a} = 10^7$.

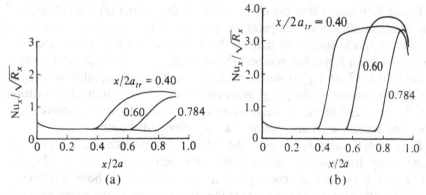

Figure 6.18. Effect of transition on wall heat flux parameter g'_w for (a) $R_{2a} = 10^6$ and (b) $R_{2a} = 10^7$.

tion, we compute $R_\theta (\equiv u_e \theta / \nu)$ as the laminar boundary-layer equations are solved at a specified surface distance s, which corresponds to $R_x \equiv u_e s / \nu$. When R_x and R_θ satisfy Eq. (6.111), the turbulent-flow calculations can be started by activating the eddy-viscosity formulas. Sometimes, either in calculation or in real life, laminar separation occurs, that is, $(\partial u / \partial y)_w$ becomes zero, before transition. In such situations it is reasonable to assume the laminar separation point to be the transition point; in practice, laminar separation is soon followed by transition in the mixing layer and then (at least in some cases) by turbulent reattachment.

The transition calculations done at two Reynolds numbers show that at $R_{2a} = 10^6$, laminar separation takes place at $x/2a = 0.784$ before a pair of values of R_θ and R_x can be found to satisfy Eq. (6.111). On the other hand, at $R_{2a} = 10^7$, calculations indicate transition to be at $x/2a = 0.658$ according to Eq. (6.111).

Figure 6.15 shows the effect of Reynolds number on the variation of dimensionless wall shear parameter $f''_w (\equiv c_f / 2\sqrt{R_x})$ with axial distance $x/2a$. We note that both flows start as stagnation-point flow at $x/2a = 0$ with $m = 1$ and $f''_w = 1.23259$ (Table 4.1) and that f''_w is independent of Reynolds number for the laminar portion of both flows. The flow with $R_{2a} = 10^6$ separates at $x/2a = 0.784$, and so the turbulent flow calculations are started at that location. The flow with $R_{2a} = 10^7$, which would otherwise separate at $x/2a = 0.784$ (because laminar separation point is independent of Reynolds number for a given pressure distribution), becomes turbulent at $x/2a = 0.658$. The wall shear parameter increases sharply with x, reaching a maximum at $x/2a = 0.85$, and then begins to decrease, becoming zero and thus indicating turbulent-flow separation at $x/2a = 0.98$. The flow with the lower Reynolds number, on the other hand, separates early at $x/2a = 0.91$, because the boundary layer is thicker, principally because of the greater growth rate in the laminar region.

From these calculations we see that increasing Reynolds number moves the transition location forward and delays the turbulent-flow separation.

Figure 6.16 shows that the effect of Reynolds number on the variation of the dimensionless wall temperature gradient $g'_w (\equiv Nu_x /\sqrt{R_x})$ with axial distance $x/2a$ is similar to that on the dimensionless wall shear stress f''_w. Again, for both Reynolds numbers the value of g'_w at the stagnation point is equal to 0.5017, and g'_w is independent of the Reynolds number so long as the flow is laminar. Like f''_w, g'_w increases sharply at the transition location for $R_a = 10^7$ and begins to decrease at $x/2a = 0.85$. However, unlike f''_w, it does not become zero at separation at any Reynolds number.

Figures 6.17 and 6.18 show the effect of transition on f''_w and g'_w, respectively, for two Reynolds numbers. We see from the results of Figures 6.17a and 6.18a, which correspond to $R_{2a} = 10^6$, that both parameters increase with decreases in the x-location of transition in the range of $x/2a$ from 0.784 to 0.40. The separation location is, however, nearly unaffected. The results in Figures 6.17b and 6.18b, which correspond to $R_{2a} = 10^7$, show similar increases in wall shear and heat flux rates as the transition location is moved from 0.784 to 0.40. However, this also shows that the flow separation is delayed with the increase in Reynolds number.

We now consider three zero-pressure-gradient flows with different wall-temperature distributions to show the extent to which the numerical calculations agree with experiment, and thereby to validate the accuracy of the numerical scheme and the turbulence model.

Figure 6.19 allows comparison of calculated distribution of local Stanton number St with measurements by Seban and Doughty [19] obtained on an isothermal heated plate for Pr = 0.70. The symbols denote the experimental data, and the solid line represents the numerical prediction. The dashed lines correspond to the predictions of the empirical formulas given by Eqs. (6.67) and (6.70).

Figures 6.20 and 6.21 allow a comparison of calculated and experimental local Stanton numbers for the experimental data obtained by Reynolds et al. [17] for a single-step and double-step wall-temperature variation, respectively. As can be seen, the predictions of the differential method as well as the predictions of the empirical formulas resulting from Eqs. (6.79) and

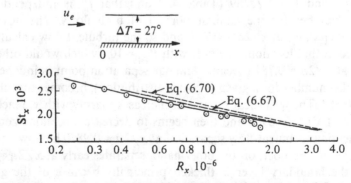

Figure 6.19. Comparison of calculated (solid line) and experimental Stanton numbers for the flat plate boundary layer. Symbols denote experimental data of Seban and Doughty [23].

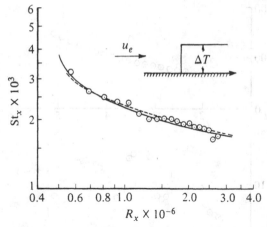

Figure 6.20. Comparison of calculated (solid line) and experimental (symbols) Stanton numbers for the flat-plate boundary-layer measured by Reynolds et al. [17]. Dashed line denotes the predictions of Eq. (6.79).

(6.83) agree well with experiment for both cases. Note that Eq. (6.83) for double-step wall temperature can be written as

$$St = St_T(x)\left\{1 + \frac{T_{w_2} - T_{w_1}}{T_{w_1} - T_e}\left[1 - \left(\frac{x_0}{x}\right)^{9/10}\right]^{-1/9}\right\}.$$

As a final example, we consider a turbulent flow on a smooth surface with longitudinal pressure gradient and compare the predictions of the differential method with experiment. Figure 6.22 shows the results for the accelerating flow of Moretti and Kays [24]. The experimental temperature difference

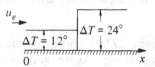

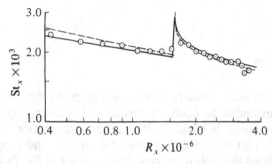

Figure 6.21. Comparison of calculated (solid line) and experimental (symbols) Stanton numbers for the flat plate boundary layer measured by Reynolds et al. [17]. Note the rather gradual change of experimental St indicating a poor approximation to a step in ΔT. Dashed line denotes the predictions of Eq. (6.83).

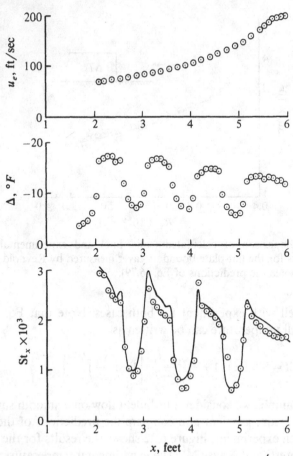

Figure 6.22. Comparison of calculated (solid line) and experimental (symbols) Stanton numbers for the accelerating boundary layer measured by Moretti and Kays [24].

between wall and freestream, $\Delta T(x)$, and experimental velocity distribution $u_e(x)$ were used in the calculations. Again, good agreement between experiment and numerical predictions is revealed.

Rough Surfaces

The Cebeci–Smith eddy-viscosity formulation given by Eqs. (6.102)–(6.110) has also been extended to compute turbulent boundary layers on rough surfaces. The extension is based on the earlier contribution of Rotta [25], which recognizes that the velocity profiles for smooth and rough walls can be similar, provided that y is displaced. We rewrite l in Eq. (6.103) as

$$l = \kappa(y + \Delta y)[1 - \exp\{-(y + \Delta y)/A\}] \tag{6.112}$$

and express Δy as a function of an equivalent sand-grain-roughness parameter $k_s^+ (\equiv k_s u_\tau / \nu)$, that is,

$$\frac{\Delta y u_\tau}{\nu} = \Delta y^+ = \begin{cases} 0.9\left[\sqrt{k_s^+} - k_s^+ \exp\left(-\frac{k_s^+}{6}\right)\right], & 5 \le k_s^+ \le 70 \\ 0.7(k_s^+)^{0.58}, & 70 \le k_s^+ \le 2000 \end{cases}$$

(6.113)

for a hydraulically smooth surface; for smaller k_s^+ we put $\Delta y = 0$ to recover smooth-surface results.

Figure 6.23 allows a comparison of calculated and experimental results for the experimental data obtained by Pimenta et al. [26], who measured zero-pressure-gradient flows on rough surfaces composed of densely packed spheres of uniform size with a diameter of 0.05 in. The dashed lines in Fig. 6.23 correspond to the empirical formulas given by Pimenta et al. [26].

$$\frac{c_f}{2} = 0.00328\left(\frac{\theta}{r}\right)^{-0.175}, \tag{6.114a}$$

$$\mathrm{St}_x = 0.00317\left(\frac{\theta_t}{r}\right)^{-0.175}. \tag{6.114b}$$

This arrangement is the same as that discussed by Schlichting [13] in connection with pipe flows. The equivalent sand-roughness height $k_s (\equiv 1.25r$, with r denoting the radius of the roughness element) suggested by Schlichting was used in the calculations of Cebeci and Chang [27]. We do not expect the accuracy of rough-surface calculations to be significantly affected by pressure gradient, provided that the calculation method is adequate for pressure gradients on smooth surfaces. In effect, roughness just changes the surface boundary condition and should not affect the formulas used in the outer layer.

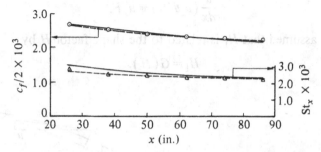

Figure 6.23. Skin friction and heat transfer on a rough surface of densely-packed spheres of diameter $d = 0.05$ in.; $u_\tau d / \nu \approx 120$. Solid lines denote calculations, dashed lines the predictions of Eq. (6.114) and the symbols denote the data of Pimenta et al. [26]. Case 2, $u_\infty = 89$ ft/sec.

Integral Methods

As in laminar flows with pressure gradient, there are several integral methods for calculating momentum transfer in turbulent boundary layers and a more limited number for heat transfer. This disparity arises because of the difficulty of incorporating possible rapid changes in wall temperature or heat flux into the temperature profiles used in the solution of the energy integral equation. In the following, we discuss integral methods, first for momentum transfer and then for heat transfer.

The momentum integral equation

$$\frac{d\theta}{dx} + \frac{\theta}{u_e}\frac{du_e}{dx}(H+2) = \frac{c_f}{2} \tag{3.68}$$

contains the three unknowns θ, H, and c_f, and assumed relationships between these integral parameters are required. There are several approaches to the achievement of this objective, as discussed, for example, by Reynolds [28]. One approach that we shall consider here adopts the notion that a turbulent boundary layer grows by a process of "entrainment" of nonturbulent fluid at the outer edge and into the turbulent region. It was first used by Head [29], who assumed that the mean-velocity component normal to the edge of the boundary layer (which is known as the entrainment velocity v_E as shown on Fig. 6.24) depends only on the mean-velocity profile, specifically on H. He assumed that the dimensionless entrainment velocity v_E/u_e is given by

$$\frac{v_E}{u_e} \equiv \frac{1}{u_e}\frac{d}{dx}\int_0^\delta u\,dy = \frac{1}{u_e}\frac{d}{dx}[u_e(\delta - \delta^*)] = F(H), \tag{6.115}$$

where we have used the definition of δ^*, Eq. (3.69b). If we define

$$H_1 = \frac{\delta - \delta^*}{\theta}, \tag{6.116}$$

then the right-hand equality in Eq. (6.115) can be written as

$$\frac{d}{dx}(u_e\theta H_1) = u_e F. \tag{6.117}$$

Head also assumed that H_1 is related to the shape factor H by

$$H_1 = G(H). \tag{6.118}$$

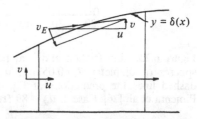

Figure 6.24. Definition of entrainment velocity.

The functions F and G were determined from experiment, and a best fit to several sets of experimental data showed that they can be approximated by

$$F = 0.0306(H_1 - 3.0)^{-0.6169},\tag{6.119}$$

$$G = \begin{cases} 0.8234(H - 1.1)^{-1.287} + 3.3 & H \le 1.6, \\ 0.5501(H - 0.6778)^{-3.064} + 3.3 & H \ge 1.6. \end{cases}\tag{6.120}$$

With F and G defined by Eqs. (6.119) and (6.120), Eq. (6.117) provides a relationship between θ and H. Another equation relating c_f to θ and/or H is needed, and Head used the semiempirical skin-friction law given by Ludwieg and Tillmann [30],

$$c_f = 0.246 \times 10^{-0.678H} R_\theta^{-0.268},\tag{6.121}$$

where $R_\theta = u_e \theta / \nu$. The system [Eqs. (3.68) and (6.115)–(6.121)], which includes two ordinary differential equations, can be solved numerically for a specified external velocity distribution to obtain the boundary-layer development on a two-dimensional body with a smooth surface (see Appendix D). To start the calculations, say at $x = x_0$, we note that initial values of two of the three quantities θ, H, and c_f must be specified, with the third following from Eq. (6.121). When turbulent-flow calculations follow laminar calculations for a boundary layer on the same surface, Head's method is often started by assuming continuity of momentum thickness θ and taking the initial value of H to be 1.4, an approximate value corresponding to flat-plate flow.

This model, like most integral methods, uses a given value of the shape factor H as the criterion for separation. [Equation (6.121) predicts c_f to be zero only if H tends to infinity.] It is not possible to give an exact value of H corresponding to separation, and values between the lower and upper limits of H makes little difference in locating the separation point since the shape factor increases rapidly close to separation.

A more refined integral method for computing momentum transfer in turbulent flows is Green's "lag-entrainment" method [31], which is an extension of Head's method in that the momentum integral equation and the entrainment equation are supplemented by an equation for the streamwise rate of change of entrainment coefficient F. This additional equation allows for more realistic calculations in rapidly changing flows and is a significant improvement over Head's method. In effect this is an "integral" version of the "differential" method of Bradshaw et al. [32]. It requires the solution of Eqs. (3.68) and (6.115) as before and also considers the "rate of change of entrainment coefficient" equation given by

$$\theta(H_1 + H)\frac{dF}{dx} = \frac{F(F + 0.02) + 0.2667c_{f_0}}{F + 0.01} \times \left\{ 2.8\left[(0.32c_{f_0} + 0.024F_{eq} + 1.2F_{eq}^2)^{1/2}\right.\right.$$

$$\left.\left. - (0.32c_{f_0} + 0.024F + 1.2F^2)^{1/2}\right] + \left(\frac{\delta}{u_e}\frac{du_e}{dx}\right)_{eq} - \frac{\delta}{u_e}\frac{du_e}{dx}\right\},$$

$$\tag{6.122}$$

where the numerical coefficients arise from curve fits to experimental data and the empirical functions of Bradshaw et al. [32]. Here c_{f_0} is the flat-plate skin-friction coefficient calculated from the empirical formula

$$c_{f_0} = \frac{0.01013}{\log R_\theta - 1.02} - 0.00075. \tag{6.123}$$

The subscript eq in Eq. (6.122) refers to equilibrium flows, which are defined as flows in which the shape of the velocity and shear-stress profiles in the boundary layer do not vary with x. The functional forms of the equilibrium values of F_{eq}, $[(\theta/u_e)(du_e/dx)]_{eq}$, and $[(\delta/u_e)(du_e/dx)]_{eq}$ are given by

$$F_{eq} = H_1 \left[\frac{c_f}{2} - (H+1) \left(\frac{\theta}{u_e} \frac{du_e}{dx} \right)_{eq} \right], \tag{6.124}$$

$$\left(\frac{\theta}{u_e} \frac{du_e}{dx} \right)_{eq} = \frac{1.25}{H} \left[\frac{c_f}{2} - \left(\frac{H-1}{6.432H} \right)^2 \right], \tag{6.125}$$

and an obvious consequence of the definitions of H and H_1,

$$\left(\frac{\delta}{u_e} \frac{du_e}{dx} \right)_{eq} = (H + H_1) \left(\frac{\theta}{u_e} \frac{du_e}{dx} \right)_{eq}. \tag{6.126}$$

The skin-friction formula and the relationship between the shape factors H and H_1 complete the number of equations needed to solve the system of ordinary differential equations (3.68), (6.115), and (6.122). The skin-friction equation is given by

$$\left(\frac{c_f}{c_{f_0}} + 0.5 \right) \left(\frac{H}{H_0} - 0.4 \right) = 0.9, \tag{6.127a}$$

where

$$1 - \frac{1}{H_0} = 6.55 \left(\frac{c_{f_0}}{2} \right)^{1/2}, \tag{6.127b}$$

so that Eqs. (6.123), (6.127), and (6.128) give c_f as a function of H and R_θ with values close to Eq. (6.121).

The shape-factor relation is

$$H_1 = 3.15 + \frac{1.72}{H-1} - 0.01(H-1)^2 \tag{6.127c}$$

and gives values close to Eq. (6.120).

Comparisons with experiment show good accuracy in incompressible boundary-layer flows and also in wakes. The method has also been extended to represent compressible flows [31].

Two-dimensional turbulent boundary layers can also be computed by simple methods such as Thwaites' method discussed in Chapter 4. Although these methods are limited and are not as accurate as the finite-difference and integral methods discussed in this chapter, they are nevertheless useful in estimating boundary-layer parameters without the use of computers. According to a method discussed in [33], the momentum thickness θ is computed from

$$\theta\left(\frac{u_e\theta}{\nu}\right)^{1/5}u_e^4(x) = 0.0106\int_{x_{tr}}^{x}u_e^4(x)\,dx + \text{const.} \qquad (6.128)$$

The lower limit of integration is determined by the initial conditions. If, for example, the flow is turbulent everywhere, then we can set the lower limit of integration and the constant equal to zero. On the other hand, if turbulent flow is preceded by a laminar flow, then the constant is determined by assuming continuity in θ, with θ computed from a laminar boundary-layer method. Then the constant is the value of $\theta(u_e\theta/\nu)^{1/5}u_e^4$ at transition.

The shape factor H is computed from another empirical formula given by

$$u_e^2\left(\frac{1}{H-1}-4.762\right) = \text{const}+0.00307\int_{x_{tr}}^{x}u_e^2\theta^{-1}\left(\frac{u_e\theta}{\nu}\right)^{-1/5}dx \qquad (6.129a)$$

The value of the constant is again determined from the initial conditions, which usually are taken to be the value of H at transition given as a function of R_θ for a zero-pressure-gradient turbulent flow (see Fig. 6.25). The local skin-friction coefficient c_f is calculated from another empirical formula

$$c_f = \frac{G(H)}{R_\theta^{1/5}} \qquad (6.129b)$$

where $G(H)$ is defined for a Reynolds number range 10^6 to 10^8 by Table 6.1.

The use of integral procedures to predict heat transfer in turbulent boundary layers generally requires the solution of the integral forms of the energy and momentum equations, although solutions of the integral form of one equation and the differential form of the other have, on occasions, been used. Empirical information is, of course, required to allow the solution of the energy equation, and this usually involves a relationship between the wall heat flux and known integral quantities together with an equation to link the thicknesses of the temperature and velocity boundary layers. It is difficult to provide empirical relationships that can be used for more than the simplest flows; as a consequence, integral procedures are not widely used, and differential methods are generally to be preferred.

Where an approximate heat-transfer result is required in relation to a simple flow, expressions derived from integral procedures can be useful. The method of Ambrok [20], for example, assumes the Reynolds analogy and, with the integral energy equation, arrives at the approximate equation

$$\text{St} = \frac{\dot{q}_w}{\rho c_p u_e(T_w - T_e)} = \frac{\text{Pr}^{-0.4}R_L^{-0.2}(T_w - T_e)^{0.25}}{\left[\int_0^{x^*}u_e^*(T_w - T_e)^{1.25}\,dx^*\right]^{0.2}}, \qquad (6.130)$$

u_e^*, and R_L denote dimensionless quantities defined by Eq. (4.77),

$$x^* = \frac{x}{L}, \qquad u_e^* = \frac{u_e}{u_{\text{ref}}}, \qquad R_L = \frac{u_{\text{ref}}L}{\nu}.$$

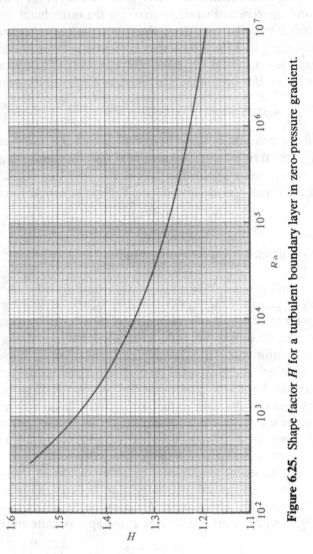

Figure 6.25. Shape factor H for a turbulent boundary layer in zero-pressure gradient.

Table 6.1 The universal function $G(H)$

H	$G(H)$	H	$G(H)$
1.2	0.0108	2.0	0.0027
1.3	0.0092	2.1	0.0022
1.4	0.0079	2.2	0.0017
1.5	0.0067	2.3	0.0012
1.6	0.0057	2.4	0.0008
1.7	0.0048	2.5	0.0004
1.8	0.0040	2.6	0
1.9	0.0033		

It is useful to note that Eq. (6.130) does represent, albeit approximately, the effect of variable surface temperature.

The above procedure is similar to that described previously for laminar boundary layers with pressure gradient, Eq. (4.86). Further details of this and other integral procedures are given by Kays and Crawford [20].

6.7 Wall Jets and Film Cooling

In turbulent flows, the flow properties in film cooling, as in laminar flows, depend on the ratio of slot to freestream velocity, the ratio of slot to freestream temperature T_c/T_e, the boundary conditions on the surface to be protected, and the initial profile shapes. In addition, the geometric possibilities are so great that the solution of wall-jet and film-cooling problems must be obtained by using a differential method such as the case discussed in Section 4.5 and as will be discussed later in this section.

For simple flow configurations dealing with turbulent film cooling on flat plates, several empirical formulas have been suggested. A recent study conducted by Ko and Liu [34] indicates that for the flow configuration shown in Fig. 4.18, depending on the blowing parameter $F \equiv \rho_c u_c / \rho_e u_e$, the film-cooling effectiveness η^* can be predicted satisfactorily by the following empirical formulas:

$$\eta^* = \begin{cases} 2.73 F^{0.4}\left(\dfrac{x}{h}\right)^{-0.38} & \dfrac{x}{h} \leq 20, \\[2mm] 5.44 F^{0.4}\left(\dfrac{x}{h}\right)^{-0.58} & 20 \leq \dfrac{x}{h} \leq 150, F <1, \\[2mm] 2.04\left(\dfrac{x}{h}\right)^{-0.38} & \end{cases} \qquad (6.131a)$$

In the region $x/h < 65$ and $0.45 < F < 1$, the lower value of η^* from the last two equations is recommended.

$$\eta^* = \begin{cases} 1.96 F^{0.55}\left(\dfrac{x}{h}\right)^{-0.38} & \dfrac{x}{h} <150, 1 \leq F < 2, \\[2mm] 2.71\left(\dfrac{x}{h}\right)^{-0.38} & \dfrac{x}{h} <150, 2 \leq F < 3.5. \end{cases} \qquad (6.131b)$$

The authors also recommend the following formula for Nusselt number $\mathrm{Nu}_c(\equiv \hat{h}x/k_c)$, which is based on the thermal conductivity evaluated at the coolant temperature T_c and the Reynolds number $R_x(\equiv u_e x/\nu_e)$ based on the mainstream conditions u_e and T_e for $F \leq 1$:

$$\mathrm{Nu}_c = 0.004 R_x^{0.9} F^{-0.1}, \qquad 8 \leq \frac{x}{h} \leq 60. \tag{6.131c}$$

For the higher blowing rate, $1 < F < 2$, the authors find that the coolant-jet Reynolds number $R_{x_c}(\equiv u_c x/\nu_c)$ has a stronger influence on heat transfer; so we correlate the Nusselt number as

$$\mathrm{Nu}_c = \begin{cases} 0.057 R_{x_c}^{0.7}, & \dfrac{x}{h} < 10, \\[2ex] 6.39 \times 10^{-5}\left(\dfrac{R_{x_c}}{F}\right)^{1.3}, & 10 \leq \dfrac{x}{h} \leq 35. \end{cases} \tag{6.131d}$$

The calculation procedure for turbulent wall jets and film cooling is similar to that of laminar wall jets and film cooling discussed in Section 4.5 except that now we need to use a model for the Reynolds stresses and turbulent heat fluxes appearing in the momentum and energy equations. For a given model, we can again solve the governing equations for specified initial velocity and temperature profiles and boundary conditions.

The velocity profiles that occur in the near-slot region (see Fig. 4.18 for laminar flow) possess a maximum and a minimum (in addition to the asymptotic maximum freestream value of flows considered previously). This complex profile poses a problem in the use of eddy-viscosity and mixing-length turbulence models, such as the Cebeci–Smith model discussed in the previous section, since the locations of zero velocity gradient correspond, with this approach, to the locations of zero eddy viscosity or mixing length. Even if a laminar viscosity is retained, regions of low and zero shear will occur and will present a barrier to the transport of momentum and thermal energy or species.

Figure 6.26 shows two wall-jet velocity profiles, corresponding to high and low jet velocity. Points of infinite or zero eddy viscosity occur in these and all asymmetric flows, independent of whether length scales are obtained from algebraic assumptions, as here, or from the solution of dissipation-rate or related length-scale equations. The corresponding difficulty can be overcome by the use of a turbulence model based on the solution of Reynolds-stress equations, where eddy-viscosity assumptions are not used, or by ad hoc changes to the eddy-viscosity model. The latter approach is chosen here since it is considerably simpler and, as shown by Pai and Whitelaw [35], leads to results with accuracy adequate for most engineering purposes. It may be described as follows:

For a given wall-jet profile, Fig. (6.26), we first determine the values of y for which the values of u are maximum ($u_{\max}$) and minimum ($u_{\min}$). Let us denote them, say, by $y_{\max}$ and $y_{\min}$, respectively. Then we compute

$$y_+ = y_{\min} + 0.04(x - x_0), \tag{6.132a}$$

$$y_- = y_{\min} - 0.04(x - x_0), \tag{6.132b}$$

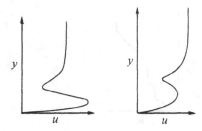

Figure 6.26. Wall-jet profiles for high and low jet velocities.

which assume a spreading rate based on experimentally determined knowledge. Then a modified version of the algebraic eddy-viscosity formulation of Cebeci and Smith discussed in the previous section is computed as follows:

1. In the near-wall region, $0 \le y \le y_{max}$, the usual formulation given by Eqs. (6.102) and (6.109), without the γ term, is used; that is,

$$(\varepsilon_m)_i = l^2 \left| \frac{\partial u}{\partial y} \right|, \qquad 0 \le y \le y_c, \tag{6.133a}$$

$$(\varepsilon_m)_0 = 0.0168 \int_0^{y_{max}} (u_{max} - u)\, dy, \qquad y_c \le y \le y_{max}, \tag{6.133b}$$

with l given by Eq. (6.103). (In the literature, there are several papers purporting to show that the logarithmic law of the wall is different in wall jets and in boundary layers. This is true, but it is simply a consequence of the large negative value of $\partial \tau / \partial y$ near the wall; Eq. (6.133a) is still valid.)

2. In the central wake region, $y_+ \le y \le y_-$, we define

$$\varepsilon_m = \kappa \frac{u_\tau + (u_\tau)_{b.l.}}{2} (y_+ - y_-) \tag{6.134a}$$

or

$$\varepsilon_m = \kappa [u_\tau + (u_\tau)_{b.l.}] 0.04(x - x_0) \tag{6.134b}$$

with u_τ denoting the friction velocity at $y = 0$ and $(u_\tau)_{b.l.}$ denoting the friction velocity of the boundary layer on the slot "splitter plate"; both values of u_τ are *fixed* by the velocity profile specified at $x = x_0$.

3. In the region $y_+ \le y \le \delta$, the eddy-viscosity distribution is again defined by expressions similar to those given by Eqs. (6.102) and (6.109) except that now

$$(\varepsilon_m)_i = l^2 \left| \frac{\partial u}{\partial y} \right|, \qquad y_+ \le y \le y_0, \tag{6.135a}$$

$$(\varepsilon_m)_0 = 0.0168 \int_{y_+}^{\delta} |(u_e - u)|\, dy, \qquad y_0 \le y \le \delta, \tag{6.135b}$$

with l given by $l = \kappa(y - y_{min})$ and y_0 corresponding to the same y_c. definition used in Eqs. (6.113).

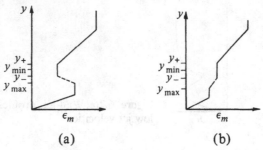

(a)   (b)

Figure 6.27. Distribution of eddy viscosity in wall jets according to Eqs. (6.133) through (6.135).

Figure 6.27 shows the eddy-viscosity distribution across the layer determined by Eqs. (6.132)–(6.135). In the region between y_{max} and y_- the eddy-viscosity distribution is obtained by drawing a straight line as shown by the dashed lines in Fig. 6.27(a).

We note that with changing wall-jet profile, the value of ε_m determined by Eq. (6.134) can exceed the value of $(\varepsilon_m)_0$ determined by Eq. (6.133b). In such cases, in the region $y_{max} \le y \le y_-$ the eddy-viscosity distributions are connected as shown in Fig. 6.27(b). If the value of ε_m computed by Eq. (6.134) exceeds the value of $(\varepsilon_m)_0$ computed by Eq. (6.135b), then in the region $y_- \le y \le \delta$ we use the eddy-viscosity distribution given by Eq. (6.135), thus abandoning the use of Eq. (6.134).

In order to solve the boundary-layer equations for the configuration shown in Fig. 4.18, we again use Eqs. (4.123)–(4.128). Since the flow is turbulent and the eddy-viscosity and turbulent-Prandtl-number concepts are being used, Eqs. (4.123) and (4.124) are slightly different than those for laminar flows; that is,

$$\left(bf''\right)' + \frac{1}{2}ff'' = \xi\left(f'\frac{\partial f'}{\partial \xi} - f''\frac{\partial f}{\partial \xi}\right), \tag{6.136}$$

$$\left(eg'\right)' + \frac{1}{2}fg' = \xi\left(f'\frac{\partial g'}{\partial \xi} - g''\frac{\partial f}{\partial \xi}\right), \tag{6.137}$$

where

$$b = 1 + \varepsilon_m^+, \qquad e = \frac{1}{Pr} + \frac{\varepsilon_m^+}{Pr_t}. \tag{6.138}$$

Figures 6.28 and 6.29 show the variation of the local skin-friction coefficient and the variation of adiabatic-wall effectiveness η^* as a function of dimensionless distance ξ, respectively, for four values of the velocity ratio u_c/u_e and for $\delta/y_c = 2.0$. The initial velocity and temperature profiles were generated for a length Reynolds number corresponding to 10^6 at $\xi_0 = 1$ and, as in laminar flows, the calculations used the numerical method and the computer program discussed in Chapters 13 and 14. We note from these

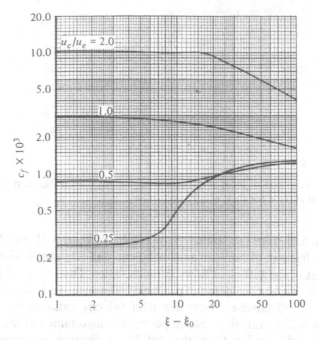

Figure 6.28. Variation of local skin-friction coefficient c_f as a function of $\xi - \xi_0$ for four velocity ratios and for $\delta/y_c = 2$ and $Pr = 0.72$. Calculations are for an adiabatic wall.

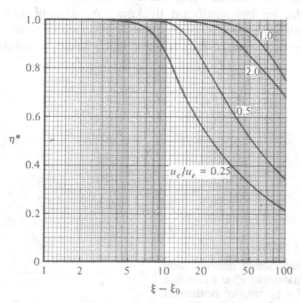

Figure 6.29. Adiabatic wall effectiveness η^* as a function of $\xi - \xi_0$ for four velocity ratios u_c/u_e, $Pr = 0.72$ and for $\delta/y_c = 2.0$.

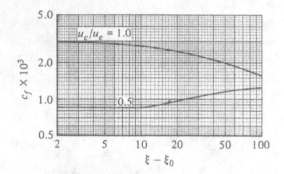

Figure 6.30. Variation of local skin-friction coefficient c_f as a function of $\xi - \xi_0$ on an isothermal surface for $\delta/y_c = 0.95$ and $Pr = 0.72$.

two figures, and from the laminar flow results of Figs. 4.21 and 4.22, that the general features of laminar and turbulent wall jets are similar, but the development of the turbulent flows is much faster. The results of Fig. 6.29, for example, have considerable practical relevance and show that the adiabatic wall effectiveness increases with velocity ratio in the range of u_c/u_e from 0.25 to 1.0: the wall temperature is maintained at the temperature of the fluid issuing from the slot for a distance corresponding to approximately 3 to 20 times the slot height as the velocity ratio increases from 0.25 to 1.0. It can be expected that this trend will not always continue to higher velocity ratios since, at least when the boundary layers on the two sides of the slot lip are of equal thickness, equal velocities should lead to the least mixing. The thickness of the boundary layer on the upper surface of the slot lip, δ, is of secondary importance in turbulent flows and especially where the slot lip has significant thickness. A value of δ/y_c of 2.0 is consistent with some practical configurations since, with a value of y_c of, say 5 mm, it implies a value of δ of 10 mm, which would be obtained with a

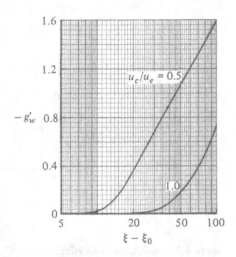

Figure 6.31. Variation of dimensionless wall heat flux parameter g'_w as a function of distance $\xi - \xi_0$ on an isothermal surface for $\delta/y_c = 0.95$ and $Pr = 0.72$.

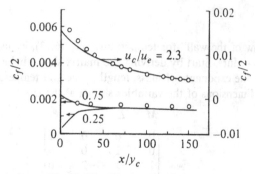

Figure 6.32. Variations of local skin-friction coefficient with x/y_c for three values of u_c/u_e.

free-stream velocity of 10 ms^{-1} and a plate length of 0.5 m, or shorter if the boundary layer is artificially thickened.

Figures 6.30 and 6.31 show the results for an isothermal flat plate. The calculations were made for two velocity ratios with Pr = 0.72 and $\delta/y_c = 0.95$ by using the same initial velocity and temperature profiles as in previous calculations. As before, the relation between the dimensionless wall-temperature gradient g'_w and heat flux and the Reynolds number is given by Eq. (4.130).

The accuracy of the numerical results presented in these two figures depends on the accuracy of the turbulence model. To see how well the calculations agree with experiment, we consider the experimental data of Kacker and Whitelaw [36] and compare them with the numerical calculations obtained with the modified Cebeci–Smith model as given by Eqs. (6.133)–(6.135). The results shown in Figs. 6.32 and 6.33 are in close agreement with the measurements of Kacker and Whitelaw [36] and with the calculations of Pai and Whitelaw [35] which made use of a mixing-length assumption. It is to be expected that higher-order turbulence models, such as those proposed by Launder, Reece and Rodi [37] will lead to similar results, although assumptions such as those of Fig. 6.27 would be replaced by assumptions about the nature of the equations with Reynolds stresses and turbulence dissipation rate as dependent variables.

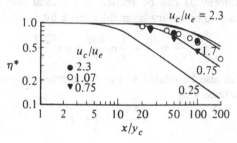

Figure 6.33. Comparison between computed and experimental results as a function of u_c/u_e for turbulent flows.

Problems

6.1 Derive the "law of the wall" for temperature, Eq. (6.12), by using the matrix elimination procedure. Start by deriving the matrix shown below, in which the rows represent the exponents of mass, length, time and temperature (M, L, T and θ) in the dimensions of the variables shown at the left

$$
\begin{array}{c|cccc}
 & M & L & T & \theta \\
\hline
T_w - T & 0 & 0 & 0 & 1 \\
y & 0 & 1 & 0 & 0 \\
\tau_w & 1 & -1 & -2 & 0 \\
\dot{q}_w & 1 & 0 & -3 & 0 \\
\rho & 1 & -3 & 0 & 0 \\
\mu & 1 & -1 & -1 & 0 \\
k & 1 & 1 & -3 & -1 \\
c_p & 0 & 2 & -2 & -1
\end{array}
$$

6.2 Using the matrix elimination procedure, show that the law of the wall for velocity on a rough surface is

$$u^+ = \phi(y^+, k^+).$$

6.3 Show that the continuity equation requires that $\overline{u'v'}$ should vary as at least the third power of y in the viscous sublayer, whereas the Van Driest formula for mixing length, Eq. (6.17) implies $\overline{u'v'} \sim y^4$ for small y.

6.4 Show that the "kinematic heating" parameter $\rho u_\tau^3 / \dot{q}_w$ is equal to

$$
\frac{(\gamma - 1)(u_\tau / a_w)^2}{\dot{q}_w / (\rho c_p u_\tau T_w)}
$$

where $a_w = \sqrt{\gamma R T_w}$.

6.5 Using the inner-layer formula, Eq. (6.13), show that the logarithmic law for solid surfaces, Eq. (6.11), namely, can be generalized to flows with a uniform transpiration velocity v_w and can be written in the form

$$\frac{2}{v_w}\left[(u_\tau^2 + u v_w)^{1/2} - u_\tau\right] = \frac{1}{\kappa}\ln\frac{y u_\tau}{\nu} + c. \tag{P6.1}$$

Here c is a dimensionless quantity that is a function of v_w / u_τ.

6.6 Show that Eq. (P6.1) reduces to Eq. (6.11) as $v_w \to 0$.

6.7 If the expression for the whole velocity profile, Eq. (6.39), is evaluated at $y = \delta$, the profile parameter Π can be related to the local skin-friction coefficient $c_f = 2\tau_w / \rho u_e^2$ and to boundary-layer thickness δ by

$$\frac{\sqrt{2}}{c_f} \equiv \frac{u_e}{u_\tau} = \frac{1}{\kappa}\ln\frac{\delta u_\tau}{\nu} + c + \frac{2\Pi}{\kappa}. \tag{P6.2}$$

Show that it can also be related to the displacement thickness δ^* and to the momentum thickness θ by

$$\kappa\frac{\delta^* u_e}{\delta u_\tau} = 1 + \Pi \tag{P6.3}$$

and

$$\kappa^2\frac{(\delta^* - \theta)u_e^2}{\delta u_\tau^2} = 2 + 2\left[1 + \frac{1}{\pi}Si(\pi)\right]\Pi + \frac{3}{2}\Pi^2 \tag{P6.4}$$

Also, show that

$$\frac{H}{H-1}\frac{u_\tau}{\kappa u_e} \equiv \frac{1}{\kappa G} = F(\Pi) \tag{P6.5}$$

where $Si(\pi) = \int_0^\pi [\sin u / u] \, du = 1.8519$ and G is the Clauser shape parameter.

$$G = \int_0^\infty \left(\frac{u-u_e}{u_\tau}\right)^2 d\left(\frac{y}{\Delta}\right),$$

$$\Delta = -\int_0^\infty \left(\frac{u-u_e}{u_\tau}\right) dy.$$

6.8 Use Eqs. (P6.2) and (P6.3) to find the skin friction on a flat-plate boundary layer at $R_{\delta^*} = u_e \delta^*/v = 20,000$ and then use Eq. (P6.5) to calculate $R_\theta = u_e\theta/v$. Take $c = 5.0$ and $\kappa = 0.41$.

6.9 Air at 25°C and 1 atm flows at 30 m s^{-1} past a flat plate of length 5 m. Assume $R_{x_{tr}} = 3 \times 10^6$ and

(a) Find the effective origin x_0 of the turbulent boundary layer.
Hint: To estimate x_0 neglect the transitional region, assume that the momentum thickness is continuous at transition, and replace x in Eq. (6.59) by $x_{tr} - x_0$.
(b) With Reynolds number based on the effective origin, calculate the local and average skin-friction coefficients at $x = 5$ m.
(c) At $x = 4$ m, calculate the distances from the surface at which y^+ is equal to 5, 50, 100, 500 and 1000.

6.10 (a) If in problem 6.9 the surface temperature of the plate is maintained at 75°C, calculate the rate of cooling of the plate per unit width. Use the arithmetic-mean film temperature T_f to evaluate the fluid properties.
(b) What error is involved if the boundary layer is assumed to be turbulent from the leading edge?
(c) Repeat (a) and (b) for a velocity of 15 m s^{-1} with all the other data remaining the same. Discuss the results.

6.11 Liquid ammonia at $-25°C$ flows at 2 m s^{-1} past a 2 m-long flat plate whose surface temperature is maintained at $-10°C$. If transition from laminar to turbulent flow occurs at 4×10^5, calculate the local and average heat transfer coefficient $\hat{h}$ at the end of the plate. Evaluate the fluid properties at the arithmetic-mean film temperature T_f.

6.12 Using the momentum integral equation (3.68) and Eq. (6.54), derive Eqs. (6.58), (6.59), (6.56) and (6.57). Represent the local skin-friction in Eq. (3.68) by

$$c_f = \frac{0.0456}{(u_e\delta/v)^{1/4}}$$

which is a modified version of the Blasius friction equation (7.17) for friction factors in circular pipes.

6.13 Air at $u_e/v = 10^7$ m s^{-1} flows past a 3 m long plate covered with spanwise square-bar roughness elements. Determine the local skin-friction coefficient at $x = 1$ m and the average skin-friction coefficient of the plate. As a simplication, assume that roughness causes the transition to be at the leading edge so that the contribution of laminar flow can be neglected, and take $k = 0.0005$ m.

Hint: First determine the equivalent sand-grain height of the square-bar roughness distribution tested by Moore and shown in Fig. 6.7.

6.14 Consider problem 6.13 where the plate surface is (a) covered with camouflage paint with equivalent sand roughness $k_s = 1 \times 10^{-3}$ cm and (b) cast iron with $k_s = 25 \times 10^{-3}$ cm.

Calculate the momentum thickness, boundary-layer thickness, local skin-friction coefficient and average-skin friction coefficient at $x = 3$ m.

Hint: In calculating θ and δ for rough wall boundary layers on a flat plate, remember that

$$\frac{\bar{c}_f}{2} = \frac{\theta}{x}$$

and a power-law assumption for the velocity profile is a good approximation for δ^*, δ and θ.

6.15 Using the power-law profile for velocity, derive Eq. (6.71b). Hint: First show that the momentum integral equation for a flat plate can be written as

$$-\frac{\partial}{\partial x} \int_0^y u(u_e - u) \, dy + (u_e - u) \frac{\partial}{\partial x} \int_0^y u \, dy = \frac{\tau - \tau_w}{\rho}.$$

6.16 Using Eqs. (6.71) and the definition of eddy viscosity and eddy conductivity, derive Eq. (6.72) for $Pr_t = 1$.

6.17 Air at $u_e/\nu = 3 \times 10^6$ m^{-1} flows past a 3 m-long flat plate. Consider the plate: (a) heated at uniform wall temperature T_w, and (b) the heated portion preceded by an unheated portion x_0 of 1 m. Calculate the Stanton number distribution along the plate for both cases. What role does the term $(T_w/T_e)^{0.4}$ in Eq. (6.79) play in the results. Assume the flow to be turbulent from the leading edge with $T_w/T_e = 1.1$ and $Pr = 0.7$.

6.18 Use Eq. (6.82) to derive an expression for wall heat flux on a flat plate for which the difference between wall temperature and freestream temperature varies linearly with x, that is,

$$T_w - T_e = A + Bx.$$

Hint: Note that there is a temperature jump at the leading edge of the plate where $T_w - T_e = A$.

6.19 Use Eq. (6.82) to obtain an expression for the heat transfer rate on a flat plate for $x > x_2$ and with $T_w = T_{w_1}$ for $0 < x < x_1$, $T_w = T_{w_2}$ for $x_1 < x < x_2$ and $T_w = T_{w_3}$ for $x > x_2$.

6.20 Air at 20°C and with $u_e/\nu = 5 \times 10^5$ m^{-1} flows over a 4 m long flat plate. Assume that the flow is turbulent at the leading edge and has the following temperature distribution: 0 to 1 m, 50°C; 1 to 1.5 m, 60°C; 1.5 to 4 m, 80°C: Find the local heat flux at $x = 4$ m for $Pr = 1$.

6.21 Air at 20°C and with $u_e/\nu = 5 \times 10^5$ m^{-1} flows over a 5 m long flat plate. Assume that the heating starts at $x_0 = 2$ m and is applied uniformly in the range $x_0 < x \le 5$, and find the wall temperature distribution along the plate. Assume also that the flow is turbulent from the leading edge.

6.22 Show that the integral in Eq. (6.86) can be expressed by Eq. (6.87). Hint: To put the integral in the form of the Beta function,

$$B_r(m, n) = \int_0^r z^{m-1}(1-z)^{n-1} \, dz \qquad (m, n > 0)$$

first multiply and divide the integral by x (x is treated as a parameter) and let $z = 1 - (\xi/x)^{9/10}$.

6.23 Use the law of the wall, Eq. (6.4) and the continuity equation (3.5), to derive Eq. (6.97).

6.24 Show that the entrainment velocity v_E, which is the rate at which the volume flow rate per unit span changes with x, namely,

$$v_E = \frac{d}{dx} \int_0^\delta u \, dy$$

can be written in the form given by Eq. (6.115).

Hint: First use the continuity equation and Leibnitz's rule for differentiation to show that the v-component of the velocity at the boundary-layer edge, $y = \delta$, can be written as

$$v_e = -\frac{d}{dx} \int_0^\delta u \, dy + u_e \frac{d\delta}{dx}.$$

Then use the definition of displacement thickness δ^* to show that

$$v_e = \frac{d}{dx} (u_e \delta^*) - \delta \frac{du_e}{dx}$$

and

$$\int_0^\delta u \, dy = u_e (\delta - \delta^*).$$

6.25 Show that the entrainment velocity v_E in axisymmetric flow is given by

$$v_E = \frac{1}{r_\delta} \frac{d}{dx} \int_0^\delta ru \, dy = \frac{1}{r_\delta} \frac{d}{dx} \left[u_e \left(\delta \frac{r_0 + r_\delta}{2} - \delta^* \right) \right]$$

where

$$r_\delta = r_0 + \delta \cos \phi$$

$$\delta^* = \int_0^\delta r \left(1 - \frac{u}{u_e} \right) dy, \qquad \theta = \int_0^\delta r \frac{u}{u_e} \left(1 - \frac{u}{u_e} \right) dy.$$

6.26 To prevent icing on a windshield, air is blown tangentially to the surface at a velocity and temperature of 25 m s^{-1} and 325°K, respectively. If the vehicle is travelling at 50 m s^{-1} and the surrounding air is at 275°K,

(a) Calculate the required slot width if the windshield is to be maintained at a temperature in excess of 300°K for a length of 1 m.

(b) Determine if the value of δ/y_c used in the calculation of the results of figure 6.28 is appropriate and suggest the likely effect of the difference between the calculated value of δ/y_c and the value of the figure. Assume that the external stream is directed tangential to the windshield and that the leading edge of the related boundary layer occurs 2 m upstream of the slot exit.

6.27 The assumption of an adiabatic wall is appropriate to a glass windshield but less so to lifting surfaces of aircraft. Suppose that it is required to maintain a surface downstream of a blown slot at a uniform temperature of 300°K by controlling the heat transfer to the surface with the assistance of electrical heaters. With the same parameters as those of Problem 6.26, plot the distribution of heat flux over a 1 m length in dimensional terms. Note that g_w' is defined by Eq. (4.130).

6.28 Air at atmospheric pressure and 15°C is heated as it flows over a two-dimensional smooth surface with an elliptic cross section of 4 to 1 thickness ratio.

Calculate the wall shear parameter $c_f/2\sqrt{R_x}$ and wall heat flux parameter $Nu_x/\sqrt{R_x}$ as a function of dimensionless axial distance $x/2a$ for both laminar and turbulent flows with transition computed from Eq. (6.111) for $R_{2a} = 10^7$ and for a uniform wall temperature of 25°C and for a Prandtl number 0.72. Perform these calculations using Thwaites' method and the Spalding-Smith method for laminar flows, and with Eqs. (6.128) to (6.130) for turbulent flows. Compare your results with those given in Figures 6.15 and 6.16.

6.29 A gas turbine blade is based on an NACA 2412 airfoil and has the geometry and external velocity distribution given below. It operates at the following conditions:

$$\begin{aligned} \text{total temperature,} \quad & T_0 = 850°C \\ \text{kinematic viscosity,} \quad & \nu = 6 \times 10^{-5} \text{ m}^2 \text{ s}^{-1}. \\ \text{freestream velocity,} \quad & u_\infty = 200 \text{ m s}^{-1}. \end{aligned}$$

We wish to determine the amount of cooling (kW m^{-1}) needed to maintain the surface temperature at 550°C from the leading-edge to the transition point, then, decreasing linearly to 450°C at the trailing edge. We are also interested in determining the local skin-friction distribution and displacement thickness distributions on the upper and lower surfaces of the blade. Suggested procedure is:

1. Determine momentum and heat transfer for laminar flow.
2. Compute transition location.
3. Determine momentum and heat transfer for turbulent flow.

Take the chord length to be 0.1 m. To calculate the heat transfer in turbulent flows, use Eq. (6.130).

Coordinates and external (inviscid) velocity distribution of NACA 2412 airfoil for a zero degree angle of attack

x/c	y/c	u_e/u_∞
0.975017	−0.002574	−0.923056
0.925000	−0.006500	−0.951147
0.850000	−0.011600	−0.967792
0.750002	−0.018235	−0.989543
0.650002	−0.024535	−1.006718
0.550005	−0.030586	−1.026400
0.450007	−0.035862	−1.046163
0.350006	−0.039787	−1.064401
0.275001	−0.041775	−1.084731
0.225000	−0.042390	−1.103499
0.174994	−0.041869	−1.122426
0.124977	−0.039573	−1.138186
0.087484	−0.036186	−1.154834
0.062454	−0.032608	−1.166511
0.037354	−0.026895	−1.159470
0.018586	−0.019930	−1.124105
0.004651	−0.009461	−0.729720
0.003899	0.012117	0.603138
0.018541	0.026010	1.075167

Coordinates and external (inviscid) velocity
distribution (*Continued*)

x/c	y/c	u_e/u_∞
0.037283	0.036076	1.156801
0.062418	0.045696	1.198170
0.087457	0.053112	1.219485
0.124906	0.061678	1.241882
0.174955	0.069695	1.249947
0.224978	0.074922	1.249534
0.274990	0.077978	1.244678
0.350006	0.079106	1.227716
0.450027	0.075688	1.194753
0.550034	0.068384	1.162275
0.650039	0.058038	1.129654
0.750043	0.044950	1.091903
0.850053	0.029465	1.046916
0.925026	0.016242	1.000793
0.975055	0.005940	0.923055

6.30 In Appendix D we describe a computer program for calculating the development of turbulent boundary layers in two-dimensional flows. It is based on
Head's method and on the solution of two ordinary differential equations given
by Eqs. (3.68) and (6.117) by a fourth-order Runge-Kutta method. Show that
with minor modifications to Eqs. (3.68) and (6.117), this code can be extended
to axisymmetric turbulent boundary layers.

Hint: Note that:

$$B(2) = u_e r_0 \theta H_1$$

$$C(1) = \frac{c_f}{2} - (H+2)\frac{\theta}{u_e}\frac{du_e}{dx} - \frac{\theta}{r_0}\frac{dr_0}{dx}$$

$$C(2) = r_0 u_e F.$$

In addition we need to read in $r_0(x)/L$ and compute dr_0/dx or $d/d(x/L)$
(r_0/L). The latter can again be done with a three-point Lagrange-interpolation
formula.

References

[1] Rodi, W.: *Turbulence Models and Their Application in Hydraulics*. International
Association for Hydraulics Research, Delft, Netherlands, 1980.

[2] Bradshaw, P., Cebeci, T., and Whitelaw, J. H.: *Engineering Calculation Methods
for Turbulent Flows*. Academic, London, 1981.

[3] Klebanoff, P. S. and Diehl, Z. W.: Some features of artificially thickened fully
developed turbulent boundary layers with zero pressure gradient. NACA Rep.
1110, 1952.

[4] Jacobs, W.: Unformung eines turbulenten Geschwindigkeitsprofils, *Z. Angew.
Math. Mech.* **19**:87 (1939).

[5] Taylor, E. S.: *Dimensional Analysis for Engineers*. Clarendon, Oxford, 1973.

[6] Van Driest, E. R.: On turbulent flow near a wall. *J. Aero. Sci.* **23**:1007 (1956).

[7] Cebeci, T.: A model for eddy conductivity and turbulent Prandtl number. *J. Heat Transfer* **95**:227 (1973).

[8] Na, T. Y. and Habib, J. S.: Heat transfer in turbulent pipe flow based on a new mixing-length model. *Appl. Sci. Res.* **28**: (1973).

[9] Clauser, F. H.: The turbulent boundary layer. *Advan. Appl. Mech.* **4**:1 (1956).

[10] Owen, P. R. and Thomson, W. R.: Heat transfer across rough surfaces. *J. Fluid Mech.* **15**:321–334 (1963).

[11] Hoffmann, P. H. and Perry, A. E.: The development of turbulent thermal layers on flat plates. *Int. J. Heat Mass Transfer* **22**:39 (1979).

[12] Cebeci, T. and Smith, A. M. O.: *Analysis of Turbulent Boundary Layers*. Academic, New York, 1974.

[13] Schlichting, H.: *Boundary-Layer Theory*. McGraw-Hill, New York, 1981.

[14] Simonich, J. C. and Bradshaw, P.: Effect of freestream turbulence on heat transfer through a turbulent boundary layer. *J. Heat Transfer* **100**:671 (1978).

[15] Kader, B. A. and Yaglom, A. M.: Heat and mass transfer laws for fully turbulent wall flows. *Int. J. Heat Mass Transfer* **15**:2329 (1972).

[16] Reynolds, W. C., Kays, W. M., and Kline, S. J.: Heat transfer in the turbulent incompressible boundary layer. I. Constant wall temperature. NASA MEMO 12-1-58W, 1958.

[17] Reynolds, W. C., Kays, W. M., and Kline, S. J.: Heat transfer in the turbulent incompressible boundary layer. II. Step wall-temperature distribution. NASA MEMO 12-22-58W, 1958.

[18] Reynolds, W. C., Kays, W. M., and Kline, S. J.: Heat transfer in the turbulent incompressible boundary layer. III. Arbitrary wall temperature and heat flux. NASA MEMO 12-3-58W, 1958.

[19] Hildebrand, F. B.: *Advanced Calculus for Applications*. Prentice-Hall, Englewood Cliffs, NJ, 1962.

[20] Kays, W. M. and Crawford, M. E.: *Convective Heat and Mass Transfer*. McGraw-Hill, New York, 1980.

[21] Chen, K. K. and Thyson, N. A.: Extension of Emmons' spot theory to flows on blunt bodies. *AIAA J.* **5**:821 (1971).

[22] Michel, R.: Etude de la transition sur le profiles d'aile; établissement d'un critere de détermination de point de transition et calcul de la trainee de profile incompressible. ONERA Rep. 1/157A, 1951.

[23] Seban, R. A. and Doughty, D. L.: Heat transfer to turbulent boundary layers with variable freestream velocity. *J. Heat Transfer* **78**:217 (1956).

[24] Moretti, P. M. and Kays, W. M.: Heat transfer to a turbulent boundary layer with varying freestream velocity and varying surface temperature, an experimental study. *Int. J. Heat Mass Transfer* **8**:1187 (1965).

[25] Rotta, J. C.: Turbulent boundary layers in incompressible flows. *Prog. Aero. Sci.* **2**:1 (1962).

[26] Pimenta, M. M., Moffat, R. J., and Kays, W. M.: The turbulent boundary layer: An experimental study of the transport of momentum and heat with the effect of roughness. Rep. No. HMT-21, Stanford University, Dept. of Mech. Eng., Stanford, CA, 1975.

[27] Cebeci, T. and Chang, K. C.: Calculation of incompressible rough-wall boundary-layer flows. *AIAA J.* **16**:730 (1978).

[28] Reynolds, W. C.: Computation of turbulent flows–state of the art, 1970. Rep. MD-27, Stanford University, Stanford, CA, 1970.

[29] Head, M. R.: Entrainment in the turbulent boundary layers. ARC R&M 3152, 1958.

[30] Ludwieg, H. and Tillmann, W.: Untersuchungen über die Wandschubspannung in turbulenten reibungsschichten. *Ing. Arch.* **17**:288 (1949). English translation in NACA TM 1285, 1950.

[31] Green, J. E., Weeks, D. J., and Brooman, J. W. F.: Prediction of turbulent boundary layers and wakes in incompressible flow by a lag-entrainment method. ARC R&M 3791, 1973.

[32] Bradshaw, P., Ferriss, D. H., and Atwell, N. P.: Calculation of boundary-layer development using the turbulent energy equation. *J. Fluid Mech.* **28**:593 (1967).

[33] Thwaites, B. (ed): *Incompressible Aerodynamics*. Clarendon, Oxford, 1960.

[34] Ko, S.-Y. and Liu, D.-Y: Experimental investigation of effectiveness, heat transfer coefficient and turbulence of film cooling. *AIAA J.* **18**:907–913 (1980).

[35] Pai, B. R. and Whitelaw, J. H.: The prediction of wall temperature in the presence of film cooling. *Int. J. Heat Mass Transfer* **14**:409 (1971).

[36] Kacker, S. C. and Whitelaw, J. H.: The effect of slot height and of slot turbulence intensity on the effectiveness of the uniform density, two-dimensional wall jet. *J. Heat Transfer* **90**:469 (1969).

[37] Launder, B. E., Reece, G. J., and Rodi, W.: Progress in the development of a Reynolds-stress turbulence closure. *J. Fluid Mech.* **68**:537 (1975).

CHAPTER 7

Uncoupled Turbulent Duct Flows

The calculation of the momentum and heat-transfer properties of turbulent internal flows is similar to that for the laminar flows of Chapter 5, with the obvious differences implied by the addition of turbulent stresses. Again we concentrate our attention on three different conditions in which momentum and heat transfer may occur in duct flows. Section 7.1 deals with fully developed velocity and temperature fields in ducts with smooth and rough surfaces. Section 7.2 considers heat transfer starting far downstream of the duct entry, so that the velocity field is fully developed ($\partial u / \partial x = 0, v = 0$), and the temperature field is developing. In this case, the energy equation can be written in a form like Eq. (5.3), using the eddy-viscosity and turbulent-Prandtl-number concepts, as

$$u\frac{\partial T}{\partial x} = \frac{\nu}{\Pr}\frac{1}{r^K}\frac{\partial}{\partial y}\left[r^K\left(1 + \frac{\Pr}{\Pr_t}\varepsilon_m^+\right)\frac{\partial T}{\partial y}\right], \qquad (7.1)$$

where y is measured from the duct wall as usual. We discuss the solutions of this equation for flows on smooth and rough surfaces with suitable formulas for ε_m^+ and $\Pr_t$.

Section 7.3 describes the solution of the third problem in which both velocity and temperature fields are developing in the entry region of a heated duct.

7.1 Fully Developed Duct Flow

To solve the energy equation (7.1) for the case in which the temperature profile is fully developed, as in Section 5.1 for laminar flow, we make use of

216

Eq. (5.14) and write Eq. (7.1) as

$$u\left[(1-G)\frac{dT_w}{dx}+G\frac{dT_m}{dx}\right]=\frac{\nu}{\text{Pr}}\frac{1}{r^K}\frac{\partial}{\partial y}\left[r^K\left(1+\frac{\text{Pr}}{\text{Pr}_t}\varepsilon_m^+\right)\frac{\partial T}{\partial y}\right], \quad (7.2)$$

Velocity Profile and Friction Factor

To obtain the velocity profile u for a fully developed flow, we consider the x-momentum equation. Using the definition of eddy viscosity and recalling that the inertia terms on the left-hand side are equal to zero, we can write this equation as

$$-\frac{1}{\rho}\frac{dp}{dx}+\frac{\nu}{r^K}\frac{d}{dy}\left[r^K(1+\varepsilon_m^+)\frac{du}{dy}\right]=0. \quad (7.3)$$

If we let

$$b=r^K(1+\varepsilon_m^+), \qquad \bar{p}=\frac{p}{\rho} \quad (7.4)$$

and use the definition of r ($\equiv r_0 - y$), we can write Eq. (7.3) as

$$\frac{d}{dr}\left(b\frac{du}{dr}\right)=\frac{r^K}{\nu}\frac{d\bar{p}}{dx}. \quad (7.5)$$

Integrating Eq. (7.5) and noting that $du/dr = 0$ at $r = 0$, we get

$$b\frac{du}{dr}=\frac{r^{K+1}}{K+1}\frac{1}{\nu}\frac{d\bar{p}}{dx}. \quad (7.6)$$

Evaluating Eq. (7.6) at the wall to relate dp/dx to $(du/dr)_{r=r_0}=-\tau_w/\mu \equiv -u_\tau^2/\nu$, we can write Eq. (7.6), after some algebra, as

$$\left(\frac{r}{r_0}\right)\frac{u_\tau^2}{\nu}=(1+\varepsilon_m^+)\frac{du}{dy}. \quad (7.7)$$

In terms of dimensionless variables $u^+ = u/u_\tau$, $y^+ = yu_\tau/\nu$, and $r_0^+ = r_0 u_\tau/\nu$, Eq. (7.7) can be written as

$$\left(1-\frac{y^+}{r_0^+}\right)=(1+\varepsilon_m^+)\frac{du^+}{dy^+}. \quad (7.8)$$

There are several expressions that can be used for ε_m^+ for a fully developed flow. One popular formula is based on Nikuradse's mixing-length expression

$$l=r_0\left[0.14-0.08\left(1-\frac{y}{r_0}\right)^2-0.06\left(1-\frac{y}{r_0}\right)^4\right] \quad (7.9)$$

which tends to $l = 0.4y$ for small y, and Van Driest's damping factor $1-e^{-y/A}$ discussed in Section 6.2. These expressions give

$$\varepsilon_m=l^2(1-e^{-y/A})^2\frac{du}{dy}, \quad (7.10)$$

which in terms of dimensionless quantities can be written as

$$\varepsilon_m^+ = \{ l^+ [1 - \exp(-y^+/A^+)] \}^2 \frac{du^+}{dy^+}, \tag{7.11}$$

where $l^+ = l u_\tau / \nu$ and $A^+ = 26$.

Another useful expression is a two-layer formula similar to that used by Cebeci and Smith for boundary-layer flows discussed in Section 6.6. In this case, the eddy viscosity for the inner region is represented by Eq. (7.10) with l given by Eq. (6.83); that is,

$$l = \kappa y \left[1 - \exp\left(-\frac{y}{A}\right) \right]. \tag{6.83}$$

In the outer region $(\varepsilon_m)_0$ is represented by

$$(\varepsilon_m)_0 = \alpha \int_0^L (u_c - u)\, dy. \tag{7.12}$$

Here L represents the radius of the pipe or the half-width of the two-dimensional duct, depending on the problem, u_c is the centerline velocity, and α is an empirical constant, which is a function of Reynolds number as we shall see later.

For a demonstration, we choose Eqs. (7.9) and (7.11) rather than the two-layer model. Inserting Eq. (7.11) into Eq. (7.8) and following a procedure similar to that used to obtain Eq. (6.19), we can obtain the whole turbulent velocity profile by integrating the following expression, which is an axisymmetric version of Eq. (6.20):

$$u^+ = \int_0^{y^+} \frac{2(1 - y^+/r_0^+)^K}{1 + \sqrt{1 + 4(1 - y^+/r_0^+)^K (L^+)^2}}\, dy^+, \tag{7.13}$$

where

$$L^+ = l^+ \left[1 - \exp\left(-\frac{y^+}{A^+}\right) \right]. \tag{7.14}$$

Once u^+ is known, u immediately follows as a function of y provided that u_τ is known. That is, we need expressions for friction factor to start the calculation.

A relation between the friction factor and the Reynolds number R_d can be obtained by making use of a velocity profile expression similar to that given for a boundary layer by Eq. (6.39) but with r_0 replacing δ; that is,

$$u^+ = \frac{1}{\kappa} \ln \frac{y u_\tau}{\nu} + c + \frac{\Pi}{\kappa} w\left(\frac{y}{r_0}\right). \tag{7.15}$$

Π is negligibly small in pipe flow, by coincidence and not because the inner-layer arguments are valid throughout the pipe. Then evaluating Eq.

(7.15) at the centerline gives a formula for u_{max}/u_τ. After rearrangement to replace u_{max} with the bulk-average velocity u_m (see Problem 7.2), we get

$$\frac{1}{\sqrt{f}} = 0.87 \ln R_d\sqrt{f} - 0.8, \tag{7.16}$$

where f is the pipe friction factor $\tau_w/(1/8\rho u_m^2)$. This equation is known as Prandtl's friction law for smooth pipes. The corresponding law for two-dimensional ducts, where $u_m = u_{max} - 2.64u_\tau$ and Π is slightly positive, has an additive constant of -0.41 instead of -0.8.

A wholly empirical formula that can be used to calculate the friction factor in smooth pipes has been obtained by Blasius [1]; it is given by

$$f = \frac{0.3164}{R_d^{0.25}}, \qquad R_d \leq 10^5. \tag{7.17}$$

A lower limit of R_d, about 2×10^3, is set by the requirement that the flow be turbulent. At $R_d = 10^5$, Eq. (7.17) gives $f = 0.0178$, whereas the more accurate relation gives 0.0179; at $R_d = 2 \times 10^3$, the values are 0.0473 and 0.0497, respectively.

To set up the velocity field (implying the eddy conductivity $\varepsilon_m/\mathrm{Pr}_t$) for insertion into Eq. (7.2), we first calculate f from Eq. (7.16) or (7.17) to compute y_{max}^+ ($\equiv r_0 u_\tau/\nu$) and then calculate u^+ from Eq. (7.13), then use it to get ε_m^+ as a function of y from Eq. (7.11), converting from y^+ to y using u_τ from $u_m\sqrt{f/8}$.

In calculating the velocity profile u^+ and f we must ensure that the principle of conservation of mass flow rate in the duct is not violated. To explain this, let us consider Eq. (5.2). For a circular pipe this equation can be written in the present notation as

$$1 = 2\int_0^1 \bar{r}\bar{u}\,d\bar{r}. \tag{7.18}$$

With the velocity profile computed from Eqs. (7.13) and (7.14), it is possible that the above relation may not be satisfied exactly since the generation of the velocity profile and the calculation of the friction factor are based on empirical data that may not be compatible. One possible way to avoid the discrepancy in mass flow rate, small though it may be, is to use the two-layer eddy-viscosity model given by Eqs. (7.10), (6.83), and (7.12).

To use the two-layer eddy-viscosity model, first we use Eq. (7.13) to generate an initial dimensionless velocity profile, and then we use this profile to compute the eddy-viscosity formulas for the inner and outer regions. Since α in Eq. (7.12) is not known initially, we assume a value for it, say 0.0168. With ε_m^+ computed in this way, we integrate Eq. (7.8) to obtain a new velocity profile u^+, and use Eq. (7.17) to get $\bar{u}$. If this dimensionless

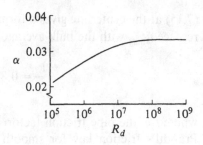

Figure 7.1. Variation of α with R_d.

velocity does not satisfy Eq. (7.18), a new value of outer eddy-viscosity parameter α is computed. This process is repeated until Eq. (7.18) is satisfied.

Figure 7.1 shows the variation of α with R_d obtained in this way. We note from this figure that at first the value of α increases with increasing Reynolds number, ranging from a value of 0.0114 at $R_d = 10^4$ to 0.0270 at $R_d = 10^6$. However, it tends to a constant value equal to 0.0333 around $R_d = 10^8$.

Heat-Transfer Calculations in Smooth Pipes

To present results for a circular pipe to illustrate the calculation of heat transfer in a fully developed turbulent flow, we write Eq. (7.2) in terms of dimensionless quantities as

$$\bar{u}\left[(1-G)\frac{dT_w}{d\bar{x}} + G\frac{dT_m}{d\bar{x}}\right] = \frac{1}{\bar{r}}\frac{\partial}{\partial \bar{r}}\left[\bar{r}\left(1 + \frac{\Pr}{\Pr_t}\varepsilon_m^+\right)\frac{\partial T}{\partial \bar{r}}\right]. \qquad (7.19)$$

Here

$$\bar{u} = \frac{u}{u_m}, \qquad \bar{x} = \frac{2x/r_0}{R_d\Pr}, \qquad R_d = \frac{u_m d}{\nu}, \qquad \bar{r} = \frac{r}{r_0} \qquad (7.20)$$

with the boundary conditions given by

$$\bar{r} = 0, \qquad \frac{\partial T}{\partial \bar{r}} = 0; \qquad \bar{r} = 1, \qquad T = T_w. \qquad (7.21)$$

Figure 7.2 shows the variation of Nusselt number defined by Eq. (5.23a), that is,

$$\mathrm{Nu} = \frac{\dot{q}_w}{T_w - T_m}\frac{d}{k} \qquad (5.23a)$$

with Reynolds number R_d at three Prandtl numbers, 0.02, 0.72, and 14.3. The calculations were done by using the two-layer eddy-viscosity formula-

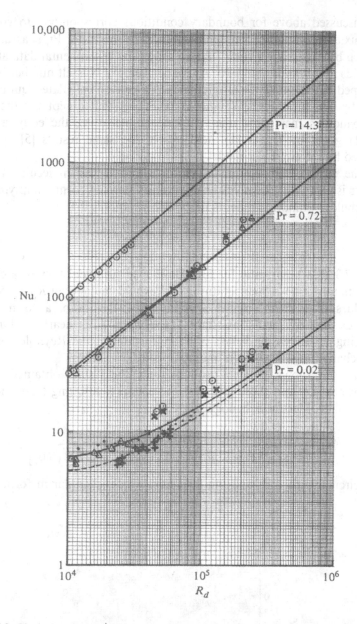

Figure 7.2. Variation of Nusselt number with Reynolds number at three Prandtl numbers in a fully developed flow in a smooth pipe with uniform wall temperature (dashed lines) and heat flux (solid lines). Symbols denote the experimental data: McAdams [2], for $Pr = 0.02$; Lawn [3], for $Pr = 0.72$; Gowen and Smith [4], for $Pr = 14.3$.

tion discussed above for boundary conditions corresponding to constant heat flux and to uniform wall temperature. The model for Pr_t is assumed to be given by Eq. (6.26). Figure 7.2 also shows the experimental data at these Reynolds and Prandtl numbers. As we can see, the Nusselt number in fully developed turbulent flows in circular pipes can be calculated quite accurately by this procedure. We should point out that except for the small discrepancy in mass flow rate, the calculations using the eddy-viscosity formula given by Eq. (7.11) yield nearly the same results [5] as those obtained by the two-layer model.

If the surface temperature is uniform and if the molecular Prandtl number is unity, the Reynolds analogy in its simplest form—implying that the turbulent Prandtl number is unity also—requires that

$$\frac{St}{f/8} = 1. \tag{7.22}$$

Figure 7.3 shows this ratio for fully developed turbulent flow in a circular pipe with uniform wall temperature at a Reynolds number of $R_d = 1.5 \times 10^5$. At values of Prandtl number around unity, this ratio, a form of the Reynolds analogy factor introduced in Chapter 6, is nearly 1, but with increasing or decreasing values of Prandtl number, the Reynolds analogy factor changes significantly from unity.

Figure 7.4 shows a comparison between the results obtained by the two-layer model and those given by the equation resulting from the Karman–Boelter–Martinelli analogy [1],

$$Nu = \frac{R_d Pr\sqrt{f/8}}{0.833\left[5\,Pr + 5\ln(5\,Pr + 1) + 2.5\ln\left(R_d\sqrt{f/8}\,/60\right)\right]}, \tag{7.23}$$

for a circular pipe with uniform heat flux, which is similar in form to Eq.

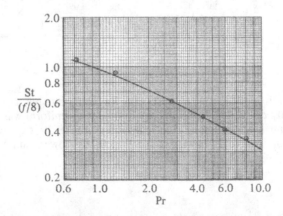

Figure 7.3. Prandtl number influence on Reynolds analogy ratio in a circular pipe with uniform wall temperature at $R_d = 1.5 \times 10^5$. Circles denote experimental data reported in Dipprey and Sabersky [6] and solid line denotes calculations.

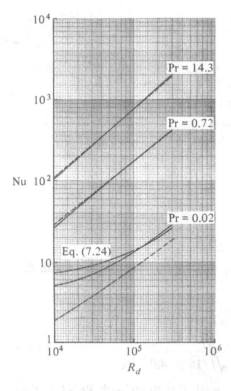

Figure 7.4. Comparison of the predictions from the two-layer model (solid lines), Eq. (7.23), (dashed lines), and Eq. (7.24) for a smooth pipe with uniform wall heat flux.

(6.69) for the Stanton number in a turbulent boundary layer. As we can see, the predictions of Eq. (7.23) agree well with the predictions of the two-layer eddy-viscosity model at Pr = 0.72 and 14.3 but not at Pr = 0.02. This is not surprising because Eq. (7.23) assumes that Pr_t is constant across the layer including the sublayer very close to the surface, and such an assumption, though reasonable at Prandtl numbers near or above unity, is not satisfactory at lower values.

Figure 7.4 also shows the predictions of another empirical formula,

$$\text{Nu} = 6.3 + 0.0167 R_d^{0.85} \text{Pr}^{0.93}, \tag{7.24}$$

proposed by Sleicher and Rouse [7] for low-Prandtl-number turbulent flows in circular pipes with uniform heat flux. As we can see, the predictions of Eq. (7.24) agree better with those of the two-layer eddy-viscosity model at Pr = 0.02 than with Eq. (7.23). The agreement at low Reynolds numbers is clearly not as good as that at high Reynolds numbers.

A similar procedure can be used to calculate the Nusselt number in two-dimensional ducts as discussed in Problem 7.4. Note that to use Eq. (7.9) for two-dimensional duct flows, we set $r_0 = h$.

According to an extensive study conducted by Kader and Yaglom [8], for fully developed turbulent flows with constant surface temperature, the

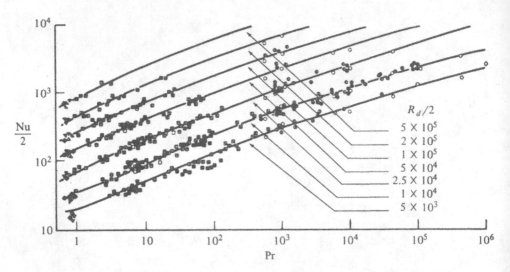

Figure 7.5. Comparison of heat and mass transfer data for pipe flows with Eq. (7.26).

empirical formula

$$Nu = \frac{R_d\sqrt{f}}{8.6\ln\left(R_d\sqrt{f/16}\right)-4.0} \qquad (7.25)$$

fits the existing data on air (Pr = 0.7) well. For fluids with Pr ≥ 0.5, they recommend

$$Nu = \frac{R_d Pr\sqrt{f/2}}{4.24\ln\left(R_d\sqrt{f/16}\right)+25.0\,Pr^{2/3}+4.24\ln Pr-20.2}. \qquad (7.26)$$

Equations (7.25) and (7.26) utilize Prandtl's friction law, Eq. (7.16), to compute the friction factor. Kader and Yaglom have compared Eq. (7.26) with numerous data on heat transfer in smooth pipes. The data covered more than 6 orders of magnitude of Prandtl-number variations (Pr = 1 to 10^6) and 2 orders of Reynolds-number variations ($R_d = 10^4$ to 10^6). In general, Eq. (7.26) fitted the data satisfactorily, although there was some scatter at moderate values of Reynolds number, $R_d < 2 \times 10^4$ (see Fig. 7.5).

Velocity Profile and Friction Factor in Rough Pipes

The calculation of momentum and heat-transfer properties of fully developed turbulent flows in rough pipes can be obtained by following procedures similar to those used for smooth pipes. Again we must start by establishing the velocity profile and friction factor. The friction-factor formulas for such flows are obtained in a manner similar to that discussed for a smooth surface. Again, the variation of friction factor depends on the three distinct flow regimes discussed in Section 6.2. In the hydraulically smooth regime, the variation of f is the same as that for a smooth surface. In the transitional region, f depends on the relative roughness k/d as well as on the Reynolds number R_d. In the fully rough regime, f varies only with relative roughness and is independent of the Reynolds number. Here we describe how the friction factor for the fully rough regime can be obtained and present results for the other regimes.

The law of the wall for a surface with uniform roughness is given by Eq. (6.32). For a fully rough regime with sand-grain roughness, this equation can be written as

$$u = u_\tau \left(\frac{1}{\kappa} \ln \frac{y}{k_s} + 8.5 \right) = u_\tau \left(\frac{2.303}{\kappa} \log \frac{y}{k_s} + 8.5 \right). \qquad (7.27)$$

Evaluating Eq. (7.27) plus the unaltered "wake" contribution $\Pi(x)w(y/r_0)u_\tau$ at the pipe centerline $y = r_0$ and making use of the relations given by Eq. (7.18), we obtain

$$f = \frac{1}{[2\log(r_0/k_s) + 1.74]^2}, \qquad (7.28)$$

a relation of similar form to Eq. (7.10), first obtained by von Karman.

The velocity profile in a rough circular pipe can also be computed, as in the smooth case, by integrating the expression given by Eq. (7.8) in which the eddy viscosity is computed by using the two-layer model. The inner-region eddy viscosity is computed from Eq. (7.10) with the mixing length l given by Eq. (6.102). The outer-region eddy viscosity is computed from Eq. (7.12) with the values of parameter α given in Fig. 7.1. For a given Reynolds number R_d and for a roughness height k, we can calculate the friction factor f and the roughness Reynolds number k^+ to find the Δy from Eq. (6.113). Then we use the same procedure as in the case of the smooth pipe to calculate the velocity profile.

Heat-Transfer Calculations in Rough Pipes

To obtain the variation of Nusselt number in a fully developed rough circular pipe, we again follow the procedure used for the smooth pipe. With f and u computed by the procedure discussed above, we integrate Eq. (7.19) as before. Figures 7.6 and 7.7 show the results obtained in this manner. Figure 7.6 shows the effect of molecular Prandtl number on Nusselt number at different Reynolds numbers in a rough pipe with $r_0/k = 208$ and uniform wall heat flux. All the data, shown by symbols, are due to Dipprey and Sabersky [6] except those for $Pr = 0.72$, which are due to Nunner [9]. We see that for a given Reynolds number, local Nusselt number increases with

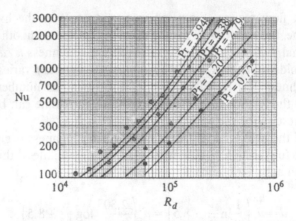

Figure 7.6. Effect of Prandtl number on Nusselt number in circular rough pipe ($r_0/k = 208$) with uniform wall heat flux. Symbols denote experimental data and solid lines the calculations.

increasing Prandtl number and that the solutions agree well with the experimental data.

Figure 7.7 shows the effect of roughness on Nusselt number at different Reynolds numbers in a circular rough pipe with $r_0/k = 60$ and 252 for uniform wall heat flux and $Pr = 0.72$. The experimental data, due to Nunner [9], agree reasonably well with the calculations and indicate that Nusselt number increases significantly with increasing roughness.

Figure 7.8 shows the effect of roughness on Nusselt number at different Reynolds numbers and at three Prandtl numbers. The figure also shows the results for a smooth pipe. We see that the effect of roughness for all three Prandtl numbers is relatively small at lower Reynolds numbers and becomes quite pronounced at higher Reynolds numbers.

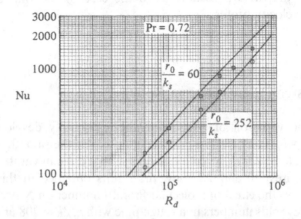

Figure 7.7. Effect of roughness on Nusselt number in a circular rough pipe with uniform wall heat flux at $Pr = 0.72$. Symbols denote experimental data and solid lines the calculations.

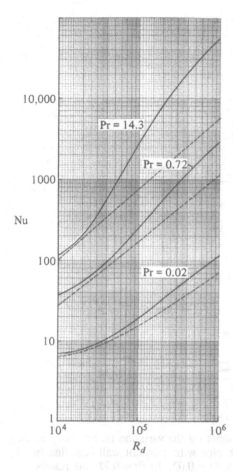

Figure 7.8. Effect of roughness on Nusselt number in a circular rough pipe ($r_0/k_s = 100$) with uniform wall heat flux at different Reynolds numbers and Prandtl numbers. Solid lines denote roughness and dashed lines smooth surfaces.

7.2 Thermal Entry Length for a Fully Developed Velocity Field

The prediction of heat transfer in turbulent flows in constant-area ducts with a fully developed velocity profile can be obtained by following a procedure similar to that described in Section 5.2 for laminar flows. Again we express Eq. (7.1) in terms of dimensionless quantities. For the case of specified wall temperature, we first solve Eqs. (5.38) and (5.39) in terms of transformed variables and then change over to solving Eqs. (5.41) and (5.42) in primitive variables before the thermal boundary layer exceeds the half-width of the duct or the radius of the pipe. The coefficients a_2 and a_3 in Eq. (5.40a) remain unchanged; a_1 now becomes

$$a_1 = (\hat{r})^K \left(1 + \frac{Pr}{Pr_t} \varepsilon_m^+ \right). \tag{7.29}$$

A similar procedure also applies for the case of specified wall heat flux.

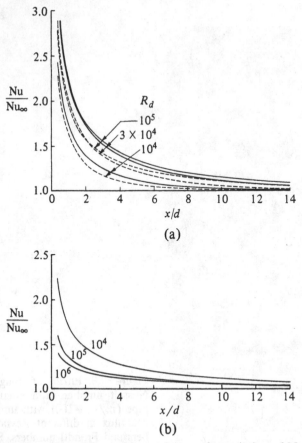

Figure 7.9. (a) The effect of Reynolds number on the variation of Nu/Nu_∞ in the thermal entry region of a circular smooth pipe with constant wall heat flux (solid lines) and wall temperature (dashed lines). $Pr = 0.02$. (b) $Pr = 0.72$; the results are for uniform wall temperature only.

Figures 7.9 through 7.11 show the solutions of Eq. (7.1) for a turbulent flow in a pipe, both with uniform wall temperature and with uniform wall heat flux. The calculations were made by using a modified version of the Fortran program described in Section 13.5. The velocity profile for the smooth wall case was obtained by integrating Eq. (7.13) using the two-layer eddy-viscosity model discussed in Section 7.1. The friction factor was calculated from Eq. (7.17), and as in the fully developed heat-transfer calculations, a variable turbulent Prandtl number given by Eq. (6.26) was used.

Figure 7.9 shows the effect of Reynolds number on the variation of Nu/Nu_∞ in the thermal entry region of a circular smooth duct with different boundary and flow conditions. Figure 7.9(a) shows that thermal entry length decreases with increasing Reynolds number when the pipe is subject to either uniform wall temperature or heat-flux boundary conditions at $Pr = 0.02$. We observe the same behavior for $Pr = 0.72$ in Figure 7.9(b), for the wall boundary condition corresponding to uniform temperature,

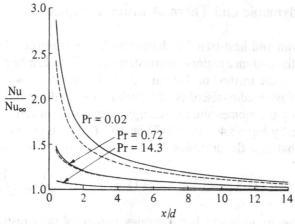

Figure 7.10. The effect of Prandtl number on the variation of Nu/Nu_∞ in the thermal entry region of a circular smooth pipe with uniform wall heat flux (solid lines) and wall temperature (dashed lines) at $R_d = 3 \times 10^5$.

though thermal energy length decreases less. Figure 7.10 shows the effect of Prandtl number on the thermal entry length in a circular smooth pipe at a Reynolds number of $R_d = 3 \times 10^5$. We see that as the Prandtl number increases, the thermal entry length decreases. This applies to both the uniform wall-heat-flux and uniform wall-temperature cases.

Figure 7.11 shows the effect of roughness on the thermal entry length of a circular pipe with uniform heat flux at three Prandtl numbers. These calculations, for $R_d = 3 \times 10^5$ at $k^+ = 93$, show that the roughness effect is rather pronounced and that the thermal entry length of a rough pipe is smaller than that of a smooth pipe, because of the general increase in turbulent mixing.

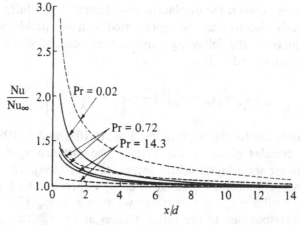

Figure 7.11. The effect of roughness (solid lines) in the variation of Nu/Nu_∞ in the thermal entry region of a circular pipe with uniform wall heat flux at $R_d = 3 \times 10^5$. The dashed lines denote the results for a smooth pipe.

7.3 Hydrodynamic and Thermal Entry Lengths

The momentum and heat-transfer characteristics of transitional and turbulent flows in the entrance regions of constant-area ducts can be predicted by an extension of the method of Section 5.3. As in external flows, this can be done by employing eddy-viscosity and turbulent-Prandtl-number concepts, thus expressing the momentum and energy equations in a form very similar to those given by Eqs. (5.46) and (5.47) or Eqs. (5.53) and (5.54) for laminar flow except that now the definitions of b and e in Eq. (5.50a) become

$$b = (1-t)^{2K}(1+\varepsilon_m^+), \qquad e = (1-t)^{2K}\left(1 + \frac{\text{Pr}}{\text{Pr}_t}\varepsilon_m^+\right)\frac{1}{\text{Pr}}. \quad (7.30)$$

Turbulent flows in ducts, like laminar ones, consist of three distinct regions. Near the entrance the shear layers develop in the region close to the wall and the core velocity increases continuously because of the growth in displacement thickness. The flow in this region, called the *displacement-interaction region* [10], is physically no more than a boundary-layer flow in a mild favorable pressure gradient, and it is known that pressure gradients have much less effect on turbulent boundary layers than on laminar ones. Therefore we can confidently use the eddy-viscosity formulation presented in Eqs. (6.102)–(6.109) in this region, replacing u_e by the centerline velocity u_c.

As the flow moves downstream, the shear layers grow, and finally they meet on the centerline. The interaction between the eddies in the two shear layers—considering a two-dimensional duct—increases in strength until a steady state of mutual eddy intrusion and fine-scale mixing is achieved, where the flow is fully developed.

In the fully developed region, we can use Nikuradse's empirical mixing length expression given by Eq. (7.9), together with Van Driest's damping factor, and represent the eddy viscosity for the entire layer by Eq. (7.10).

For the region between the displacement-interaction and fully-developed regions, the eddy viscosity can be represented with acceptable accuracy in pipes or in ducts by the following empirical expression, which combines Eqs. (6.102), (6.109), and (7.10):

$$\varepsilon_m = \varepsilon_m^I + \left(\varepsilon_m^F - \varepsilon_m^I\right)\left[1 - \exp\left(-\frac{x - x_0}{\lambda r_0}\right)\right]. \quad (7.31)$$

Here ε_m^I denotes the formulas given by Eqs. (6.102) and (6.109), and ε_m^F denotes the formulas given by Eq. (7.10). The distance x_0 denotes the location where the shear layers nearly merge, say $\delta \leq 0.99 r_0$. The parameter λ is an empirical constant taken equal to 20 by Cebeci and Chang [11]. We use ε_m^I for ε_m until $x = x_0$; at this point we switch to Eq. (7.31) with the value of ε_m^I corresponding to the value it takes at $x = x_0$. This simple but arbitrary interpolation function gives good results (for details see Cebeci and Chang [11]).

Sample Calculations

Here we present results for two turbulent flows in the entrance region of a circular smooth pipe. Figure 7.12 allows a comparison of calculated and experimental results for the isothermal experimental data obtained by Barbin and Jones [12] and provides an example of a developing pipe flow that asymptotes to the classical measurements of Laufer [13]. It should be emphasized that the fully developed conditions are achieved asymptotically. At a Reynolds number $R_d = 3.88 \times 10^5$, Barbin and Jones concluded that these conditions had not been achieved at $x/d = 40$. The pressure-drop

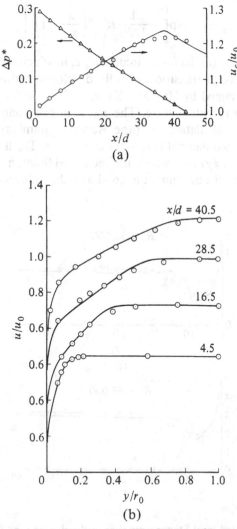

(a)

(b)

Figure 7.12. Comparison of calculated (solid line) and experimental (symbols) results for the pipe flow measured by Barbin and Jones [12]. (a) Centerline velocity and pressure drop. (b) Velocity profile.

results of Fig. 7.12(a), which are admittedly not a sensitive check, would suggest that their flow was, however, very close to the asymptote. The computed nondimensional centerline velocity distribution is also shown in Fig. 7.12(a) and, like the pressure-drop data, is in close agreement with measurements. In the region where the boundary layers fill the pipe, however, the calculations show a more rapid change in shape than do the measurements. This slight discrepancy between calculations and measurements can be attributed to the form of the eddy-viscosity correlation.

In cases where the location of the onset of transition is known, the flow may be calculated with fair accuracy through the transition region and in the fully turbulent region by using the empirical intermittency-factor expression defined by Eq. (6.105), with Eq. (6.106), which for a pipe becomes

$$\gamma_{tr} = 1 - \exp\left[-\frac{1}{1200} R_{x_{tr}}^{0.66}\left(\frac{x}{x_{tr}} - 1\right)^2\right]. \qquad (7.32)$$

Figure 7.13 shows the results for a circular pipe, in which the flow goes from a laminar state through transition to fully developed turbulent flow. Measurements were obtained by Hall and Khan [14] for uniform wall heat flux and for uniform wall temperature. The figure allows comparison between measurements and calculations at pipe Reynolds numbers of 31,000 and 33,000 and for the two thermal boundary conditions. The flow was assumed to begin transition at $x/r_0 = 8$ where the measured Stanton number starts to rise. As can be seen, the agreement is good and the influence of the thermal

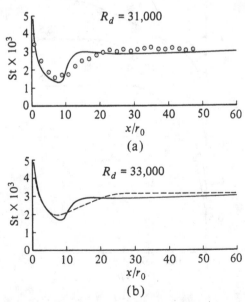

Figure 7.13. Computed local Stanton number in developing pipe flow. Symbols and dotted line are data due to Hall and Khan [14], and the solid line is the calculation of Cebeci and Chang [11]. (a) Constant heat flux. (b) Constant wall temperature.

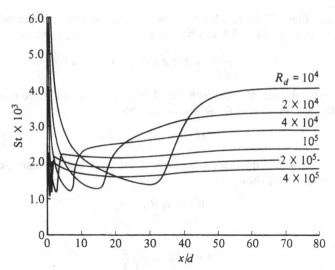

Figure 7.14. Local Stanton number distribution in developing pipe flow as a function of Reynolds number ($Pr = 0.72$ and $R_{\theta_{tr}} = 320$).

boundary condition is much greater in the region of transition (where of course the comparatively crude calculations are not expected to be quantitatively accurate) than in the laminar and turbulent regions. The results of Hall and Khan show that the thermal entry length diminishes with increasing Reynolds number, because transition moves forward, and present calculations support this finding as shown in Figure 7.14.

Problems

The solutions of Problems 7.1 to 7.19 can be obtained without the use of the computer program of Section 13.3. For those wishing to become familiar with the use of computer programs for solving problems of fluid mechanics and heat transfer, we recommend that Problems 7.20 and 7.23 be solved with this program.

7.1 Deduce Eq. (7.7) from Eq. (7.6).

7.2 Use the velocity defect law for a pipe, that is

$$\frac{u_{max} - u}{u_\tau} = \phi\left(\frac{y}{r_0}\right)$$

and the empirically obtained relation that

$$u_m = u_{max} - 4.07u_\tau$$

and show that the friction factor f can be written as

$$u_{max} = u_m\left(1 + 1.44\sqrt{f}\right). \tag{P7.1}$$

7.3 Evaluate Eq. (7.15) at the centerline of the pipe and, with Eq. (P7.1), show that Prandtl's friction law for smooth pipes can be written in the form given by Eq. (7.16).

7.4 Show that Prandtl's friction law for a two-dimensional duct, corresponding to the pipe law in Problem 7.3, can be written, with $(R = u_m h / \nu)$, as

$$\frac{1}{\sqrt{f}} = 0.87 \ln\left(R\sqrt{f} \right) - 0.41$$

7.5 Assume that the heat transfer coefficient $\hat{h}$ in fully developed forced convection in a tube is a function of the following variables

$$\hat{h} = f\left(u_0, \rho, \mu, c_p, k, d \right)$$

where k denotes the thermal conductivity of the fluid. Use the matrix-elimination procedure and show that

$$Nu = \phi\left(R_d, Pr \right)$$

where

$$Nu = \frac{\hat{h} d}{k} = \frac{\dot{q}_w}{\Delta T} \frac{d}{k}.$$

7.6 Integrate Eq. (7.19) with respect to r twice subject to the boundary conditions given by Eq. (7.21) for the case of turbulent flow with constant heat flux condition at the wall. Use the mixing-length formula given by Eqs. (7.9) and (7.10) and the turbulent Prandtl number expression given by Eq. (6.26) for $R_d = 10^4$ and 10^5 with $Pr = 0.02, 0.72$ and 14.3. Compute the friction factor by using Eq. (7.17). Compare your results with those given in Fig. 7.2.

7.7 Repeat problem 7.6 for uniform wall temperature.

7.8 Compare the values of Nusselt number obtained from Eqs. (7.23), (7.24) and (7.26) for the following pipe flows with
   (a) $Pr = 0.7$, $R_d = 10^4$ (Compare Eq. (7.25) also.)
   (b) $Pr = 0.2$, $R_d = 10^4$.
   (c) $Pr = 100$, $R_d = 10^4$.
   (d) $Pr = 0.7$, $R_d = 10^6$.

7.9 Liquid mercury enters a 0.02 m-diameter tube at 50°C and is heated to 95°C as it passes through the tube at a mass flow rate of 2 kg s^{-1}. If a constant heat flux is maintained along the tube and the surface temperature of the tube is 15°C higher than the liquid mercury bulk temperature, calculate the length of the tube.

7.10 Engine oil enters a 0.05 m-diameter tube 10 m long at a temperature of 30°C and at a velocity of 1 m s^{-1}. The tube wall temperature is 10°C higher than the oil bulk temperature and is maintained at 50°C. Calculate the total heat transfer to the oil and the exit temperature of the oil.

7.11 Water at a rate of 1 kg s^{-1} and at an average temperature of 20°C flows in a 0.05 m-diameter tube 15 m long. The pressure drop is measured as 5×10^{-6} Pa s. A uniform heat flux condition is maintained at the wall and the average wall temperature is 50°C and is 10°C higher than the water bulk temperature. Calculate the exit temperature of the water.

7.12 Air at atmospheric pressure and 100°C is heated as it flows through a 0.01 m-diameter tube 15 m long at a velocity of 20 m s^{-1}. A constant wall heat flux condition is maintained and the wall temperature is 7°C above the bulk air temperature. Calculate the heat transfer.

7.13 Repeat Problem 7.6 for a turbulent flow in a pipe with roughness, $r_0/k_s = 100$ for $R_d = 10^5$, Pr = 0.02, 0.72 and 14.3. Compute the friction factor from Eq. (7.28) and compare your results with those given in Fig. 7.8.

Hint: To calculate the velocity profile, use the two-layer eddy-viscosity model in which the inner-region eddy viscosity is obtained from Eq. (7.10) with the mixing length l given by Eq. (6.102), and the outer-region eddy viscosity from Eq. (7.12) with the value of parameter α obtained either in Fig. 7.1 or by making use of Eq. (7.18). Since the establishment of inner and outer regions requires a velocity profile, obtain the initial profile from Eq. (7.13) with the modified mixing-length given by Eqs. (7.9) and (7.14). Then compute the eddy-viscosity distribution from the two-layer model and the velocity profile from Eq. (7.8). Repeat this procedure until the computed velocity profiles converge.

7.14 Use Eq. (7.16) or (7.28) for friction factor f and the results of Fig. 7.8 for Nusselt number to evaluate the Reynolds analogy factor $St/(f/8)$ for smooth and rough ($r_0/k_s = 100$) pipes at $R_d = 10^5$ and Pr = 0.02, 0.72, 14.3.

7.15 Water at a rate of 3 kg s^{-1} and at 40°C enters a 5 cm-diameter pipe with a relative roughness of $k_s/d = 0.005$. If the wall temperature is maintained at 60°C and is 10°C higher than the bulk water temperature, and the pipe is 20 m long, calculate the total heat transfer. Use Reynolds analogy.

7.16 0.5 kg s^{-1} of air at atmospheric pressure and 10°C is heated as it flows through a 2 cm-diameter pipe with a relative roughness of $k_s/d = 0.002$. Calculate the heat transfer per unit length of pipe if a constant-heat-flux condition is maintained at the wall and the wall temperature is 5°C above the average air temperature.

7.17 Air at atmospheric pressure and 10°C flows in a 0.05 m-diameter, 1 m long pipe at a velocity of 10 m s^{-1}. The heating starts at 5 cm from the entrance. A constant heat flux is imposed such that the wall temperature is 5°C above the average air temperature. Calculate the local heat transfer rates at 13, 25 and 45 cm from the entrance of the tube. Assume velocity field is fully developed.

7.18 2 kg s^{-1} of air at atmospheric pressure flows in a 0.1 m-diameter, 1.5 m long pipe. A constant uniform wall temperature of 140°C is applied for 1 m starting 0.5 m from the entrance. The average air temperature in the pipe is 200°C. Calculate
(a) the local heat transfer rates at 0.3, 0.4 and 0.6 m.
(b) the decrease in temperature of the air as it passes through the duct.

7.19 As a crude approximation, the thermal entry length of a turbulent flow in a circular duct with uniform surface temperature and $R_d = 10^4$, Pr = 100 can be estimated from the Leveque analysis discussed in Problem 4.20. To do this, assume that the Leveque solution is valid up to the fully-developed limit, for which the Nusselt number may be obtained from appropriate fully-developed analysis. Put your results in terms of number of diameters from the start of heating.

7.20 Use the computer program of Section 13.3, which provides a solution of Eq. (7.1) for a pipe with specified uniform wall temperature, to obtain solutions for Pr = 14.3, $R_d = 10^5$. Plot
(a) the local Nusselt number variation, and
(b) the temperature profiles at $x/d = 2, 4, 6$ and 8.

7.21 Repeat problem 7.20 for Pr = 0.02.

7.22 Use the computer program of Section 13.3 to obtain solutions for a turbulent flow in a pipe with roughness and with uniform wall temperature. Take $r_0/k_s = 100$, $R_d = 10^5$, Pr = 0.02, 0.72 and 14.3. Compare the temperature profile and the Nusselt number corresponding to fully-developed flow conditions with those computed in Problem 7.13. Again use the two-layer eddy-viscosity model.

7.23 For an incompressible turbulent flow of gases in a circular pipe with constant heat flux, Reynolds, Swearingen and McEligot [15] show that the axial variation of Nusselt number is correlated by

$$\frac{\text{Nu}}{(\text{Nu})_\infty} = 1 + 0.8\left(1 + 70{,}000\,R_d^{-1.5}\right)\left(\frac{x}{d}\right)^{-1} \tag{P7.2}$$

to within ± 5 percent for $x/d \geq 2$ and for a range of Reynolds numbers from 3,000 to 50,000. Here Nu is given by Eq. (5.23a) and $(\text{Nu})_\infty$ denotes the Nusselt number corresponding to fully developed conditions obtained from the empirical formula of Dittus and Boelter,

$$(\text{Nu})_\infty = 0.021\,R_d^{0.8}\text{Pr}^{0.4} \tag{P7.3}$$

Modify the computer program of Section 13.3 and compare the numerical results with those given by Eqs. (P7.2) and (P7.3). Take Pr = 0.72.

References

[1] Schlichting, H.: *Boundary-Layer Theory*,. McGraw-Hill, New York, 1981.

[2] McAdams, W. H.: *Heat Transmission*, 3d ed.. McGraw-Hill, New York, 1954.

[3] Lawn, C. J.: Turbulent heat transfer at low Reynolds number. *J. Heat Transfer* **91**:532 (1969).

[4] Gowan, R. A. and Smith, J. W.: The effect of the Prandtl number on temperature profiles for heat transfer in turbulent pipe flow. *Chem. Eng. Sci.* **22**:1701 (1967).

[5] Habib, I. S. and Na, T. Y.: Prediction of heat transfer in turbulent pipe flow with constant wall temperature. *J. Heat Transfer* **96**:253 (1974).

[6] Dipprey, D. F. and Sabersky, R. M.: Heat and momentum transfer in smooth and rough tubes at various Prandtl numbers. *Int. J. Heat Mass Transfer* **6**:329 (1963).

[7] Sleicher, C. A. and Rouse, M. W.: A convenient correlation for heat transfer to constant and variable property fluids in turbulent pipe flow. *Int. J. Heat Mass Transfer* **18**:677–683 (1975).

[8] Kader, B. A. and Yaglom, A. M.: Heat and mass transfer laws for fully turbulent wall flows. *Int. J. Heat Mass Transfer* **15**:2329–2351 (1972).

[9] Nunner, W.: Heat transfer and pressure drop in rough tubes. *VDI-Forschungschaft* 455, *Ser. B* **22**:5 (1956). Also Atomic Energy Research Establishment Doc. AERE Lib/Trans. 786 (translated by F. Hudswell). Harwell, England, 1958.

[10] Johnston, J. P.: Turbulence, in *Topics in Applied Physics* (P. Bradshaw, ed.), Vol. 12. Springer-Verlag, Berlin, 1978.

[11] Cebeci, T. and Chang, K. C.: A general method for calculating momentum and heat transfer in laminar and turbulent duct flows. *Num. Heat Transfer* **1**:39 (1978).

[12] Barbin, A. R. and Jones, J. B.: Turbulent flow in the inlet region of a smooth pipe. *J. Basic Eng.* **85**:29 (1963).

[13] Laufer, J.: The structure of turbulence in fully developed pipe flow. NACA Rep. No. 1174, 1954.

[14] Hall, W. B. and Khan, S. A.:Experimental investigation into the effect of the thermal boundary condition on heat transfer in the entrance region of a pipe. *J. Mech. Eng. Sci.* **6**:250 (1964).

[15] Reynolds, H. C., Swearingen, T. B., and McEligot, D. M.: Thermal entry for low Reynolds number turbulent flow. *J. Basic Eng.* **91**:87 (1969).

CHAPTER 8

Free Shear Flows

In this chapter we discuss the momentum and heat-transfer properties of uncoupled laminar and turbulent free shear flows, far from solid walls. As in the case of flows over walls, the thin-shear-layer equations admit similarity solutions for some laminar free shear flows, and the corresponding similarity variables can be found by a number of methods. In Section 4.1 we discussed the group-theoretic method. Here in Sections 8.1 and 8.2 we shall use a different approach to find the similarity variables of a two-dimensional laminar jet and a mixing layer between two uniform streams at different temperatures. Later, in Sections 8.3 and 8.4, we shall use the same approach for similar turbulent free shear flows. It should be noted that the similarity solutions become valid only at large distances from the origin because the initial conditions at, for example, a jet nozzle will not match the similarity solution. To illustrate the slow approach to similarity, we shall obtain the solutions of free shear layers for nonsimilar flows and compare them with similarity solutions. It should also be noted that while in practical cases free shear flows are nearly always turbulent, the turbulent-flow solutions are closely related to laminar ones, as we shall see in Sections 8.2 and 8.4. In fact the analysis for free shear flows below is carried out for a general shear stress τ, which can be purely laminar or include a Reynolds (turbulent) shear stress, and for a similarly general heat-flux rate $\dot{q}_y$. This is an illustration of one of the main points of the book; in spite of considerable differences in physical behavior, laminar and turbulent flows can be analyzed and computed in essentially the same way, provided that a model for the Reynolds stresses and turbulent heat-flux rates is available. The present

238

chapter is also a convenient place to introduce coupling between velocity and temperature (density) fields and to discuss the differences and similarities between low-speed and high-speed flows. To do this we take the simplest possible kind of shear layer, the free mixing layer between two parallel streams; although a very special case, it is easier to discuss than the coupled boundary layers and duct flows that we shall meet in later chapters, and Section 8.5 is a convenient introduction to coupled flows for readers with insufficient time to go through the more detailed chapters on coupled flows, Chapters 9 to 12.

8.1 Two-Dimensional Laminar Jet

In this section we discuss a two-dimensional laminar jet that represents a good approximation for many real jets such as high-aspect-ratio rectangular ("slot") nozzles. Since round (axisymmetric) jets are also important, some comments on them are included.

Figure 8.1 shows a two-dimensional heated jet emerging from a slot nozzle and mixing with the surrounding fluid, which is at rest and at another (uniform) temperature. Let the x direction coincide with the jet axis with the origin at the slot. Since the streamlines are nearly parallel within the jet, although the streamlines in the entraining flow are more nearly normal to the axis, the pressure variation in the jet is small and can be neglected. The relevant equations follow from Eqs. (3.5), (3.12), and (3.14) and can be written as

$$\frac{\partial u}{\partial x} + \frac{\partial v}{\partial y} = 0, \tag{8.1}$$

$$u\frac{\partial u}{\partial x} + v\frac{\partial u}{\partial y} = \frac{1}{\rho}\frac{\partial \tau}{\partial y}, \tag{8.2}$$

$$u\frac{\partial T}{\partial x} + v\frac{\partial T}{\partial y} = -\frac{1}{\rho c_p}\frac{\partial \dot{q}}{\partial y}. \tag{8.3}$$

Here in general

$$\tau = \mu\frac{\partial u}{\partial y} - \rho\overline{u'v'}, \tag{8.4}$$

$$\dot{q} \equiv \dot{q}_y = -k\frac{\partial T}{\partial y} + \rho c_p\overline{T'v'}. \tag{8.5}$$

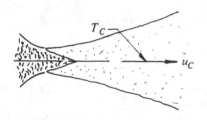

Figure 8.1. The two-dimensional thermal jet.

These equations are subject to the symmetry and boundary conditions

$$y=0, \qquad v=0, \qquad \frac{\partial u}{\partial y}=0, \qquad \frac{\partial T}{\partial y}=0, \qquad (8.6a)$$

$$y=\infty, \qquad u=0, \qquad T=T_e. \qquad (8.6b)$$

Because the pressure is constant in the jet and the motion is steady, the total momentum in the x direction is constant; that is,

$$J=\rho\int_{-\infty}^{\infty} u^2\,dy \equiv 2\rho\int_0^{\infty} u^2\,dy = \text{const.} \qquad (8.7)$$

The heat flux (rate of transport of enthalpy of the mean flow) in the x direction is independent of x and equal to its value at the orifice; that is,

$$K=2\rho c_p\int_0^{\infty} u(T-T_e)\,dy = \text{const}, \qquad (8.8)$$

K being equal to the product of the initial mass flow rate and the mean enthalpy per unit mass.

Similar Flows

To find the similarity solution for the above system, we define dimensionless velocity and temperature ratios by

$$f'(\eta)=\frac{u(x,y)}{u_c(x)}, \qquad (8.9)$$

$$g(\eta)=\frac{T(x,y)-T_e}{T_c(x)-T_e}. \qquad (8.10)$$

Here $u_c(x)$ and $T_c(x)$ denote the velocity and temperature, respectively, along the centerline $y=0$, and η denotes the similarity variable defined by

$$\eta=\frac{y}{\delta(x)}, \qquad (8.11)$$

where δ is the shear-layer thickness, to be defined quantitatively below. We assume, as was done explicitly or implicitly in all the "similar" flows discussed earlier in this book, that the stream function $\psi(x,y)$ is related to a dimensionless stream function $f(\eta)$, independent of x, by

$$\psi(x,y)=u_c(x)\delta(x)f(\eta). \qquad (8.12)$$

Note that since $\psi(x,y)$ has the units (length)^2/time, and since $f(\eta)$ is dimensionless, the product $u_c(x)\delta(x)$ has the same units as ψ. Our interest here is to find the functional form of $\delta(x)$.

Using Eqs. (8.9)–(8.11), we can write Eqs. (8.7) and (8.8) as

$$J=2\rho M\int_0^{\infty}(f')^2\,d\eta, \qquad (8.13)$$

$$K=2\rho c_p N\int_0^{\infty} f'g\,d\eta, \qquad (8.14)$$

where

$$M = u_c^2 \delta, \qquad N = u_c \delta (T_c - T_e).$$ (8.15)

We note that since the total momentum J and the heat flux K are constant, then M and N must be constant, since the integrals in Eqs. (8.13) and (8.14) are pure numbers. By using Eqs. (8.9)–(8.12) and (8.15), together with the chain rule, we can write Eqs. (8.2) and (8.3) as

$$\frac{u_c^2}{2} \frac{d\delta}{dx} \left[(f')^2 + ff'' \right] = -\frac{\tau'}{\rho},$$ (8.16)

$$\delta u_c \frac{dT_c}{dx} (fg)' = -\frac{1}{\rho c_p} (\dot{q})',$$ (8.17)

$$\eta = 0, \qquad f = f'' = 0, \qquad g' = 0,$$ (8.18a)
$$\eta = \eta_e, \qquad f' = 0, \qquad g = 0.$$ (8.18b)

Equations (8.16) and (8.17), with τ and $\dot{q}$ defined by Eqs. (8.4) and (8.5), apply to both laminar and turbulent jets. For laminar flows, using the definitions of τ and $\dot{q}$, we can write them as

$$\frac{u_c \delta}{2\nu} \frac{d\delta}{dx} \left[(f')^2 + ff'' \right] = -f''',$$ (8.19)

$$\frac{1}{\nu} \frac{\delta^2 u_c}{T_c - T_e} \frac{dT_c}{dx} (fg)' = \frac{g''}{\mathrm{Pr}}.$$ (8.20)

Let us first consider Eq. (8.19). We see that for a two-dimensional laminar jet, the momentum equation will have a similarity solution if the coefficient of the left-hand side is independent of x, that is, if

$$\frac{u_c \delta}{2\nu} \frac{d\delta}{dx} = \mathrm{const} = c_1,$$ (8.21)

where we expect $c_1 > 0$. With this restriction and with $c_1 = 1$ (which actually fixes the definition of δ), Eq. (8.19) becomes

$$f''' + (f')^2 + ff'' = 0.$$ (8.22)

Likewise for a heated two-dimensional laminar jet, the energy equation will also have a similarity solution provided that (see Problem 8.3)

$$\frac{\delta^2 u_c}{\nu (T_c - T_e)} \frac{dT_c}{dx} = \mathrm{const} = -1.$$ (8.23)

Eq. (8.20) can be written as

$$g'' + \mathrm{Pr}(fg)' = 0$$ (8.24)

To find the variation of δ with x, so that we can find the similarity variable η from Eq. (8.11), and to find the variation of $u_c(x)$ and $T_c(x)$ with x required for similarity, we use Eqs. (8.21) and (8.23) and the relation given by Eq. (8.15). From Eqs. (8.21) and (8.15) we can write

$$\frac{u_c \delta}{2\nu} \frac{d\delta}{dx} = c_1 = 1 = \frac{\sqrt{\delta M}}{2\nu} \frac{d\delta}{dx} \sim \delta^{1/2} \frac{d\delta}{dx},$$

and upon integration we find, again using the relation between u_c and δ given by the definition of M,

$$\delta = \left(\frac{9\nu^2 x^2}{M}\right)^{1/3} \sim x^{2/3} \quad \text{and} \quad u_c = \left(\frac{M^2}{3\nu x}\right)^{1/3} \sim x^{-1/3}. \quad (8.25a)$$

Similarly, noting the relations given by Eq. (8.25a), we can integrate Eq. (8.23) to get

$$(T_c - T_e) \simeq x^{-1/3}. \quad (8.25b)$$

To find the solution of Eq. (8.22) subject to the boundary conditions given in Eqs. (8.18), we integrate it once to get

$$f'' + ff' = \text{const.} \quad (8.26)$$

It follows from Eq. (8.18a) that the constant of integration is zero. Noting this, integrating Eq. (8.26) once more, and using the boundary condition in Eq. (8.18b) to evaluate the integration constant, we can write the similarity solution for a two-dimensional laminar jet as

$$f = \sqrt{2} \tanh \frac{\eta}{\sqrt{2}}, \quad (8.27)$$

so that

$$f' = \text{sech}^2 \frac{\eta}{\sqrt{2}}. \qu(8.28)$$

Inserting Eq. (8.28) into Eq. (8.13) and integrating, we find

$$M = \frac{3}{4\sqrt{2}} \left(\frac{J}{\rho}\right).$$

Therefore the half-width δ [which we now see to be fixed, our choice of $c_1 = 1$ in Eq. (8.21), as the point where $u/u_c = \text{sech}^2(1/\sqrt{2}) = 0.63$] and the centerline velocity u_c of a two-dimensional similar jet are

$$\delta = \left(\frac{12\sqrt{2}\,\nu^2 x^2}{J/\rho}\right)^{1/3}, \quad (8.29a)$$

$$u_c = \left[\frac{3}{32}\left(\frac{J}{\rho}\right)^2 \frac{1}{\nu x}\right]^{1/3}. \quad (8.29b)$$

The mass flow rate $\dot{m} = 2\rho \int_0^\infty u\, dy$ is

$$\dot{m} = 3.302\rho \left(\frac{J}{\rho}\right)^{1/3} (\nu x)^{1/3}. \quad (8.29c)$$

Similarly the solution of the energy equation (8.24) can be written as

$$g = \text{sech}^{2\,\text{Pr}} \frac{\eta}{\sqrt{2}}. \quad (8.30)$$

Inserting the velocity and temperature profiles given by Eqs. (8.28) and (8.30) into Eq. (8.14), we get

$$K = 2\sqrt{2}\,\rho c_p N \int_0^\infty \frac{d\eta/\sqrt{2}}{\cosh^{2\mathrm{Pr}+2}\left(\eta/\sqrt{2}\right)}. \tag{8.31}$$

This expression can be integrated in closed form when $2\,\mathrm{Pr}+2$ is an integer. For example, when $\mathrm{Pr}=1$, we find that

$$N = \frac{3}{4\sqrt{2}}\frac{K}{\rho c_p} \tag{8.32a}$$

and that the difference between the centerline and the ambient fluid temperature of a two-dimensional laminar jet is

$$T_c - T_e = \frac{3^{2/3}}{4\sqrt{2}}\frac{K}{\rho c_p}\left(\nu x M\right)^{-1/3}. \tag{8.32b}$$

Figure 8.2 shows the similar velocity profile of a two-dimensional jet according to Eq. (8.28), as well as the similar velocity profile of a circular jet (see Problem 8.1). Figure 8.3 shows the corresponding temperature profiles of a two-dimensional jet and a circular jet at different Prandtl numbers. Note that the η scales for the two-dimensional and axisymmetric cases are not directly comparable since they depend on arbitrary choices of δ. Nor can the growth rates be compared because $\delta \approx x$ for a circular jet and $\delta \approx x^{2/3}$ for a two-dimensional jet. For a circular jet, starting with uniform velocity u_0 from an orifice of diameter d_0, $d\delta/dx = 3.27(u_0 d_0/\nu)^{-1}$. If the two-dimensional and axisymmetric profiles were scaled so as to coincide where $u/u_c = 0.5$ (say), it would be seen that the velocity in the axisymmetric jet decreased to zero at large η more slowly than in the two-dimensional

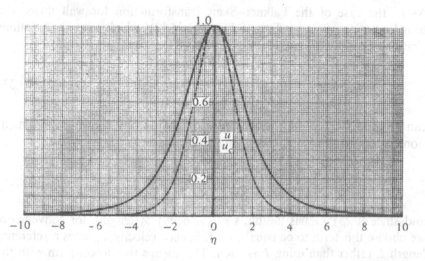

Figure 8.2. Velocity profiles for laminar two-dimensional (dashed lines) and circular (solid lines) jets.

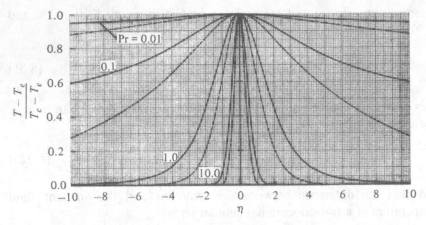

Figure 8.3. Effect of Prandtl number on the temperature profiles of laminar two-dimensional (dashed lines) and circular (solid lines) jets.

jet. In fact, Eq. (8.28) shows that in the two-dimensional jet $f' \equiv u/u_c$ goes to zero as $\exp(-\eta/\sqrt{2})$ at large η, and the solution for an axisymmetric jet (Problem 8.1) shows that $u/u_c \sim \eta^{-4}$ at large η. The temperature profiles in Fig. 8.3 coincide with the velocity profiles if $Pr = 1$, and as usual they become wider if $Pr < 1$ and thinner if $Pr > 1$. The low-Pr temperature profiles for the circular jet are much wider than for the two-dimensional jet, an exaggerated version of the behavior of the velocity profiles.

Nonsimilar Flows

As in the case of the Falkner–Skan transformation for wall flows, the transformation given by Eqs. (8.11) and (8.12), which with the relations given by Eqs. (8.29a) and (8.29b) can be written as

$$\eta \equiv \frac{y}{\delta} = \frac{y}{x^{2/3}} \left(\frac{J/\rho}{12\sqrt{2}\,\nu^2} \right)^{1/3}, \qquad \psi = \left(\frac{9\sqrt{2}}{8} \frac{J}{\rho} \nu x \right)^{1/3} f(\eta), \quad (8.33)$$

can be used for two-dimensional nonsimilar flows by allowing the dimensionless stream function to vary with x. Since

$$u = \frac{\partial \psi}{\partial y} = \frac{\partial \psi}{\partial \eta} \frac{\partial \eta}{\partial y} = \left[\frac{3}{32} \frac{(J/\rho)^2}{\nu} \right]^{1/3} \frac{f'}{x^{1/3}}$$

and since $(J/\rho)^2/\nu$ has the units of (velocity)$^3 \times$(length), for convenience we choose this term to be equal to a reference velocity u_0^3 times a reference length L rather than using J as such. This means that to conform with the form of transformations used for two-dimensional jets, we must replace the constant $(12\sqrt{2})^{1/3}$ in Eq. (8.33) by 3 and the constant $(9\sqrt{2}/8)^{1/3}$ by unity, which is just a change of scaling factors, and rewrite Eq. (8.33) as

$$\eta = \sqrt{\frac{u_0}{\nu L}} \frac{y}{3\xi^{2/3}}, \qquad \xi = \frac{x}{L}, \qquad \psi = \sqrt{u_0 \nu L}\, \xi^{1/3} f(\xi, \eta). \qquad (8.34)$$

Introducing the relations given by Eq. (8.34) into Eqs. (8.2) and (8.3), and allowing g to vary with ξ also, we can write the transformed momentum and energy equations for two-dimensional nonsimilar flows as

$$f''' + (f')^2 + ff'' = 3\xi\left(f'\frac{\partial f'}{\partial \xi} - f''\frac{\partial f}{\partial \xi} \right), \qquad (8.35)$$

$$\frac{1}{\mathrm{Pr}} g'' + (fg)' = 3\xi\left(f'\frac{\partial g}{\partial \xi} - g'\frac{\partial f}{\partial \xi} \right). \qquad (8.36)$$

We see that for similar flows these equations reduce to Eqs. (8.22) and (8.24).

While the solution of the momentum and energy equations for laminar flow can readily be obtained analytically for similar jet flows, solution for nonsimilar flows require the use of numerical methods.

Figure 8.4 shows the variation of the centerline velocity u_c/u_0 with nondimensional streamwise distance ξ for a similar flow far from a two-dimensional nozzle and for a nonsimilar flow with realistic conditions at the nozzle exit, $x = 0$. The latter solutions, which are due to Lai and Simmons [1], are obtained by solving Eq. (8.35) subject to its boundary conditions given in Eq. (8.18) by a laminar-flow version of a Fortran program similar to that described in Section 14.3. Two separate expressions are used for the initial velocity profiles at the exit of the duct. The first one corresponds to a parabolic velocity profile given by Eq. (P5.4) of Problem 5.3, and the second one corresponds to a uniform velocity profile approximated by an expres-

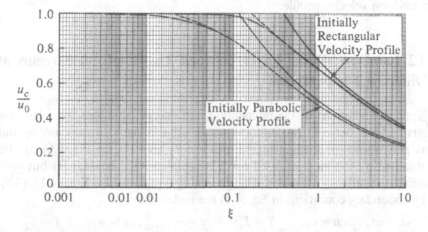

Figure 8.4. Variation of dimensionless jet centerline velocity with dimensionless streamwise distance $\xi(=x/h)$ from nozzle. (——) similarity solution with virtual origin at nozzle exit; (---) similarity solution with virtual origin at $\xi_v = -0.222$ for initially rectangular velocity profile. Here h denotes the channel half-width.

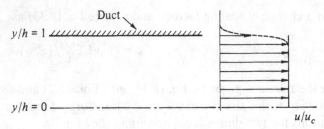

Figure 8.5. Uniform velocity profile (solid line) at the exit of a duct and the fairing (dashed line) according to Eq. (8.37), with $\zeta_c = 1$ and $\beta = 20$.

sion similar to that given by Eq. (4.129),

$$\frac{u}{u_c} = \frac{1}{2}\left[1 - \tanh\beta\left(\frac{y}{h} - \zeta_c\right)\right]. \tag{8.37}$$

Here β and ζ_c are specified constants. Figure 8.5 shows a comparison between the exact and approximate expressions. We see that by choosing $\zeta_c = 1$ and β equal to, say, 20, we can approximate a uniform velocity profile with a thin nozzle-wall boundary layer by Eq. (8.37).

Figure 8.4 also shows the similarity solutions computed according to Eq. (8.29b), with the virtual origin taken at the nozzle exit or computed by plotting the similarity solution for $(u_c/u_0)^{-3}$ against ξ and extrapolating it back to zero ordinate. Lai and Simmons find that the virtual origin is at $\xi \equiv \xi_v = -0.088$ for an initial parabolic velocity profile and at $\xi_v = -0.222$ for a uniform velocity profile. The similarity solution using the virtual origin computed this way indicates that similarity is actually attained earlier than would have been predicted by assuming the virtual origin to be at the nozzle exit. According to the results with adjustment of virtual origin, similarity starts at $\xi = 0.07$ with an initial parabolic velocity profile and at $\xi = 0.3$ with a uniform velocity profile.

8.2 Laminar Mixing Layer between Two Uniform Streams at Different Temperatures

Similarity solutions of the momentum and energy equations for a laminar mixing layer between two uniform streams that move with velocities u_1 and u_2 and whose (uniform) temperatures are T_1 and T_2 (see Fig. 8.6) can be obtained by the methods used for a two-dimensional laminar jet but with different similarity variables. The governing equations are Eqs. (8.1)–(8.5). The boundary conditions in Eq. (8.6) are replaced by

$$y = \infty, \quad u = u_1, \quad T = T_1; \quad y = -\infty, \quad u = u_2, \quad T = T_2. \tag{8.38}$$

Sometimes the velocity of one uniform stream may be zero. If we use the definition of stream function ψ and relate it to a dimensionless stream

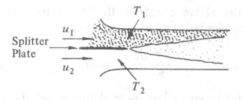

Figure 8.6. The thermal mixing layer.

function f by

$$\psi(x, y) = u_1 \delta(x) f(\eta) \tag{8.39}$$

on the usual argument that ψ must be made dimensionless by (velocity scale)×(flow width), then we can write

$$u = u_1 f', \qquad v = u_1 \frac{d\delta}{dx}(f'\eta - f) \tag{8.40}$$

with $\eta = y/\delta(x)$. Here $y = 0$, defined as the line on which $v = 0$, is *not* in general parallel to the splitter plate dividing the two streams for $x < 0$. The lateral location of the profile is determined by the boundary conditions applied by the external flow. If there is a solid boundary, parallel to the splitter plate, at the upper edge of the high-velocity stream, then $v = 0$ for large positive y (where $f' = 1$), which requires $f = \eta$ for large η. If we now define a dimensionless temperature by

$$g(\eta) = \frac{T - T_2}{T_1 - T_2}, \tag{8.41}$$

then using the definition of η and the definition of dimensionless stream function given by Eq. (8.39), together with the chain rule, we can write the momentum and energy equations and their boundary conditions as

$$u_1^2 \frac{d\delta}{dx} ff'' = -\frac{1}{\rho}\tau', \tag{8.42}$$

$$(T_1 - T_2)\delta u_1 \frac{d\delta}{dx} fg' = \frac{1}{\rho c_p}q', \tag{8.43}$$

$$\eta = \eta_e, \quad f' = 1, \quad g = 1; \quad \eta = -\eta_e, \quad f' = \frac{u_2}{u_1} \equiv \lambda, \quad g = 0, \tag{8.44a}$$

$$\eta = 0, \quad f = 0 \quad \text{or} \quad f' = \tfrac{1}{2}(1 + \lambda). \tag{8.44b}$$

Equations (8.42) and (8.43) with τ and $\dot{q}$ defined by Eqs. (8.4) and (8.5) apply to both laminar and turbulent flows; for laminar flows, they become

$$f''' + \frac{u_1 \delta}{\nu} \frac{d\delta}{dx} ff'' = 0, \tag{8.45}$$

$$g'' + \frac{u_1 \delta}{\nu} \frac{d\delta}{dx} \Pr fg' = 0. \tag{8.46}$$

We see from the above equations that both momentum and energy equations will have similarity solutions provided that

$$\frac{u_1 \delta}{\nu} \frac{d\delta}{dx} = \text{const.}$$

Taking the constant equal to $\frac{1}{2}$ so as to define δ [see the comment on Eq. (8.21)], we see that

$$\delta = \left(\frac{\nu x}{u_1}\right)^{1/2}, \tag{8.47}$$

and Eqs. (8.45) and (8.46) become

$$f''' + \tfrac{1}{2}ff'' = 0. \tag{8.48}$$

$$g'' + \tfrac{1}{2}\Pr fg' = 0. \tag{8.49}$$

No closed-form solutions are known for the system given by Eqs. (8.44), (8.48), and (8.49). Hence the solutions shown in Figs. 8.7 and 8.8 were obtained numerically by the procedure described in Section 14.4 for the centerline boundary condition $f = 0$.

Figure 8.7 shows the velocity profiles for $\lambda \equiv u_2/u_1 = 0$, $\frac{1}{4}$, and $\frac{1}{2}$, showing a decrease in growth rate with increasing λ, mainly on the low-velocity side, and Fig. 8.8 shows the temperature profiles for different values of Prandtl number and for the case $\lambda = 0$. We see that the Prandtl number has a significant effect on the solutions. It is interesting to note that the centerline temperature (at $\eta = 0$), as well as the width, changes with Pr; the reason is that enthalpy conservation requires $\int_{-\infty}^{\infty} u(T - T_2)\,dy$ to be constant and equal to $\int_{0}^{\infty} u_1(T_1 - T_2)\,dy$, which is its value at $x = 0$, so that if the temperature decrease extends further in the positive η direction where u is high, there must be a compensating increase of temperature in the medium u region (regions of low u do not contribute much to the integral).

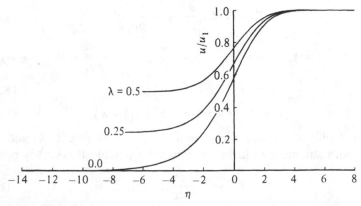

Figure 8.7. Velocity profiles for the mixing of two uniform laminar streams at different velocities, $\lambda = u_2/u_1$.

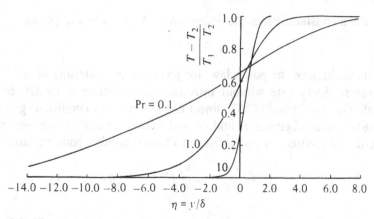

Figure 8.8. The temperature profile of a mixing layer for different Prandtl numbers with $\lambda = 0$. The velocity profile u/u_e is identical with the temperature profile for $Pr = 1$.

8.3 Two-Dimensional Turbulent Jet

Hot turbulent jets appear frequently in engineering practice. Here, as in Section 8.1 for laminar flow, we consider the simplest case of a two-dimensional turbulent hot jet in still, constant-temperature air, and examine the possibility of profile similarity.

Equations (8.16) and (8.17) apply to both laminar and turbulent two-dimensional jets. For turbulent jets the contributions of the laminar shear stress and heat transfer to τ and $\dot{q}$, defined by Eqs. (8.4) and (8.5), respectively, are small, just as they are outside the sublayer in a wall flow, and can be neglected. Assuming that the turbulent shear stress and the heat flux scale on similarity variables, so that

$$\frac{\tau}{\rho} = -\overline{u'v'} = u_c^2 G(\eta),$$

(8.50)

$$-\frac{\dot{q}}{\rho c_p} = -\overline{v'T'} = u_c(T_c - T_e)H(\eta),$$

(8.51)

where $\eta = y/\delta$, we can write Eqs. (8.16) and (8.17) as

$$\frac{1}{2}\frac{d\delta}{dx}\left[(f')^2 + ff''\right] + G' = 0,$$

(8.52)

$$\frac{\delta}{T_c - T_e}\frac{d}{dx}(T_c - T_e)(fg)' - H' = 0.$$

(8.53)

For similarity, the coefficients $d\delta/dx$ and

$$\frac{\delta}{T_c - T_e}\frac{d}{dx}(T_c - T_e)$$

must be constant so that

$$\delta \sim x, \qquad T_c - T_e \sim x^s,$$

(8.54a)

where s is a constant. From the definition of M given in Eq. (8.15),

$$u_c \sim x^{-1/2}. \tag{8.54b}$$

We have obtained the power laws for growth rate, centerline velocity, and temperature decay rate without introducing a turbulence model, but to integrate Eqs. (8.52) and (8.53) subject to the boundary conditions given by Eq. (8.6), relations between f' and G' and between g and H' are needed. If we use the eddy-viscosity and turbulent-Prandtl-number concepts and let

$$\frac{\tau}{\rho} = \varepsilon_m \frac{\partial u}{\partial y} = \varepsilon_m \frac{u_c}{\delta} f'' = u_c^2 G(\eta) \tag{8.55}$$

and

$$-\frac{\dot{q}}{\rho c_p} = \varepsilon_h \frac{\partial T}{\partial y} = \frac{\varepsilon_m}{\mathrm{Pr}_t}(T_c - T_e)\frac{g'}{\delta} = u_e(T_c - T_e)H(\eta) \tag{8.56}$$

and if we assume that it is accurate enough to take ε_m and Pr_t to be independent of η, we can write Eqs. (8.52) and (8.53) as

$$\frac{u_c \delta}{2\varepsilon_m} \frac{d\delta}{dx}\left[(f')^2 + ff''\right] + f''' = 0, \tag{8.57}$$

$$\frac{\mathrm{Pr}_t}{\varepsilon_m} \frac{u_c \delta^2}{T_c - T_e} \frac{d}{dx}(T_c - T_e)(fg)' - g'' = 0. \tag{8.58}$$

If we define δ as the y distance where $u/u_c = \frac{1}{2}$, then experimental data [2] suggest

$$\varepsilon_m = 0.037 u_c \delta. \tag{8.59}$$

If we write the first relation in Eq. (8.54a) as

$$\delta = Ax \tag{8.60a}$$

and use Eq. (8.59), the coefficient in Eq. (8.57) becomes

$$\frac{u_c \delta}{2\varepsilon_m} \frac{d\delta}{dx} = \text{const} = \frac{A}{2(0.037)} = c_1, \tag{8.60b}$$

as required for similar solution of Eq. (8.57), and Eq. (8.57) can be written as

$$f''' + c_1\left[(f')^2 + ff''\right] = 0. \tag{8.61}$$

After integrating it three times and using at first the boundary conditions that at $\eta = \eta_e$, $f' = f'' = 0$ and then the condition that at $\eta = 0$, $f' = 1$, $f = 0$, we find the solution to be

$$f = \sqrt{\frac{2}{c_1}} \tanh\sqrt{\frac{c_1}{2}}\,\eta. \tag{8.62}$$

Requiring that f' ($\equiv u/u_c$) = $\frac{1}{2}$ at $y = \delta$, that is, $\eta = 1$, we find the value of c_1 to be 1.5523. Then it follows from Eq. (8.60b) that $A = 0.115$. As a result, the similarity solution for the dimensionless velocity profile of a two-dimensional turbulent jet can be written as

$$f' = \frac{u}{u_c} = \text{sech}^2 0.881\eta, \qquad (8.63)$$

the dimensionless stream function f can be written as

$$f = 1.135 \tanh 0.881\eta, \qquad (8.64)$$

and Eq. (8.60a) for the width of the jet becomes

$$\delta = 0.115x. \qquad (8.65)$$

As in laminar flows we now insert Eq. (8.63) into Eq. (8.13), and upon integration we get

$$u_c = 2.40\sqrt{\frac{J/\rho}{x}}. \qquad (8.66)$$

The mass flow rate $\dot{m}$ is

$$\dot{m} = 0.625\sqrt{\rho J x}. \qquad (8.67)$$

To obtain the similarity solution of the energy equation (8.58), we denote

$$\frac{\text{Pr}_t}{\varepsilon_m} \frac{u_c \delta^2}{T_c - T_e} \frac{d}{dx}(T_c - T_e) = \text{const} = -C\text{Pr}_t. \qquad (8.68)$$

and we write Eq. (8.58) as

$$g'' + C\text{Pr}_t(fg)' = 0. \qquad (8.69)$$

Letting $C = c_1$ ($\equiv 1.5523$), we integrate Eq. (8.69) to get

$$g' + c_1\text{Pr}_t fg = c_2. \qquad (8.70)$$

Noting that the constant of integration $c_2 = 0$ according to the centerline boundary condition imposed on g, and using the relation for f obtained from Eq. (8.64), we integrate Eq. (8.70) once more to get

$$g = \frac{T - T_e}{T_c - T_e} = \frac{c_3}{[\cosh 0.881\eta]^{2\text{Pr}_t}} = c_3[\text{sech} 0.881\eta]^{2\text{Pr}_t}, \qquad (8.71)$$

where $c_3 = 1$ because $g(0) = 1$.

Clearly, if $\text{Pr}_t = 1$, the velocity profile of Eq. (8.63) and the temperature profile of Eq. (8.71) are identical. The profile shapes are also identical with those given for a laminar jet by Eqs. (8.28) and (8.30) because the eddy viscosity is assumed to be independent of y. Since the eddy viscosity ε_m depends on x, the growth rate is different; the jet width varies linearly with x in turbulent flow and as $x^{2/3}$ in laminar flow.

8.4 Turbulent Mixing Layer between Two Uniform Streams at Different Temperatures

Turbulent mixing layers between two streams of different (but uniform) speed and temperature are also common. In both cases the mass-transfer problem (mixing of one fluid with another) is at least as important as the heat-transfer problem; combustion of a stream of gaseous fuel in a gaseous oxidant involves both heat and mass transfer.

We recall that Eqs. (8.42) and (8.43) apply to both laminar and turbulent mixing layers between two uniform streams whose temperatures are T_1 and T_2. For turbulent flows, the contributions of laminar momentum and heat transfer to τ and $\dot{q}$ are small and can be neglected. As before, if we use the eddy-viscosity, eddy-conductivity, and turbulent-Prandtl-number concepts and let

$$-\overline{u'v'} = \varepsilon_m \frac{\partial u}{\partial y} = \varepsilon_m \frac{u_1}{\delta} f'' \tag{8.72a}$$

and

$$-\overline{T'v'} = \varepsilon_h \frac{\partial T}{\partial y} = \frac{\varepsilon_m}{\mathrm{Pr}_t} \frac{\partial T}{\partial y} = \frac{\varepsilon_m}{\mathrm{Pr}_t} \frac{T_0}{\delta} g'', \tag{8.72b}$$

where $T_0 = T_1 - T_2$, we can write Eqs. (8.45) and (8.46) as

$$f''' + \frac{u_1 \delta}{\varepsilon_m} \frac{d\delta}{dx} ff'' = 0 \tag{8.73}$$

and

$$g'' + \mathrm{Pr}_t \frac{u_1 \delta}{\varepsilon_m} \frac{d\delta}{dx} fg' = 0. \tag{8.74}$$

For similarity of the velocity field, we must have

$$\frac{u_1 \delta}{\varepsilon_m} \frac{d\delta}{dx} = \text{const.} \tag{8.75}$$

With Pr_t also assumed to be a constant, this requirement for similarity then applies to both velocity and temperature fields. If we take the constant in Eq. (8.75) to be $\frac{1}{2}$, as in laminar flows—which implies a specific definition of δ in terms of the velocity profile width—then Eqs. (8.73) and (8.74) become

$$f''' + \tfrac{1}{2}ff'' = 0, \tag{8.76}$$

$$g'' + \tfrac{1}{2}\mathrm{Pr}_t fg' = 0. \tag{8.77}$$

We observe that, as in the case of the jet, these equations, which are subject to the boundary conditions given by Eq. (8.44), are identical to those for laminar flows (Eqs. (8.48) and (8.49)] if we replace Pr_t by Pr. In fact, if we assume Pr_t to be, say, 0.9, then the laminar-flow profile for $\mathrm{Pr} = 0.9$ will be the same as the turbulent-flow profile; as usual, if $Pr_t = 1.0$, the velocity and temperature profiles will be identical.

The difference between the solutions of Eqs. (8.45) and (8.76) is due to the definition of δ. For laminar flows it is given by Eq. (8.47), and for turbulent flows it is given by a solution of Eq. (8.75), which requires an expression for ε_m. Several expressions can be used for this purpose. Here we use the one given by Prandtl. Assuming that $\varepsilon_m \sim \delta$, we expect that ε_m will be determined by the velocity and length scales of the mixing layer:

$$\varepsilon_m = \kappa_1 \delta (u_{max} - u_{min}) = \kappa_1 \delta (u_1 - u_2), \qquad (8.78)$$

where κ_1 is an empirical factor, nominally dependent on y but usually taken as constant.

If we assume that $-\overline{u'v'} = u_1^2 H(\eta)$, then from similarity arguments it follows that δ is proportional to x; in laminar flow it is proportional to $x^{1/2}$. Denoting δ by cx, we can write Eq. (8.78), with $\kappa_1 c = C$, as

$$\varepsilon_m = C x u_1 (1 - \lambda). \qquad (8.79)$$

For uniformity with the existing literature on turbulent mixing layers, we now introduce a parameter σ used by Görtler and defined by him as

$$\sigma = \frac{1}{2} \sqrt{\frac{1 + \lambda}{(1 - \lambda)C}}.$$

This can be written as

$$C = \frac{(1 + \lambda)}{4\sigma^2 (1 - \lambda)}. \qquad (8.80)$$

Substituting Eq. (8.80) into (8.79) and the resulting expression into Eq. (8.75), and taking the constant in Eq. (8.75) to be $\frac{1}{2}$, we get

$$\delta = \frac{x}{\sigma} \sqrt{\frac{1 + \lambda}{8}}. \qquad (8.81)$$

With δ given by this equation, we can now plot the solutions of Eqs. (8.76) and (8.77) in terms of f' ($\equiv u / u_1$) and g' [$\equiv (T - T_2)(T_1 - T_2)$] as a function of

$$\eta = \frac{y}{x} \sigma \sqrt{\frac{8}{1 + \lambda}} \qquad (8.82)$$

for a given value of λ. The Görtler parameter σ, a numerical constant, must be determined empirically. For a turbulent "half jet" (mixing layer in still air) for which $\lambda = 0$, experimental values are mostly between 11 and 13.5. For mixing layers with arbitrary velocity ratios λ, Abramovich [3] and Sabin [4] proposed that

$$\sigma = \sigma_0 \left(\frac{1 + \lambda}{1 - \lambda} \right) \qquad (8.83)$$

for flows with and without pressure gradient. In Eq. (8.83) σ_0 is, of course, the value of σ for the half jet, $\lambda = 0$. This relation was later confirmed by Pui and Gartshore [5] to be a good fit to data.

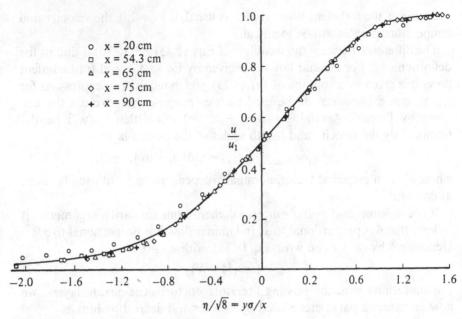

Figure 8.9. A comparison between the numerical solutions of Eq. (8.67) for a turbulent mixing layer (shown by solid line) and the experimental data of Liepmann and Laufer, ref. 6, for a half jet with $\sigma \sim 12$.

Figure 8.9 shows a comparison between the numerical solutions of Eq. (8.76) and the experimental data of Liepmann and Laufer [6] for a half jet, with $\sigma = 12.0$ and taking $y = 0$ where $u/u_1 = 0.5$.

Data for thermal mixing layers are rare but suggest a turbulent Prandtl number of the order of 0.5. This does not necessarily imply that the temperature profile is wider than the velocity profile but merely that the two shapes are different. However, good agreement with experiment near the edges would probably require Pr_t to depend on y.

8.5 Coupled Flows

We have already defined coupled flows as those in which changes in density or viscosity, induced by temperature changes, result in changes in the velocity field, which, of course, control the temperature field. As usual, we can replace "temperature" by "concentration" in qualitative discussion, and the quantitative changes in analysis to deal with concentration changes in fluid mixtures, rather than temperature changes in a homogeneous fluid, are small.

In low-speed flows with heat transfer, density changes are attributable almost entirely to temperature changes because—virtually by definition —pressure changes in low-speed flows are negligibly small compared with the absolute pressure, and it therefore follows from the gas law that pressure changes induce negligibly small changes in absolute temperature. Density variations across the thickness of a shear layer do not have a very large effect on flow behavior, either in laminar flow or in turbulent flow; a change of 10 percent in absolute density will typically produce changes in spreading rate of 1 percent, although actual values vary from flow to flow. The same applies to changes in viscosity; in gases the viscosity μ is very roughly proportional to the absolute temperature, whereas in liquids μ *decreases* rapidly with increase in temperature (while the change of density with temperature is negligible) so that small changes in temperature can have a noticeable effect on laminar flows in liquids.

In high-speed laminar flows, the only additional effect of pressure changes is to produce a density change of the same order; the direct effect of a streamwise pressure gradient in accelerating the flow is likely to be much larger than the indirect effect of the density changes it induces. Therefore, as in low-speed flows, coupling via density or viscosity variations produces effects that have to be taken into account in quantitative calculations but which do not lead to striking changes in physical behavior. An exception is the appearance of shock waves in supersonic flows; this is not a shear-layer phenomenon and the basic equations will not be discussed in this book, but the interaction between shock waves and laminar shear layers is treated in Sections 10.3 and 10.4 with extension to turbulent flow in Section 11.5.

In high-speed turbulent flows, the question arises whether the temperature and pressure *fluctuations* that accompany velocity fluctuations are a large enough fraction of the absolute temperature and pressure to couple with the velocity fluctuations and thus influence the transfer of momentum and enthalpy. It was shown in Chapter 3 that the effect of density fluctuations on the terms in the mean-flow equations is small, but perhaps not negligible, in boundary-layer flows; in fact, it is masked by the effect of mean density variations. The effect of pressure fluctuations seems to be even smaller in nonhypersonic boundary layers (Mach number less than 5) but increases rapidly with Mach number.

It happens that the mixing layer between two parallel streams is an excellent illustration of the two main features of coupled flows: (1) the small, if not negligible, effects of mean-density variations and mean density fluctuations and (2) the large effects of pressure fluctuations at very high Mach numbers, which provide the only serious difference between low-speed and high-speed coupled flows. We therefore discuss the (turbulent) mixing layer in more detail.

Low-Speed Flows

The interpretation of experimental data obtained in free turbulent shear layers with large density differences is made difficult by the influence of initial conditions and the absence of flows that can be regarded as "fully developed." Indeed, in the case of the mixing layer between two streams of different densities and unequal (subsonic) speeds, even the direction of the change in spreading rate with density ratio is uncertain. The measurements covering the widest range of density ratio appear to be those of Brown and Roshko [7], who varied the ratio of low-speed stream density ρ_2 to high-speed stream density ρ_1 between 7 and $\frac{1}{7}$. In the former case the spreading rate was about 0.75 of that of a constant-density mixing layer, and in the latter case it was about 1.35 times as large as in the constant-density case. Other experiments over smaller ranges of density ratio are inconsistent, but it is clear that in most practical cases, such as the mixing of air and gaseous hydrocarbon fuel, the density ratio will be sufficiently near unity for the change in spreading rate to be negligible. Furthermore, the effect of density ratio on the percentage change of spreading rate with velocity ratio u_2/u_1 is also small.

In the case of a jet of one fluid emerging into another fluid of different density, the density ratio has inevitably fallen to a value fairly near unity at the location where the jet has become fully developed (say, $x/d = 20$), and the change of jet spreading rate with density ratio is effectively negligible. The case of low-speed wakes with significant density differences is not of great practical importance except for buoyant flows in the ocean, and there appear to be no data available.

High-Speed Flows

The effect of Mach number on the spreading rate of a mixing layer is extremely large. Most data refer to the case in which the total temperatures of both streams are the same, so that the temperature ratio and density ratio are uniquely related to the Mach numbers of the two streams. In the most common case, the mixing layer between a uniform stream and still air ($u_2 = 0$), the density ratio is given by

$$\frac{\rho_2}{\rho_1} = \frac{T_1}{T_2} = \frac{1}{1+(\gamma-1)M_1^2/2}, \tag{8.84}$$

where M_1 is the Mach number of the uniform stream. The usual measure of spreading rate is the Görtler parameter σ, related to the standard deviation of the "error function" that fits the velocity profiles at all Mach numbers to adequate accuracy. Figure 8.10 shows the data plotted by Birch and Morrisette [8] with a few later additions. Measurements at a Mach number of 19 are reported by Harvey and Hunter [11] and show a spreading parameter σ in the region of 50, which suggests that the trend of σ with Mach number flattens out considerably above the range of the data shown in Fig. 8.10. However, even the data in Fig. 8.10 show considerable scatter, mainly due to the effect of initial conditions (possibly including shock waves in the case where the pressure of the supersonic jet at exit was not adjusted to be accurately atmospheric). In cases where the exit pressure is significantly different from atmospheric pressure, the pattern of shock waves

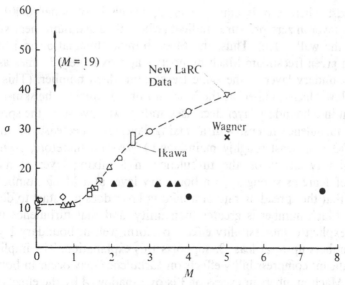

Figure 8.10. Variation of σ with Mach number in single-stream turbulent mixing layers. Symbols □ and X denote data of Ikawa and Kubota [9] and Wagner [10], respectively: for other symbols see [8].

and/or expansions (Fig. 10.9) considerably affects the spreading rate of the mixing layer.

Equation (8.84) implies that the density ratio across a mixing layer at a Mach number of 5 is roughly equal to the factor of 7 investigated in a low-speed flow by Brown and Roshko, who found an increase in spreading rate of about 35 percent compared with the *decrease* of almost a factor of 3 indicated by Fig. 8.10. Clearly, the high Mach number implies an effect of compressibility on the turbulence, as well as on the mean density gradient. Indeed it is easy to show that the Mach-number *fluctuation* in a mixing layer is considerably higher than in a boundary layer at the same mean Mach number. A typical velocity fluctuation can be expressed in terms of the shear stress, so that a representative maximum root-mean-square (rms) velocity fluctuation can be written as

$$\sqrt{\frac{\tau_m}{\rho}},$$

where τ_m is the maximum shear stress within the layer. The square of the speed of sound, a^2, is $\gamma p/\rho$, and we can see that the Mach number based on the above-mentioned representative velocity fluctuation and the local speed of sound can be written in terms of the external stream Mach number and a shear-stress coefficient based on external stream parameters, that is,

$$M_1\sqrt{\frac{\tau_m}{\rho_1 u_1^2}}.$$

The quantity under the square root sign is of order 0.01 in a mixing layer at low speeds, whereas it is equal to $c_f/2$, which is of order 0.001, in a boundary layer in zero pressure gradient (where the maximum shear stress is equal to the wall value). Thus, the Mach-number fluctuation in a mixing layer at a given freestream Mach number is approximately 3 times as large as in a boundary layer at the same freestream Mach number. (This result refers to low Mach number; as the Mach number increases, the skin-friction coefficient in a boundary layer decreases and, as we have seen, the spreading rate and turbulence intensity in a mixing layer also decrease, so that the factor of 3 is at least roughly maintained.) We can, therefore, argue that compressibility effects on the turbulence in a mixing layer at a Mach number of 1 are as strong as in a boundary layer at a Mach number of 3. The fact that the spreading rate of a mixing layer does not start to decrease until the Mach number is greater than unity, and that turbulence models with no explicit compressibility effects perform well in boundary layers at Mach numbers up to at least 3, supports this explanation. The implication that significant compressibility effects on turbulence may occur in boundary layers at Mach numbers in excess of 3 is overshadowed by the effects of the very large heat-transfer rates found in practice at hypersonic speeds and the fact that the viscous sublayer becomes extremely thick in hypersonic boundary layers.

No convincing explanation of the compressibility effects exists. Clearly, pressure fluctuations are in some way responsible because large density differences at low speeds have very little effect. Pressure fluctuations within a turbulent flow are of the order of the density multiplied by the mean-square velocity fluctuation, which, we argued above, is in turn of the same order as the shear stress. In fact the ratio of the maximum shear stress to the absolute pressure is, except for a factor of γ, equal to the square of the Mach-number fluctuation derived above. Since the root mean square of the Mach-number fluctuation in a mixing layer is about 0.1 of the stream Mach number, this suggests that the ratio of the rms pressure fluctuation to the absolute pressure is of order $\frac{1}{4}$ at a Mach number of 5. It is not necessary that these large pressure fluctuations are caused by shock waves, although the latter may well occur, nor is it necessary to suppose that the main reason for the decrease in spreading rate with increase in Mach number is the increasing loss of turbulent kinetic energy by acoustic radiation ("eddy Mach waves"), although the latter may have some effect. It is known that pressure fluctuations play a large part in the generation and destruction of shear stress in turbulent flow, and this effect of pressure fluctuations is certain to alter if the pressure fluctuations become a significant fraction of the absolute pressure. However, this approach to the role of pressure fluctuations does not explain why the spreading rate should *decrease* with increasing Mach number.

In jets and wakes, the Mach number based on the maximum velocity difference between the shear layer and the external flow falls rapidly with increasing distance downstream, and the density ratio returns rapidly toward unity. As in the case of low-speed jets and wakes with significant density differences, it is difficult to establish general effects of compressibility on spreading rate, independent of the initial conditions. There is considerable interest in the wakes of axisymmetric bodies moving at high speeds, with reference to the detection of reentering missiles. In this case, the most important variables are the temperature and the electron density in the partly ionized gas. Wake data for moderate freestream Mach numbers are given by Demetriades [12, 13].

Problems

8.1 As in a two-dimensional jet, the pressure gradient term in the momentum equation for an axisymmetric jet issuing into still air can be neglected. The differential equations for a heated laminar jet follow from Eqs. (3.15) to (3.17) and can be written as

$$\frac{\partial}{\partial x}(ur) + \frac{\partial}{\partial r}(vr) = 0, \tag{P8.1}$$

$$u\frac{\partial u}{\partial x} + v\frac{\partial u}{\partial r} = \frac{\nu}{r}\frac{\partial}{\partial r}\left(r\frac{\partial u}{\partial r}\right), \tag{P8.2}$$

$$u\frac{\partial T}{\partial x} + v\frac{\partial T}{\partial r} = \frac{\nu}{\mathrm{Pr}}\frac{1}{r}\frac{\partial}{\partial r}\left(r\frac{\partial T}{\partial r}\right). \tag{P8.3}$$

The boundary conditions of these equations are

$$r = 0, \qquad v = 0, \qquad \frac{\partial u}{\partial r} = 0, \qquad \frac{\partial T}{\partial r} = 0 \tag{P8.4a}$$

which express symmetry about $r = 0$ and

$$r \to \infty, \qquad u \to 0, \qquad T \to T_e. \tag{P8.4b}$$

In addition to the above equations, the total momentum denoted by J, and the heat flux denoted by K (both in the x-direction) remain constant and are independent of the distance x from the orifice. Hence

$$J = 2\pi\rho \int_0^\infty u^2 r \, dr = \text{const.} \tag{P8.5}$$

$$K = 2\pi\rho c_p \int_0^\infty ur(T - T_e) \, dr = \text{const.} \tag{P8.6}$$

In Eq. (P8.6), K is equal to the product of the initial mass flow rate and mean enthalpy at the orifice.

(a) Using the procedure of Sec. 4.1, show that the similarity variable η and dimensionless stream function for continuity and momentum equations are

$$\eta = \frac{r}{x}, \qquad f(\eta) = \frac{\psi}{x}. \tag{P8.7}$$

Note that the second expression in Eq. (P8.7) is dimensionally incorrect. It can easily be corrected by rewriting it as

$$\psi = \nu x f(\eta). \tag{P8.8}$$

(b) From the definitions of η and stream function,

$$ru = \frac{\partial \psi}{\partial r}, \qquad rv = -\frac{\partial \psi}{\partial x}$$

and from Eq. (P8.8), show that

$$\frac{u}{u_c} = \frac{1}{(u_c x / v)} \frac{f'}{\eta}. \tag{P8.9}$$

Note that the right-hand side of Eq. (P8.9) is independent of x by virtue of

$$u_c x = \text{const.} \tag{P8.10}$$

As a result we can redefine η as

$$\eta = \left(\frac{u_c x}{v} \right)^{1/2} \frac{r}{x}. \tag{P8.11}$$

(c) Use the transformation defined by Eqs. (P8.8) and (P8.11), observe the chain-rule and show that Eqs. (P8.1) to (P8.4) can be written as

$$\left[\eta \left(\frac{f'}{\eta} \right)' \right]' + f \left(\frac{f'}{\eta} \right)' + \frac{(f')^2}{\eta} = 0, \tag{P8.12}$$

$$\left(\frac{\eta}{\mathrm{Pr}} G' + fG \right)' = 0, \tag{P8.13}$$

$$\eta = 0, \quad f = G' = 0, \quad f'' = 0 \tag{P8.14a}$$
$$\eta = \eta_e, \quad f' = G = 0 \tag{P8.14b}$$

where

$$G(\eta) = \frac{T - T_e}{T_c - T_e}.$$

(d) Note that

$$\lim_{\eta \to \infty} \frac{f'(\eta)}{\eta} \to 0 \quad \text{and} \quad \lim_{\eta \to \infty} f''(\eta) \to 0 \tag{P8.15}$$

and show that the solutions of Eq. (P8.12) subject to $f(0) = 0$ are given by

$$f(\eta) = \frac{1/2 \eta^2}{1 + 1/8 \eta^2} \tag{P8.16}$$

and

$$\frac{f'(\eta)}{\eta} = \frac{1}{\left[1 + 1/8 \eta^2 \right]^2}. \tag{P8.17}$$

(e) With $f(\eta)$ given by Eq. (P8.15), show that the solution of Eq. (P8.13) subject to the boundary conditions given by Eq. (P8.14) is

$$G = \frac{1}{\left[1 + \eta^2 / 8 \right]^{2\,\mathrm{Pr}}}. \tag{P8.18}$$

8.2 Compute u_c, δ/h, $\dot{m}/\rho$ for a two-dimensional laminar jet issuing into still air assuming a half duct width, $h = 5$ cm, $v = 1.5 \times 10^{-5}$ m^2 s^{-1} and duct velocity profile to be:

(a) parabolic; that is,

$$u = u_{max}\left[1 - (y/h)^2\right];$$

(b) uniform

$$u = u_{max} = \text{const}.$$

In each case, take $u_{max} = 0.5$ m s^{-1}.

8.3 Show that the center-line temperature excess, $T_c - T_e$, in a heated plane turbulent jet in still air varies as $x^{-1/2}$.

8.4 For laminar mixing of two streams of nearly equal velocity ($u_1 - u_2 \ll u_1$), linearize the momentum equation. Discuss the solution, the velocity profile, the relation of the free shear layer to the external flow, and the shear stress on the dividing streamline. Compare it with the numerical solutions given in Fig. 8.7 for the cases $u_2/u_1 = 0$ and $u_2/u_1 = 0.5$.

8.5 Integrate Eq. (8.31) for very small values of Pr (~ 0) and for Pr = 0.5 and obtain expressions for the difference betwen the centerline and the ambient fluid temperature of a two-dimensional laminar jet, similar to Eq. (8.32b) which was obtained for Pr = 1.0.

References

[1] Lai, J. C. S. and Simmons, J. M.: Numerical solution of periodically pulsed laminar free jets. *AIAA J* **19**:813 (1981).

[2] Schlichting, H.: *Boundary-Layer Theory*. McGraw-Hill, New York, 1981.

[3] Abramovich, G. N.: *The Theory of Turbulent Jets*. M.I.T., Cambridge, MA, 1963.

[4] Sabin, C. M.: An analytical and experimental study of the plane, incompressible, turbulent free-shear layer with arbitrary velocity ratio and pressure gradient. *J. Basic Eng.* **87**:421 (1965).

[5] Pui, N. K. and Gartshore, I. S.: Measurement of the growth rate and structure in plane turbulent mixing layers. *J. Fluid Mech.* **91**:111 (1979).

[6] Liepmann, H. W. and Laufer, J.: Investigations of free turbulent mixing. NACA TN 1257, 1947.

[7] Brown, G. L. and Roshko, A.: On density effects and large structure in turbulent mixing layers. *J. Fluid Mech.* **64**:775 (1974).

[8] Birch, S. F. and Eggers, J. M.: A critical review of the experimental data for developed free turbulent shear layers, in *Free Turbulent Shear Flows*. NASA SP-321, 1972, p. 11.

[9] Ikawa, H. and Kubota, T.: Investigation of supersonic turbulent mixing layer with zero pressure gradient. *AIAA J.* **13**:566 (1975).

[10] Wagner, R. D.: Mean flow and turbulence measurements in a Mach 5 free shear layer. NASA TN D-7366, 1973.

[11] Harvey, W. D. and Hunter, W. W.: Experimental study of a free turbulent shear flow at Mach 19 with electron-beam and conventional probes. NASA TN D-7981, 1975.

[12] Demetriades, A.: Turbulence measurements in a supersonic two-dimensional wake. *Phys. Fluids* **13**:1672 (1970).

[13] Demetriades, A.: Turbulence correlations in a compressible wake. *J. Fluid Mech.* **74**:251 (1976).

CHAPTER 9

Buoyant Flows

Buoyancy effects occur, in principle, in any variable-density flow in a gravitational field. Their importance increases with density gradient and may be characterized by the size of the Richardson number, the simplest form of which is

$$Ri = \frac{\Delta\rho}{\rho}\frac{gh}{u^2},$$ (9.1)

where $\Delta\rho$ is the density difference that occurs over a typical (usually vertical) length scale h in a flow of velocity u. As this Richardson number tends to zero, so does the importance of buoyancy. Equation (9.1) implies that significant buoyancy effects can occur either if $\Delta\rho/\rho$ is significant, as in small-scale plumes, or if $gh/\rho u^2$ is significant, as in large-scale geophysical flows. Flows created entirely by buoyancy forces are referred to as *natural* or *free*-convection flows to distinguish them from *forced*-convection flows, which are driven primarily by imposed pressure differences but can still suffer buoyancy effects. The range of buoyant flows that can occur in nature and in engineering practice is large and has been extensively considered (see Jaluria [1]). Simple examples of natural convection are the two-dimensional plume above a long horizontal heated body, and the two-dimensional boundary layer on a vertical plate heated at its lower edge. These plumes can be regarded as having line sources, as can, for example, the flame front of a forest fire. Similarly an axisymmetric plume in still air can be regarded as having a point source of buoyancy and behaves qualitatively like a round jet.

Many flows are subject to a combination of natural and forced convection. Heated jets or diffusion flames created by blowing combustible gas from a vertical pipe are controlled by forced convection in the initial region and by buoyancy forces far from the jet or pipe exit. Industrial smokestacks usually have a significant imposed momentum flux to assist the initial rise of the contaminant plume, and the effluent from chemical and power plants usually enters a river with sufficient momentum to carry it away from the bank and toward the center of the stream before buoyancy forces carry the contaminant to the surface or bottom of the river. Examples of this type are considered in Sections 9.2 to 9.4, with the simpler cases of natural-convection wall boundary layers and free shear flows in Sections 9.1 and 9.5, respectively.

Buoyancy is of considerable importance in the environment where differences between land (or sea) and air temperatures can give rise to complicated flow patterns and in enclosures such as ventilated and heated rooms or reactor configurations. The calculation of these flows is beyond the scope of this book, but important parts of them (for example, the flow around and above a room-heating radiator panel) can be estimated with the present methods that also provide a starting point for the representation of the more complicated flows. It should be appreciated that, with some arrangements including, for example, the fluids between two horizontal plates with the lower plate at the higher temperature, cellular flow patterns can exist and are particularly difficult to calculate.

The equations which represent natural or free convection in Newtonian fluids in two-dimensional laminar flow stem from those used to represent momentum and heat transfer in general and, with boundary-layer assumptions appropriate in the plume or plate flows just mentioned, are different from the forced-convection equations only insofar as a body force term is included. With the x axis chosen *vertically upward*, the momentum equation with the thin-shear-layer approximation becomes

$$\rho\left(u\frac{\partial u}{\partial x} + v\frac{\partial u}{\partial y}\right) = -\frac{dp}{dx} + \frac{\partial}{\partial y}\left(\mu\frac{\partial u}{\partial y}\right) - \rho g, \qquad (9.2)$$

where the body force f_x appearing in the general form of the equation, Eq. (3.23), is now equal to $-g$, turbulent terms have been neglected, and the flow index k has been set to zero.

The continuity equation is

$$\frac{\partial}{\partial x}(\rho u) + \frac{\partial}{\partial y}(\rho v) = 0, \qquad (9.3)$$

and the energy (enthalpy) equation, Eq. (3.24), simplifies to

$$\rho c_p\left(u\frac{\partial T}{\partial x} + v\frac{\partial T}{\partial y}\right) = \frac{\partial}{\partial y}\left(k\frac{\partial T}{\partial y}\right) \qquad (9.4)$$

on neglecting turbulent terms. In the absence of shear or acceleration of the fluid, Eq. (9.2) reduces to

$$\frac{dp}{dx} = -\rho g, \tag{9.5}$$

and if we denote the density of the unaccelerated (e.g., stationary) fluid outside the heated shear layer by ρ_e and remember the thin-shear-layer approximation that the pressure within the shear layer equals the pressure outside it, we can replace dp/dx in Eq. (9.2) by $-\rho_e g$ and get

$$\rho\left(u\frac{\partial u}{\partial x} + v\frac{\partial u}{\partial y}\right) = \frac{\partial}{\partial y}\left(\mu\frac{\partial u}{\partial y}\right) - g(\rho - \rho_e). \tag{9.6}$$

If the flow outside the shear layer is accelerating, we subtract the so-called hydrostatic pressure gradient $-\rho_e g$ from the actual pressure gradient [so that the whole effect of buoyancy still appears as $-g(\rho - \rho_e)$] and write Eq. (9.6) as

$$\rho\left(u\frac{\partial u}{\partial x} + v\frac{\partial u}{\partial y}\right) = -\frac{dp}{dx} + \frac{\partial}{\partial y}\left(\mu\frac{\partial u}{\partial y}\right) - g(\rho - \rho_e). \tag{9.7}$$

From now on the word "pressure" and the symbol p *exclude* the hydrostatic contribution, so that $dp/dx = 0$ in still fluid.

If the density changes are *small*, we can make the *Boussinesq approximation* and put $\rho = \rho_e$ everywhere in the equations except in the buoyancy term itself, and we can write the buoyancy term as

$$\rho - \rho_e = -\rho_e \beta(T - T_e), \tag{9.8}$$

where T_e is the ambient temperature and β is the coefficient of thermal volumetric expansion of the fluid defined by

$$\beta = -\frac{1}{\rho}\left(\frac{\partial \rho}{\partial T}\right)_p. \tag{9.9}$$

For an ideal gas $\beta = 1/T$. Note that Eq. (9.8) requires only that the pressure within the heated shear layer equals the pressure outside it, as above, not that dp/dx is small. If $T - T_e$ is small, we can also assume $\mu = \mu_e$, $k = k_e$.

Using Eq. (9.8), we can generalize Eq. (9.7) as

$$\rho_e\left(u\frac{\partial u}{\partial x} + v\frac{\partial u}{\partial y}\right) = -\frac{dp}{dx} + \mu_e\frac{\partial^2 u}{\partial y^2} + \rho_e g\beta(T - T_e) + \rho_e\frac{\partial}{\partial y}(-\overline{u'v'})$$

$$\tag{9.10}$$

for turbulent flows; the continuity equation (9.3) reduces to its incompressible flow form, that is,

$$\frac{\partial u}{\partial x} + \frac{\partial v}{\partial y} = 0, \tag{9.11}$$

and the energy equation, Eq. (9.4), becomes

$$u\frac{\partial T}{\partial x} + v\frac{\partial T}{\partial y} = \frac{k_e}{\rho_e c_{p_e}}\frac{\partial^2 T}{\partial y^2} + \frac{\partial}{\partial y}(-\overline{T'v'}).$$

Dropping the subscript e and introducing the definition of Prandtl number, the above equation can also be written as

$$u\frac{\partial T}{\partial x} + v\frac{\partial T}{\partial y} = \frac{v}{\Pr}\frac{\partial^2 T}{\partial y^2} + \frac{\partial}{\partial y}(-\overline{T'v'}). \qquad (9.12)$$

In the *absence* of a pressure gradient, with $v_e \equiv \mu_e/\rho_e$ and denoted by v for convenience, Eq. (9.10) becomes

$$u\frac{\partial u}{\partial x} + v\frac{\partial u}{\partial y} = v\frac{\partial^2 u}{\partial y^2} + g\beta(T - T_e) + \frac{\partial}{\partial y}(-\overline{u'v'}). \qquad (9.13)$$

Let us consider the application of Eqs. (9.11)–(9.13) to the flow up a heated vertical flat plate [Fig. 9.1(a)]. It is easy to observe the dust particles that follow the flow upward along the surface of a heated flat iron held in the vertical position. Far from the surface, the air in the room is still, and there is consequently no freestream velocity; that is, $u_e = 0$. The boundary layer grows from the bottom (leading edge of the flat surface, $x = 0$) to the top (trailing edge, $x = L$), where it separates and forms a plume. The driving force stems from the term $\rho_e g\beta(T - T_e)$, and the absolute velocities therefore increase with $(T_w - T_e)$.

It is easy to envisage that the flow and heat-transfer processes of the previous paragraph can take place in the presence of a flow caused by a pressure gradient. Thus, air can be blown up (or down) the surface and will act with (or against) the buoyancy-driven flow. Thus buoyancy can aid or oppose the forced-convection flow. This process is represented by Eqs. (9.10)–(9.12); in this case a freestream velocity u_e exists, and dp/dx in Eq. (9.10) may be nonzero.

The dimensionless parameters governing the flow can be obtained either by Taylor's formal procedure (Chapter 6) or by inspection. For a typical velocity in the flow we have

$$u = f(L, T_w - T_e, v, \Pr, g\beta), \qquad (9.14)$$

where the dimensions of $g\beta$ can be deduced from Eq. (9.9) as $L/(T^2\theta)$,

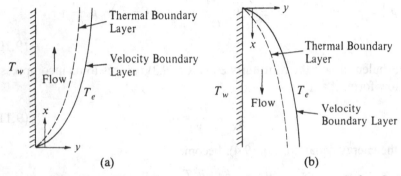

(a)   (b)

Figure 9.1. Coordinate system for natural convection flow over a vertical surface. (a) $T_w > T_e$. (b) $T_w < T_e$: y-coordinate exaggerated.

writing the dimensions of temperature as $[\theta]$. Thus we have

$$\frac{u}{\sqrt{g\beta L(T_w - T_e)}} = f\left[\text{Pr}, \frac{L\sqrt{g\beta L(T_w - T_e)}}{\nu}\right],\tag{9.15}$$

deriving by inspection a quantity

$$u_c \equiv \sqrt{g\beta L(T_w - T_e)}\tag{9.16}$$

with the dimensions of velocity and using it both to nondimensionalize u and to form a "Reynolds number"; the latter is usually presented as its square, the Grashof number

$$\text{Gr}_L = \frac{u_c^2 L^2}{\nu^2} = \frac{g\beta L^3(T_w - T_e)}{\nu^2}.\tag{9.17}$$

In physical terms, the velocity that results from buoyancy stems from the force per unit mass, $g\beta(T_w - T_e)$, which acts over a distance related to L. The resulting kinetic energy per unit mass is proportional to $g\beta L(T_w - T_e)$, and the natural-convection velocity is proportional to u_c in Eq. (9.16).

In combined forced- and free-convection flows where a separate velocity scale, u_e say, is available, the ratio

$$\frac{\sqrt{g\beta L(T_w - T_e)}}{u_e}$$

is a measure of the relative importance of free and forced convection; its square is a form of the Richardson number (see Problem 9.2) and can be written as

$$\text{Ri} = \frac{u_c^2}{u_e^2} = \frac{\text{Gr}_L}{R_L^2}.\tag{9.18}$$

The following sections discuss specific flow configurations where natural convection is important. In most cases the results presented have been obtained by the numerical method and the Fortran program described in Sections 13.4 and 13.5 and in Chapter 14. Most calculations have been done for laminar flow, for simplicity. Similar calculations could be done for turbulent flow, but turbulence models for buoyant flows are less well developed than for the simpler flows discussed in Chapters 6 to 8.

9.1 Natural-Convection Boundary Layers

Let us consider a vertical plate placed in a fluid at rest. If the surface temperature T_w of the plate is greater than the ambient temperature T_e, the fluid adjacent to the surface is heated, becomes lighter, and rises, leading to the formation of velocity and temperature boundary layers as shown in Fig. 9.1a. A similar boundary-layer development occurs, this time with the flow as shown in Fig. 9.1(b), if the wall temperature is less than the ambient temperature.

In this section we shall only discuss the steady-state problem, in which the plate temperature is independent of time. The reader interested in the time-dependent problem is referred to a recent text by Jaluria [1]; it is of considerable practical interest because if heat is transferred to a solid surface from within, starting at $t = 0$, heat transfer to the fluid initially takes place only by conduction, and the large heat transfer rates in buoyant convection take some time to build up, leaving the possibility of transient overheating.

For a steady two-dimensional laminar or turbulent natural-convection flow with zero external velocity and with small temperature differences, the governing equations are given by Eqs. (9.11)–(9.13). These equations, in the absence of mass transfer at the surface, are subject to the following boundary conditions:

$$y = 0, \qquad u = v = 0, \qquad T = T_w \quad \text{or} \quad \left(\frac{\partial T}{\partial y}\right)_w = -\frac{\dot{q}_w}{k}, \qquad (9.19a)$$

$$y = \delta, \qquad u = 0, \qquad T = T_e. \qquad (9.19b)$$

As in conventional (forced-convection) boundary layers discussed in Chapter 4, it is convenient and useful to express the governing equations and their boundary conditions in terms of similarity variables. For laminar buoyant flows, the appropriate similarity variable η and dimensionless stream function f for specified wall temperature (see Problem 9.3) are

$$\eta = \left[\frac{g\beta(T_w - T_e)}{\nu^2 x}\right]^{1/4} y, \qquad \psi = \left[g\beta(T_w - T_e)\nu^2 x^3\right]^{1/4} f(x, \eta). \qquad (9.20)$$

Here we have allowed f to vary with x also, in order to extend the transformation to nonsimilar flows.

As discussed in Chapter 6, we can use eddy-viscosity and turbulent-Prandtl-number concepts to write the momentum and energy equations for turbulent flows with buoyancy effects in the same form as in laminar flows. Using the definitions of ε_m and Pr_t given by Eqs. (6.1) and (6.3) and the transformation given by Eq. (9.20), we can express Eqs. (9.12), (9.13), and (9.19) as

$$(bf'')' + \tfrac{3}{4}ff'' - \tfrac{1}{2}(f')^2 + \theta - \tfrac{1}{2}n\left[(f')^2 - \tfrac{1}{2}ff''\right] = x\left(f'\frac{\partial f'}{\partial x} - f''\frac{\partial f}{\partial x}\right), \qquad (9.21)$$

$$(e\theta')' + \tfrac{3}{4}f\theta' + n\left[\tfrac{1}{4}f\theta' - f'\theta\right] = x\left(f'\frac{\partial \theta}{\partial x} - \theta'\frac{\partial f}{\partial x}\right), \qquad (9.22)$$

$$\eta = 0, \qquad f = f' = 0, \qquad \theta = 1 \qquad (9.23a)$$

$$\eta = \eta_e, \qquad f' = 0, \qquad \theta = 0. \qquad (9.23b)$$

Here primes denote differentiation with respect to η and

$$b = 1 + \varepsilon_m^+, \qquad e = \frac{1}{\text{Pr}} + \frac{\varepsilon_m^+}{\text{Pr}_t}. \qquad (9.24)$$

The parameter θ is a dimensionless temperature defined by

$$T = T_e + T_e(\theta)\phi(x) \qquad (9.25)$$

similar to the g function defined by Eq. (4.25) and $\phi(x)$ is defined by

$$\phi(x) = \frac{T_w - T_e}{T_e}, \qquad (9.26)$$

and n is again a dimensionless temperature-gradient parameter defined by

$$n = \frac{x}{(T_w - T_e)} \frac{d}{dx} (T_w - T_e). \qquad (9.27)$$

In general, n is a function of x in forced-convection boundary layers. For a similar flow it is a constant. For the case of specified wall heat-flux rate, Eq. (9.20) should be modified. A convenient transformation is

$$\eta = \left(\frac{g\beta T_e}{\nu^2 x} \right)^{1/4} y, \qquad \psi = \left(g\beta T_e \nu^2 x^3 \right)^{1/4} f(x, \eta)$$

The resulting equations are slightly different than those defined by Eqs. (9.21) and (9.22).

Equations (9.21) and (9.22), subject to the boundary conditions given by Eq. (9.23) and with a turbulence model where appropriate, can be solved for various values of n and Pr, for both laminar and turbulent natural-convection flows, as discussed below.

Laminar Flow on Vertical Plates

For laminar flows we set the Reynolds-shear-stress and heat-flux terms equal to zero. This amounts to redefining b and e in Eq. (9.24) as

$$b = 1, \qquad e = \frac{1}{\text{Pr}}. \qquad (9.28)$$

In the same manner in which the requirements for similarity of laminar boundary layers were established for forced-convection flows, it can be shown that the laminar natural-convection boundary layer on a vertical plate will be similar if

$$T_w - T_e = c_2 x^n \qquad (9.29)$$

for specified wall temperature with c_2 and n as constants. The similarity equations follow from Eqs. (9.21) and (9.22) and, with boundary conditions still being given by Eqs. (9.23), can be written as

$$f''' + \tfrac{3}{4}ff'' - \tfrac{1}{2}(f')^2 + \theta - \tfrac{1}{2}n\left[(f')^2 - \tfrac{1}{2}ff''\right] = 0, \tag{9.30}$$

$$\theta'' + \Pr\left(\tfrac{3}{4}f\theta' - nf'\theta + \tfrac{1}{4}nf\theta'\right) = 0. \tag{9.31}$$

Equations (9.30) and (9.31), subject to the boundary conditions given by Eqs. (9.23), have been solved for various values of n and Pr. Here we first consider the case of specified wall temperature only and present solutions obtained with the numerical method described in Chapter 13.

The local Nusselt number Nu_x is defined by

$$\mathrm{Nu}_x = \frac{\dot{q}_w}{T_w - T_e}\frac{x}{k} = -\theta'_w \mathrm{Gr}_x^{1/4}. \tag{9.32}$$

Here Gr_x is a local Grashof number based on the local convective velocity scale $u_c(x)$,

$$\mathrm{Gr}_x = \frac{u_c^2(x)x^2}{\nu^2} \equiv \frac{g\beta x^3(T_w - T_e)}{\nu^2}, \tag{9.33}$$

where

$$u_c(x) = \sqrt{g\beta x(T_w - T_e)}. \tag{9.34}$$

Similarly the local skin friction can be written, in terms of primitive variables and transformed variables, respectively, as

$$c_f = \frac{\tau_w}{\tfrac{1}{2}\rho u_c^2} = \frac{2f''_w}{\sqrt{\mathrm{Gr}_x}}. \tag{9.35}$$

Table 9.1 presents the solutions of the momentum and energy equations for uniform surface temperature ($n = 0$) at different Prandtl numbers. As we see, Prandtl number has a pronounced effect on the heat-transfer parameter θ'_w and on the wall shear parameter f''_w. The results of Table 9.1 can be

Table 9.1 Solutions of Eqs. (9.30), (9.31), and (9.23) for uniform wall temperature

Pr	$-\theta'_w = \dfrac{\mathrm{Nu}_x}{\mathrm{Gr}_x^{1/4}}$	$f''_w = \dfrac{c_f}{2}\sqrt{\mathrm{Gr}_x}$
0.1	0.1637	1.2104
0.72	0.3567	0.9558
1.0	0.4009	0.9081
10	0.8266	0.5930
100	1.5495	0.3564

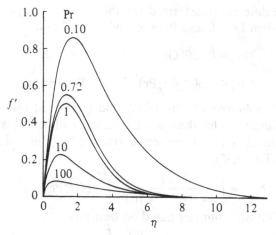

Figure 9.2. Velocity profiles for a natural convection boundary layer on a vertical plate with uniform wall temperature.

satisfactorily correlated by the following equation given by Ede [2]:

$$\text{Nu}_x = \frac{3}{4}\left[\frac{2\,\text{Pr}}{5(1+2\,\text{Pr}^{1/2}+2\,\text{Pr})}\right]^{1/4}(\text{Gr}_x\,\text{Pr})^{1/4}. \tag{9.36}$$

Figures 9.2 and 9.3 show the dimensionless velocity and temperature profiles. In contrast to forced convection in uncoupled flows where the velocity profiles are independent of Prandtl number, the velocity profiles in natural convection vary significantly as Prandtl number varies. The temperature profiles are strong functions of the Prandtl number in both cases.

When the Prandtl number of the fluid becomes very small (say less than 0.1) or very large (say greater than 100), the numerical solutions of Eqs. (9.30) and (9.31) gradually become more difficult, as in forced-convection problems. In these cases, the similarity equations (9.30) and (9.31) need to

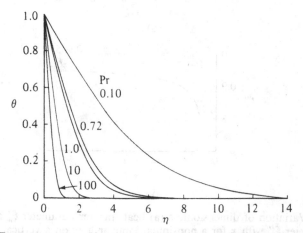

Figure 9.3. Temperature profiles for a natural convection boundary layer on a vertical plate with uniform wall temperature.

be expressed in different transformed variables. Le Fevre [3] considered the extreme cases when $\text{Pr} \to 0$ and $\text{Pr} \to \infty$ and showed that

$$\text{Nu}_x = 0.600\left(\text{Gr}_x\text{Pr}^2\right)^{1/4}, \qquad \text{Pr} \to 0, \tag{9.37a}$$

$$\text{Nu}_x = 0.503\left(\text{Gr}_x\text{Pr}\right)^{1/4}, \qquad \text{Pr} \to \infty. \tag{9.37b}$$

The similarity solutions of Eqs. (9.30) and (9.31) can also be obtained with slight modifications for the case of specified wall heat flux. The results for the local Nusselt number have been correlated by Fujii and Fujii [4] in a form similar to Eq. (9.36),

$$\text{Nu}_x = \left(\frac{\text{Pr}}{4+9\,\text{Pr}^{1/2}+10\,\text{Pr}}\right)^{1/5}\left(\text{Gr}_x^*\text{Pr}\right)^{1/5}, \tag{9.38}$$

where Gr_x^* is a Grashof number based on heat flux,

$$\text{Gr}_x^* = \frac{g\beta\dot{q}_w x^4}{k\nu^2}. \tag{9.39}$$

For general variations of $T_w - T_e$ with x, we need to solve the nonsimilar form of the equations given by Eqs. (9.21)–(9.23). Results are presented here for unity Prandtl number and for the following variation of n:

$$n = \begin{cases} 1 & 0 \leq \dfrac{x}{L} \leq 1, \text{ that is } T_w - T_e \sim x, \\[2mm] 0 & \dfrac{x}{L} > 1, \text{ that is, } T_w - T_e = \text{const.} \end{cases} \tag{9.40}$$

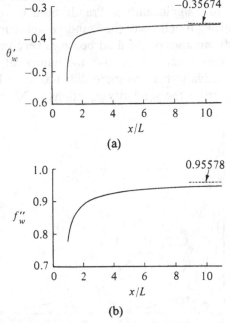

(a)

(b)

Figure 9.4. Variation of dimensionless (a) heat transfer parameter θ'_w and (b) wall shear parameter f''_w with x for a nonsimilar laminar flow on a vertical plate whose temperature distribution is given by Eq. (9.40) and for which $\text{Pr}=1$.

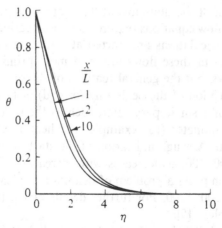

Figure 9.5. Temperature profiles for a nonsimilar laminar flow on a vertical plate whose temperature distribution is given by Eq. (9.40) and for which $Pr = 1$.

Since the flow is similar in the range $0 \leq x/L \leq 1$ and becomes nonsimilar for $x/L > 1$, we can start the calculations at $x = 0$ or at $x = L$ using the similarity solution for $n = 1$ to provide the initial condition (and the results for $0 < x/L \leq 1$) in the latter case. For $x/L > 1$, small increments in x/L ($\Delta x/L = 0.01$) were taken at first, and the solution procedure was continued to $x/L = 2$. For values of $x/L > 2$, larger increments in x/L were taken. The results presented in Fig. 9.4 show that the dimensionless heat transfer θ_w' ($\equiv -Nu_x/Gr_x^{1/4}$) approaches its similarity value of -0.35674 a little faster than the dimensionless wall shear parameter f_w'' ($\equiv c_f/2\sqrt{Gr_x}$) does. In either case, however, the length required to reach similarity values is more than $x/L = 10$. Figure 9.5 shows the temperature profiles at different x/L values.

Turbulent Flow on Vertical Plates

Turbulent natural-convection boundary layers can also be calculated by using the eddy-viscosity and turbulent-Prandtl-number concepts used for nonbuoyant flows. Again the Cebeci–Smith eddy-viscosity formulation Eqs. (6.102) and (6.109) and the turbulent-Prandtl-number expressions given by Eq. (6.26) can be used for this purpose, at least for demonstration calculations. Because of the difficulty of defining an outer eddy-viscosity expression based on the velocity defect, however, it is more appropriate to use an outer eddy-viscosity expression given by

$$(\varepsilon_m) = (0.075\delta)^2 \left| \frac{\partial u}{\partial y} \right| \tag{9.41}$$

corresponding to $l = 0.075\delta$, a typical value used in nonbuoyant boundary-layer problems.

As in the solution of the nonbuoyant-flow equations, the solution of the natural-convection-flow equations requires the specification of the transition location when the calculations are started at a leading edge. Methods for predicting transition in these flows are in a more primitive state than in unheated shear flows, but the general features of the process are similar. It appears that the location of the beginning (or end) of transition cannot be specified in terms of a single parameter, such as the Grashof number, but requires another parameter (for example, the heat flux across the solid surface) to specify it. A rough indication of the start of transition in air or water is $Gr_x^{1/4} = 500$–600; it decreases with increasing heat flux $\dot{q}_w$ in air. The end of transition poses a problem of definition, typical values for $Gr_x^{1/4}$ being in the range 900–1300. For further details of the transition process, see Jaluria and Gebhart [5].

Figures 9.6 through 9.8 show the results for turbulent natural-convection flow over a vertical flat plate with uniform wall temperature. Calculations were made by Cebeci and Khattab [6] by solving equations very similar to

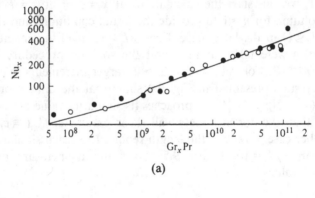

(a)

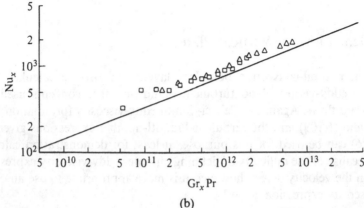

(b)

Figure 9.6. Comparison of calculated values of local Nusselt number defined by Eq. (9.32) for a laminar and turbulent natural convection boundary-layer flow on a vertical plate for air and oil. Solid line shows the calculated values and the symbols denote the experimental data, ref. 7. (a) Air, Pr = 0.72. (b) Oil, Pr = 58.7. Data from [8].

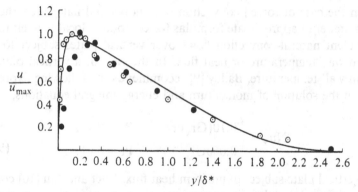

Figure 9.7. Comparison of calculated (solid line) and experimental (open and solid symbols denote different data) velocity profiles for a turbulent natural convection boundary-layer flow on a vertical flat plate, $Pr = 0.72$.

those given by Eqs. (9.21) and (9.22) and by using the turbulence model given by Eqs. (6.26), (6.102), and (9.41). The calculations were started as laminar at the leading edge and were made turbulent at a specified local Rayleigh number, $Ra_x = Gr_x Pr$. In Fig. 9.6(a) values of local Nusselt number for both laminar and turbulent flows are presented for air ($Pr = 0.72$) and with transition specified at $Ra_x = 6 \times 10^8$. The results in Fig. 9.6(b) correspond to oil with a Prandtl number of 58.7 and have a trend similar to those of Fig. 9.6(a); the variation of Nusselt number is reasonably well predicted with this turbulence model. Velocity and temperature profiles are presented in Figs. 9.7 and 9.8 with the latter showing a significant variation with Prandtl number.

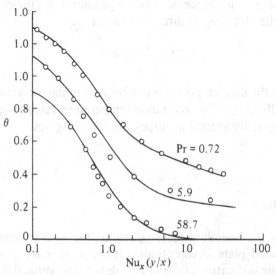

Figure 9.8. Comparison of calculated (solid lines) and experimental (symbols) temperature profiles for a turbulent natural convection boundary-layer flow on a vertical flat plate.

As in the case of forced convection over horizontal flat plates, there are also a number of approximate formulas for computing local Nusselt number in turbulent natural-convection flows over vertical plates subject to either uniform wall temperature or heat flux. In the case of a vertical plate with uniform wall temperature, Bailey [9] recommends the following expressions based on the solution of momentum and energy integral equations,

$$\mathrm{Nu}_x = \begin{cases} 0.10(\mathrm{Gr}_x\mathrm{Pr})^{1/3} & \mathrm{Pr} = 0.72, \qquad (9.42a) \\ 0.060\mathrm{Gr}_x^{1/4} & \mathrm{Pr} = 0.01. \qquad (9.42b) \end{cases}$$

For a vertical plate subject to uniform heat flux, Vliet and Liu [10] correlate the experimental data by the expression

$$\mathrm{Nu}_x = 0.568(\mathrm{Gr}_x^*\mathrm{Pr})^{0.22} \qquad (9.43)$$

over a $\mathrm{Gr}_x^*\mathrm{Pr}$ range of 2×10^{13} to 10^{16}, where Gr_x^* is given by Eq. (9.39).

An important consideration, especially for gases, is that of fluid property variation. A study conducted by Sparrow and Gregg [11] showed that for gases, where β may be taken as $1/T_e$, the constant-property solutions can satisfactorily be corrected by using a *reference temperature* T_r given by

$$T_r = T_w - 0.38(T_w - T_e). \qquad (9.44)$$

For liquid metals they again suggested that all properties be evaluated at T_r. It was also found that the "film temperature" T_f given by

$$T_f = \frac{T_w + T_e}{2}$$

is an adequate reference temperature for most cases. The relationship given by Eq. (9.44) yields a difference of only 0.6 percent, from the constant-property results at the film temperature, over the range

$$0.25 \le \frac{T_w}{T_e} \le 4.0. \qquad (9.45)$$

For this reason, the film temperature can be used in most cases of significant temperature differences. The reference temperature relation given by Eq. (9.44) may be used for greater accuracy, mainly for gases.

Inclined Plate

The solution of the boundary-layer equations for natural convection from an inclined heated plate is essentially similar to that on a vertical plate provided that the inclination angle of the plate to the vertical, γ, is less than about 45°. The momentum equation corresponding to Eq. (9.10), with the Boussinesq assumption and boundary-layer approximations and with x

measured along the plate, can be written as

$$u\frac{\partial u}{\partial x} + v\frac{\partial u}{\partial y} = g\beta(T - T_e)\cos\gamma - \frac{1}{\rho_e}\frac{\partial p}{\partial x} + \nu\frac{\partial^2 u}{\partial y^2} - \frac{\partial}{\partial y}(\overline{u'v'}), \quad (9.46)$$

and the y-component equation is, to the same approximation,

$$g\beta(T - T_e)\sin\gamma = \frac{1}{\rho_e}\frac{\partial p}{\partial y}. \quad (9.47)$$

The continuity and energy equations remain the same as Eqs. (9.11) and (9.12). It is clear from Eq. (9.46) that except for the modification of the buoyancy term due to the inclination and the retention of the (full) pressure gradient term, the resulting x-momentum equation is the same as Eq. (9.13). The pressure gradient in the x direction is retained here because of its importance in horizontal flows and will be discussed later. Equation (9.47) describes the balance between the pressure gradient in the y direction and the buoyancy force.

For small angles γ, i.e., a near-vertical plate, the problem and the solution are the same as those of the vertical plate except for the replacement of g by $g\cos\gamma$. This implies using $Gr_x\cos\gamma$ instead of Gr_x. It also implies equal rates of heat transfer on both sides of the surface. Jaluria [1] reports that the empirical formula for laminar flows on vertical plates,

$$Nu_x = 0.55(Gr_x^*Pr)^{0.2}, \quad (9.48)$$

can be used by replacing Gr_x^* with $Gr_x^*\cos\gamma$ for both upward- and downward-facing inclined surfaces with uniform wall heat flux. For turbulent flows, the empirical equation

$$Nu_x = 0.17(Gr_x^*Pr)^{0.25} \quad (9.49)$$

is suggested, with Gr_x^* the same as that for a vertical surface for the upward-facing heated surface and with Gr_x^* replaced by $Gr_x^*\cos^2\gamma$ for the downward-facing surface. For additional correlations for surfaces with various inclination angles, the reader is referred to Fujii and Imura [12].

Horizontal Plate

For a semi-infinite horizontal plate with a horizontal external stream (see Fig. 9.9), the momentum equations are

$$u\frac{\partial u}{\partial x} + v\frac{\partial u}{\partial y} = -\frac{1}{\rho}\frac{\partial p}{\partial x} + \nu\frac{\partial^2 u}{\partial y^2} - \frac{\partial}{\partial y}\overline{u'v'}, \quad (9.50)$$

$$g\beta(T - T_e) = \frac{1}{\rho}\frac{\partial p}{\partial y}, \quad (9.51)$$

where the x coordinate is measured along the horizontal plate and the buoyancy force is normal to the flow direction. Physically, the upper surface

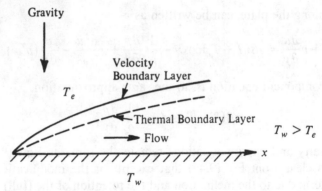

Figure 9.9. Coordinate system for natural convection boundary-layer flow on a semi-infinite horizontal surface.

of the plate surface heats up the fluid adjacent to it, which, being lighter, tends to rise. This results in a pressure gradient normal to the plate and tends to thicken the boundary layer. Similar considerations apply for the lower side of a cooled plate.

Even in still air, a boundary layer can still form on a horizontal semi-infinite plate as a consequence of buoyancy. In this case, the laminar-flow equation may be reduced to a similarity form (Stewartson [13]). We write Eqs. (9.50) and (9.51) with $\bar{p} \equiv (p - p_e)/\rho$ as

$$u\frac{\partial u}{\partial x} + v\frac{\partial u}{\partial y} = -\frac{\partial \bar{p}}{\partial x} + v\frac{\partial^2 u}{\partial y^2}, \tag{9.52}$$

$$g\beta(T - T_e) = \frac{\partial \bar{p}}{\partial y} \tag{9.53}$$

and introduce the following dimensionless quantities and similarity variables

$$\xi = \frac{x}{L}, \qquad \frac{T - T_e}{T_w - T_e} = H(\eta),$$

$$\eta = \mathrm{Gr}_L^{1/5}\left(\frac{y}{L}\right)\frac{1}{\xi^{2/5}}, \qquad \psi = \xi^{3/5}f(\eta), \qquad \bar{p} = \frac{v^2}{L^2}\mathrm{Gr}_L^{4/5}\xi^{2/5}G(\eta).$$
$$\tag{9.54}$$

Using the definition of stream function, it then follows that

$$u = \frac{v}{L}\mathrm{Gr}_L^{2/5}\xi^{1/5}f', \tag{9.55a}$$

$$v = -\frac{1}{5}\frac{v}{L}\frac{\mathrm{Gr}_L^{1/5}}{\xi^{2/5}}(3f - 2\eta f'). \tag{9.55b}$$

Note that here v/L has the dimensions of velocity and can be taken as a reference velocity.

In terms of the similarity variables defined in Eqs. (9.54), the momentum and energy equations, namely, Eqs. (9.12), (9.52), and (9.53), and their boundary conditions can be written as

$$f''' + \tfrac{3}{5}ff'' - \tfrac{1}{5}(f')^2 - \tfrac{2}{5}(G - \eta G') = 0,$$ (9.56)

$$G' = H,$$ (9.57)

$$H'' + \tfrac{3}{5}\Pr fH' = 0,$$ (9.58)

$$\eta = 0, \qquad f = f' = 0, \qquad H = 1,$$ (9.59a)

$$\eta = \eta_e, \qquad f = 0, \qquad G = H = 0.$$ (9.59b)

Rotem and Claassen [14] have solved these equations and the principal results are shown in Table 9.2. The Nusselt number for this solution is

$$\mathrm{Nu} = \frac{\dot{g}_w}{T_w - T_e}\frac{L}{k} = -\frac{\mathrm{Gr}_L^{1/5}}{\xi^{2/5}}H'(0).$$ (9.60)

As might be expected, this natural-convection boundary layer is very unstable and, at higher Grashof numbers, can give rise to a cellular pattern of longitudinal (x-wise) vortices, as also found for nonzero but small plate angles. On a plate of finite length, boundary layers can form from each of the two edges and merge into a vertical plume where they meet in the middle of the plate.

For horizontal plates with finite length, several empirical formulas are summarized by Jaluria [1]. For heated surfaces facing downward or cooled surfaces facing upward, the correlation due to McAdams [15] is

$$\overline{\mathrm{Nu}} = 0.27(\mathrm{Gr}_L\Pr)^{1/4}, \qquad 3\times10^5 < \mathrm{Gr}_L\Pr < 3\times10^{10}.$$ (9.61)

Fujii and Imura [12] give the corresponding correlation as

$$\overline{\mathrm{Nu}} = 0.58(\mathrm{Gr}_L\Pr)^{1/5}, \qquad 10^6 < \mathrm{Gr}_L\Pr < 10^{11}.$$ (9.62)

Here $\overline{\mathrm{Nu}}$ denotes an average Nusselt number defined by

$$\overline{\mathrm{Nu}} = \frac{1}{L}\int_0^L \mathrm{Nu}_x\,dx.$$ (9.63)

Table 9.2 Similarity solutions of Eqs. (9.56)–(9.59)

Pr	$f(\eta_e)$	$f''(0)$	$G(0)$	$H'(0)$
0.10	7.04147	2.03014	-3.3648	-0.19681
0.30	3.77414	1.36178	-2.2939	-0.27868
0.50	2.84050	1.12619	-1.9421	-0.32396
0.72	2.33450	0.97998	-1.7290	-0.35909
1.00	1.97860	0.86611	-1.5658	-0.39204
2.00	1.43923	0.66616	-1.2832	-0.46901
5.00	1.00826	0.47366	-1.0134	-0.58816
10.00	0.79423	0.36638	-0.8592	-0.69069

For the heated horizontal surface facing upward, Fishender and Saunders [16] provided the following correlations:

$$\overline{Nu} = 0.54(Gr_L Pr)^{1/4}, \qquad 10^5 < Gr_L Pr < 2 \times 10^7 \qquad (9.64)$$

and

$$\overline{Nu} = 0.14(Gr_L Pr)^{1/3}, \qquad 2 \times 10^7 < Gr_L Pr < 3 \times 10^{10}. \qquad (9.65)$$

Note that Eq. (9.64) is valid for laminar flow without separation and Eq. (9.65) is identical to that of Fujii and Imura [12].

9.2 Combined Natural- and Forced-Convection Boundary Layers

We now consider the situation where buoyancy forces influence the development of a forced-convection boundary layer growing in the presence of a freestream directed vertically upward, on a vertical plate that is either heated or cooled.

In the case of the heated plate, the buoyancy forces act in the direction of the freestream, and in the case of cooled plate, they oppose the freestream. In both cases, the development of the boundary layer near the leading edge, where Reynolds number R_x is low, is dominated by the large viscous forces, and the buoyancy force may be neglected. Further along the plate these viscous forces diminish, and the buoyancy then makes a significant difference to the evolution of the boundary layer, as we shall see later in this section.

We shall first discuss the case where the freestream and the buoyancy forces are in the same direction. The fluid in the boundary layer is accelerated by the buoyancy forces, which act somewhat like a favorable pressure gradient. For simplicity we shall only consider laminar flow with *uniform* wall temperature.

To obtain the solution of the governing equations for this case, we first direct our attention to the region near the leading edge of the plate. We introduce a slightly modified version of the Falkner–Skan transformation discussed in Section 4.1. We define the similarity variable η, the stream function f, and the dimensionless temperature θ by expressions similar or identical to those defined by Eqs. (4.19a), (4.19b), and (4.24); that is,

$$\eta = \left(\frac{u_e}{\nu x}\right)^{1/2} y, \qquad \theta = \frac{T - T_e}{T_w - T_e}, \qquad \psi = (u_e \nu x)^{1/2} f(x, \eta). \qquad (9.66a)$$

Also, with Ri given by Eq. (9.18) and with L denoting a reference length, ξ is a scaled x variable defined by

$$\xi = \left(\frac{x}{L}\right) Ri. \qquad (9.66b)$$

Using the transformation given by Eqs. (9.66) with Reynolds-stress terms set equal to zero, we can write Eqs. (9.12) and (9.13) and their boundary conditions corresponding to

$$y = 0, \qquad u = v = 0, \qquad T = T_w, \tag{9.67a}$$

$$y = \delta, \qquad u = u_e, \qquad T = T_e \tag{9.67b}$$

for laminar flows as

$$f''' + \frac{1}{2} ff'' = \xi \left(\theta + f' \frac{\partial f'}{\partial \xi} - f'' \frac{\partial f}{\partial \xi} \right), \tag{9.68}$$

$$\frac{\theta''}{Pr} + \frac{1}{2} f\theta' = \xi \left(f' \frac{\partial \theta}{\partial \xi} - \theta' \frac{\partial f}{\partial \xi} \right), \tag{9.69}$$

$$\eta = 0, \qquad f = f' = 0, \qquad \theta = 1, \tag{9.70a}$$

$$\eta = \eta_e, \qquad f' = 1, \qquad \theta = 0. \tag{9.70b}$$

We note from Eq. (9.68) that the flow has the Blasius solution (Section 4.2) at $\xi = 0$. With increasing ξ, however, the effect of buoyancy forces increases, and when ξ is large, it is helpful to change the similarity variables, either to those for pure natural convection or to the primitive variables in which Eqs. (9.12) and (9.13) are expressed in terms of the dimensionless y distance Y and the stream function $F(\xi, Y)$. Here we make the first choice and define the dimensionless y distance η and the dimensionless stream function f by Eq. (9.20). To separate these new parameters from those already used in Eqs. (9.66)–(9.70), we denote the dimensionless y distance by ζ and the dimensionless stream function by $F(\xi, \zeta)$. These new variables, with primes denoting differentiation with respect to ζ, enable Eqs. (9.12) and (9.13) and their boundary conditions to be written as

$$F''' + \frac{3}{4} FF'' - \frac{1}{2} (F')^2 + \theta = \xi \left(F' \frac{\partial F'}{\partial \xi} - F'' \frac{\partial F}{\partial \xi} \right), \tag{9.71}$$

$$\frac{\theta''}{Pr} + \frac{3}{4} F\theta' = \xi \left(F' \frac{\partial \theta}{\partial \xi} - \theta' \frac{\partial F}{\partial \xi} \right), \tag{9.72}$$

$$\zeta = 0, \qquad F = F' = 0, \qquad \theta = 1, \tag{9.73a}$$

$$\zeta = \zeta_e, \qquad F' = \xi^{-1/2}, \qquad \theta = 0. \tag{9.73b}$$

When the freestream and the buoyancy force are in opposite directions, the buoyancy forces retard the fluid on the boundary layer, acting somewhat like an adverse pressure gradient. This situation can lead to flow separation, as was demonstrated by Merkin [17]. Equations (9.68)–(9.73) also apply to this case, provided that we replace the plus sign in front of θ in Eqs. (9.68) and (9.71) by a minus sign.

We again define the local Nusselt number Nu_x by

$$Nu_x = \frac{\hat{h}x}{k} = \frac{\dot{q}_w}{T_w - T_e} \frac{x}{k}. \tag{9.74a}$$

In terms of similarity parameters and dimensionless quantities defined by Eq. (9.66a), this expression can be written as

$$\mathrm{Nu}_x = -\theta'_w\sqrt{R_x}. \qquad (9.74b)$$

Now using the relations given by Eqs. (9.18) and (9.66b) and with R_L denoting $u_e L/\nu$, we can also write Eq.(9.74b) as

$$\mathrm{Nu}_x \frac{\sqrt{\mathrm{Gr}_L}}{R_L^{3/2}} = -\theta'_w\sqrt{\xi}. \qquad (9.74c)$$

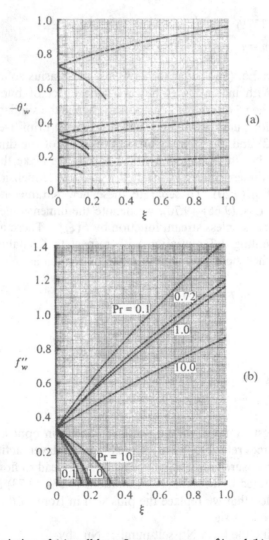

Figure 9.10. Variation of (a) wall heat flux parameter $-\theta'_w$ and (b) wall shear stress parameter f''_w on a vertical flat plate with natural and forced convection. Solid lines denote boundary conditions corresponding to a cooled surface and dashed lines those for a heated surface.

Similarly the local skin-friction coefficient defined by

$$c_f = \frac{\tau_w}{\frac{1}{2}\rho u_e^2} \qquad (9.75a)$$

can be written as

$$c_f \frac{R_L^{3/2}}{\sqrt{\mathrm{Gr}_L}} = \frac{2 f_w''}{\sqrt{\xi}}. \qquad (9.75b)$$

Similar expressions for Nu_x and c_f can also be written in terms of the variable defined by Eqs. (9.20).

Figures 9.10(a) and (b) show the variation of wall-heat-flux parameter $-\theta_w'$ and wall-shear-stress parameter f_w'' with reduced nondimensional dis-

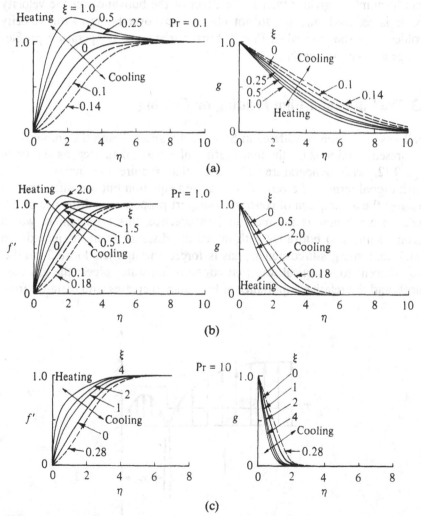

Figure 9.11. Effect of buoyancy on heated and cooled vertical flat plates with natural and forced convection at different Prandtl numbers. (a) $\mathrm{Pr} = 0.1$, (b) $\mathrm{Pr} = 1.0$, (c) $\mathrm{Pr} = 10$.

tance ξ for different values of Prandtl number. Both parts of Fig. 9.10 allow comparison between heated and cooled mixed convection.

As can be seen from Fig. 9.10(a), cooling the wall, with consequently opposed natural and forced convection, results in separation of the boundary layer at a value of ξ that decreases with decreasing Prandtl number. Thus, as the value of Prandtl number decreases from 10 to 0.1, the separation location moves upstream from $\xi = 0.35$ to 0.15. We note that, as expected, although the value of f_w'' is zero at separation, the value of θ_w' is finite, indicating that the Reynolds analogy is not realistic near separation.

We also see from the velocity and temperature profiles shown in Fig. 9.11 that buoyancy plays a very pronounced effect. At Prandtl numbers less than or equal to 1, with heating, the velocity in the boundary layer can exceed the external velocity ($f' > 1$) as the buoyancy effect becomes stronger. At Prandtl numbers greater than 1, the effect of the buoyancy on the velocity profile is reduced, and we do not observe any overshoots in the velocity profiles as is the case when Pr ≤ 1. Furthermore, the temperature profiles change less with increasing ξ.

9.3 Wall Jets and Film Heating or Cooling

Wall jets have been considered in Sections 4.5 and 6.7, and the emphasis of the present section is on the film heating of a vertical surface, as shown in Fig. 9.12, with temperature differences that require the inclusion of a gravitational term in the vertical momentum equation but are small enough to allow the assumption of constant transport properties and the use of Eq. (9.8). Laminar flow is assumed, and in practice, this arrangement would become subject to buoyancy with increasing distance from the jet exit. In many domestic gas-fired boilers, gas is forced through jets located parallel and adjacent to a vertical wall that contains hot-water pipes; the gas then burns, and the relative importance of buoyancy increases with distance from

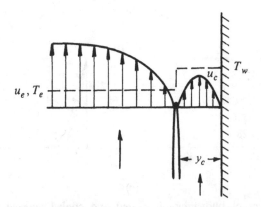

Figure 9.12. Initial velocity (solid lines) and temperature (dashed line) profiles for a wall-jet on a vertical flat plate. u_c denotes the average slot velocity.

the jet exit. If the space into which the wall jet exhausts is large, it is possible to neglect the motion of the fluid outside the jet. If forced convection is present and causes flow opposed to that of the jet, separation and recirculation may occur. Here we consider an example in which there is an external laminar flow up a uniformly heated vertical surface so that the wall jet, the freestream, and buoyancy-driven flow are all in the same direction.

The appropriate equations for the conservation of mass, momentum, and energy correspond to those given by Eqs. (9.68) and (9.69). To show the effect of buoyancy, however, we write them in a slightly different form. Using Eq. (9.66b) and letting $z = x/L$, they become

$$f''' + \frac{1}{2}ff'' = \text{Ri } z\theta + z\left(f'\frac{\partial f'}{\partial z} - f''\frac{\partial f}{\partial z}\right), \qquad (9.76a)$$

$$\frac{\theta''}{\text{Pr}} + \frac{1}{2}f\theta' = z\left(f'\frac{\partial \theta}{\partial z} - \theta'\frac{\partial f}{\partial z}\right). \qquad (9.76b)$$

As in the case of a wall jet on a horizontal surface (see Section 4.5), these equations require boundary and initial conditions. For a uniform wall temperature, the dimensionless boundary conditions are given by Eq. (9.70), and the initial conditions are the same as those for a laminar flow over a horizontal surface.

The results shown in Figs. 9.13 and 9.14 were obtained for $\text{Pr} = 0.72$, for one value of $u_c/u_e(\equiv 1.0)$, for one value of $\delta/y_c(\equiv 0.95)$, and for several values of Richardson number Ri to demonstrate the effect of buoyancy on

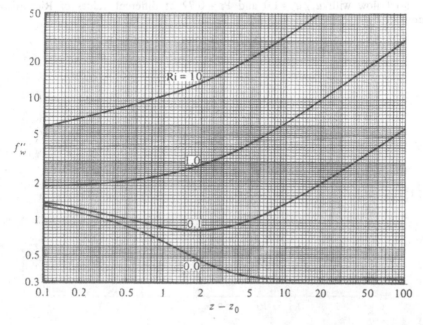

Figure 9.13. Variation of the dimensionless wall shear stress parameter f_w'' in buoyant wall-jet flow with $u_c/u_e = 1$ and $\text{Pr} = 0.72$ at different values of Richardson number.

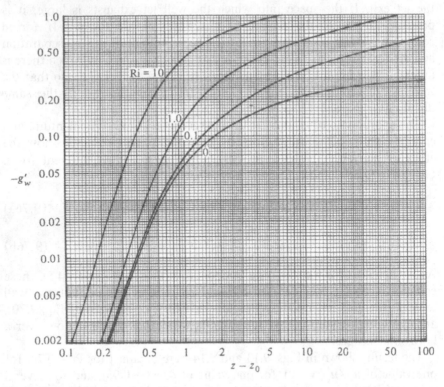

Figure 9.14. Variation of the dimensionless wall heat flux parameter, g'_w, in buoyant, wall-jet flow with $u_c/u_e = 1.0$ and $Pr = 0.72$ at different values of Richardson number.

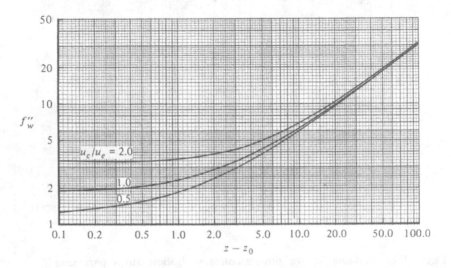

Figure 9.15. Variation of dimensionless wall shear parameter f''_w as a function of distance $z - z_0$ on an isothermal vertical surface for $\delta/y_c = 0.95$, $Pr = 1$ and $Ri = 1$.

skin friction and heat transfer. We see from Fig. 9.13 that with no buoyancy (Ri = 0) the dimensionless wall-heat-transfer parameter f_w'' decreases with increasing dimensionless surface distance $z - z_0$. With Ri = 0.1 the skin-friction parameter does not change appreciably from the results at Ri = 0 at small values of $z - z_0$, and this is to be expected since the buoyancy effect is small near the leading edge. With increasing z, however, the influence of buoyancy increases. Since the effect of buoyancy on a heated surface is that of a flow with favorable pressure gradient, the trend of variation of f_w'' with increasing z changes and, rather than decreasing with z, it begins to increase. As the Richardson number increases further, the buoyancy effect becomes more pronounced. Figure 9.14 shows a similar effect on the dimensionless wall-heat-flux parameter g_w', which, with increasing Ri, begins to vary substantially from the value corresponding to Ri = 0 as z increases.

Figures 9.15 and 9.16 show the effect of u_c/u_e on f_w'' and g_w' with Ri fixed at 1.0 and may be compared with Figs. 4.23 and 4.24. In the range of values of z, up to 10, the influence of the wall jet and its velocity ratio is significant, and the buoyancy effect has caused the increase in the skin-friction parameter. For values of z greater than 10, the boundary layer increasingly "forgets" its origin, and the influence of the wall jet diminishes to zero, at which point the result is identical to that of Table 9.1 (see Problem 9.26). The wall heat flux varies in a corresponding manner, with buoyancy increasing the value of g_w' for each velocity ratio. Again, for z greater than 10, the results correspond clearly to those for a boundary layer with natural convection and without a wall jet.

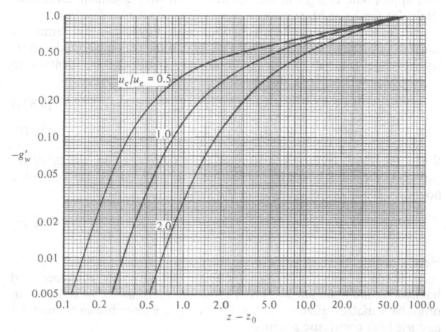

Figure 9.16. Variation of dimensionless wall heat flux parameter g_w' as a function of distance $z - z_0$ on an isothermal vertical surface for $\delta/y_c = 0.95$, Pr = 1 and Ri = 1.

9.4 Natural and Forced Convection in Duct Flows

Solutions for laminar and turbulent flows in ducts with buoyancy effects can be obtained by using procedures similar to those discussed in Chapters 5 and 7 for flows with no buoyancy effects. A wide range of flows corresponding to different boundary conditions may be calculated in regions before the shear layers merge and in the regions after they merge. One common example is flow in a vertical heated duct in which the entire flow is due to natural convection. In this case the boundary layers at the entrance of the duct begin to grow due to the higher temperature of the wall; at some distance away from the entrance, these shear layers merge and continue to accelerate the flow in the duct. Another example is one in which the forced-convection flow in the vertical duct is influenced by the surface temperature of the duct. In this case, if the ambient fluid is at a temperature higher than that of the surface of the duct, then the buoyancy effect opposes the flow and acts in roughly the same way as an adverse pressure gradient; this situation can cause the flow to separate (as in boundary layers on vertical plates discussed in Section 9.3). On the other hand, if the ambient temperature of the fluid is lower than the duct wall temperature, then the buoyancy effect aids the flow like a favorable pressure gradient.

To illustrate the calculation of heat and momentum transfer in vertical ducts, we shall consider a laminar flow with natural and forced convection between two parallel plates at constant temperature. Obviously these calculations can also be extended to circular vertical pipes and to turbulent flows with appropriate changes in the equations and with a suitable turbulence model for the Reynolds stresses.

The governing laminar-flow equations with Reynolds-stress terms set equal to zero are given by Eqs. (9.11)–(9.13). In the region before the shear layers merge, the wall and "edge" boundary conditions are identical to those given by Eqs. (9.67); that is,

$$y = 0, \qquad u = v = 0, \qquad T = T_w, \tag{9.67a}$$

$$y = \delta, \qquad u = u_e, \qquad T = T_e. \tag{9.67b}$$

When the shear layers grow so that it is convenient to use primitive variables, the edge boundary conditions are replaced by the centerline boundary conditions resulting from the symmetry,

$$y = L, \qquad \frac{\partial u}{\partial y} = 0, \qquad \frac{\partial T}{\partial y} = 0. \tag{9.77}$$

Here L denotes the half-width of the duct.

As in ducts with no buoyancy effects, the governing equations and boundary conditions can be solved either in transformed variables or in primitive variables. Here we follow the procedure described in Section 5.3 and use both coordinate systems.

In the early stages of the flow when the shear layers are thin, we solve the governing equations in transformed variables. For this purpose we use the modified Falkner–Skan transformation defined by Eq. (5.45), which for a two-dimensional flow is

$$\eta = \sqrt{\frac{u_0}{\nu x}}\, y, \qquad \psi(x, y) = \sqrt{u_0 \nu x}\, f(x, \eta). \tag{9.78}$$

We also use the dimensionless temperature θ defined in Eq. (9.66a), that is,

$$\theta = \frac{T - T_e}{T_w - T_e},$$

and with primes denoting differentiation with respect to η, we can write the momentum and energy equations and their boundary conditions as

$$f''' + \frac{1}{2}ff'' = \xi \frac{dp^*}{d\xi} + \xi\left(f'\frac{\partial f'}{\partial \xi} - f''\frac{\partial f}{\partial \xi}\right) \pm \xi\theta, \tag{9.79}$$

$$\frac{\theta''}{\text{Pr}} + \frac{1}{2}f\theta' = \xi\left(f'\frac{\partial\theta}{\partial\xi} - \theta'\frac{\partial f}{\partial\xi}\right), \tag{9.80}$$

$$\eta = 0, \qquad f = f' = 0, \qquad \theta = 1, \tag{9.81a}$$

$$\eta = \eta_e, \qquad f' = \frac{u_e}{u_0} \equiv \bar{u}_e, \qquad \theta = 0. \tag{9.81b}$$

The $\pm$ sign in Eq. (9.79) denotes aided (heating) or opposed (cooling) flow conditions respectively, ξ denotes a scaled x variable, given by Eq. (9.66b), and p^* denotes a dimensionless pressure $p/\rho u_0^2$.

To obtain the additional boundary condition due to the presence of the pressure gradient term, we use Eq. (5.57), that is,

$$u_0 L = \int_0^L u\, dy, \tag{5.57}$$

and write it in terms of transformed variables as

$$f(\xi, \eta_{sp}) = \eta_{sp}, \tag{9.82a}$$

where $\eta_{sp} = \sqrt{(\text{Gr}_L/R_L)(1/\xi)}$. Noting that $f' = \bar{u}_e$ for $\eta_e < \eta < \eta_{sp}$, we can write this equation as

$$\bar{u}_e(\eta_{sp} - \eta_e) + f_e = f_{sp}. \tag{9.82b}$$

Except for the buoyancy term in Eq. (9.79) and for the term Gr_L/R_L in the definition of η_{sp}, we note that the momentum and energy equations for a vertical duct are identical in form (as they should be) to those for a horizontal duct. More important, although the equations and their solutions for a horizontal duct are Reynolds-number independent, for a vertical duct they are both Grashof-number and Reynolds-number dependent.

When the shear layers become sufficiently thick, it is necessary to abandon the transformed variables in favor of primitive variables at some $\xi = \xi_0$. Furthermore, with increasing ξ, the effect of buoyancy force increases, and it is helpful to change the similarity variables either to those for pure natural convection or to the primitive variables. Here we make the second choice and define the dimensionless distance Y and stream function $F(x, Y)$ by Eq. (5.52), which for a two-dimensional flow becomes

$$Y = \left(\frac{u_0}{\nu L}\right)^{1/2} y, \qquad \psi = (u_0 \nu L)^{1/2} F(x, Y). \qquad (9.83)$$

These new variables enable the conservation equations and their boundary conditions, with primes now denoting differentiation with respect to Y, to be written as

$$F''' = \frac{dp^*}{d\xi} + F' \frac{\partial F'}{\partial \xi} - F'' \frac{\partial F}{\partial \xi} \pm \xi\theta, \qquad (9.84)$$

$$\frac{\theta''}{\mathrm{Pr}} = F' \frac{\partial \theta}{\partial \xi} - \theta' \frac{\partial F}{\partial \xi}, \qquad (9.85)$$

$$Y = 0, \qquad F = F' = 0, \qquad \theta = 1, \qquad (9.86a)$$

$$Y_c = \sqrt{R_L}, \qquad F'' = 0, \qquad F = \sqrt{R_L}, \qquad \theta' = 0. \qquad (9.86b)$$

As in the solution of the equations for a horizontal duct (Chapter 5), the above system of equations for a vertical duct can be solved by the nonlinear eigenvalue method as described in Section 14.1 or by the Mechul-function method as described by Bradshaw et al. [18]. While both methods can be used satisfactorily for flows in which wall shear is positive (no separation), only the Mechul-function method can be used for flows with negative wall shear. For details, see Cebeci et al. [19].

Figures 9.17 to 9.19 present the variations of the dimensionless wall shear-stress parameter f_w'' ($\equiv c_f/2R_L\sqrt{\xi}$), the dimensionless pressure drop Δp^*, and the dimensionless wall heat-flux parameter θ_w' ($\equiv -\mathrm{Nu}_x/R_L\sqrt{\xi}$), respectively, with reduced downstream distance ξ for $\mathrm{Pr} = 0.72$, $R_L = 10^3$. Here c_f is defined by Eq. (9.75a) with u_e replaced by u_0 and Nu_x by Eq.

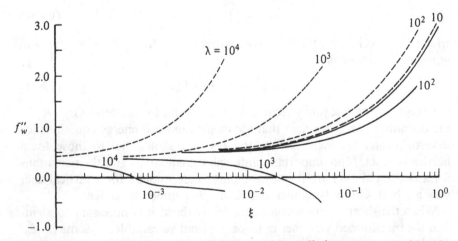

Figure 9.17. Variation of transformed dimensionless wall shear parameter f_w'' with downstream reduced distance, ξ. Here solid lines denote boundary conditions corresponding to a cooled surface and dashed lines those for a heated surface.

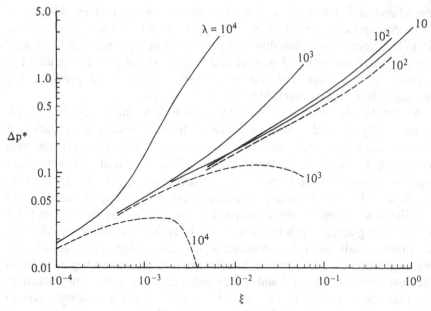

Figure 9.18. Variation of dimensionless pressure drop with reduced downstream distance, ξ. Solid lines denote boundary conditions corresponding to a cooled surface and dashed lines those for a heated surface.

(9.74a). In all cases, the velocity and temperature boundary layers started at the duct entrance, i.e., at $\xi = 0$, for a given ratio of Gr_L/R_L. The calculations were terminated at $\xi = 6$, which encompasses the range of interest for most practical problems. Results are presented for values of λ ($\equiv Gr_L/R_L^2$ $\equiv Ri$) of 10, 10^2, 10^3, and 10^4; the first value implies negligible effect of natural convection, and the result of each figure corresponds to laminar

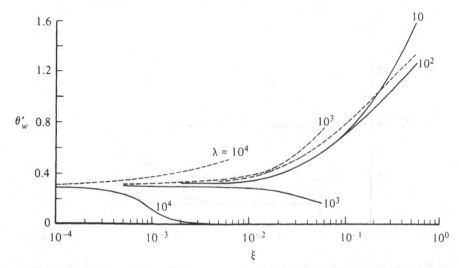

Figure 9.19. Variation of transformed dimensionless wall temperature gradient θ'_w with reduced downstream distance, ξ. Solid lines denote boundary conditions corresponding to a cooled surface and dashed lines those for a heated surface.

forced-convection flow in a horizontal two-dimensional duct. All figures allow comparison between aided and opposed natural and forced convection. The forced-convection flow is presumed to be upward, with heating of the wall resulting in aided natural convection; if the pressure gradient is assumed to act downward, the results will remain valid provided the buoyancy force is also reversed by cooling the wall.

As can be seen from Fig. 9.17, the cold-wall boundary condition with consequently opposed natural and forced convection results in separation of the boundary layer and reversed flow at values of ξ that decrease with increasing λ. Thus as the value of $T_w - T_e$ is increased, the separation location moves upstream, and the reversed flow persists at downstream locations. The corresponding pressure drop (Fig. 9.18) increases more rapidly downstream of the separation locations, but as can be seen for $\lambda = 10^4$, the gradient tends to reduce slightly as the dimensionless wall shear parameter tends toward a maximum negative value. The variation of pressure drop with λ is substantial as the onset of separation begins to affect the flow. Where the natural and forced convection act in the same direction, the pressure drop falls rapidly as the buoyancy force increasingly acts to overcome the increasing wall shear parameter f_w''.

Figure 9.20 shows the influence of Prandtl number on the dimensionless wall heat flux parameter θ_w' for $\lambda = 10^3$. It can be seen that, as would be expected, as the ratio of thermal to momentum boundary layer decreases, θ_w' increases and the magnitude of the difference between the aided and opposed situations also increases. Prandtl number has a comparatively small effect on the location of separation for the opposed situation. With Pr = 0.1 and 10, separation occurs at $\xi = 0.017$ and 0.032, respectively.

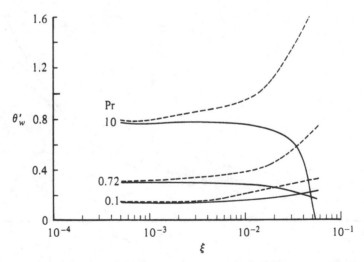

Figure 9.20. Effect of Prandtl number on θ_w' for $\lambda = 10^3$. Solid lines denote boundary conditions corresponding to a cooled surface and dashed lines those for a heated surface.

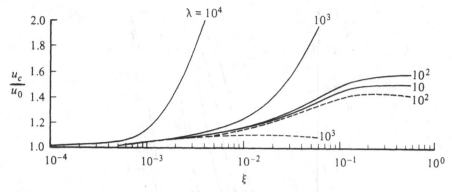

Figure 9.21. Variation of dimensionless center-line velocity u_c/u_0 with reduced downstream distance, ξ. Solid lines denote boundary conditions corresponding to a cooled surface and dashed lines those for a heated surface.

Figure 9.21 show the variation of the nondimensional centerline velocity with reduced distance ξ for several values of λ, corresponding to $\mathrm{Pr} = 0.72$. With $\lambda = 10$, which corresponds to negligible influence of natural convection, the nondimensional centerline velocity changes from unity to 1.5 as the profile changes from Blasius to its fully developed plane Poiseuille shape. The effect of natural convection acting with the forced convection is to accelerate the flow in the near-wall region, and as shown for $\lambda = 10^4$, this may happen to an extent that results in a very small increase in centerline velocity or even a reduction as fluid is entrained into a thin-wall boundary

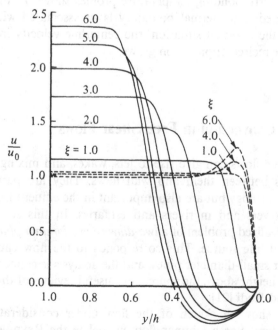

Figure 9.22. Dimensionless velocity profiles with heating (dashed lines) and cooling (solid lines) at different values of ξ for $\mathrm{Pr} = 0.72$, $\lambda = 10^4$.

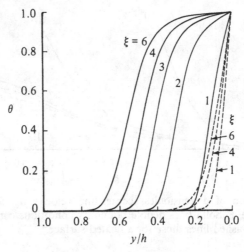

Figure 9.23. Dimensionless temperature profiles with heating (dashed lines) and cooling (solid lines) at different values of ξ for Pr = 0.72, $\lambda = 10^4$.

layer, with a substantial velocity maximum, from the central region of the flow.

The velocity profiles of Fig. 9.22, shown for aided (heating) and opposed (cooling) natural and forced convection and for Pr = 0.72, $\lambda = 10^4$, confirm the arguments of the previous paragraph. For the aided situation, the velocity in the central region does decrease as the accelerated wall layer develops. The corresponding temperature profiles shown in Fig. 9.23 confirm the expected thin thermal boundary layer associated with the aided situation. For the opposed situation, the centerline velocity increases substantially as the recirculation region grows.

9.5 Natural Convection in Free Shear Flows

Buoyancy-driven flows can give rise to jets, wakes, and mixing regions and to interactions between them and wall flows. They are particularly important in meteorology but are also important in the effluent from chimneys and cooling towers and in rivers and estuaries. In this section, we will consider the idealized problem of a *two-dimensional laminar plume* generated by a horizontal line source. This corresponds to the flow caused by long heated wires or small-diameter pipes, and the analysis is concerned with the flow at a large height above the source. For useful reviews of the subject, see Turner [20] and Rodi [21].

Figure 9.24 shows a sketch of the flow under consideration and the coordinate system. For a laminar flow we delete the Reynolds-stress and heat-flux terms from Eqs. (9.11)–(9.13) and impose the following boundary conditions on them:

$$y = 0, \qquad v = \frac{\partial u}{\partial y} = \frac{\partial T}{\partial y} = 0, \qquad (9.87a)$$

$$y \to \infty, \qquad T \to T_e, \qquad u \to 0. \qquad (9.87b)$$

In addition, since there are no other heat sources in the flow, the total thermal energy convected in the boundary layer, K_c, is constant; that is,

$$K_c = \rho c_p \int_{-\infty}^{\infty} u(T - T_e)\, dy = \text{const.} \qquad (9.88)$$

For a power-law centerline temperature distribution, such as the one given by Eq. (9.29a), Eqs. (9.11)–(9.13), (9.87), and (9.88) have similarity solutions. To demonstrate this, we define the similarity variable η and the stream function ψ, derivable by the approach of Section 4.1, by

$$\eta = \left(\frac{x}{L}\right)^{(n-1)/4} \sqrt{\frac{u_c}{\nu L}}\, y, \qquad \psi = \left(\frac{x}{L}\right)^{(n+3)/4} \sqrt{u_c \nu L}\, f(\eta) \qquad (9.89)$$

and let

$$T_c - T_e = \left(\frac{x}{L}\right)^{n} N. \qquad (9.90)$$

Here N is a constant, T_c denotes the centerline temperature, and u_c a reference velocity. We also define a dimensionless temperature,

$$\theta = \frac{T - T_e}{T_c - T_e}. \qquad (9.91)$$

Using Eqs. (9.89)–(9.91) and noting that

$$u = \frac{\partial \psi}{\partial y} = u_c \left(\frac{x}{L}\right)^{(2n+2)/4} f', \qquad (9.92)$$

we can write Eq. (9.88) as

$$K_c = 2\rho c_p \sqrt{u_c L \nu}\, N \left(\frac{x}{L}\right)^{(5n+3)/4} \int_0^{\infty} f'\theta\, d\eta. \qquad (9.93)$$

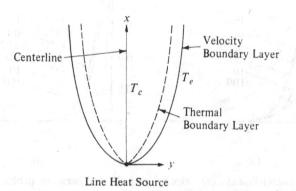

Line Heat Source

Figure 9.24. Coordinate system for a two-dimensional plume rising from a horizontal line heat source. y-coordinate is exaggerated.

Since the heat flux K_c is independent of x, the exponent n must be equal to $-\frac{3}{5}$. Using this value of n in Eqs. (9.89) and (9.90), we can now transform Eqs. (9.12) and (9.13) and their boundary conditions given by Eq. (9.87) to obtain

$$f''' + \tfrac{12}{5}ff'' - \tfrac{4}{5}(f')^2 + \theta = 0, \tag{9.94}$$

$$\theta'' + \tfrac{12}{5}\Pr(f\theta)' = 0, \tag{9.95}$$

$$\eta = 0, \quad f = f'' = 0, \quad \theta' = 0, \tag{9.96a}$$

$$\eta \to \eta_e, \quad f' \to 0, \quad \theta \to 0. \tag{9.96b}$$

Several numerical methods can be used to solve the coupled system given by Eqs. (9.94)–(9.96). One simple procedure (not the most efficient) is to solve them separately. Integrating Eq. (9.95) and writing the result as

$$\theta(\eta) = \exp\left[-\tfrac{12}{5}\Pr \int_0^\eta f(\eta)\,d\eta \right], \tag{9.97}$$

we can obtain its solution by inserting any expression for $f(\eta)$ that satisfies the boundary conditions for f and its derivatives given in Eq. (9.96). The resulting expression is then substituted into Eq. (9.94) to obtain a new solution of f. This solution is then inserted into Eq. (9.95) to get a new distribution of $\theta(\eta)$ that can be inserted into Eq. (9.94) to yield a new function $f(\eta)$. This procedure is repeated until convergence is achieved. Figure 9.25 shows the velocity and temperature profiles obtained by this procedure for different values of Prandtl numbers and indicates the pronounced effect of Prandtl number on both sets of profiles. For additional solutions, see Gebhart et al. [22].

With f and θ known, the factor N can be determined from Eq. (9.93) and the mass flow rate $\dot{m}$ in the plume can be calculated.

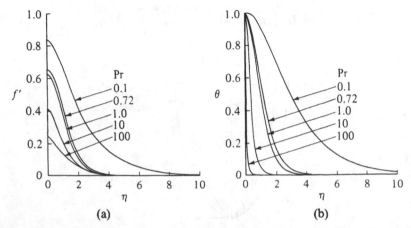

Figure 9.25. Dimensionless (a) velocity and (b) temperature profiles for a two-dimensional laminar plume at different Prandtl numbers.

In terms of dimensionless variables, with

$$I_1 = \int_0^\infty f'\theta \, d\eta,$$

Eq. (9.93) can be written as

$$K_c = 2kN \Pr\sqrt{Gr_L}\, I_1. \qquad (9.98)$$

Since the heat flux K_c along the plume is also equal to the total heat input by the line source, Eq. (9.98) can be solved to give

$$N = \frac{K_c}{2k \Pr\sqrt{Gr_L}\, I_1}. \qquad (9.99)$$

The mass flow rate in the plume is

$$\dot{m} = 2\int_0^\infty \rho u \, dy = 2\rho u_c L \left(\frac{x}{L}\right)^{3/5} \frac{f(\eta_e)}{\sqrt{Gr_L}}. \qquad (9.100)$$

We see from Eq. (9.100) that the mass flow rate increases with height above the line source as $x^{3/5}$. The value of $f(\eta_e)$ depends on the Prandtl number of the fluid. Its variation for different values of Prandtl number is given in Table 9.3.

Problems

9.1 The heat transfer coefficient $\hat{h}$ in laminar natural convection depends on the following parameters

$$\hat{h} = f(\rho, \mu, c_p, k, L, g\beta, \Delta T).$$

Using the matrix-elimination procedure discussed in Section 6.1, show that

$$Nu = f(Gr_L, \Pr).$$

9.2 Show that the square of the ratio

$$\frac{u_c}{u_e} = \frac{\sqrt{g\beta L(T_w - T_e)}}{u_e}$$

is a form of the Richardson number defined in Eq. (9.1).

9.3 Show that for a two-dimensional laminar natural convection flow over a vertical flat plate, the similarity variable η and the dimensionless stream

Table 9.3 Values of dimensionless stream function $f(\eta_e)$ as a function of Prandtl number

Pr	0.1	0.72	1.0	10	100
$f(\eta_e)$	2.127	0.920	0.827	0.499	0.353

function $f(\eta)$ given by Eq. (9.20), can also be written as

$$\eta = \left(\frac{u_c}{\nu L}\right)^{1/2} y \left(\frac{L}{x}\right)^{1/4}, \qquad \psi = (u_c \nu L)^{1/2} \left(\frac{x}{L}\right)^{3/4} f(\eta).$$

Use the method discussed in Section 4.1.

9.4 Derive the similarity equations for two-dimensional laminar natural convection flow over a vertical flat plate subject to (a) specified surface temperature and (b) heat flux. Use the transformation defined by Eq. (9.20).

9.5 Obtain an expression for the magnitude and location of the maximum velocity of air (Pr = 0.72) in a natural convection laminar boundary layer on a vertical plate with uniform wall temperature.

9.6 Two uniformly heated vertical flat plates at $50°C$ are placed in air whose temperature is $15°C$. If the plates are 0.5 m high, what is the minimum spacing which will prevent merging of the natural convection boundary layers?

9.7 Calculate the local skin-friction coefficient c_f and Nusselt number Nu_x at $x = 0.25$ m for a laminar flow on a vertical flat plate whose surface temperature is maintained at a uniform temperature of $65°C$ with ambient temperature corresponding to $15°C$. Use film temperature T_f to evaluate the fluid properties. Take the fluid to be: (a) air, (b) glycerin, (c) steam.

9.8 A horizontal plate whose surface temperature is maintained at $250°C$ is exposed to atmospheric air at $15°C$. Find the heat lost per foot of length for laminar flow when the plate is: (a) semi-infinite in extent, and (b) finite. Calculate all the fluid properties at film temperature T_f. Discuss the comparison.

9.9 Air at $20°C$ and 1 atm flows at a velocity of 10 ms^{-1} past a 1.5m-long vertical plate whose surface is maintained at a uniform temperature of $120°C$. Use film temperature T_f to evaluate the fluid properties.

(a) Calculate the local Nusselt number and skin friction at $x = 0.5$, 1, 1.5 m. Compare your results with those that would result on a horizontal flat plate under the same conditions. Discuss this comparison.

(b) If the width of the plate is 0.5 m, calculate the total heat transfer from the plate.

9.10 Consider a wall jet, with a slot height of 0.01 m, in which the slot and freestream velocities of air are equal and the freestream temperature is $300°K$.

(a) Determine the value of Richardson number for a freestream velocity of 0.1 m/s and a uniform wall temperature of $320°K$.

(b) Determine the value of Richardson number for a freestream velocity of 1.0 m/s, and a uniform wall temperature of $320°K$.

(c) Determine the value of Richardson number for a freestream velocity of 1.0 m/s, and a uniform wall temperature of $1000°K$ and consider the validity of the analysis for this case.

9.11 Determine, with the help of Figs. 9.13 and 9.14, the values of $c_f/2$ and Nu at a location 0.1 m downstream of a wall jet of air in which the freestream velocity and slot velocity are equal to 0.1 m/s, the slot height is 0.01 m and the freestream and wall temperatures are equal to 300 and $500°K$, respectively. What do you expect to happen if the values of the freestream and wall temperature are reversed?

9.12 Use Figs. 9.13–9.16 to assess the values of $c_f/2$ and Nu for the conditions of problem 9.2 but with a freestream velocity twice the slot velocity.

References

[1] Jaluria, Y.: *Natural Convection Heat and Mass Transfer*, Pergamon, Oxford, 1980.

[2] Ede, A. J.: Advances in free convection, in *Advances in Heat Transfer*, Vol. 4, Academic, New York, 1967, p. 1.

[3] LeFevre, E. J.: Laminar free convection from a vertical plane surface. *Proc. 9th Int. Cong. Appl. Mech., Brussels*, Vol. 4, 1956, p. 168.

[4] Fujii, T. and Fujii, M.: The dependence of local Nusselt number on Prandtl number in the case of free convection along a vertical surface with uniform heat flux. *Int. J. Heat Mass Transfer* **19**:121 (1976).

[5] Jaluria, Y. and Gebhart, B.: On transition mechanisms in vertical natural convection flow. *J. Fluid Mech.* **66**:309 (1974).

[6] Cebeci, T. and Khattab, A. A.: Prediction of turbulent-free-convective-heat transfer from a vertical flat plate. *J. Heat Transfer* **97**:469 (1975).

[7] Warner, C. Y. and Arpaci, V. S.: An experimental investigation of turbulent natural convection in air along a vertical heated flat plate. *Int. J. Heat Mass Transfer* **11**:397 (1968).

[8] Fujii, T., Takeuchi, M., Fujii, M., Suzaki, K., and Vehara, H.: Experiments on natural convection heat transfer from the outer surface of a vertical cylinder to liquids. *Int. J. Heat Mass Transfer* **13**:753 (1970).

[9] Bailey, F. J.: Analysis of turbulent free convection heat transfer. *Proc. Inst. Mech. Eng.* **169**:361 (1955).

[10] Vliet, G. C. and Liu, C. K.: An experimental study of turbulent natural convection boundary layers. *J. Heat Transfer* **91**:517 (1969).

[11] Sparrow, E. M. and Gregg, J. L.: Variable fluid-property problem in free convection. *J. Heat Transfer* **80**:879 (1958).

[12] Fujii, T. and Imura, H.: Natural-convection heat transfer from a plate with arbitrary inclination. *Int. J. Heat Mass Transfer* **15**:755 (1972).

[13] Stewartson, K.: On free convection from a horizontal plate. *ZAMP* **9**:276 (1958).

[14] Rotem, A. and Claassen, L.: Natural convection above unconfined horizontal surfaces. *J. Fluid Mech.* **39**:173 (1969).

[15] McAdams, W. H.: *Heat Transmission*, 3rd ed. McGraw-Hill, New York, 1954.

[16] Fishenden, M. and Saunders, O. A.: *An Introduction to Heat Transfer*, Oxford. London, 1950.

[17] Merkin, J. H.: The effect of buoyancy forces on the boundary-layer flow over a semi-infinite vertical flat plate in a uniform freestream. *J. Fluid Mech.* **35**:439 (1969).

[18] Bradshaw, P., Cebeci, T., and Whitelaw, J. H.: *Engineering Calculation Methods for Turbulent Flows*. Academic, London, 1981.

[19] Cebeci, T. Khattab, A. A., and Lamont, R.: Combined natural and forced convection in vertical ducts. *Proc. Seventh International Heat Transfer Conf.* Munich, Germany, Hemisphere Publishing Co., Washington, D.C. 1982.

[20] Turner, J. S.: *Buoyancy Effects in Fluids*. University Press, Cambridge, 1973.

[21] Rodi, W.: *Vertical Turbulent Buoyant Jets: a Review of Experimental Data.* Pergamon, Oxford, 1980.

[22] Gebhart, B., Pera, L., and Schorr, A. W.: Steady laminar natural convection plumes above a horizontal line heat source. *Int. J. Heat Mass Transfer* **13**:161 (1970).

CHAPTER 10

Coupled Laminar Boundary Layers

If the typical temperature difference in a gas flow is an appreciable fraction of the absolute temperature, the typical density difference will be an appreciable fraction of the absolute density, and the density appearing in the velocity-field equations can no longer be taken as constant. Instead, as in the buoyant flows of Chapter 9, the momentum and energy equations *must* be solved simultaneously since they are "coupled," i.e., density appears in the momentum equation and is linked through an equation of state to the dependent variable of the energy equation.

There is no important difference between the solution procedure needed for low-speed laminar shear layers with large density differences (introduced by large heat-transfer rates or large concentration changes) and that required for high-speed flows where temperature differences are generated by viscous dissipation of kinetic energy. The complication introduced by the variation of pressure, with consequent variations in density, is minor, and high-speed flows without shock waves can, therefore, be considered by the procedures of this chapter. A constant-pressure high-speed shear layer is identical to a low-speed flow in which distributed heat sources within the fluid replace the viscous dissipation of kinetic energy.

As in uncoupled laminar boundary layers, it is useful and desirable to express the governing equations in transformed variables before they are solved. In this chapter, we use two totally different types of transformations for coupled flows. The first, used in simpler form in constant-property flow, is the Falkner–Skan transformation and is intended to *reduce* (but not usually eliminate) the dependence of the equations on x. The second transformation, which is due to Illingworth [1] and Stewartson [2], is intended to eliminate the dependence of the equations on the Mach number, the ratio of wall temperature to freestream temperature, or other variable-property parameters. In fact, the effect of variable properties can be eliminated in only a few special laminar-flow cases, but both in laminar flow and in turbulent flow, a *reduction* of variable-property effects makes the remaining effects easier to calculate. However, calculations could be done in primitive untransformed variables, although less efficiently.

We define the compressible version of the Falkner–Skan transformation by

$$d\eta = \left(\frac{u_e}{\nu_e x}\right)^{1/2} \frac{\rho}{\rho_e} dy, \qquad \psi(x, y) = (\rho_e \mu_e u_e x)^{1/2} f(x, \eta). \quad (10.1)$$

We use a procedure similar to that described in Section 4.1, with the usual definition of stream function replaced by

$$\rho u = \frac{\partial \psi}{\partial y}, \qquad \rho v = -\frac{\partial \psi}{\partial x}. \quad (10.2)$$

The momentum and energy equations for a laminar flow with negligible body forces are Eqs. (3.38), omitting turbulence and body-force terms, and (3.47a); that is,

$$\rho u \frac{\partial u}{\partial x} + \rho v \frac{\partial u}{\partial y} = -\frac{dp}{dx} + \frac{\partial}{\partial y}\left(\mu \frac{\partial u}{\partial y}\right), \quad (3.38)$$

$$\rho u \frac{\partial H}{\partial x} + \rho v \frac{\partial H}{\partial y} = \frac{\partial}{\partial y}\left[\frac{\mu}{Pr} \frac{\partial H}{\partial y} + \mu\left(1 - \frac{1}{Pr}\right)u \frac{\partial u}{\partial y}\right]. \quad (3.47a)$$

After transforming and replacing the $-dp/dx$ term with $\rho_e u_e (du_e/dx)$, the equations can be written, in notation to be explained below, as

$$(bf'')' + m_1 ff'' + m_2\left[c - (f')^2\right] = x\left(f' \frac{\partial f'}{\partial x} - f'' \frac{\partial f}{\partial x}\right), \quad (10.3)$$

$$(eS' + df'f'')' + m_1 fS' = x\left(f' \frac{\partial S}{\partial x} - S' \frac{\partial f}{\partial x}\right). \quad (10.4)$$

In the absence of mass transfer the boundary conditions for these equations are

$$y = 0, \qquad v = 0, \qquad u = 0, \qquad H = H_w(x) \quad \text{or} \quad \left(\frac{\partial H}{\partial y}\right)_w = -\frac{c_{p_w}}{k_w} \dot{q}_w,$$

$$(10.5a)$$

$$y = \delta, \qquad u = u_e(x), \qquad H = H_e. \quad (10.5b)$$

In terms of transformed variables these boundary conditions can be written as

$$\eta = 0, \qquad f = 0, \qquad f' = 0, \qquad S = S_w(x) \quad \text{or} \quad S'_w = -\frac{c_{p_w} C_w x \dot{q}_w}{k_w H_e \sqrt{R_x}},$$

$$(10.6a)$$

$$\eta = \eta_e, \qquad f' = 1, \qquad S \equiv H/H_e = 1. \quad (10.6b)$$

As before, in Eqs. (10.3), (10.4), and (10.6) the primes denote differentiation with respect to η, and f' and S denote the dimensionless velocity and total-enthalpy ratios, u/u_e and H/H_e, respectively. The other parameters b, C, c, e, d, m_1, m_2, and R_x are defined by

$$b = C, \qquad C \equiv \frac{\rho\mu}{\rho_e\mu_e}, \qquad c = \frac{\rho_e}{\rho}, \qquad d = \frac{Cu_e^2}{H_e}\left(1 - \frac{1}{Pr}\right), \qquad e = \frac{b}{Pr},$$

$$\text{(10.7a)}$$

$$m_1 = \frac{1}{2}\left[1 + m_2 + \frac{x}{\rho_e\mu_e}\frac{d}{dx}(\rho_e\mu_e)\right], \qquad m_2 = \frac{x}{u_e}\frac{du_e}{dx}, \qquad R_x = \frac{u_e x}{\nu_e}.$$

$$\text{(10.7b)}$$

Because the Prandtl number of the fluid may assume a wide range of values, and because the viscosity-temperature relationship may vary greatly from one fluid to another, it is useful to consider a model fluid, for which $Pr = 1$ and for which there is a simple relationship between μ and T. The Illingworth Stewartson transformation, used with these simplifications for the fluid properties, simplifies the form of the governing equations and allows us to study the solutions of coupled external laminar boundary layers with and without heat transfer through the surface. Further, if the heat transfer through the surface is zero (adiabatic flow), the equations reduce to their incompressible form.

We define the Illingworth–Stewartson transformation by

$$dX = \frac{a_e}{a_\infty}\frac{p_e}{p_\infty}dx, \qquad \text{(10.8a)}$$

$$dY = \frac{a_e}{a_\infty}\frac{\rho}{\rho_\infty}dy. \qquad \text{(10.8b)}$$

where a is the speed of sound and subscript ∞ denotes—say—conditions at $y = \infty$. We also relate the dimensional stream function ψ to an incompressible stream function $F(X, Y)$ by

$$\psi = \rho_\infty F. \qquad \text{(10.9)}$$

Then, using the definition of stream function given by Eq. (10.2) and the relations given by Eqs. (10.8) and (10.9), and employing the usual chain rule for differentiation, we can write the momentum equation, Eq. (10.3), with primes denoting differentiation with respect to Y, as

$$F'\frac{\partial F'}{\partial X} - F''\frac{\partial F}{\partial X}$$
$$= \frac{a_\infty}{a_e}\left[-(F')^2\frac{d}{dX}\left(\frac{a_e}{a_\infty}\right) + \frac{p_e}{p_\infty}\frac{a_\infty}{a_e}u_e\frac{du_e}{dX}\right] + \frac{T_\infty}{T_e}\frac{\mu_e}{\mu_\infty}\nu_\infty(CF'')'.$$

$$\text{(10.10)}$$

Here the C denotes the density-viscosity parameter defined in Eq. (10.7a). In general, it is a function of x and y, but in the model fluid we assume that C is only a function of x. This assumption leads to useful simplifications in the solution of boundary-layer equations by approximate methods. Since viscous effects are most important near the wall, we may approximate C with good accuracy as

$$C(x) = \frac{\rho_w \mu_w}{\rho_e \mu_e} = \frac{\mu_w T_e}{T_w \mu_e} \tag{10.11}$$

by virtue of the perfect-gas assumption and the y independence of pressure. This empirical relationship is known as the *Chapman–Rubesin viscosity law*. Note that Eq. (10.11) would be exact, with $C=1$, if μ were proportional to T; actually, for air at temperatures near atmospheric, μ is approximately proportional to $T^{0.76}$.

Let us consider the first expression in Eq. (10.2) and write it as

$$\rho u = \frac{\partial \psi}{\partial y} = \rho_\infty \frac{\partial F}{\partial Y} \frac{\partial Y}{\partial y} = \rho \frac{a_e}{a_\infty} F'. \tag{10.12}$$

If we define $\bar{u}$ to be the incompressible u velocity ($\equiv F'$), then it follows from the above expression that the relationship between compressible and incompressible u velocities is

$$u = \frac{a_e}{a_\infty} \bar{u}. \tag{10.13}$$

(The speed of sound in the incompressible flow is simply assumed to be very large compared with u; the type of fluid need not be specified.) From the definition of speed of sound for an isentropic flow we can write

$$a_e^2 = \gamma R T_e = \frac{\gamma R}{c_p} c_p T_e = \frac{\gamma R}{c_p} \left[H_e - \frac{u_e^2}{2} \right]. \tag{10.14}$$

Differentiating this expression with respect to X, we get

$$2 a_e \frac{da_e}{dX} = -\frac{\gamma R}{c_p} u_e \frac{du_e}{dX}. \tag{10.15}$$

The first term on the right-hand side of Eq. (10.10) is the pressure-gradient term. Noting that $\partial F / \partial Y = \bar{u} = a_\infty / a_e u$ and that $H = c_p T + u^2/2$, we can write this pressure-gradient term as

$$\frac{1}{c_p T_e} \left(\frac{a_\infty}{a_e} \right)^2 H u_e \frac{du_e}{dX}. \tag{10.16}$$

Further, after writing $u_e = a_e / a_\infty \bar{u}_e$, we can show, by making use of (10.15), that

$$\frac{du_e}{dX} = \frac{c_p T_e}{H_e} \frac{a_e}{a_\infty} \frac{d\bar{u}_e}{dX}. \tag{10.17}$$

If we now substitute du_e / dX into Eq. (10.16) and again note the relationship between u_e and $\bar{u}_e$, we may write Eq. (10.16) as

$$S \bar{u}_e \frac{d\bar{u}_e}{dX}, \tag{10.18}$$

where $S \equiv H/H_e$. Also for a perfect gas and with $C = 1$, the transformed momentum equation (10.10) in Illingworth–Stewartson variables can be written as

$$v_\infty F''' + S\bar{u}_e \frac{d\bar{u}_e}{dX} = F' \frac{\partial F'}{\partial X} - F'' \frac{\partial F}{\partial X}, \qquad (10.19)$$

where the primes now denote differentiation with respect to Y. If the somewhat more general Chapman–Rubesin formula (10.11) is used for C, Eq. (10.19) is still valid, provided X is replaced by $\hat{x}$, where

$$d\hat{x} = C(X)\, dX = \frac{a_e p_e}{a_\infty p_\infty} C(x)\, dx. \qquad (10.20)$$

In order to express the energy equation in terms of Illingworth–Stewartson variables, we assume the Prandtl number to be unity. Then the total-enthalpy equation is given by Eq. (3.47b), which is

$$\rho u \frac{\partial H}{\partial x} + \rho v \frac{\partial H}{\partial y} = \frac{\partial}{\partial y}\left(\mu \frac{\partial H}{\partial y}\right). \qquad (3.47b)$$

By the use of the relations given by Eqs. (10.2), (10.8), and (10.9) and by a procedure similar to that used for the momentum equation, Eq. (3.47b) can be written as

$$v_\infty S'' = F' \frac{\partial S}{\partial X} - S' \frac{\partial F}{\partial X} \qquad (10.21)$$

with primes now denoting differentiation with respect to Y.

Similarly, the boundary conditions given by Eqs. (10.5) become

$$Y = 0, \qquad F = F' = 0, \qquad S' = 0 \quad \text{(adiabatic flow)}$$

or $\qquad S = S_w(x) \quad \text{(prescribed wall temperature)}, \qquad (10.22a)$

$$Y \to \infty, \qquad F' \to \bar{u}_e, \qquad S \to 1.0. \qquad (10.22b)$$

It is easy to see that the total-enthalpy equation for $\mathrm{Pr} = 1$, Eq. (3.47b), has the solution $H = H_e$ (that is, $S = 1$), which of course corresponds to an adiabatic flow. We can therefore see that $S = 1$ must also be a solution of Eq. (10.21), which then reduces Eq. (10.19) exactly to the standard form for an incompressible flow. We should remember, however, that when this equation is solved subject to the boundary conditions given by Eq. (10.22), the x and y coordinates are distorted according to Eq. (10.8), and the external velocity distribution of the incompressible flow ($\bar{u}_e$) is distorted according to $u_e(x) = \{[a_e(x)]/a_\infty(x)\}\bar{u}_e(X)$.

The adiabatic-wall solution $H = H_e$ of the total-enthalpy equation is called the *Crocco integral*. Note that since (1) the velocity u in a zero-pressure-gradient flow satisfies the same equation as Eq. (3.47b) but with H replaced by u, and (2) if H is a solution of Eq. (3.47b), so is $H - H_w$, it follows that Eq. (3.47b) has a solution $H - H_w + (\text{const})u$. The constant is evaluated by requiring $H = H_e$ when $u = u_e$ (external-flow conditions), giving

$$H = H_w - (H_w - H_e)\frac{u}{u_e}, \tag{10.23a}$$

$$T = T_w - \frac{1}{c_p}(H_w - H_e)\frac{u}{u_e} - \frac{u^2}{2c_p} \tag{10.23b}$$

for *arbitrary* wall temperature $T_w \equiv H_w/c_p$ and $Pr = 1$.

10.1 Similar Flows

For similar flows the dimensionless stream function f and dimensionless total-enthalpy ratio S in the compressible version of the Falkner–Skan transformation are functions of η only, by definition, and as a result, Eqs. (10.3), (10.4), and (10.6) reduce to

$$(bf'')' + m_1 ff'' + m_2\left[c - (f')^2\right] = 0, \tag{10.24}$$

$$(eS' + df'f'')' + m_1 fS' = 0, \tag{10.25}$$

$$\eta = 0, \quad f = 0, \quad f' = 0, \quad S = S_w \quad \text{or} \quad S' = S'_w \tag{10.26a}$$

$$\eta = \eta_e, \quad f' = 1, \quad S = 1, \tag{10.26b}$$

and for similarity all the parameters b, e, d, c, m_1, and m_2 defined by Eqs. (10.7) as well as the boundary conditions are independent of x.

To obtain the similarity solutions of the momentum and energy equations expressed in Illingworth–Stewartson transformed variables, we use the Falkner–Skan transformation for uncoupled flows discussed in Section 4.1, namely,

$$\eta = \left(\frac{\bar{u}_e}{\nu_\infty X}\right)^{1/2} Y, \quad F(X, Y) = (\bar{u}_e \nu_\infty X)^{1/2}\phi(\eta). \tag{10.27}$$

Using this transformation, we can write Eqs. (10.19) and (10.21) as

$$\phi''' + \frac{m+1}{2}\phi\phi'' + m\left[S - (\phi')^2\right] = 0, \tag{10.28}$$

$$S'' + \frac{m+1}{2}\phi S' = 0, \tag{10.29}$$

where in Eq. (10.27) we have changed the notation used for the stream function from $f(\eta)$ to $\phi(\eta)$ in order to distinguish it from the definition used in Eq. (4.19). In Eqs. (10.28) and (10.29), m is the usual definition of the dimensionless pressure-gradient parameter,

$$m = \frac{X}{\bar{u}_e}\frac{d\bar{u}_e}{dX}. \tag{10.30}$$

The same formation is obtained when the Chapman–Rubesin approximation is used except that X is replaced by $\hat{x}$.

Similarly the boundary conditions given by Eqs. (10.22) can be written as

$$\eta = 0, \qquad \phi = \phi' = 0 \qquad S = S_w \quad \text{or} \quad S' = 0, \qquad (10.31a)$$

$$\eta = \eta_e, \qquad \phi' = 1, \qquad S = 1. \qquad (10.31b)$$

Again, the solutions of Eqs. (10.28) and (10.29) are independent of X provided that the above boundary conditions are independent of X. These requirements are satisfied if S_w and m are constants.

The requirement that S_w is constant leads to uniform wall temperature, since H_e is constant in an isentropic external stream, and the requirement that m is constant, as in uncoupled flows, leads to

$$\bar{u}_e = cX^m, \qquad (10.32)$$

where c is a constant. Thus, we get similarity solutions in coupled laminar boundary layers on impermeable walls only when the distorted external velocity $\bar{u}_e$ varies with the distorted surface distance X as prescribed by Eq. (10.32) and when the surface temperature is constant. As in uncoupled laminar flows, the importance of similar flows is quite evident; we solve two ordinary differential equations, rather than two partial differential equations. However, as in uncoupled flows, values of m other than 0 or 1, corresponding to flat-plate flow and stagnation-point flow, respectively, are rare in practice.

Relation between Coupled and Uncoupled Boundary-Layer Parameters

As in uncoupled similar flows, the boundary-layer parameters such as δ^*, θ, and c_f can be written in simplified forms by using either the compressible version of the Falkner–Skan transformation or the Illingworth–Stewartson transformation. Formulas that connect coupled boundary-layer parameters to the solutions of Eqs. (10.28) and (10.29), subject to Eq. (10.31) with constant S_w, can also be derived. For an *adiabatic* flow their solutions reduce to uncoupled flow solutions; in this case Table 4.1 can be used.

As an example, let us consider a flow with zero pressure gradient, governed by the incompressible version of the compressible-flow equations given by Eqs. (10.28) and (10.29). Regardless of heat transfer, the latter can be written as

$$\phi''' + \tfrac{1}{2}\phi\phi'' = 0, \qquad (10.33)$$

$$S'' + \tfrac{1}{2}\phi S' = 0, \qquad (10.34)$$

which are identical to the incompressible-flow equations, (4.31) and (4.32) with $m = 0$, $n = 0$, and $\text{Pr} = 1$. Solving Eqs. (10.33) and (10.34) for ϕ and equating the two expressions, we get

$$\frac{S''}{S'} = \frac{\phi'''}{\phi''}. \qquad (10.35)$$

Integrating this expression twice and using the boundary conditions given by Eq. (10.31), with S_w now being constant for similarity, we get

$$S = S_w - (S_w - 1)\phi', \tag{10.36}$$

which is the same as Eq. (10.23) when that equation is written in terms of S. From Eq. (10.36) we see that the static-temperature distribution across the boundary layer is

$$\frac{T}{T_e} = -\frac{\bar{u}_e^2}{2c_p T_e}(\phi')^2 + \left(1 + \frac{\bar{u}_e^2}{2c_p T_e}\right)[\phi' + S_w(1 - \phi')]. \tag{10.37}$$

By definition, the displacement thickness for a compressible flow is

$$\delta_c^* = \int_0^\infty \left(1 - \frac{\rho u}{\rho_e u_e}\right) dy, \tag{10.38a}$$

which in terms of Illingworth–Stewartson variables can be written as

$$\delta_c^* = \frac{a_\infty \rho_\infty}{a_e \rho_e}\left[\int_0^\infty \left(\frac{\rho_e}{\rho} - \frac{\bar{u}}{\bar{u}_e}\right) dY\right]. \tag{10.38b}$$

For a perfect gas, the above expression becomes

$$\delta_c^* = \frac{a_\infty \rho_\infty}{a_e \rho_e}\left[\int_0^\infty \left(\frac{T}{T_e} - \frac{\bar{u}}{\bar{u}_e}\right) dY\right]. \tag{10.38c}$$

The temperature distribution for an adiabatic flow ($S \equiv 1$) can be obtained by recalling the definition of total enthalpy, which can be written as

$$H = c_p T + \frac{u^2}{2} = c_p T_e + \frac{u_e^2}{2},$$

or in terms of Illingworth–Stewartson variables

$$\frac{T}{T_e} = 1 + \frac{\gamma - 1}{2}M_\infty^2\left[1 - \left(\frac{\bar{u}}{\bar{u}_e}\right)^2\right]. \tag{10.39}$$

Inserting Eq. (10.39) into (10.38c), rearranging, and recalling that for a constant-pressure flow $a_\infty \rho_\infty / a_e \rho_e = 1$, we can write δ_c^* as

$$\delta_c^* = \int_0^\infty \left(1 - \frac{\bar{u}}{\bar{u}_e}\right) dY + \frac{\gamma - 1}{2}M_\infty^2 \int_0^\infty \left[1 - \left(\frac{\bar{u}}{\bar{u}_e}\right)^2\right] dY. \tag{10.40}$$

Noting that for any variable z,

$$\int_0^\infty (1 - z^2)\, dY = \int_0^\infty z(1 - z)\, dY + \int_0^\infty (1 - z)\, dY,$$

we can write Eq. (10.40) as

$$\delta_c^* = \delta_i^* + \sigma\left(\delta_i^* + \theta_i\right), \tag{10.41}$$

where δ_i^* and θ_i are dimensionless incompressible displacement and momentum-thickness parameters, respectively (see Table 4.1), and σ is a

Mach-number parameter defined by

$$\sigma = \frac{\gamma - 1}{2} M_\infty^2. \tag{10.42}$$

By following a procedure similar to that described for displacement thickness, we can also derive relations between compressible and incompressible values of momentum thickness and local skin-friction coefficients. For an adiabatic flat-plate flow they are

$$\theta_c = \theta_i, \qquad c_{f_c} = \frac{\rho_w}{\rho_\infty} \frac{\mu_w}{\mu_\infty} c_{f_i}. \tag{10.43}$$

Here c_{f_i} denotes the incompressible value of the local skin-friction coefficient at the same value of $u_e \theta / \nu_e$.

Since $\theta_c = \theta_i$, we can rewrite Eq. (10.41) as

$$H_c = H_i + \sigma(H_i + 1), \tag{10.44}$$

where H is the shape parameter δ^*/θ and *not* the total enthalpy. This shows that the shape parameter H_c of a compressible laminar boundary layer increases with Mach number. The physical reason is that viscous dissipation

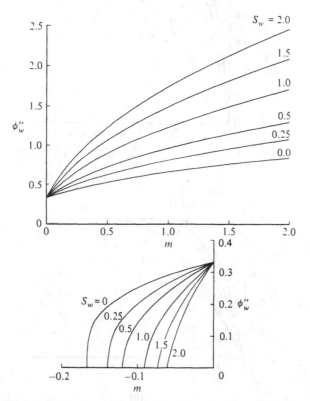

Figure 10.1. Variation of the wall shear parameter f_w'' with pressure gradient parameter m and heat transfer parameter S_w for compressible similar flows: Note different scales for $m > 0$ or $m < 0$.

of kinetic energy into heat increases T and reduces ρ near the surface so that the integrand $\rho u/\rho_e u_e(1 - u/u_e)$ in the definition of θ is a smaller fraction of the integrand $1 - \rho u/\rho_e u_e$ in the definition of δ^* than if ρ/ρ_e were equal to unity.

If we assume that $\mu \sim T$, then it follows from the second relation in Eq. (10.43) that for an adiabatic flat plate, the compressible local skin-friction coefficient is the same as in the transformed (incompressible) flow since $T_e = T_\infty$.

Solution with Heat Transfer

For an adiabatic flow, $S \equiv 1$, and the solution of the momentum equation (10.28) is independent of the energy equation (10.29). This means that we can now use the incompressible flow solutions of Eq. (10.28) for coupled

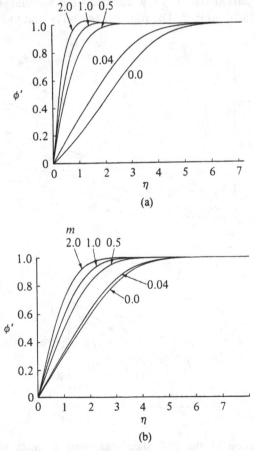

Figure 10.2. The dimensionless velocity, f' as a function of η for various values of m (a) $S_w = 2.0$. (b) $S_w = 0$.

adiabatic flows. For a flow with heat transfer, $S_w \neq 1$, the solutions of the momentum and energy equations are coupled. In this case it is necessary to solve Eqs. (10.28) and (10.29) simultaneously for specified values of S_w and m. Note that for each value of the dimensionless pressure gradient m, we can get different solutions corresponding to different values of $S_w \neq 1$.

Figure 10.1 shows the variation of the wall shear parameter ϕ_w'', with pressure-gradient parameter m and with heat-transfer parameter S_w. We note that although in zero pressure gradient ($m = 0$) the Illingworth–Stewartson transformation makes the solution of the momentum equation in transformed variables independent of S_w [see Eq. (10.33)], heat transfer has a large effect on the solution in pressure-gradient flows. From Fig. 10.1 we see that for a given pressure gradient, heating the wall ($S_w > 1$) increases the value of ϕ_w'' and cooling the wall ($S_w < 1$) decreases the value of ϕ_w''.

Figures 10.2 and 10.3 show the velocity profiles and enthalpy profiles across the boundary layer for various values of S_w and m. From Fig. 10.2(a) we see that, with heating, the velocity in the boundary layer can exceed the external velocity ($\phi' > 1$) when the pressure gradient is *favorable*. This

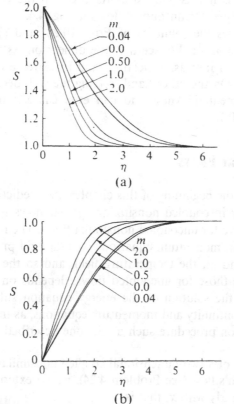

(a)

(b)

Figure 10.3. The dimensionless enthalpy, S, as a function of η for various values of m (a) $S_w = 2.0$. (b) $S_w = 0$.

Table 10.1. Solutions of Eqs. (10.28) and (10.29) for $m = 1$

S_w	ϕ_w''	$\delta_1^* = \int_0^\infty (S - \phi')\,d\eta$	$\theta_1 = \int_0^\infty \phi'(1 - \phi')\,d\eta$	H	S_w'
2	1.7361	1.3855	0.1753	7.904	−0.6154
1.5	1.4911	1.0235	0.2338	4.378	−0.2972
1.0	1.2326	0.6479	0.2919	2.216	0
0.5	0.9545	0.2567	0.3489	0.736	0.2710
0.25	0.8060	0.0529	0.3766	0.141	0.3041
0	0.6487	−0.1577	0.4032	−0.391	0.5066

feature occurs at all positive values of m, $0 < m < \infty$, if the wall is heated, but it is less marked as m becomes smaller. Physically the reason is that when $m > 0$, the favorable pressure gradient tends to accelerate the fluid; heating the wall diminishes ρ, from the equation of state, and so augments the effect of the pressure gradient on the acceleration of the fluid, increasing ϕ' and, as remarked above, ϕ_w''. Also, an inflexion appears even in the zero-pressure-gradient profile ($m = 0$) because μ varies with y.

From Fig. 10.3 we see that the effect of the pressure gradient on the enthalpy profiles is not as marked as its effect on the velocity profiles, as already demonstrated for uncoupled flows in Fig. 4.4.

Table 10.1 shows the solutions of Eqs. (10.28) and (10.29) for several values of S_w when $m = 1$. We see that the dimensionless momentum thickness decreases with increasing heating, and in extreme cases (if $\phi' > 1$) it may be negative. On the other hand, *cooling* the wall decreases the dimensionless displacement thickness and can even cause it to be negative, as shown in Table 10.1.

10.2 Nonsimilar Flows

As mentioned at the beginning of this chapter, the prediction of momentum and heat transfer in coupled nonsimilar laminar flows is slightly different from the procedure for uncoupled nonsimilar flows. The main reason is that the continuity and momentum equations contain fluid properties such as ρ and μ that depend on the temperature field, and so the solution of these equations, unlike those for uncoupled flows, depends on the energy equation. Of course, the solution of the energy equation still depends on the solution of the continuity and momentum equations, as in uncoupled flows, and so an iteration procedure such as the one described in Section 13.4 is needed.

As an example of a two-dimensional coupled nonsimilar laminar flow, we consider Howarth's flow (see Problem 4.24), whose external velocity distribution varies linearly with x, that is,

$$u_e = u_\infty \left(1 - \frac{x}{8}\right). \tag{10.45}$$

This flow can be started with a flat-plate profile of negligible thickness (at $x = 0$), but the pressure gradient is adverse, and separation eventually occurs. For an uncoupled flow with no heat transfer, calculations (Problem 4.24) indicate separation at $x = 0.96$.

To show the effect of heat transfer on separation point, we first consider the momentum and energy equations expressed in Illingworth–Stewartson variables, namely, Eqs. (10.19), (10.21), and (10.22). Using the Falkner–Skan transformation given by Eq. (10.27) and allowing the dimensionless stream function ϕ to vary with X as well, we can write Eqs. (10.19), (10.21), and (10.22) as

$$\phi''' + \frac{m+1}{2}\phi\phi'' + m\left[S - (\phi')^2\right] = X\left(\phi'\frac{\partial\phi'}{\partial X} - \phi''\frac{\partial\phi}{\partial X}\right), \quad (10.46)$$

$$S'' + \frac{m+1}{2}\phi S' = X\left(\phi'\frac{\partial S}{\partial X} - S'\frac{\partial\phi}{\partial X}\right), \quad (10.47)$$

$$\eta = 0, \qquad \phi = \phi' = 0, \qquad S = S_w(x) \quad \text{or} \quad S' = 0, \quad (10.48a)$$

$$\eta = \eta_e, \qquad \phi' = 1, \qquad S = 1. \quad (10.48b)$$

We solve the above equations by using the numerical method and Fortran program discussed in Sections 13.4 and 13.5 for values of S_w ($\equiv H_w/H_e$) equal to 2, 1, 0.25, and 0, thus covering the case in which the wall is heated ($S_w = 2$), cooled ($S_w = 0.25$), or cooled to effectively zero enthalpy ($S_w = 0$) and the case in which there is no heat transfer ($S_w = 1$). In the latter case, we expect the solutions of the momentum equation to indicate the flow separation to be at $X = 0.96$, as in uncoupled flow.

Figures 10.4 and 10.5 show the variation of dimensionless wall shear and heat-transfer parameters ϕ_w'' and S_w' with X for four different values of S_w. As anticipated, for $S_w = 1$, we find flow separation at $X = 0.96$ (Fig. 10.4). With heating the flow separation moves forward, to $X = 0.6$ if $S_w = 2$. On the other hand, cooling stabilizes the boundary layer and delays the separation. The calculations indicate separation at $X = 1.86$ for $S_w = 0.25$ and at $X = 2.74$ for $S_w = 0$. We also note from Fig. 10.5 that at separation, the wall heat-transfer parameter S_w' is finite, indicating, as in the uncoupled calculations presented in Table 4.3, that the Reynolds analogy is not realistic near separation.

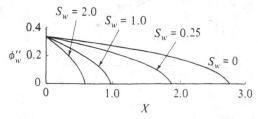

Figure 10.4. Variation of the wall shear parameter for a compressible Howarth flow with dimensionless wall temperature.

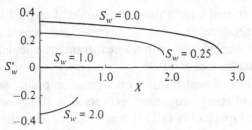

Figure 10.5. Variation of the wall heat transfer parameter for a compressible Howarth flow with dimensionless wall temperature.

Figures 10.6 and 10.7 show the effect of Mach number and Prandtl number on the solutions of momentum and energy equations. These calculations are made for the same external velocity distribution, Howarth's flow, by solving the momentum and energy equations given by Eqs. (10.3) and (10.4) for values of S_w equal to 2.0 and 0.25. The results in Figs. 10.6 and 10.7 are for a Prandtl number of 0.72 and for freestream Mach numbers of 0.05, 1.0, and 2.0. Shown in these figures are the incompressible results obtained by the solution of Eqs. (10.46) and (10.47).

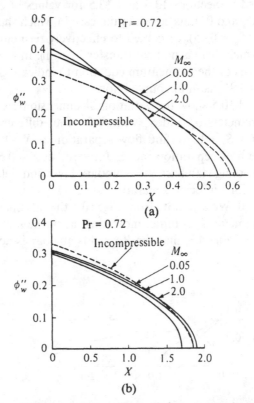

Figure 10.6. Effect of Mach number on the wall shear parameter ϕ_w'' (a) $S_w = 2.0$. (b) $S_w = 0.25$.

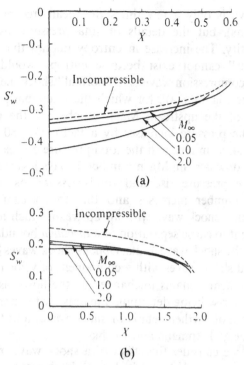

Figure 10.7. Effect of Mach number on the wall heat transfer parameter S_w'. (a) $S_w = 2.0$. (b) $S_w = 0.25$.

The results in Fig. 10.6 show that with either heating or cooling, the Mach number has a considerable influence on the separation point. In the case of heating ($S_w = 2$), the separation point moves from $X \simeq 0.61$ for $M_\infty = 0.05$ to $X \simeq 0.55$ for $M_\infty = 1.0$ and to $X \simeq 0.425$ for $M_\infty = 2.0$. In the case of cooling ($S_w = 0.25$), the separation point moves from $X \simeq 1.9$ for $M_\infty = 0.05$ to $X \simeq 1.86$ for $M_\infty = 1.0$ and to $X \simeq 1.7$ for $M_\infty = 2.0$. We see that for a given S_w, the effect of Mach number on the location of separation point is similar to the effect of heating for a given M_∞; that is, the separation point moves forward with increase of M_∞ or S_w. In Figs. 10.6 and 10.7 we also observe that the wall shear parameter ϕ_w'' and the location of separation are affected little by compressibility if M_∞ is less than about 1.0. We also observe that the Mach number has a smaller effect on the heat-transfer parameter S_w' when the surface is cooled than when it is heated.

10.3 Shock-Wave/Shear-Layer Interaction

A shock wave is a sudden increase of pressure and density across a very thin "wave" front in supersonic flow. It is, of course, governed by the equations of conservation of mass, momentum, and energy outlined above, but it results in a sudden decrease in total pressure and increase in entropy; the dissipation is effected by viscosity (as modified by the fact that the thickness of the shock wave is only a few times the mean free path between gas molecule collisions), but the details of total-pressure loss, etc., do not depend on viscosity. The increase in entropy implies that a concentrated "rarefaction wave" cannot exist because entropy would decrease. The simplest kind of compression wave is a "normal" shock, i.e., normal to the flow direction [see Fig. 10.8(a)], for which the fractional pressure rise is a unique function of the upstream Mach number. If the upstream Mach number is 1.3, the pressure increases by a factor of 1.80 (if the ratio of specific heats is 1.4, as in air), and the total pressure decreases to 0.98 of its initial value. The downstream Mach number, always less than unity, is 0.79 in this case. The pressure rise and total-pressure loss increase as the upstream Mach number increases and the downstream Mach number decreases. A normal shock wave with an upstream Mach number of 1.3 is just about sufficient to cause separation of a *turbulent* boundary layer from a surface on which the shock wave abuts. *Oblique* shock waves [Fig. 10.8(b), (c)] are simply normal shock waves with a component of velocity parallel to the shock; this component remains unchanged as the flow passes through the shock, the pressure rise being determined mainly by the normal component of Mach number. For further details of shock waves and the formulas for pressure rise, see, e.g., Liepmann and Roshko [3].

In this section we consider the effect of a shock wave on a shear layer. The two main cases are that in which the shock wave is imposed from outside (an "incident" shock) and that in which the shock springs from a solid surface carrying a boundary layer; note that in the latter case a shock system of some sort would be present even in the absence of a boundary layer. Figure 10.9(a) shows the pattern of shock waves and expansion fans that occurs in a jet exhausting into a still atmosphere at a speed greater than that of sound when the atmospheric pressure is greater than that in the jet nozzle; an example is the lift-off of a launch vehicle whose rocket nozzle is optimized for flight in the upper atmosphere. Figure 10.9(b) shows the case of an underexpanded jet, where the atmospheric pressure is less than that in the nozzle. The obvious difference between the two cases is that the overexpanded flow starts with a shock wave, through which the pressure rises so that the pressure in the downstream part of the jet boundary is equal to atmospheric pressure, whereas the underexpanded flow starts with an expansion fan, through which the pressure *falls* to the atmospheric value. The expansion takes place over a finite distance because the speed of sound decreases through the expansion and the Mach number M increases, so that the angle of a weak-disturbance wave to the flow direction, $\sin^{-1}(1/M)$, increases.

In a compression, on the other hand, the Mach number decreases downstream, the inclination of a weak-disturbance wave is greater than upstream of the shock, and all the weak-compression waves merge into a shock wave. Where a shock wave intersects a constant-pressure boundary, such as the atmospheric boundary at point A in Fig. 10.9(a), it has to reflect as an expansion (crudely speaking, the pressure rise caused by the shock has

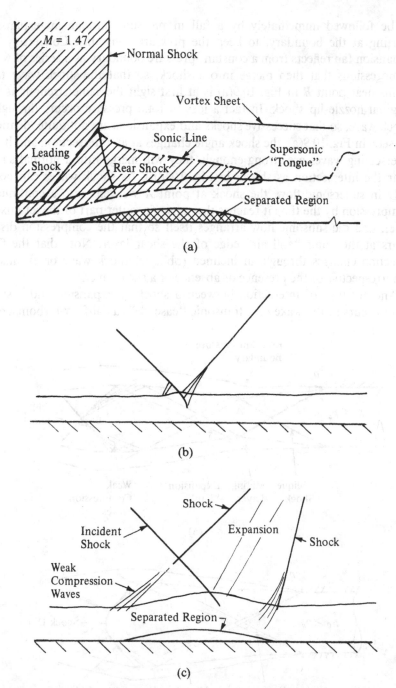

(a)

(b)

(c)

Figure 10.8. (a) Interaction between normal shock wave and turbulent boundary layer (after [4]). (b) Weak incident oblique shock. (c) Strong oblique shock (the reflected shocks and expansion merge to form one shock).

to be followed immediately by a fall in pressure through an expansion starting at the boundary, to keep the pressure constant). Conversely, an expansion fan reflects from a constant-pressure boundary as a series of weak compressions that then merge into a shock, so that the shock wave that forms near point B in Fig. 10.9(a) is at first sight the same strength as the original nozzle lip shock. In fact a loss in total pressure occurs through a shock wave, so that successive shocks and expansions are not quite identical. As seen in Fig. 10.9(a), the shock angle changes at an intersection, and if the intersecting waves are strong enough, they will merge into a normal shock near the intersection as shown in Fig. 10.9(b). As shock waves can occur only in supersonic flow, the shock at point A weakens into a continuous compression by the time it reaches the subsonic, outer part of the jet mixing layer, and the subsonic flow arranges itself so that the compression disappears at the outer "still air" edge of the shear layer. Note that the flow direction changes through an inclined (oblique) shock wave or expansion fan irrespective of the presence or absence of a shear layer.

Another type of interaction between a shock or expansion and a shear layer occurs in the wake of a transonic "cascade" (axial-flow turbomachine

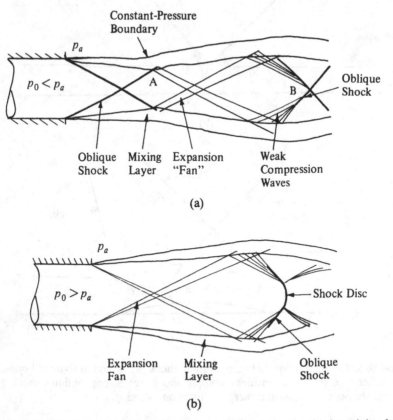

Figure 10.9. Interactions between shock or expansion waves and a jet mixing layer. (a) "Over-expanded" jet: exit pressure below ambient. (b) "Under-expanded" jet: exit pressure above ambient.

blade row) where waves shed from one blade—say, at the trailing edge—intersect the wake of another blade. Here the wave passes right through the wake, and only if the speed near the wake centerline is subsonic will there be a large spread of the pressure disturbance upstream and downstream of the intersection point.

Figure 10.8(b) shows the interaction of a shock wave, generated somewhere upstream, with a boundary layer. At a solid surface the boundary condition is one of constant flow direction rather than constant pressure, so that the incoming shock, which deflects the flow downward, is followed by an outgoing shock—not an expansion—which deflects the external flow upward to something near its original direction. As in the case of the jet mixing layer, the shock cannot penetrate into the subsonic part of the flow, near the wall, so that a distributed pressure rise occurs upstream and downstream of the shock intersection position.

Whether the shock wave is an incident (usually oblique) shock or a shock wave generated at the surface—and therefore necessarily normal to the surface so that the flow remains nominally parallel to the surface—a large positive ("adverse") pressure gradient is imposed on the boundary layer so that the flow is retarded, especially near the surface. If the shock is strong enough, the boundary layer may separate, as shown in Fig. 10.8(a). The separated region is often fairly short because of the effects of upstream influence. As can be seen in Fig. 10.8(c), the retardation of the flow at the front of the interaction region causes the streamlines to move away from the surface, generating compression waves in the supersonic part of the flow that then merge into an outgoing shock. The first and second outgoing shocks and the expansion fan from the point of intersection of the incoming shock merge further out in the flow into a shock wave of the right strength to cancel the incoming shock's deflection of the flow. If the incident shock is strong enough, a nearly normal shock may extend outward from the shear layer to the intersection of the normal and reflected shocks, as in the shock intersection near the centerline of a jet (Fig. 10.9). Similar behavior can occur, less spectacularly, even if the flow does not separate, because the slow-moving fluid near the wall tends to simulate a constant-pressure boundary whether it is moving downstream or upstream.

Other forms of shock-wave shear-layer interaction include the flow over a backward-facing step, the prototype of many more complicated flows that reattach to a solid surface after an extended separation region (shock induced or otherwise). As seen in Fig. 10.10, an expansion occurs at the edge of the step, and since the pressure in the separated region behind the step is nearly constant, the external flow is also nearly straight, according to the usual rules of supersonic flow in which flow angle and pressure are simply connected. Thus, a well-behaved free-mixing layer (Section 8.2 or 8.4) develops, and unless the initial boundary-layer thickness is a large fraction of the step height, the mixing layer is nearly independent of the initial conditions. Fluid is entrained into the underside of the mixing layer, as shown by the streamlines in Fig. 10.10, and is supplied by the upstream

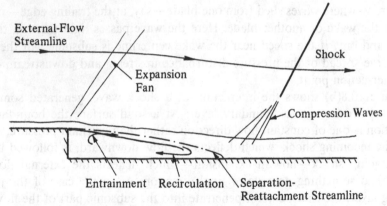

Figure 10.10. Supersonic flow down a backward-facing step.

deflection of fluid near the reattachment point—indeed without this entrainment the flow would not reattach. The pressure gradient near reattachment is large because the downstream-going flow has to be turned to follow the surface again, and the distributed pressure rise near the surface leads to the formation of a shock wave in the outer part of the flow, nominally canceling the expansion at the edge of the step. That is, the shock wave is caused by the shear layer rather than being imposed on it from outside.

Free Interaction and Upstream Influence

A striking feature of the interaction between shock waves and (especially laminar) boundary layers is that when the shock is of moderate strength, typically with pressure rises of 20 percent or more, the boundary layer is disturbed by the shock at considerable distances ahead of it. At first sight this phenomenon is surprising since the inviscid flow above the boundary layer is supersonic and cannot permit disturbances to travel upstream; moreover the boundary-layer equations are parabolic, and so, unless separation occurs, disturbances can only travel downstream. It would seem that upstream disturbances are simulated only if the full Navier–Stokes equations are used since these are elliptic and permit upstream disturbances. However, the expected distance upstream that disturbances within the shear layer would be propagated in this way is only one or two boundary-layer thicknesses and much smaller than the extent actually observed in laminar flow.

The explanation for this dramatic phenomenon was provided by Oswatitsch and Wieghardt [5]. They pointed out that the interaction between the boundary layer and the external supersonic inviscid stream is of such a character as to permit the spontaneous evolution of the flow at any position $x = x_0$, as follows. Suppose the pressure in the external stream at x_0 increased by a small amount. Then the boundary layer for $x < x_0$ would be under the influence of an adverse pressure gradient and would be expected to thicken, leading to an increase in the displacement thickness and an increase in the equivalent blowing velocity $v_w = (1/\rho_e)d(\rho_e u_e \delta^*)/dx$ [see Eq. (10.68)]. Once this happens, the external stream starts to turn away from the wall, and then Ackeret's formula [3]

$$\Delta p = \frac{u_e \rho_e v_w}{\sqrt{M_e^2 - 1}} \qquad (10.49)$$

implies that the pressure also increases. Hence the boundary layer will continue to develop until something happens to change the mutual interaction. It turns out that if the development carries on far enough, separation occurs, and thereafter an increase in the boundary-layer thickness is associated with a decrease in the pressure gradient. Eventually the pressure reaches a constant value, the so-called plateau pressure, while the separated boundary layer continues to grow. Part of the reason for the decrease in the adverse pressure gradient in separated flows is that the reverse flow at the bottom of the boundary layer is moving very slowly and needs only a weak pressure gradient to sustain it.

In a similar way the assumption of a small pressure *fall* leads to a boundary-layer change that encourages a further fall in the pressure so that the process is amplified. Now, however, an acceptable ultimate form of the interaction is not possible, and the development continues until the assumptions underlying the division of the flow field into inviscid and boundary-layer flows is no longer tenable. These two ways in which the interaction can evolve are commonly referred to in the literature as *compressive free interaction* and *expansive free interactions*, respectively, the first being relevant to shock-wave interactions and concave corners, while the second is relevant to the flows just upstream of convex corners.

If the external flow was subsonic, this boundary-layer behavior would be impossible since perturbations in the boundary layer and the external stream would tend to act in opposition to each other. For example, a small pressure *increment* thickens the boundary layer and induces a positive normal velocity v_w, but positive values of v_w induce small pressure *decrements*.

The principal reason why the growth of free interactions in boundary layers is limited is that the pressure changes with x and hence must be compensated for by a geometrical or physical feature of the flow field. Otherwise, there will be an imbalance with the pressure at large distances from the body.

The scaling laws of the evolutionary part of the free interaction were given by Lighthill [6]. As might be expected, the length scale along the body over which it occurs is short, being formally $O(\delta^* R^{1/8})$, where δ^* is the displacement thickness of the boundary layer and R is the Reynolds number equal to $(u_e x_0)/\nu_w$, and amounts in practice to only a few boundary-layer thicknesses. The distinguishing feature of compressible free interactions that gives them special importance in practical calculations is that the final state of the evolution is self-preserving. As already explained, it consists of a wedge-shaped separation region over which the oncoming boundary layer and inviscid stream are deflected. This flow deflection leads to a pressure increment that for insulated walls is

$$\frac{1}{2} \frac{cC^{1/4}\lambda_0^{1/2}\rho_e u_e^2}{R^{1/4}\left(M_e^2 - 1\right)^{1/4}},$$

(10.50)

where λ_0 is Chapman's constant equal to $(T_e\mu_e/T_w\mu_w)$ and c is a quantity that appears to take an almost constant value of about 1.6. Below this forward-moving fluid, there is a slowly recirculating region of "back-flow."

Since the occurrence of the free interaction is surprising at first sight, it is worthwhile describing at length the arguments that lead to Lighthill's scaling laws so that the reader may appreciate in detail how they arise.

The starting point for our analysis is a remarkable and simple result originally due to Prandtl, known as the *transposition theorem*. This states that if $u(x, y), v(x, y), T(x, y)$ is any solution of the boundary-layer equations of steady flow, then so is $U(x, Y), V(x, Y), \theta(x, Y)$, where

$$Y = y + A(x), \qquad U(x, Y) = u(x, y), \qquad V = v - A'(x)u, \qquad \theta = T,$$

(10.51)

and $A(x)$ is an arbitrary function of x. The proof is by direct substitution into the boundary-layer equations and is immediate. It is an exact theorem, there being no restriction on A, and in particular we shall be able to allow $A'(x) = O(R^{-1/4})$, where R is the Reynolds number of the flow, $u_e x_0/\nu_w$. Then the contribution to V from $A'(x)u$ is much larger than that from v, which is only $O(R^{-1/2}u)$. Hence we may neglect v in Eq. (10.51).

For simplicity we consider a boundary layer on a flat plate under a uniform external stream. Let the free interaction be centered at a station x_0 of the boundary layer, and let the distance over which it evolves be small compared with x_0 but large compared with $\delta^* = O(R^{-1/2}x_0)$. Further let it be described by a displacement of the streamlines and characterized by the function $A(x)$, which we have to determine. To a first approximation, therefore, we are assuming that any induced pressure gradient does not modify the flow properties of the majority of the boundary layer in the interaction region. It follows from Eq. (10.51) that the normal component of the velocity just outside the boundary layer is given by

$$v_e(x) = v(x_0, \infty) - A'(x)u_e(x_0),$$

(10.52)

and, as pointed out earlier, we shall neglect $v(x_0, \infty)$ in Eq. (10.52). Notice that $x - x_0$ is assumed to be small so that in the *unperturbed* boundary layer, the flow properties do not change to first order. The induced pressure is then given by

$$\Delta p = -\frac{u_e^2(x_0)\rho_e A'(x)}{\left(M_e^2 - 1\right)^{1/2}}$$

(10.53)

from Eq. (10.49).

Although this pressure rise is not significant over most of the boundary layer, it does become important near the plate because the fluid is moving slowly there. In addition to this induced pressure, the simple solution given by Eq. (10.51) must itself be modified near the plate because $U \neq 0$ when $y = 0$. A velocity sub-boundary layer must be introduced near $y = 0$, which is driven by the pressure Δp and matches the solution given by Eq. (10.53), i.e., has the property

$$u = \lambda[y + A(x)], \qquad (10.54)$$

where λ is the value of the undisturbed velocity gradient at $y = 0$, when y is small on the scale of the main boundary layer and large on the scale of the sub-boundary layer. A similar sub-boundary layer is needed in general to adjust the thermal properties of the fluid, but this does not affect the velocity sub-boundary layer.

The governing equations for the sub-boundary layer are essentially the same as those for the main boundary layer except that the density and kinematic viscosity may be assigned their wall values. They are given by

$$\frac{\partial u}{\partial x} + \frac{\partial v}{\partial y} = 0, \qquad (10.55)$$

$$u \frac{\partial u}{\partial x} + v \frac{\partial u}{\partial y} = -\frac{1}{\rho_w} \frac{\partial p}{\partial x} + v_w \frac{\partial^2 u}{\partial y^2}, \qquad (10.56)$$

where we note that the scale of y is now much smaller than in the main boundary layer. Indeed we shall see that the scale of y here is $O(R^{-5/8})$, whereas in the main boundary layer the scale of y is $O(R^{-1/2})$. The boundary conditions are

$$u = v = 0 \qquad \text{at } y = 0, \qquad (10.57a)$$

$$u \rightarrow \lambda y \qquad \text{upstream of the disturbance,} \qquad (10.57b)$$

$$u \rightarrow \lambda[y + A(x)] \qquad \text{as } y \rightarrow \infty \qquad (10.57c)$$

as the sub-boundary layer merges with the main boundary layer. Here ρ_w and v_w are the undisturbed density and kinematic viscosity of the fluid at the plate. The boundary condition given by Eq. (10.57b) reflects the fact that the interaction is spontaneous; i.e., there is no external agency forcing it in the neighborhood of $x = x_0$. Finally Eq. (10.57c) is necessary since u takes this form near the base of the main boundary layer. Recall that $\lambda = O(R^{1/2})$ and $v_w = O(R^{-1})$, and bear in mind that we shall have to assume y and $A(x)$ are of the same size in the sub-boundary layer—otherwise, a contradiction is obtained.

The set of equations given by Eqs. (10.55) and (10.56) is sufficient to determine A as a function of x, although it is clear that a considerable numerical effort is required. It is straightforward to determine the scaling laws of the free interaction, however. We write, in the sub-boundary layer,

$$y = \lambda^{-5/4}\alpha z, \qquad x = x_0 + \frac{\lambda^{-3/4}\alpha^3 X}{v_w \lambda^2},$$

$$u = \lambda^{-1/4}\alpha\tilde{u}(X,z), \qquad v = \lambda^{-3/4}\alpha^{-1}\tilde{v}(X,z)(v_w\lambda^2),$$

$$A(x) = \lambda^{-5/4}\alpha\tilde{A}(X), \qquad p = p_e(x_0) + \frac{\lambda^{-1/2}\rho_e u_e^2 \alpha^{-2} P(X)(v_w\lambda^2)}{\left(M_e^2 - 1\right)^{1/2}},$$

$$\tag{10.58}$$

where

$$\alpha = \left(\frac{\rho_e u_e^2 v_w \lambda^2}{\rho_w \left(M_e^2 - 1\right)^{1/2}}\right)^{1/4}, \tag{10.59}$$

and note that since $\lambda^2 v_w$ is independent of the Reynolds number, so is α. The governing equations to determine $A(x)$ now reduce to

$$\frac{\partial \tilde{u}}{\partial X} + \frac{\partial \tilde{v}}{\partial z} = 0, \tag{10.60}$$

$$\tilde{u}\frac{\partial \tilde{u}}{\partial X} + \tilde{v}\frac{\partial \tilde{u}}{\partial z} = \frac{d^2\tilde{A}}{dX^2} + \frac{\partial^2 \tilde{u}}{\partial z^2} \tag{10.61}$$

with boundary conditions

$$\tilde{u} = \tilde{v} = 0 \qquad \text{at } z = 0, \tag{10.62a}$$

$$\tilde{u} \to z \qquad \text{as } X \to -\infty, \tag{10.62b}$$

$$\tilde{u} \to z + A(X) \quad \text{as } z \to \infty. \tag{10.62c}$$

It now follows, from the fact that $\lambda = O(R^{1/2})$, that the pressure rise in the compressible free interaction is $O(R^{-1/4})$, as indicated in Eq. (10.50), and that the longitudinal extent of the interaction is $O(R^{-3/8}) = O(R^{1/8}\delta^*)$. Further discussion of the scaling laws depends on the nature of the fluid, but if, for example, the fluid satisfies the Chapman–Rubesin relation,

$$\lambda = 0.332\frac{\rho_w u_e}{\rho_e x_0} R^{1/2} C^{-1/2},$$

then Eq. (10.50) follows at once.

We now demonstrate that a solution to Eq. (10.61) can be found when A is small and X is large and negative. Assume that

$$\tilde{A} = a_1 e^{\kappa X}, \qquad \tilde{u} = z + a_1 e^{\kappa X} f'(z),$$

$$v = -a_1 \kappa e^{\kappa X} f(z), \tag{10.63}$$

where κ is a positive constant to be found and $e^{2\kappa X}$ is neglected. The constant a_1 is chosen to be equal to -1 for a compressive free interaction and $+1$ for an expansive free interaction. Then from Eq. (10.63) we have at once

$$\kappa z f' - \kappa f = \kappa^2 + f'''$$ (10.64)

with

$$f'(0) = f(0) = 0, \qquad f'(\infty) = 1.$$ (10.65)

The solution of the differential equation is

$$f(z) = \frac{-\kappa^{5/3} Ai(z\kappa^{1/3})}{Ai'(0)},$$ (10.66)

where $Ai(t)$ is Airy's function, satisfying $w'' - tw = 0$ and tending to zero as $t \to \infty$. The boundary conditions (10.65) are also satisfied if

$$-\kappa^{4/3} \int_0^\infty \frac{Ai(t)\, dt}{Ai'(0)} = 1, \quad \text{that is, } \kappa = 0.8272.$$ (10.67)

This solution, due to Lighthill, shows that the boundary layer can spontaneously change from the conventional kind to one in which interaction effects are important. For expansive interactions the pressure is decreasing, and for compressive interactions the pressure is increasing. In both cases the change in the fluid velocity is small, but the change in the skin friction is important, since it is represented by $\partial u/\partial z$. Indeed in compressive flows, the skin friction changes sign, and thereafter the boundary layer separates from the plate as explained earlier. In expansive flows the skin friction increases without limit, and eventually the velocity changes do become significant. The solution given by Eq. (10.66) can be made the basis of a series expansion in powers of $e^{\kappa X}$, but it is more usual to solve the equations numerically.

This account of free interactions forms the basis of the triple-deck theory of interactive flows, which has been reviewed by Stewartson [7] and, more recently, by Smith [8]. It is hoped that this brief introduction will give the reader a flavor of the theory and confidence that crucial features of the phenomenon can be explained by an interactive theory, built on a firm theoretical basis.

10.4 A Prescription for Computing Interactive Flows with Shocks

Formally the flow configuration described in the last section can persist indefinitely along the body and in fact is brought to an end by the feature provoking the interaction. If it is a shock wave, it strikes the boundary layer (see Fig. 10.11), which by now is behaving in essence like a mixing layer (see Fig. 6.9), is reflected as a rarefaction fan, and deflects the mixing layer back to the body. Now, however, the backflow in the boundary layer moves under quite a strong favorable pressure gradient and must also disappear at the reattachment point R_0 (Fig. 10.11). Thus if we consider the

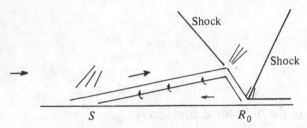

Figure 10.11. Representative sketch (not to scale) of the interaction between a shock wave and laminar boundary layer. S denotes location of separation and R_0 reattachment.

behavior of this back-flow starting from R_0 and following it to the separation point S, we see that to begin with it acquires a significant momentum between R_0 and the shock interaction that it must lose by entrainment to the mixing layer between the shock interactions and S. Since entrainment is a rather weak method of losing momentum, it is not surprising that separation often occurs at considerable distances ahead of the shock. In practice this argument needs modification because a separated boundary layer is very prone to instability, and transition to turbulence occurs between S and the shock if it is at all strong. A turbulent boundary layer entrains fluid at a much greater rate than a laminar boundary layer and consequently the pressure plateau between separation and the shock interaction is reduced, although still of importance.

If the feature provoking the free interaction is a concave corner or ramp (see Fig. 10.12), the mixing layer continues unchanged in character past the corner, but now instead of it slightly diverging from the wall, the two are coming together rather quickly, and when they meet at reattachment, the mixing layer is turned sharply to move parallel to the wall, inducing a rapid rise in pressure. This compression fan merges into a shock outside the boundary layer and also serves to turn around some of the fluid in the mixing layer to initiate the boundary-layer back-flow. It was argued by Chapman et al. [9] that the length of the separated region can be determined by computing the velocity just upstream of reattachment on the streamline that bifurcates from the wall at separation. This velocity must be reduced to zero at reattachment to preserve continuity, and this is affected by the inviscid forces associated with the pressure rise there.

At first sight one might think that a boundary-layer approach to strong interaction problems is ruled out because a basic assumption of that theory,

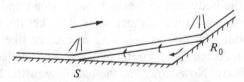

Figure 10.12. Representative sketch (not to scale) of the interaction between a shock wave and the laminar boundary layer on a ramp.

namely that the variation of the velocity across the boundary layer is much larger than that along it, is contradicted in the shock structure. Indeed the length scale in the shock is $O(R^{-1})$ and much smaller than for the boundary layer, which is $O(R^{-1/2})$. Nevertheless there are a number of arguments that suggest that a boundary-layer approach is worth considering. The obvious alternative, the Navier–Stokes equations, even if taken in their steady form, require extremely large computers.

However, if we restrict attention to a boundary-layer approach, we must recognize that it has its limitations. For example, if the separated region is thick enough, then the basic assumptions must be called into question. At the present time it is not possible to put a frame around the external conditions outside which the theory gives unacceptable errors, but we may note that in the case of wholly laminar flow, calculations at $M_\infty = 4$ are in quite good agreement with observation and solution of the Navier–Stokes equations [10–12].

The first problem to be faced in the computation is the determination of the point where the pressure begins to rise. We have already explained that this is fixed in some way by the strength of the incident shock, but it can only be found by trial and error. It will in fact be necessary to make a number of sweeps of the boundary layer in the downstream direction and to develop an iterative sequence that converges. The interaction law by which the boundary layer influences the inviscid stream may be written down as

$$v_w = \frac{1}{\rho_e} \frac{d}{dx} \int_0^\infty (\rho_e u_e - \rho u)\, dy, \tag{10.68}$$

where ρ and u are computed from the boundary-layer solution and v_w is the effective blowing velocity. The contribution to the inviscid pressure from v_w in simple cases is then given by Eq. (10.49), but in general it may be more appropriate to use Eq. (10.50) as a replacement for the boundary condition $v = 0$ on the body in the general inviscid program. Such a procedure is particularly appropriate when the external flow is transonic, when Eq. (10.49) is not correct. A further complication should be mentioned at this stage. Should the flow downstream of the main shock be subsonic, which is always a possibility in a transonic situation and must be the case if the incident shock arises from a Mach reflection, then the computation of the external flow is even more complicated. Upstream influences, in addition to fixing the initial pressure rise of the interaction, make the iterative process more difficult to treat.

A first guess is made of the position of the critical pressure rise, and this is introduced into the flow field. Typically this is of relative order 10^{-3} or 10^{-4}. The evolution of the boundary layer is then self-sustaining, and separation soon occurs. If the oncoming boundary layer is turbulent, the turbulence model which forms the basis for turbulent studies in this book needs modification, and if it is laminar, a numerical scheme must be devised to carry the integration against the direction of motion of the fluid in the back-flow. There are various ways in which this might be done, but among the more popular and effective ways is the so-called FLARE approach, after the original users Reyhner and Flügge-Lotz [13]. The idea here is simply to set $u(\partial u/\partial x)$ equal to zero in the governing equation wherever $u < 0$. As a result, the numerical instabilities that plague attempts to integrate the boundary-layer equation against the local direction of flow are avoided, and the integration is a straightforward process. Of course there is some loss of accuracy as a result of such an approximation, but it is small since the magnitude of the reversed values of u is never more than a small fraction of the external flow velocity. We note that even with the FLARE approximation, the pressure distribution in interactive studies is predicted remarkably accurately. There are also methods available for improving the numerical method if the need arises.

One point should be born in mind when endeavoring to continue the integration through separation: whether the boundary layer is laminar or turbulent, the solution of the equations must become singular if the external velocity is prescribed. In order to avoid its occurrence, and there is no reason to expect the final converged solution to have such a singularity, the external velocity must be related to the displacement thickness in some way. Thus, in each sweep along the wall solving the boundary-layer equation, the external velocity must in some sense be regarded as unknown. There are many ways in which this can be done, and well-known examples include the use of Ackeret's formula (10.49) to relate the external velocity and v_w for supersonic flow, Veldman's use of the Hilbert integral in subsonic flow [14–16], and the relaxation formula suggested for displacement thickness by Le Balleur [17, 18] and used extensively by various investigators (see, for example, [19]). At this time they all need to be studied or explored further.

Problems

10.1 For an adiabatic flow the static temperature distribution across the boundary layer is given by Eq. (10.39).

(a) Show that for a nonadiabatic flow the static temperature is given by

$$\frac{T}{T_e} = S + \frac{\gamma - 1}{2} M_\infty^2 \left(S - \frac{\bar{u}^2}{\bar{u}_e^2} \right). \tag{P10.1}$$

(b) Using the above expression and Eq. (10.36), show that an equivalent relation for Eq. (10.41) for a flow with heat transfer is

$$\delta_c^* = S_w \delta_i^* + \sigma \left(S_w \delta_i^* + \theta_i \right). \tag{P10.2}$$

Here $S_w \delta_i^*$ with δ_i^* corresponding to the incompressible value of displacement thickness, is given by

$$S_w \delta_i^* = \int_0^\infty \left(S - \frac{\bar{u}_\cdot}{\bar{u}_e} \right) d\bar{Y}. \tag{P10.3}$$

(c) Show that for a flow with heat transfer, the relations given by Eq. (10.43) remain unchanged.

(d) Show that the heat transfer at the wall can be calculated from the
following expression

$$\dot{q}_w = \frac{\rho_\infty u_\infty H_\infty}{\sqrt{R_x}} \left(\frac{\mu_w}{\mu_\infty} \right) S_w f_w'' \qquad (P10.4)$$

or from

$$\dot{q}_w = \frac{\rho_w u_\infty H_\infty}{\sqrt{R_x}} \left(\frac{\mu_w}{\mu_\infty} \right) S_w' \qquad (P10.5)$$

$$R_x = \frac{\rho_\infty u_\infty x}{\mu_\infty}.$$

10.2 Consider a cooled flat plate with $S_w = 0.5$ at $M_\infty = 3.0$ at 2300m altitude.
Assume standard atmosphere. Calculate the displacement thickness, shape
factor, the local skin-friction coefficient, and the heat transfer rate per unit
width at 3m from the leading edge. Use Sutherland's viscosity law and take
$\gamma = 1.4$.

10.3 Calculate the Chapman-Rubesin factor at the wall for laminar boundary
layers in air of total temperature 300K, at $M_e = 1, 2, 5$ and 10. Take recovery
factor as 0.9.

10.4 Show that the shape factor $H \equiv \delta^*/\theta$ is always greater than unity in
constant density flows but may be less than unity in boundary layers on
highly-cooled walls, even at low speeds.

10.5 Show that in the similar flows for a model fluid on either a heated or a
cooled wall, S takes extreme values at the wall and at infinity.

Note: If $\dot{m} < 0$, the solutions of Eqs. (10.28) and (10.29) are not unique
unless an extra condition is applied, namely that $1 - \phi'$ is exponentially
small when η is large. Without this condition velocity overshoot is possible
whether the wall is heated or not.

10.6 For Prandtl numbers Pr not too different from unity, a reasonably good
approximation to the temperature in a boundary layer near an adiabatic wall
is

$$T = T_e \left[1 + \frac{\gamma - 1}{2} Pr^{1/2} M_e^2 \left(1 - \frac{u^2}{u_e^2} \right) \right]$$

Show that the Illingworth-Stewartson transformation may be generalized to

$$dX = \left(\frac{a_e}{a_\infty} \right)^{Pr^{1/2}} \frac{p_e}{p_\infty} dx; \qquad dY = \left(\frac{a_e}{a_\infty} \right)^{Pr^{1/2}} \frac{\rho}{\rho_\infty} dy$$

for such flows, and determine the equivalent external velocity of the incompressible flow. For details see Rott [20].

10.7 A model fluid flows past an adiabatic flat plate at Mach number M_∞. If the
momentum thickness of the boundary layer at distance x from the leading
edge is θ_c, show that

$$\theta_c = \theta_i + \frac{\gamma - 1}{2} M_\infty^2 \theta_{1i}$$

where θ_i is the momentum thickness and

$$\theta_{1i} = \int_0^\infty \frac{\bar{u}}{\bar{u}_e}\left(1 - \frac{\bar{u}^2}{\bar{u}_e^2}\right) dY$$

for the equivalent incompressible flow. Obtain an approximation to θ_c in the following way. Write down the momentum integral for the equivalent incompressible flow and, assuming that

$$\bar{u} = \bar{u}_e \sin\left(\frac{\pi Y}{2\delta(x)}\right) \qquad Y < \delta$$

$$= \bar{u}_e \qquad Y > \delta$$

show that

$$\delta^2 = \frac{2x\nu_\infty}{(4-\pi)u_e}$$

Hence deduce that

$$\theta_c = \frac{2}{\pi}\left(\frac{2x\nu_\infty}{u_e}\right)^{1/2}\left[0.463 + \frac{\gamma-1}{6}M_\infty^2\right]$$

10.8 Obtain the results corresponding to problem 10.7 when the temperature of the wall is maintained at a constant value T_w. Hint: Make use of Eq. (10.36).

10.9 Instead of Eq. (10.58), assume that

$$y = \beta, z \qquad x = x_0 + \beta_2 X$$

$$u = \beta_3 \bar{u} \qquad v = \beta_4 \bar{v}$$

$$A = \beta_5 \tilde{A} \qquad p = p_e(x_0) + \beta_6 P$$

where $\beta_1 \dots \beta_6$ are constants. Show that the only choice of β's which reduces Eqs. (10.53)–(10.57) to (10.60)–(10.62) is given by Eqs. (10.58) and (10.59).

10.10 In Eqs. (10.60)–(10.62), assume that

$$\bar{u} = \frac{1}{X_s - X}F(\eta), \qquad \eta = \frac{z}{X_s - X}$$

where X_s is a constant, $X < X_s$ and F is a function, of η only, to be found. Obtain the corresponding form for $\bar{v}$ by the use of Eq. (10.60). Verify that these forms satisfy Eqs. (10.61) and (10.62) in the limit $X \to X_s$ provided

$$F = 3\tanh^2(\eta + \beta) - 2, \qquad \tanh^2\beta = \tfrac{2}{3}$$

and write down the corresponding form of the pressure. For details, see Stewartson [23].

References

[1] Illingworth, C. R.: Steady flow in the laminar boundary layer of a gas. *Proc. Roy. Soc.* **A199**:533 (1949).

[2] Stewartson, K.: Correlated compressible and incompressible boundary layers. *Proc. Roy. Soc.* **A200**:84 (1949).

[3] Liepmann, H. W. and Roshko, A.: *Elements of Gas Dynamics*, Wiley, New York, 1957.

[4] Seddon, J.: The flow produced by interaction of a turbulent boundary layer with a normal shock wave of sufficient strength to cause separation. Aero Res. Council R&M 3502 (London), 1960.

[5] Oswatitsch, K. and Wieghardt, K.: Theoretical investigations on steady potential flows and boundary layers at high speed. Aero Res. Council Rep. 10356 (London), 1946.

[6] Lighthill, M. J.: On boundary layers and upstream influence. II. Supersonic flows without separation. *Proc. Roy. Soc.* **A217**:478 (1953).

[7] Stewartson, K.: Multistructured boundary layers on flat plates and related bodies. *Adv. Appl. Mech.* **14**:145 (1974).

[8] Smith, F. T.: On the high Reynolds number theory of laminar flows. *IMA J. Appl. Math.* **28**:207 (1982).

[9] Chapman, D. R., Kuehn, D. M., and Larson H. K.: Investigation of separated flows in supersonic and subsonic streams with emphasis on the effect of transition. NACA Rep. 1356, 1958.

[10] Carter, J. E.: Numerical solution of the Navier–Stokes equations for the supersonic laminar flow over a two-dimensional compression corner. NASA Rep. TR-R-385, 1972.

[11] Werle, M. J. and Vatsa, V. N.: A new method for supersonic boundary-layer separation. *AIAA J.* **12**:1491 (1974).

[12] Lewis, J. E., Kubota, T., and Lees, L.: Experimental investigation of supersonic laminar two-dimensional boundary-layer separation in a compression corner with and without cooling. *AIAA J.* **6**:7–14 (1968).

[13] Reyhner, T. A. and Flügge-Lotz, I.: The interaction of a shock wave with a laminar boundary layer. *Int. J. Non-Linear Mech.* **3** (2):173 (1968).

[14] Veldman, A. E. P.: A numerical method for the calculation of laminar incompressible boundary layers with strong-inviscid interaction. NLR TR 79023L, 1979.

[15] Cebeci, T., Stewartson, K., and Williams, P. G.: Separation and reattachment near the leading edge of a thin airfoil at incidence. AGARD Symposium on Computation of Viscous-Inviscis Interaction Flows, Colorado Springs, 1980.

[16] Cebeci, T., and Schimke, S. M.: The calculation of separation bubbles in interactive turbulent boundary layers. *J. Fluid Mech.* **131**:305 (1983).

[17] LeBalleur, J. C.: Couplage visqueux-non visqueux: analyse du problème incluant dècollements et ondes de choc. *La Rech. Aèrosp.* n° 1977-6, 349. English translation in ESA TT476, 1977.

[18] LeBalleur, J. C.: Couplage visqueux-non visqueux: mèthode numèrique et applications aux ècoulements bidimensionnels transsoniques et supersoniques. *La Rech. Aèrosp.* n° 1978-2, 67–76. English translation in ESA TT496, 1978.

[19] Carter, J. and Wornom, S. F.: Solutions for incompressible separated boundary layers including viscous-inviscid interaction in *Aerodynamic Analysis Requiring Advanced Computers*. NASA SP-347, 1975, p. 125.

[20] Rott, N. J.: Compressible laminar boundary layer on a heat-insulated body.

J. Aero. Sci. **20**:67 (1953).

[21] Cohen, C. B. and Reshotko, E.: The compressible laminar boundary layer with heat transfer and arbitrary pressure gradient. NACA Rep. 1294, 1956.

[22] Monaghan, R. J.: Effects of heat transfer on laminar boundary-layer development under pressure gradients in compressible flow. Aero Res. Council R&M 3218, 1961. (Contains results of unpublished work by Monaghan and Crabtree.)

[23] Stewartson, K.: On supersonic laminar boundary layers near convex corners. *Proc. Roy. Soc.* **A319**:289 (1970).

CHAPTER 11
Coupled Turbulent Boundary Layers

As in Chapter 6, we begin with the statement that the main difference between laminar flows and turbulent flows is that the effective diffusivities in turbulent flow are unknown. In Chapter 6 the temperature and/or concentration differences were small enough not to affect the mean velocity field, and it was assumed without explicit comment that the *fluctuating* velocity field, which controls the turbulent transport of momentum, heat, or mass, was also unaffected.

As discussed in Section 3.3, the effect of density fluctuations on the fluctuating velocity field is quite small in most practical cases, except in buoyant flow, because density fluctuations are usually only a small fraction of the mean density. The effect of *mean*-density gradients on turbulence is not fully understood but may be significant in some cases. For many purposes the turbulence models developed for uncoupled flows may be adequate for variable-density flows of the same general type, provided that the local mean density is inserted in differential equations or algebraic formulas relating turbulent stresses and enthalpy-flux rates to the mean velocity and temperature fields. Of course, the local mean density must also be inserted in the time-average momentum and energy equations just as in the laminar flows discussed in Chapter 10.

Like the corresponding material in Chapter 10, the preceding part of this introduction, and most of the analysis below, applies both to low-speed flows (in which large density changes are introduced by high heat-transfer rates or by the mixing of dissimilar fluids) and to high-speed flows (in which large temperature changes result from viscous dissipation of kinetic energy into thermal energy or from pressure changes that are large fractions of the absolute pressure). Of course, the configurations, and the engineering hardware, are likely to be rather different in low-speed and high-speed flows, and comparatively little work has been done on low-speed coupled flows; this means that since current turbulence models are empirical and are all based on data for limited ranges of flows, the formulas or differential equations developed for turbulent transfer rates may have slightly different coefficients in the two cases. In the section below, therefore, the confidence with which high-speed flow results are applied to low-speed coupled flows is based mainly on physical arguments rather than on data. For simplicity the discussion is carried through for gas flows. Liquid flows with large fractional variations in density are rare, so that coupling arises mainly via the dependence of viscosity on temperature, which is a fairly minor complication in turbulent flow and merely leads to temperature dependence of the constants A^+ and c in the law of the wall (Section 6.1).

In Section 11.1 we discuss inner-layer similarity ideas in compressible flows, corresponding to the discussion in Section 6.1. Section 11.2 is a review of attempts to find a mathematical transformation of coordinates and variables to convert a compressible turbulent flow into an observable (or calculable) low-speed flow. The authors' view of such transformations is that since they do not have the rigorous justification of the laminar-flow transformations discussed in Chapter 10, they are just as arbitrary as turbulence models and rather less easy to check against experimental data. The exception is that the transformation used in the inner-layer analysis for coupled turbulent flows is useful for *correlating data* for velocity profiles and skin-friction coefficients.

In general it is best to obtain solutions of coupled turbulent boundary-layer flows by differential methods. For a given two-dimensional or axisymmetric body, which implies that the external velocity distribution can be determined and that the surface boundary condition is known, the momentum- and heat-transfer properties of the flow can be obtained by solution of the momentum and energy equations with accuracy sufficient for most engineering purposes. A general computer program for this purpose is presented in Section 13.5; it utilizes the eddy-viscosity formulation of Cebeci and Smith and the so-called Box numerical scheme. This and other similar methods allow the calculation of turbulent boundary layers for a wide range of boundary conditions but require the use of large computers, and, as with uncoupled turbulent boundary layers, it is sometimes desirable and useful to have simpler methods or formulas. The accuracy of correlation formulas is usually *better* than that of general methods, but only for restricted ranges of the variables. Frequently, because of the scarcity of experimental information, they are restricted to flows with negligible pressure gradient. They are, however, valuable for many practical engineering calculations and are discussed in Section 11.3.

11.1 Inner-Layer Similarity Analysis for Velocity and Temperature Profiles

In coupled turbulent flows, the velocity profile and temperature profile in the inner layer depend on all the quantities that affect the velocity *or* temperature profile in uncoupled flows (Section 6.1), and in addition the absolute temperature (at the wall, say) must be included since, by definition, the temperature differences in coupled flows are a significant fraction of the absolute temperature. Also, if the Mach number of the flow is not small compared with unity, the speed of sound, a, and the ratio of specific heats, γ, will appear; in a perfect gas $a = \sqrt{\gamma R T} = \sqrt{(\gamma - 1)c_p T}$, so that either a or T or both may be used as a variable. With these additions to the variables listed in Section 6.2, we see that

$$u = f_1(\tau_w, y, \rho_w, \nu_w, \dot{q}_w, c_p, T_w, k_w, \gamma), \tag{11.1}$$

$$T_w - T = f_2(\tau_w, y, \rho_w, \nu_w, \dot{q}_w, c_p, T_w, k_w, \gamma), \tag{11.2}$$

and dimensional analysis with $u_\tau = \sqrt{\tau_w / \rho_w}$ gives

$$\frac{u}{u_\tau} = f_3\left(\frac{u_\tau y}{\nu_w}, \frac{\dot{q}_w}{\rho c_p u_\tau T_w}, \frac{u_\tau}{a_w}, \gamma, \Pr_w \right), \tag{11.3}$$

$$\frac{T}{T_w} = f_4\left(\frac{u_\tau y}{\nu_w}, \frac{\dot{q}_w}{\rho c_p u_\tau T_w}, \frac{u_\tau}{a_w}, \gamma, \Pr_w \right), \tag{11.4}$$

where T/T_w is used instead of $(T_w - T)/T_w$ for convenience, and the speed of sound at the wall, a_w, is used instead of the dimensionally correct but less meaningful quantity $\sqrt{c_p T_w}$. The quantity u_τ / a_w is called the *friction Mach number*, M_τ. The evaluation of fluid properties at the wall is adequate if, for example, ν / ν_w can be expressed as a function of T/T_w only (the pressure being independent of y in any case); this is the case if $\nu \propto T^\omega$ for some ω, which is a good approximation for common gases over a range of, say, $2:1$ in temperature. If ν or k are more complicated functions of T, the absolute temperature must be included in the list of variables on the right-hand sides of Eqs. (11.27) and (11.28).

The arguments that led to the inner-layer formulas for velocity and temperature in uncoupled flows (Chapter 6) can be applied again to the coupled case if we are satisfied that the effects of viscosity are again small for $u_\tau y/\nu_w \gg 1$ and that the effects of thermal conductivity are again small for $(u_\tau y/\nu_w)\Pr \gg 1$. Provided that the effects of *fluctuations* in viscosity and thermal conductivity are small and that ν and ρ do not differ by orders of magnitude from their values at the wall so that $u_\tau y/\nu_w$ is still a representative Reynolds number, molecular diffusion should indeed be small compared with turbulent diffusion if $u_\tau y/\nu_w$ is large. However, some care is still needed to arrange the formulas in the most convenient form because of the variation of ρ with y. In Chapter 6 it was argued that if the shear stress τ varied with y, then the local value of $(\tau/\rho)^{1/2}$ would provide a better velocity scale than the wall value u_τ; it therefore seems logical to use the local value in coupled flows also, and in the simplest case when only ρ, and not τ, varies with y, the appropriate velocity scale is $(\tau_w/\rho)^{1/2}$. Analogously we use $\dot{q}/\rho$, rather than $\dot{q}_w/\rho_w$, in the mixing-length formula for temperature; we shall see below that $\dot{q}$ always varies with y in high-speed flows. Then elimination of ν_w and Pr from the lists of variables and the use of dimensional analysis on $\partial u/\partial y$ and $\partial T/\partial y$ instead of u and T give, instead of Eqs. (11.3) and (11.4),

$$\frac{\partial u}{\partial y} = \frac{(\tau/\rho)^{1/2}}{\kappa y} f_u \left[\frac{\dot{q}}{\rho c_p T (\tau/\rho)^{1/2}}, \frac{(\tau/\rho)^{1/2}}{a}, \gamma \right], \qquad (11.5)$$

$$\frac{\partial T}{\partial y} = -\frac{\dot{q}/\rho c_p}{(\tau/\rho)^{1/2} \kappa_h y} f_T \left[\frac{\dot{q}}{\rho c_p T (\tau/\rho)^{1/2}}, \frac{(\tau/\rho)^{1/2}}{a}, \gamma \right], \qquad (11.6)$$

where we have consistently used *local* variables in the arguments of the f functions, even for τ, as in Eqs. (6.13) and (6.14). The analysis below will be restricted to the case of a constant-stress layer, $\tau = \tau_w$, which, as was discussed in Section 6.1, is a good approximation to the inner 20 percent of a boundary layer in a small pressure gradient.

In order to equate the f functions to unity and recover effectively incompressible versions of the mixing-length formulas, we need to neglect the effect of density fluctuations discussed in Section 3.2. With this assumption, the heat-transfer parameter $\dot{q}/\rho c_p (\tau/\rho)^{1/2} T$ and the friction Mach number $(\tau/\rho)^{1/2}/a$—representing the two sources of density (temperature) fluctuations—do not appear in formulas (11.5) and (11.6) for the gradients of u and T; they will, however, remain in the full formulas for u and T, Eqs. (11.3) and (11.4), because they affect the temperature gradient in the viscous sublayer. Formulas for $\partial u/\partial y$ and $\partial T/\partial y$ in the viscous sublayer would nominally contain all the variables on the right-hand sides of Eq. (11.3) or (11.4). The assumption that turbulence processes are little affected by density fluctuations implies that γ, which is a measure of the difference between adiabatic and isothermal processes, would have a negligible effect even in the viscous sublayer, but there are not enough data to check this. With the assumption that f_u and f_T are constant outside the viscous sublayer, we can now write, exactly as in the constant-property formulas (6.13) and (6.14),

$$\frac{\partial u}{\partial y} = \frac{(\tau/\rho)^{1/2}}{\kappa y}, \qquad (11.7)$$

$$\frac{\partial T}{\partial y} = \frac{-\dot{q}/\rho c_p}{(\tau/\rho)^{1/2} \kappa_h y}. \qquad (11.8)$$

Now Eq. (11.8) still retains the local value of $\dot{q}$, and in a high-speed flow this will differ from the wall value, even if $\tau = \tau_w$, because of viscous dissipation of mean and turbulent kinetic energy into heat. The rate at which kinetic energy is extracted from a unit volume of the mean flow by work done against viscous and turbulent stresses (equal, in a thin shear layer, to total shear stress times rate of shear strain) is $\tau \, \partial u / \partial y$; the part corresponding to the viscous shear stress represents direct viscous dissipation into heat, and the part corresponding to the turbulent shear stress represents production of turbulent kinetic energy. We cannot immediately equate turbulent energy production to viscous dissipation of that turbulent energy into heat because turbulence processes include transport of turbulent kinetic energy from one place to another. However, this transport is negligible in the inner layer (outside the viscous sublayer), so that we can write a degenerate version of the enthalpy equation (3.46), with all transport terms neglected and only y derivatives retained, as

$$\frac{\partial \dot{q}}{\partial y} = \tau_w \frac{\partial u}{\partial y}, \tag{11.9}$$

which simply states that the net rate of (y-component) transfer of heat leaving a control volume in the inner layer is equal to the rate at which the fluid in the control volume does work against (shear) stress. Integrating this equation, we get

$$\dot{q} = \dot{q}_w + u\tau_w. \tag{11.10}$$

In a low-speed flow the work done is negligible, and $\dot{q} \approx \dot{q}_w$, corresponding to $\tau \approx \tau_w$. If we divide Eq. (11.8) by Eq. (11.7), we obtain

$$\frac{\partial T}{\partial u} = \frac{-(\kappa/\kappa_h)\dot{q}}{c_p \tau_w}, \tag{11.11}$$

where κ/κ_h is the turbulent Prandtl number. Substituting for $\dot{q}$ from Eq. (11.10) and integrating with respect to u, we get

$$T = -\frac{(\kappa/\kappa_h)\dot{q}_w u}{c_p \tau_w} - \frac{(\kappa/\kappa_h)u^2}{2c_p} + \text{const.} \tag{11.12}$$

Here, unlike the corresponding equation for laminar flows, Eq. (10.23b) for Pr $= 1$, the constant of integration is *not* exactly equal to T_w because the formulas (11.7) and (11.8) are not valid in the viscous or conductive sublayers, but it is conventionally written as $c_1 T_w$, where c_1 is fairly close to unity and is a function of $\dot{q}_w/\rho_w c_p u_\tau T_w$, u_τ/a_w, and the molecular Prandtl number Pr. That is,

$$T = c_1 T_w - \frac{(\kappa/\kappa_h)\dot{q}_w u}{c_p \tau_w} - \frac{(\kappa/\kappa_h)u^2}{2c_p}. \tag{11.13}$$

Noting that $\rho = \rho_w T_w / T$, we can use Eq. (11.13) to eliminate ρ from Eq. (11.7). The integral required to obtain u as a function of y from Eq. (11.7) then becomes

$$\int \frac{dy}{\kappa y} = \int \frac{du/u_\tau}{\left[c_1 - (\kappa/\kappa_h)\dot{q}_w u/c_p T_w \tau_w - (\kappa/\kappa_h)u^2/2c_p T_w\right]^{1/2}}. \quad (11.14)$$

Replacing $c_p T_w$ by $a_w^2/(\gamma - 1)$ and integrating Eq. (11.14), we obtain the law of the wall for coupled turbulent flows:

$$\frac{u}{u_\tau} = \frac{\sqrt{c_1}}{R} \sin\left(R\frac{u^*}{u_\tau}\right) - H\left[1 - \cos\left(R\frac{u^*}{u_\tau}\right)\right], \quad (11.15)$$

where

$$R = \frac{u_\tau}{a_w}\left[\frac{(\gamma-1)\kappa}{2\kappa_h}\right]^{1/2}, \qquad H = \frac{\dot{q}_w}{\tau_w u_\tau} \equiv \frac{1}{(\gamma-1)}\frac{\dot{q}_w}{\rho_w c_p u_\tau T_w}\left(\frac{a_w}{u_\tau}\right)^2,$$

(11.16a)

$$u^* = \frac{1}{\kappa}\ln y + \text{const.} \quad (11.16b)$$

Recalling from Eq. (11.3) that y appears in the group $u_\tau y/\nu_w$, we rewrite

$$\frac{u^*}{u_\tau} = \frac{1}{\kappa}\ln\frac{u_\tau y}{\nu_w} + c. \quad (11.16c)$$

If $\dot{q}_w = 0$ and $u_\tau/a_w \to 0$, then $H = 0$, $R \to 0$, and

$$\frac{u}{u_\tau} = \sqrt{c_1}\frac{u^*}{u_\tau}. \quad (11.17)$$

It can be shown that if u_τ/a_w is small, then $c_1 = 1 - 0\,(\dot{q}_w)$, for compatibility with the logarithmic law for temperature in uncoupled flows, Eq. (6.12a), so that $u = u^*$ for small $\dot{q}_w$; thus the constant c in Eq. (11.16c) can be identified with the additive constant c in the logarithmic law for constant property wall layers. In general c, like c_1, is a function of the friction Mach number M_τ and of B_q defined by

$$B_q = \frac{\dot{q}_w}{\rho_w c_p u_\tau T_w}. \quad (11.18)$$

The above analysis is originally due to Rotta [1] and is an extension of that of Van Driest [2].

Van Driest Transformation for the Inner Layer

Simpler versions of Eq. (11.15) and the accompanying temperature profile Eq. (11.13) have been proposed by many authors. Van Driest [2] assumed $c_1 = 1$ and $\kappa/\kappa_h = 1$ [recall from Eq. (6.23) that κ/κ_h is a turbulent Prandtl number in the fully turbulent part of the flow, and note that $c_1 = 1$ implies that the effective Prandtl number in the viscous and conductive sublayers is unity]. Van Driest presented the inverse of Eq. (11.5), giving u^* in terms of u. In our more general notation this is

$$\frac{u^*}{u_\tau} = \frac{1}{R}\left(\sin^{-1}\frac{R(u/u_\tau + H)}{\left(c_1 + R^2 H^2\right)^{1/2}} - \sin^{-1}\frac{RH}{\left(c_1 + R^2 H^2\right)^{1/2}}\right). \quad (11.19)$$

This formula is called the *Van Driest transformation*; it can be regarded as transforming the inner-layer part of the compressible boundary-layer profile $u(y)$ to an equivalent incompressible flow $u^*(y)$ that obeys the logarithmic formula, Eq. (11.16c). However, it is simpler to regard Eq. (11.15), with Eq. (11.16), as the direct prediction of inner-layer similarity theory for the compressible boundary layer.

If there is no heat transfer through the surface, H is zero, the second term on the right of Eq. (11.15) disappears, and Eq. (11.19) reduces to

$$\frac{u^*}{u_\tau} = \frac{1}{R}\sin^{-1}\left(\frac{R}{\sqrt{c_1}}\frac{u}{u_\tau}\right), \quad (11.20)$$

which is easy to identify as the inverse of Eq. (11.15) without the second term on the right and of course reduces to $u^* = u$ as $R \to 0$ and $c_1 \to 1$.

The basic assumptions imply that κ and κ_h are the same as in incompressible flow. As we have seen, c_1 and c, which are constants of integration in Eqs. (11.13) and (11.6c), respectively, must be expected to be functions of the friction Mach number and the heat-transfer parameter. [The fact that c is *not* necessarily equal to its low-speed value detracts from the interpretation of Eq. (11.19) as a transformation to the inner part of a realizable low-speed boundary layer.] Experimental data, reviewed in great detail by Fernholz and Finley [3], support the extension of inner-layer similarity to compressible flow but fail to provide definite evidence about the variation of c and c_1. The low-speed values $c = 5.0$ and $c_1 = 1$ fit the data as a whole to within the rather large scatter but Bradshaw [4] presents formulas for variable c and c_1 based on a selection of the more reliable data, as we shall see later.

If we replace H by the expressions given in Eq. (11.16a) and (11.13), use the definition of Mach number, $M_e = u_e/a_e$, take $c_1 = 1$ and $\kappa/\kappa_h = 1$ following Van Driest [2], and note that $\rho_w/\rho_e = T_e/T_w$, then Eq. (11.19) can be written as

$$\frac{u^*}{u_\tau} = \frac{1}{A\sqrt{(c_f/2)(T_w/T_e)}}\left(\sin^{-1}\frac{A\sqrt{(c_f/2)(T_w/T_e)}\,(u/u_\tau) + B/2A}{\sqrt{1 + (B/2A)^2}}\right.$$

$$\left. - \sin^{-1}\frac{B/2A}{\sqrt{1 + (B/2A)^2}}\right) \quad (11.21)$$

or as

$$u^* = \frac{u_e}{A}\left(\sin^{-1}\frac{2A^2(u/u_e)-B}{(B^2+4A^2)^{1/2}}+\sin^{-1}\frac{B}{(B^2+4A^2)^{1/2}}\right), \quad (11.22)$$

where

$$A^2 = \frac{\gamma-1}{2}\frac{M_e^2}{T_w/T_e}, \qquad B = \frac{1+(\gamma-1)/2M_e^2}{T_w/T_e}-1. \quad (11.23)$$

Note that the above relations assume that the recovery factor $r \equiv (T_{aw}/T_e - 1)/[(\gamma-1)/2M_e^2]$, where T_{aw} is the temperature of an adiabatic wall, is 1. To account for the fact that r is less than unity (about 0.89), we rewrite Eqs. (11.23) as

$$A^2 = \frac{[(\gamma-1)/2]\,M_e^2 r}{T_w/T_e}, \qquad B = \frac{1+[(\gamma-1)/2]\,M_e^2 r}{T_w/T_e}-1. \quad (11.24)$$

11.2 Transformations for Coupled Turbulent Flows

The Van Driest Transformation for the Whole Layer

The Van Driest transformation (11.19) applied to the fully turbulent part of the inner (constant-stress) layer of a compressible boundary layer produces the logarithmic profile (11.16c). Applying the transformation to the outer layer of a constant-pressure compressible boundary layer, we obtain a profile that looks qualitatively like that of a constant-pressure constant-density boundary layer. In particular, the transformed profile $u^*(y)$ can be described, more or less as accurately as an incompressible profile, by the wall-plus-wake formula given by Eqs. (6.39) and (6.41),

$$\frac{u^*}{u_\tau} = \frac{1}{\kappa}\ln\frac{u_\tau y}{\nu_w}+c+\frac{\Pi}{\kappa}\left(1-\cos\pi\frac{y}{\delta}\right). \quad (11.25)$$

However, this convenient data correlation is a consequence of the strong constraint on the wake profile, which has to have zero slope and zero intercept at $y = 0$, whereas the profile as a whole has zero slope at $y = \delta$ [although the crude wake formula in Eq. (11.25) does not]; also, the "wake parameter" Π and the boundary-layer thickness δ are constants that can be adjusted to optimize the fit of Eq. (11.25) to any real or transformed profile. We must, therefore, not claim that the success of Eq. (11.25) proves the validity of Van Driest's inner-layer analysis in the *outer* layer.

As was discussed in Section 6.4, Π is constant in incompressible constant-pressure flows at high Reynolds number and equal to about 0.55. The value of Π that best fits a transformed profile is expected to be a function of the friction Mach number $M_\tau\,(\equiv u_\tau/a_w)$ and of the heat-transfer parameter B_q ($\equiv \dot{q}_w/\rho_w c_p u_\tau T_w$). Evaluation of this function from experimental data is hampered by the low Reynolds number of most of the compressible-flow data and uncertainty about the definition of Reynolds number that should be used in correlating low-Reynolds-number effects on the velocity-defect profile. If it is accepted that these originate at the irregular interface between the turbulent fluid and the nonturbulent "irrotational" fluid, then the fluid properties in the Reynolds number should be evaluated at freestream conditions. Now the largest-scale interface irregularities seen in flow-visualization pictures have a length scale of order δ, being the result of the largest eddies that extend across the full thickness of the shear layer. Therefore δ is the appropriate interface scale and, since it is found that the shear-stress profile plotted as $\tau/\tau_w = f(y/\delta)$ has nearly the same shape at any Reynolds number, we can use τ_w as a shear-stress scale and $(\tau_w/\rho_e)^{1/2}$—evaluated using the *freestream* density—as a velocity scale. Therefore the appropriate Reynolds number is $(\tau_w/\rho_e)^{1/2}\delta/\nu_e \equiv (u_\tau\delta/\nu_e)$ $\cdot(\rho_w/\rho_e)^{1/2}$, rather than the Reynolds number $u_\tau\delta/\nu_w$ that arises naturally in the Van Driest transformation. However, Fernholz [5] has shown that the Reynolds number $\rho_e u_e\theta/\mu_w$ gives excellent correlations of data over a wide range of Mach numbers and Reynolds numbers.

If the physics of low-Reynolds-number effects is the same in compressible flow as in incompressible flow, then the wake parameter Π of the transformed profile should be independent of Reynolds number for $(\tau_w/\rho_e)^{1/2}\delta/\nu_e > 2000$ approximately (corresponding to $u_e\theta/\nu > 5000$ in the case of a low-speed boundary layer). Since Π is nominally a function of M_τ and B_q, it is not possible to predict the trend with Reynolds number explicitly, but it is probably adequate to assume that the ratio of Π to its high-Reynolds-number asymptotic value is the same function of the chosen Reynolds number as in incompressible flow. In fact, although the data are rather scattered by low-speed standards, it appears that Π decreases only very slowly with increasing Mach number [6] in adiabatic-wall boundary layers ($\dot{q}_w = 0$) and, therefore, that $\Pi = 0.55$ is an adequate high-Reynolds-number value for all M_τ.

If c and c_1 are known as functions of M_τ and B_q, and if Π is known as a function of M_τ, B_q, and Reynolds number, then putting $y = \delta$ and $u = u_e$ in Eq. (11.19) and using Eq. (11.25) to substitute for u^*, we obtain u_e/u_τ as a function of $u_\tau\delta/\nu_w$. In practice we require the skin-friction coefficient $\tau_w/\frac{1}{2}\rho_e u_e^2$ as a function of $u_e\theta/\nu_e$ or Fernholz's variable $\rho_e u_e\theta/\mu_w$, given M_e (and T_e) and either $\dot{q}_w$ or T_w. This requires iterative calculation, starting with an estimate of τ_w. Also, the velocity profile of the compressible flow has to be integrated at each iteration to obtain θ/δ for conversion from the "input" Reynolds number $u_e\theta/\nu_e$ to the Reynolds number $u_\tau\delta/\nu_w$ that appears in the transformation [4]. Skin-friction formulas are discussed in Section 11.3.

The Van Driest transformation could be regarded as a *solution* of the compressible-flow problem only if the coefficients c, c_1, and Π were independent of Mach number and heat-transfer parameter. However, we can use the transformation, plus *compressible-flow* data for c, c_1 and Π, to correlate the mean properties of constant-pressure compressible boundary layers. As noted above, the change in the coefficients is almost within the (large) experimental scatter.

In pressure gradients, the transformed boundary-layer profile still fits Eq. (11.25) as does its true low-speed equivalent, but, as at low speeds, there is no simple formula to relate the shape parameter Π to the local pressure gradient. Moreover, the variation of u_τ, Π, and δ with x will not generally correspond to any realizable low-speed boundary layer; that is, it may not be possible to choose a pressure distribution $p(x)$ for a low-speed flow that will reproduce, at each x, the same velocity profile as in the compressible flow. The spirit of Van Driest's transformation, although not its details, would be retained if compressible boundary layers were calculated using the mixing-length formula to predict the shear stress and the assumption of constant turbulent Prandtl number to predict the heat transfer. We consider such calculation methods in Section 11.4.

Other Transformations

The transformations between compressible and incompressible laminar boundary layers discussed in Chapter 10 are rigorous but limited in application. Transformations for turbulent flow are necessarily inexact because our knowledge of the time-averaged properties of turbulent motion is inexact. As in the case of laminar flow, the need for transformation has decreased as our ability to do lengthy numerical calculations has increased, and the assumption that density fluctuations have negligible effect on turbulence has permitted low-speed models to be cautiously extended to compressible flow. The Van Driest transformation relies on the application of this assumption to the inner-layer (mixing-length) formula. In the outer layer, where the mixing length departs from its inner-layer value κy, the Van Driest analysis is not exact, and the transformation will not eliminate compressibility effects though it certainly *reduces* them, apparently within the scatter of current experimental data.

The transformations between compressible and incompressible flow fall into two classes: (1) transformations for the complete velocity profile (and by implication the shear stress profile) and (2) transformations for integral parameters only (specifically, skin-friction formulas). It is generally recognized that transformations for compressible flows in pressure gradient do not necessarily lead to realizable low-speed flows, and we will discuss only constant-pressure flows here.

The paper by Coles [7] is a useful review of previous work and presents one of the most general transformations so far proposed. The two main assumptions are that suitably defined ratios of coordinates in the original (high-speed) and transformed (low-speed) planes are functions only of x

and not of y, and that the ratio of the stream functions in the transformed and original flows is equal to the ratio of the (constant) viscosity in the transformed flow to the viscosity evaluated at an "intermediate temperature," somewhere between T_w and T_e, in the compressible flow. The justification for the latter assumption is carefully discussed by Coles, but the choice of intermediate temperature is necessarily somewhat arbitrary. Coles chooses the temperature at the outer edge of the viscous sublayer, but in order to fit the experimental data for skin friction, it is necessary to locate the sublayer edge at $u_\tau y/\nu = 430$ in the transformed flow, whereas the thickness of the real sublayer is only about one-tenth of this. Coles conjectures that the relevant region is perhaps not the viscous sublayer as such but the whole turbulent boundary layer at the lowest Reynolds number at which turbulence can exist; this boundary layer indeed has $u_\tau \delta/\nu$ of the order of 430. Coles' transformation should not be confused with the simpler intermediate-temperature assumption that any low-speed skin friction formula can be applied to a high-speed flow if fluid properties are evaluated at a temperature T_i somewhere between T_w and T_e (see the review by Hopkins and Inouye [8]).

11.3 Two-Dimensional Boundary Layers with Zero Pressure Gradient

Skin-Friction Formulas on Smooth Surfaces

A number of empirical formulas for varying degrees of accuracy have been developed for calculating compressible turbulent boundary layers on flat plates. Those developed by Van Driest [2] and by Spalding and Chi [9] have higher accuracy than the rest (see Hopkins and Keener [10] and Cary and Bertram [11]) and cover a wide range of Mach number and ratio of wall temperature to total temperature. These two methods have similar accuracy, although the approaches followed to obtain the formulas are somewhat different. Both methods define compressibility factors by the following relation between the compressible and incompressible values:

$$c_{f_i} = F_c c_f, \tag{11.26a}$$

$$R_{\theta_i} = F_{R_\theta} R_\theta, \tag{11.26b}$$

$$R_{x_i} = \int_0^{R_x} \frac{F_{R_\theta}}{F_c} dR_x = F_{R_x} R_x. \tag{11.26c}$$

Here the subscript i denotes the incompressible values, and the factors F_c, F_{R_θ}, and F_{R_x} ($\equiv F_{R_\theta}/F_c$) defined by Eq. (11.26) are functions of Mach number, ratio of wall temperature to total temperature, and recovery factor. Spalding and Chi's method is based on the postulate that a unique relation exists between $c_f F_c$ and $F_R R_x$. The quantity F_c is obtained by means of mixing-length theory, and F_R is obtained semiempirically. According to Spalding and Chi,

$$F_c = \frac{T_{aw}/T_e - 1}{(\sin^{-1}\alpha + \sin^{-1}\beta)^2}, \qquad F_{R_\theta} = \left(\frac{T_{aw}}{T_e}\right)^{0.772}\left(\frac{T_w}{T_e}\right)^{-1.474}, \qquad (11.27)$$

where, with r denoting the recovery factor $(T_{aw} - T_e)/(T_{0e} - T_e)$,

$$\alpha = \frac{T_{aw}/T_e + T_w/T_e - 2}{\left[(T_{aw}/T_e + T_w/T_e)^2 - 4(T_w/T_e)\right]^{1/2}}, \qquad (11.28a)$$

$$\beta = \frac{T_{aw}/T_e - T_w/T_e}{\left[(T_{aw}/T_e + T_w/T_e)^2 - 4(T_w/T_e)\right]^{1/2}},$$

$$\frac{T_{aw}}{T_e} = 1 + \frac{\gamma - 1}{2} r M_e^2. \qquad (11.28b)$$

According to Van Driest's method, which is based entirely on the mixing-length theory, F_c is again given by the expression defined in Eqs. (11.27) and (11.28). However, the parameter F_{R_θ} is now given by

$$F_{R_\theta} = \frac{\mu_e}{\mu_w}. \qquad (11.29)$$

The development of Van Driest's formula for skin friction is analogous to the solution steps discussed for uncoupled flows (see Section 6.5) except that the derivation is more tedious. The solution requires the expansion of the integral into a series by means of integration by parts and a simple expression is again obtained when higher-order terms are neglected. With this procedure and with the power-law temperature-viscosity relation

$$\mu \propto T^\omega,$$

which implies $F_{R_\theta} = (T_e/T_w)^\omega$, the following relation for c_f and R_x is obtained for compressible turbulent boundary layers with and without heat transfer, with x measured from the effective origin of the turbulent flow:

$$\frac{0.242(\sin^{-1}\alpha + \sin^{-1}\beta)}{A\sqrt{c_f(T_w/T_e)}} = 0.41 + \log R_x c_f - \left(\frac{1}{2} + \omega\right)\log\frac{T_w}{T_e}, \qquad (11.30)$$

where A is defined by Eq. (11.23). This formula is based on Prandtl's mixing-length formula $l = \kappa y$. If the procedure leading to this equation is repeated with the mixing-length expression given by von Karman's similarity law

$$l = \kappa \left|\frac{\partial u/\partial y}{\partial^2 u/\partial y^2}\right|,$$

a formula similar to that given by Eq. (11.30) is obtained except that $\frac{1}{2} + \omega$ in Eq. (11.30) is replaced by ω. This formula is known as Van Driest II, in order to distinguish it from Eq. (11.30), which is known as Van Driest I, and may be written as

$$\frac{0.242(\sin^{-1}\alpha + \sin^{-1}\beta)}{A\sqrt{c_f(T_w/T_e)}} = 0.41 + \log R_x c_f - \omega \log \frac{T_w}{T_e}. \qquad (11.31)$$

The predictions of Eq. (11.31) are in better agreement with experiment than those of Eq. (11.30), and Van Driest II should therefore be used in preference to Van Driest I.

Equations (11.30) and (11.31) constitute a compressible form of the von Karman equation discussed in Section 6.5 [see Eq. (6.52)]. For an incompressible flow, they reduce to Eq. (6.52). For an incompressible adiabatic flow, $T_w/T_e \to 1$ and $B = 0$, so that with Eq. (11.28a), we can write Eq. (11.31) as

$$\frac{0.242 \sin^{-1} A}{A\sqrt{c_f}} = 0.41 + \log R_x c_f.$$

In addition, A is of the order of M_e, and since it is small, $\sin^{-1} A \simeq A$. The resulting equation is then identical to Eq. (6.52).

According to Van Driest II, the average skin-friction coefficient $\bar{c}_f$ is obtained from the expression

$$\frac{0.242(\sin^{-1}\alpha + \sin^{-1}\beta)}{A\sqrt{\bar{c}_f(T_w/T_e)}} = \log R_x \bar{c}_f - \omega \log \frac{T_w}{T_e}. \qquad (11.32)$$

Figures 11.1 and 11.2 show the variation of local and average skin-friction coefficients calculated from Eqs. (11.31) and (11.32), respectively, on an adiabatic flat plate for various Mach numbers. The recovery factor was assumed to be 0.88.

Figure 11.3 shows the effect of compressibility on the local and average skin-friction coefficients. Here, the skin-friction formulas were solved at a specified Reynolds number ($R_x = 10^7$) as functions of Mach number for fixed values of T_w/T_e. In the results shown in Fig. 11.1, the local skin-friction values for incompressible flows with heat transfer were obtained from the limiting form of Eq. (11.31). We note that as $M_e \to 0$ and when $T_w/T_e = 1$, $A \to 0$, $\alpha \to -1$, and $\beta \to 1$. It follows that the term

$$\frac{\sin^{-1}\alpha + \sin^{-1}\beta}{A}$$

is indeterminate. Using L'Hospital's rule and recalling that $B = T_e/T_w - 1$, we can write Eq. (11.31) for an incompressible turbulent flow with heat transfer, after some algebraic manipulation, as

$$\frac{2}{\sqrt{T_w/T_e} + 1} \frac{0.242}{\sqrt{c_f}} = 0.41 + \log R_x c_f - \omega \log \frac{T_w}{T_e}. \qquad (11.33)$$

The average skin-friction formula, Eq. (11.32), can also be written for an

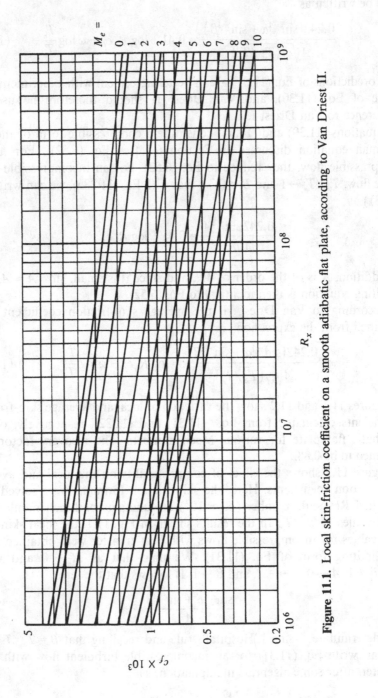

Figure 11.1. Local skin-friction coefficient on a smooth adiabatic flat plate, according to Van Driest II.

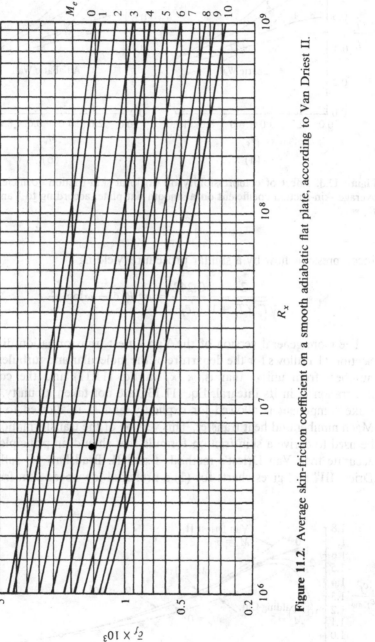

Figure 11.2. Average skin-friction coefficient on a smooth adiabatic flat plate, according to Van Driest II.

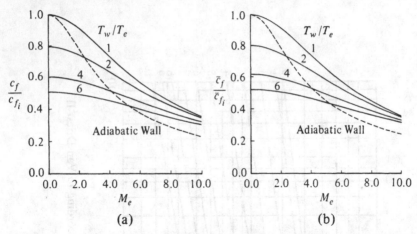

Figure 11.3. Effect of compressibility on (a) local skin-friction coefficient and (b) average skin-friction coefficient on a smooth flat plate, according to Van Driest II. $R_x = 10^7$.

incompressible flow by a similar procedure, yielding

$$\frac{2}{\sqrt{T_w/T_e}+1}\frac{0.242}{\sqrt{\bar{c}_f}} = \log R_x \bar{c}_f - \omega \log \frac{T_w}{T_e}. \qquad (11.34)$$

The more general version of the Van Driest transformation discussed in Section 11.1 allows for the departure of the molecular and turbulent Prandtl numbers from unity; that is, κ/κ_h in Eq. (11.11) and the constant of integration c_1 in its integral, Eq. (11.13), are not taken as unity. Also, the wake component is allowed for explicitly, and Π is allowed to vary with Mach number and heat transfer. This version of the transformation can also be used to derive a skin-friction formula that should, in principle, be more accurate than Van Driest's methods I and II. Bradshaw [4] calls it "Van Driest III" and gives charts for (1) a preferred low-speed skin-friction law,

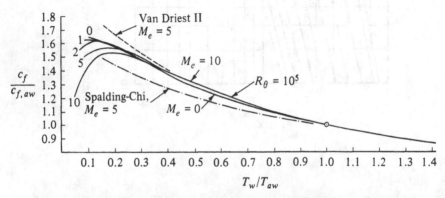

Figure 11.4. Variation of skin-friction coefficient with wall temperature [4].

(2) the ratio of adiabatic-wall skin friction to the low-speed value as a function of Mach number and Reynolds number, and (3) the ratio of skin-friction coefficient to the adiabatic value as a function of wall-temperature ratio, for a range of Mach number and Reynolds number. The chart (3) is reproduced in Fig. 11.4, with results from Van Driest II and the Spalding–Chi formula for comparison. The drop in c_f on very cold walls results from changes in the logarithmic-law additive constant c [Eq. (11.25)] with heat transfer, as inferred from a very limited amount of data; differences between Van Driest II (which assumes $c =$ const) and Van Driest III thus reflect real experimental uncertainty.

Reynolds Analogy Factor

According to the studies conducted by Spalding and Chi [9] and Cary [12] it appears that for Mach numbers less than approximately 5 and near-adiabatic wall conditions, a Reynolds analogy factor of

$$\frac{St}{c_f/2} = 1.16 \qquad (11.35)$$

adequately represents the available experimental data. However, for turbulent flow with significant wall cooling and for Mach numbers greater than 5 at any ratio of wall temperature to total temperature, the Reynolds analogy factor is ill-defined. Recent data [11] indicate that for local Mach numbers greater than 6 and T_w/T_0 less than approximately 0.3, the Reynolds analogy factor scatters around a value of 1.0. A sample of the results is presented in Fig. 11.5 for a Mach number of 11.3 and indicates that the Reynolds measured analogy factor is scattered from around 0.8 to 1.4 with no discernible trend for T_w/T_0.

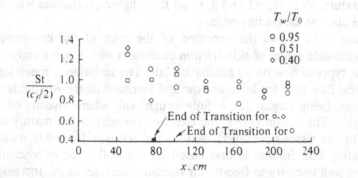

Figure 11.5. Reynolds analogy factors at $M_e = 11.3$, $R_e/m = 54 \times 10^6$, after Cary [12].

Skin-Friction Formulas on Rough Surfaces

The skin-friction formulas for a smooth flat plate, Eqs. (11.32) and (11.33), can also be used to obtain formulas for sand-grain-roughened flat plates by assuming a relation between the compressible and incompressible values such as that given by Eq. (11.26a). According to the experiments of Goddard [13] on adiabatic fully rough flat plates,

$$F_c = \frac{T_{aw}}{T_e},$$
(11.36)

and the experimental values of c_f verified the relation (11.36) for his chosen turbulent recovery factor, $r = 0.86$. It should be emphasized that this equation is for fully rough flow in which the flow on top of the roughness elements remains subsonic. It is consistent with the observation originally noted by Nikuradse for incompressible flow, namely, that the skin-friction drag for fully rough flow is the sum of the form drags of the individual roughnesses.

Fenter [14] also presented a theory for the effect of compressibility on the turbulent skin friction of rough plates with heat transfer. This gave results that agree with those of relation (11.36) only at Mach numbers close to unity and only for zero heat transfer. For $T_w = T_e$, the value of c_f given by this theory is 14 percent less than that given by Goddard's relation at $M_e = 2.0$ and 45 percent less at $M_e = 4.0$. Fenter presented experimental data for $M_c = 1.0$ and 2.0 that agreed well with this theory for the case of zero heat transfer. The difference in the experimental values of c_f of the two reports is probably within the accuracy to which the roughness heights were measured. The theory of Fenter is based on assumptions whose validity is questionable at high Mach numbers, and these assumptions may account for the difference in c_f predicted by Fenter and by Goddard for the case of $T_w = T_e$.

Figures 11.6 and 11.7 show the average skin-friction distribution for a sand-roughened adiabatic plate, and Figs. 11.8 and 11.9 show the results for a sand-roughened plate with a wall temperature equal to the freestream temperature, all at $M_e = 1$ and 2. In all these figures, transition was assumed to take place at the leading edge.

Figure 11.10 shows the variation of the ratio of the compressible to incompressible values of skin-friction coefficient with Mach number for the various types of flow on an adiabatic plate. The variation is much larger for turbulent flow than for laminar flow and increases as the Reynolds number increases, being largest for a fully rough wall where viscous effects are negligible. The reason is that the effect of viscosity is felt mainly near the wall (in the viscous sublayer), and so the relevant Reynolds number for correlating skin friction is that based on the wall value of viscosity. The ratio of wall viscosity to freestream viscosity increases as M_e increases; so a given value of $u_e L / \nu_e$ corresponds to a smaller value of $u_e L / \nu_w$ and thus a

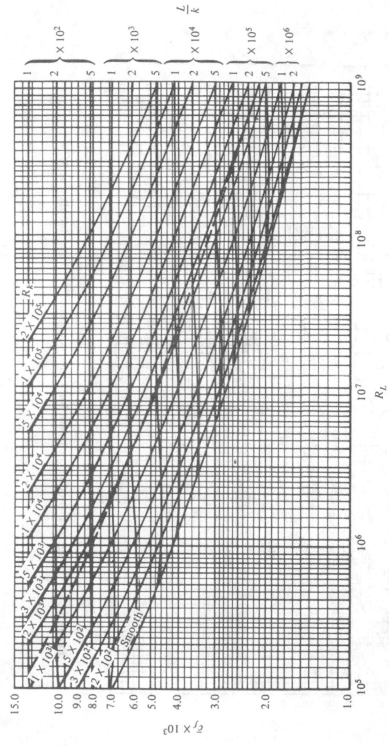

Figure 11.6. Average skin-friction coefficient for a sand-roughened adiabatic flat plate at $M_e = 1$.

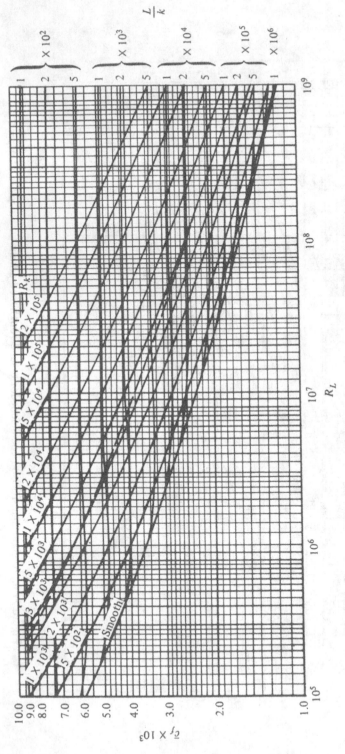

Figure 11.7. Average skin-friction coefficient for a sand-roughened adiabatic flat plate at $M_e = 2$.

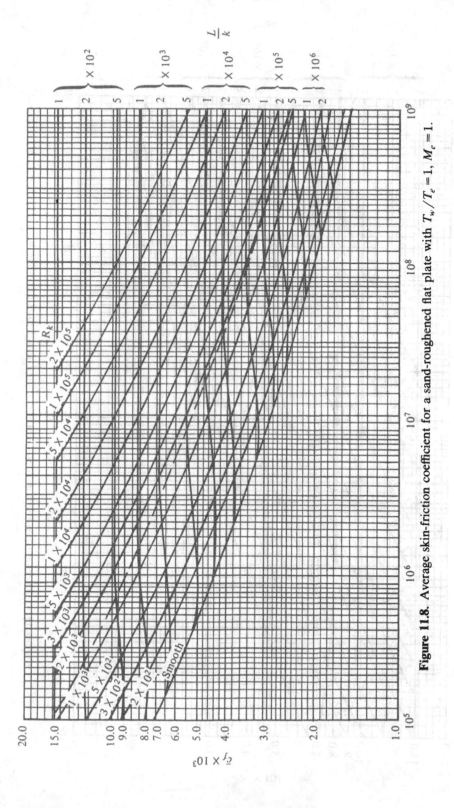

Figure 11.8. Average skin-friction coefficient for a sand-roughened flat plate with $T_w/T_e = 1$, $M_e = 1$.

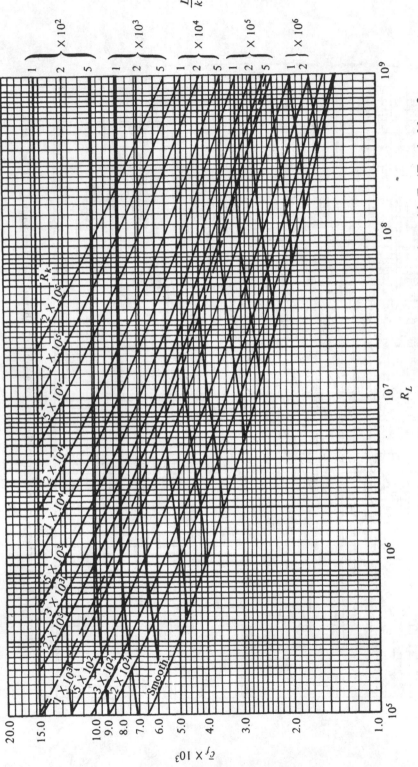

Figure 11.9. Average skin-friction coefficient for a sand-roughened flat plate with $T_w/T_e = 1$, $M_e = 2$.

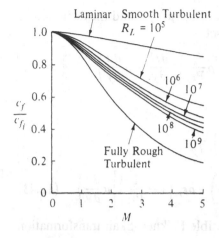

Laminar   Smooth Turbulent
$R_L = 10^5$

Figure 11.10. Mach number variation of the ratio of the compressible to incompressible values of local skin-friction coefficient for the various types of air flow on an adiabatic flat plate, for given Reynolds number $u_e L/\nu_e$.

larger c_f. The effect on c_f decreases as $u_e L/\nu_e$ increases because the change of c_f associated with R_L is smaller. The effect is absent on fully rough walls.

11.4 Two-Dimensional Flows with Pressure Gradient

In parallel with Section 6.6, we now discuss the calculation of two-dimensional flows by a differential method that employs a compressible version of the eddy-viscosity formulation described in Section 6.6.

The governing boundary-layer equations for a two-dimensional turbulent flow in the absence of body forces, neglecting the terms $\overline{\rho'u'v'}$ and $\overline{\rho'T'v'}$ involving density fluctuations, follow from Eqs. (3.38) and (3.46) and can be rewritten as

$$\frac{\partial}{\partial x}(\rho u) + \frac{\partial}{\partial y}(\overline{\rho v}) = 0, \tag{11.37}$$

$$\rho u \frac{\partial u}{\partial x} + \overline{\rho v}\frac{\partial u}{\partial y} = -\frac{dp}{dx} + \frac{\partial}{\partial y}\left(\mu\frac{\partial u}{\partial y} - \rho\overline{u'v'}\right), \tag{11.38}$$

$$\rho u \frac{\partial H}{\partial x} + \overline{\rho v}\frac{\partial H}{\partial y} = \frac{\partial}{\partial y}\left[k\frac{\partial T}{\partial y} - c_p\rho\overline{T'v'} + u\left(\mu\frac{\partial u}{\partial y} - \rho\overline{u'v'}\right)\right], \tag{11.39}$$

where, as before, $\overline{\rho v} = \rho v + \overline{\rho'v'}$. With eddy viscosity and turbulent Prandtl number defined by

$$-\overline{u'v'} = \varepsilon_m\frac{\partial u}{\partial y}, \qquad -\overline{T'v'} = \frac{\varepsilon_m}{\mathrm{Pr}_t}\frac{\partial T}{\partial y} \tag{11.40}$$

and with the use of Bernoulli's equation, $-dp/dx = \rho_e u_e(du_e/dx)$, Eqs. (11.38) and (11.39) become

$$\rho u \frac{\partial u}{\partial x} + \overline{\rho v}\frac{\partial u}{\partial y} = \rho_e u_e \frac{du_e}{dx} + \frac{\partial}{\partial y}\left[(\mu + \rho\varepsilon_m)\frac{\partial u}{\partial y}\right], \tag{11.41}$$

$$\rho u \frac{\partial H}{\partial x} + \overline{\rho v}\frac{\partial H}{\partial y} = \left[\left(k + c_p\rho\frac{\varepsilon_m}{\mathrm{Pr}_t}\right)\frac{\partial T}{\partial y} + u(\mu + \rho\varepsilon_m)\frac{\partial u}{\partial y}\right]. \tag{11.42}$$

Recalling the definition of Prandtl number $\Pr = \mu c_p / k$ and the definition of total enthalpy for a perfect gas, we can set

$$\left(k + c_p \rho \frac{\varepsilon_m}{\Pr_t}\right) \frac{\partial T}{\partial y} = \left(\frac{\mu}{\Pr} + \rho \frac{\varepsilon_m}{\Pr_t}\right)\left(\frac{\partial H}{\partial y} - u \frac{\partial u}{\partial y}\right)$$

and write Eq. (11.42) as

$$\rho u \frac{\partial H}{\partial x} + \overline{\rho v} \frac{\partial H}{\partial y} = \frac{\partial}{\partial y}\left\{\left(\frac{\mu}{\Pr} + \rho \frac{\varepsilon_m}{\Pr_t}\right)\frac{\partial H}{\partial y}\right.$$

$$\left. + \left[\mu\left(1 - \frac{1}{\Pr}\right) + \rho \varepsilon_m\left(1 - \frac{1}{\Pr_t}\right)\right] u \frac{\partial u}{\partial y}\right\}. \quad (11.43)$$

As in Chapter 10, we use the compressible Falkner–Skan transformation defined by Eq. (10.1),

$$d\eta = \sqrt{\frac{u_e}{v_e x}} \frac{\rho}{\rho_e} dy, \qquad \psi(x, y) = \sqrt{\rho_e \mu_e u_e x}\, f(x, \eta), \qquad (10.1)$$

and the definition of stream function that satisfies the continuity equation (11.37), so that

$$\rho u = \frac{\partial \psi}{\partial y}, \qquad \overline{\rho v} = -\frac{\partial \psi}{\partial x}. \qquad (11.44)$$

The momentum and energy equations, given by Eqs. (11.41) and (11.43), can now be written in a form similar to that of Eqs. (10.3) and (10.4), that is,

$$(bf'')' + m_1 ff'' + m_2\left[c - (f')^2\right] = x\left(f' \frac{\partial f'}{\partial x} - f'' \frac{\partial f}{\partial x}\right), \quad (11.45)$$

$$(eS' + df'f'')' + m_1 fS' = x\left(f' \frac{\partial S}{\partial x} - S' \frac{\partial f}{\partial x}\right), \quad (11.46)$$

except that now

$$b = C(1 + \varepsilon_m^+), \qquad e = \frac{C}{\Pr}\left(1 + \varepsilon_m^+ \frac{\Pr}{\Pr_t}\right),$$

$$d = \frac{C u_e^2}{H_e}\left[1 - \frac{1}{\Pr} + \varepsilon_m^+\left(1 - \frac{1}{\Pr_t}\right)\right], \qquad \varepsilon_m^+ = \frac{\varepsilon_m}{\nu}. \quad (11.47)$$

Including a transpiration velocity v_w at the wall, the boundary conditions given by

$$y = 0, \quad v = v_w(x), \quad u = 0, \quad H = H_w(x) \quad \text{or} \quad \left(\frac{\partial H}{\partial y}\right)_w = -\frac{c_{p_w}}{k_w} \dot{q}_w,$$

$$(11.48a)$$

$$y = \delta, \quad u = u_e(x), \quad H = H_e \qquad (11.48b)$$

can be written in terms of transformed variables as

$$\eta = 0, \quad f' = 0, \quad f_w = \frac{-1}{(u_e \mu_e \rho_e x)^{1/2}} \int_0^x \rho_w v_w \, dx,$$

$$S = S_w(x) \quad \text{or} \quad S'_w = -\frac{c_{p_w}}{k_w} \frac{c_w}{H_e} \frac{x \dot{q}_w}{\sqrt{R_x}}, \tag{11.49a}$$

$$\eta = \eta_e, \quad f' = 1, \quad S = 1. \tag{11.49b}$$

Note that in the absence of mass transfer, Eq. (11.49) is identical to the equations given by Eq. (10.6).

As in Section 6.6, we take $\text{Pr}_t = 0.9$ and use the eddy-viscosity formulation of Cebeci and Smith, which, for compressible flows, is similar to that given by Eqs. (6.102) and (6.109) except that now the damping-length parameter is constructed with local values of density and viscosity so that

$$A = 26 \frac{\nu}{N} u_\tau^{-1} \left(\frac{\rho}{\rho_w} \right)^{1/2}, \quad u_\tau = \left(\frac{\tau_w}{\rho_w} \right)^{1/2}, \quad p^+ = \frac{\nu_e u_e}{u_\tau^3} \frac{du_e}{dx}, \quad v_w^+ = \frac{v_w}{u_\tau},$$

$$\tag{11.50a}$$

$$N = \left\{ \frac{\mu}{\mu_e} \left(\frac{\rho_e}{\rho_w} \right)^2 \left(\frac{p^+}{v_w^+} \right) \left[1 - \exp\left(11.8 \frac{\mu_w}{\mu} v_w^+ \right) \right] + \exp\left(11.8 \frac{\mu_w}{\mu} v_w^+ \right) \right\}^{1/2}.$$

$$\tag{11.50b}$$

For flows with no mass transfer through the surface, N can be written as

$$N = \left[1 - 11.8 \left(\frac{\mu_w}{\mu_e} \right) \left(\frac{\rho_e}{\rho_w} \right)^2 p^+ \right]^{1/2}. \tag{11.50c}$$

Sample Calculations

As an example of a two-dimensional coupled turbulent flow with heat transfer (but no mass transfer), we consider the turbulent version of the Howarth flow discussed in Section 10.2. We assume that at first there is no pressure gradient from $0 \leq x \leq x_0$, and this is followed by a region where the external velocity distribution varies linearly with x for $x \geq x_0$; that is,

$$u_e = \begin{cases} u_\infty & 0 \leq x \leq x_0, \\ u_\infty [1 - \frac{1}{8}(x - x_0)] & x \geq x_0. \end{cases} \tag{11.51}$$

To show the effect of heat transfer on the separation point at different Mach numbers, Eqs. (11.45), (11.46), and (11.49) have been solved with the numerical method and the Fortran program to be discussed in Sections 13.4 and 13.5 for values of S_w ($\equiv H_w / H_e$) equal to 2, 1, and 0.25. Thus, as in the laminar flows of Section 10.2, the results correspond to cases where the wall is heated ($S_w = 2$), cooled ($S_w = 0.25$), and adiabatic ($S_w = 1$) for three values of the Mach number M_∞ equal to 0.05, 1, and 2. The calculations were done for given freestream pressure and temperature for different values of M_∞, giving a Reynolds number per unit x, u_∞ / ν_∞, of $7.15 \times 10^6 \, M_\infty \, \text{ft}^{-1}$ since for air we have

$$\frac{u_\infty}{\nu_\infty} = \frac{49.01 \, M_\infty \sqrt{T_\infty}}{\nu_\infty},$$

with T_∞ in deg. R. The calculations were started as laminar flow at $x = 0$, and transition was specified at $x = 0.001$, but we can assume that the flow was turbulent from the leading edge. The location of x corresponding to the beginning of the adverse pressure-gradient flow, namely x_0, was taken to be equal to 5 ft.

Figure 11.11 shows the effect of Mach number on local skin-friction coefficient $c_f \, (= \tau_w / \frac{1}{2} \rho_e u_e^2)$ for the three values of S_w. For $x < 5$ (constant u_e) the results correspond to those in Fig. 11.10 (recall that for the present calculations Reynolds number varies proportionally to Mach number). The separation point is almost independent of Mach number for the adiabatic- and cooled-wall cases ($S_w = 1.0$ and 0.25), but cooling delays separation because it increases density near the wall. Heating ($S_w = 2.0$) moves separation forward by an amount that increases with increasing Mach number.

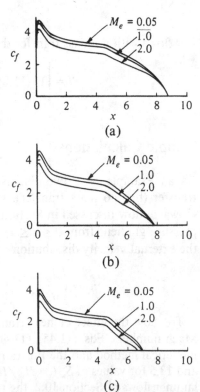

Figure 11.11. Effect of Mach number of local skin-friction coefficient c_f. (a) $S_w = 0.25$, (b) $S_w = 1.0$, (c) $S_w = 2.0$.

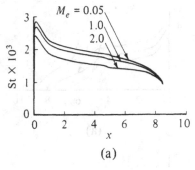

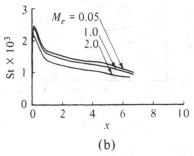

Figure 11.12. Effect of Mach number on Stanton number (a) $S_w = 0.25$ and (b) $S_w = 2.0$.

Figure 11.12 shows the effect of Mach number on Stanton number defined by

$$St = \frac{\dot{q}_w}{\rho_e u_e (H_w - H_e)} \qquad (11.52)$$

for two wall-temperature conditions and at three separate Mach numbers. We notice that at low Mach numbers heat transfer is nearly proportional to

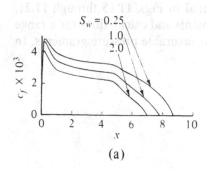

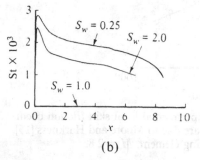

Figure 11.13. Effect of wall temperature boundary conditions on (a) local skin friction coefficient and (b) Stanton number at $M_e = 0.05$.

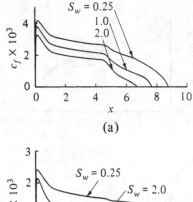

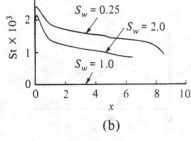

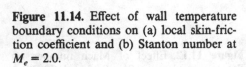

Figure 11.14. Effect of wall temperature boundary conditions on (a) local skin-friction coefficient and (b) Stanton number at $M_e = 2.0$.

surface shear stress except near separation (the ends of the curves), where c_f goes to zero but St does not. That is, the Reynolds analogy $St/\frac{1}{2}c_f \simeq$ constant works well. At the highest Mach number, the predicted heat transfer *increases* as the flow decelerates, especially on the cold wall, because deceleration increases the temperature of the fluid.

Figures 11.13 and 11.14 show the effect of wall temperature conditions on local skin-friction coefficient and Stanton number at two separate Mach numbers, cross-plotted from the above results.

A sample of additional results is presented in Figs. 11.15 through 11.21, which allow comparison between measurements and calculations for a range of boundary conditions including zero and favorable pressure gradients. In

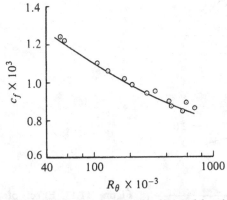

Figure 11.15. Comparison of calculated and experimental local skin-friction coefficients for an adiabatic flat-plate flow. The data are due to Moore and Harkness [15]. Skin-friction measurements were made by floating element. $M_e = 2.8$.

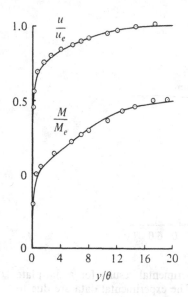

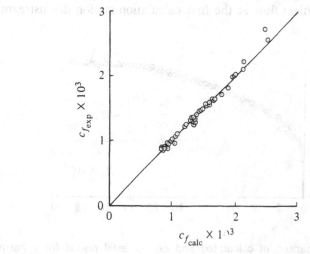

Figure 11.16. Comparison of calculated and experimental results for an adiabatic flat-plate flow. Experimental data are due to Matting, Chapman, Nyholm, and Thomas [16]. Skin friction was measured by floating element: $c_{f_{\text{exp}}} = 0.00129$; $c_{f_{\text{calc}}} = 0.00131$; $M_e = 2.95$; $R_x = 31 \times 10^6$.

the former case the results correspond to adiabatic walls with and without mass transfer and to an impervious nonadiabatic wall; in the latter the wall is impervious and adiabatic.

Figures 11.15 and 11.16 present local skin friction and velocity distributions for adiabatic flat-plate flows with Mach numbers of 2.8 and 2.95, respectively. The agreement between the measurements and calculations is clearly very satisfactory, and Fig. 11.17, which is based on local skin-friction coefficients for the range of adiabatic, turbulent flat-plate flows assembled by Cebeci et al. [17], confirms that this is so over a wide range of values of

Figure 11.17. Comparison of calculated and experimental local skin-friction coefficients on adiabatic flat-plate flows [17]. The rms error based on 43 experimental values obtained by the floating-element technique is 3.5%.

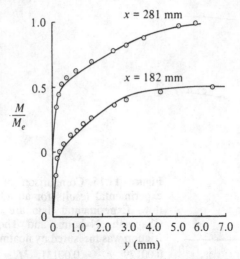

Figure 11.18. Comparison of calculated and experimental results for a flat-plate flow with heat transfer $T_w/T_e = 1.95$, $M_e = 2.57$. The experimental data are due to Michel [18].

the skin-friction coefficient. These data correspond to a Mach-number range from 0.4 to 5 and to momentum-thickness Reynolds numbers from 1.6×10^3 to 702×10^3. The rms deviation of the measurements from the straight line was 3.5 percent, which is within the measurement uncertainty.

The heat-transfer results of Figs. 11.18 and 11.19 also confirm the ability of the calculation method to represent high-Mach-number flows at least in zero pressure gradient. In Figs. 11.18 and 11.19 the values of T_w/T_e are 1.95 and 2.16, respectively, and M_e is 2.57 and 2.27, respectively. In both cases, the calculations began with laminar-flow assumptions and then assumed transition to turbulent flow at the first calculation station downstream. In

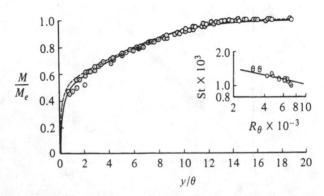

Figure 11.19. Comparison of calculated and experimental results for a flat-plate flow with heat transfer $T_w/T_e = 2.16$, $M_e = 2.27$. The experimental data are due to Pappas [19]. The solid lines denote calculations for $R_\theta = 3500$ and the dashed lines results for $R_\theta = 9500$.

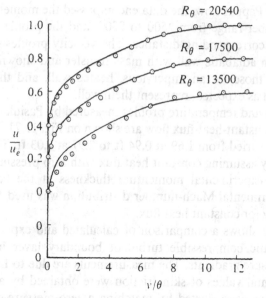

Figure 11.20. Comparison of calculated and experimental results for an adiabatic flat-plate flow with mass transfer. $(\rho v)_w / \rho_e u_e = 0.0013$, $M_e = 1.8$. The experimental data are due to Squire [20].

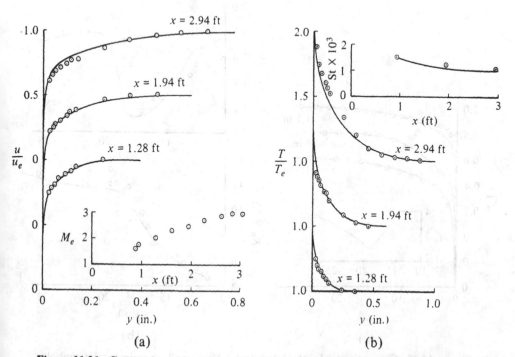

Figure 11.21. Comparison of calculated and experimental results for an accelerating flow with constant heat flux. (a) Velocity profiles, (b) temperature profiles and Stanton-number distribution. The experimental data are due to Pasiuk et al. [21].

the case of the Pappas flow, the data encompassed the momentum-thickness Reynolds-number range from 3500 to 9500, and the profile shape changes little over the corresponding distance. The velocity profiles of Squire [20], obtained on an adiabatic wall with mass transfer and shown in Fig. 11.20, are similar to those for an impervious, heated wall, and the calculations, once again and as expected, represent them well.

The velocity and temperature profiles measured by Pasiuk et al. [21] in an accelerating, constant-heat-flux flow are shown on Fig. 11.21. The freestream Mach number varied from 1.69 at 0.94 ft to 2.97 at 3.03 ft. The calculations were started by assuming constant heat flux with zero pressure gradient and matching the experimental momentum thickness at the 0.94-ft location. Then the experimental Mach-number distribution was used to compute the rest of the flow for constant heat flux.

Figure 11.22 shows a comparison of calculated and experimental results for an adiabatic compressible turbulent boundary layer in adverse and favorable pressure gradients. The measurements are due to Lewis et al. [22]. The experimental values of skin friction were obtained by a Stanton tube. The calculations were started by matching a zero-pressure-gradient profile ($R_\theta = 4870$) at $x = 11.5$ in. downstream of the leading edge of the model. Then the experimental Mach-number distribution shown in Fig. 11.22(a)

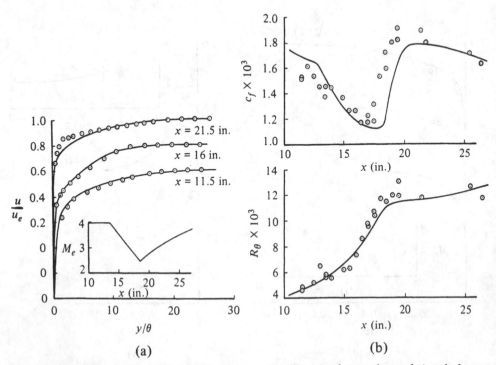

Figure 11.22. Comparison of calculated (solid lines) and experimental (symbols [22]) results for an adiabatic compressible turbulent boundary layer with an adverse and a favorable pressure gradient: (a) velocity profiles; (b) c_f and R_θ distributions. The experimental c_f values were measured by Stanton tube $M_\infty = 4$.

was used to compute the rest of the flow. Once again, comparison of measured and calculated results supports the use of the calculation method.

11.5 Shock-Wave/Boundary-Layer Interaction

We now come to one of the most important and outstanding problems in the theory of boundary layers, the response of a turbulent boundary layer to a shock wave generated either by a concave corner on the surface or externally. Normally the structure of the boundary layer has little effect on the external flow unless there are extensive regions of separation, but the situation can be quite different when shocks are involved. For example, the lifting force on airfoils at transonic speeds can be halved, and the shock position moved 20 percent of the chord, by viscous effects even when the Reynolds number is high and the boundary layer is attached. The resolution of the structure of the boundary layer in such flows is still in the development stage and far from complete. It is inappropriate here to attempt to review the enormous amount of work that has been done on this topic. Instead, we refer the reader to the general review by Adamson and Messiter [23] and to the AGARD Conference at Colorado Springs [24], and especially to the paper by Melnik [25], for an assessment of the present state of knowledge. Some comments of a general nature are in order, however, to give the reader an appreciation of the achievements.

The interaction has some general features in common with laminar interactions, principally the formation of a pressure plateau ahead of the shock intersection, but there are important differences. For example, the upstream extent of the interaction is only a few boundary-layer thicknesses, even with a large pressure rise across the shock, and much less than has been observed with laminar flows. Again the pressure rise to provoke separation in a turbulent boundary layer is higher than for laminar flow because the flow speed and dynamic pressure near the wall are larger; a rough rule is that a turbulent boundary layer passing through a normal shock will separate if the Mach number ahead of the shock exceeds 1.3 (this corresponds to a pressure rise of 0.68 of the freestream dynamic pressure, about the same as that needed to separate a low-speed boundary layer in a strong adverse pressure gradient). When the shock is oblique, the pressure rise to separation is roughly the same as for a normal shock.

The simplest experimental configuration for examining the nature of the interaction has the shock generated by a ramp on the test surface, and the resulting flow has been studied by many workers during the last 30 years or so. In Fig. 11.23 we show a comparison between the measured pressure distribution on a 25° ramp at $M_\infty = 2.96$ and the predictions of inviscid theory (Law [26]), and in Fig. 11.24 we show a photograph of hypersonic flow past a ramp. A large number of measurements of the ramp angle α_i needed to produce incipient separation have been made, and the results are

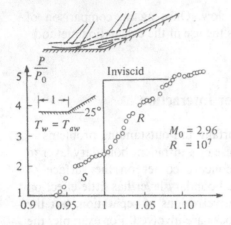

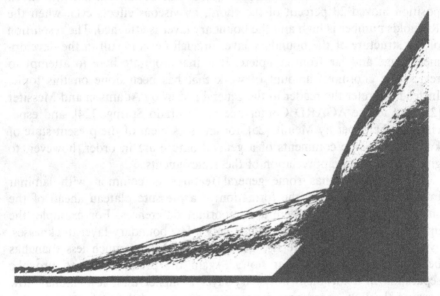

Figure 11.23. Pressure distribution along a supersonic ramp with a turbulent boundary layer. The freestream Mach number is 2.96 and the Reynolds number is 10^7 at the corner. The solid line denotes the inviscid pressure and the symbols experimental results [26].

Separated Flow

(a)

Attached Flow

(b)

Figure 11.24. Turbulent hypersonic flow at a wedge-compression corner. (a) Separated flow. (b) Attached flow. Photographs courtesy of Dr. G. M. Elfstrom, DSMA Inc., Toronto.

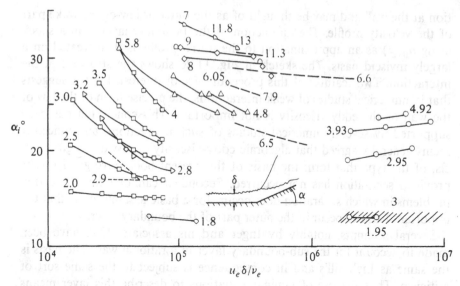

Figure 11.25. Incipient separation of a turbulent boundary layer on a ramp [27]. Labels on curves indicate Mach number.

reviewed by Roshko and Thomke [27]. The trends are displayed in Fig. 11.25. According to inviscid theory, the ramp angle that is required to produce a detached shock increases with Mach number, and so it is not surprising that the critical ramp angle α_i, for incipient separation, also increases with Mach number, from about 15° at $M_\infty = 2$ to about 30° at $M_\infty = 8$. The variation of α_i with Reynolds number R is more interesting. At low values of R, α_i decreases as R increases, apparently because of the changing outer profile (law of the wake, discussed in Sec. 11.2). At higher values the trend is reversed due to the profile becoming fuller with consequent decrease of the shape parameter H and skin-friction coefficient c_f. For recent results, see Hayakawa and Squire [28].

One of the earliest attempts to describe the details of the interaction is due to Lighthill [29], who used arguments similar to those leading to the establishment of the triple-deck theory of laminar free interactions (Sec. 10.3). The chief modification is to use the characteristic turbulent profile as initial profile instead of the laminar form, but as Lighthill immediately noticed, the resulting scaling laws [compare Eq. (10.58)] are not consistent, the interaction region being thinner than the boundary layer. The most recent precise studies of the interaction ignore the sub-boundary layer [compare Eqs. (10.55)–(10.57)] and take the velocity upstream of the interaction in the form

$$u_0(y) = u_e\left[1 + \varepsilon u_0\left(\frac{y}{\delta}\right)\right], \tag{11.52}$$

where u_e is the external velocity, ε is a small parameter $O(\log \mathrm{Re})^{-1}$, and δ is the boundary-layer thickness. This form does not satisfy the no-slip condition at the wall and may be thought of as the outer or law-of-the-wake part of the velocity profile. The interaction of the boundary layer with a shock using $u_0(y)$ as an approximation to the initial profile is then treated on a largely inviscid basis. The sketch in Fig. 11.26 shows the structure of the interaction. Two features of this procedure are of interest. First it suggests that in numerical studies of weak interactions, the precise form of the law of the wall or the eddy viscosity is not important. This conclusion has been supported by several numerical studies of such interactions, and indeed it seems generally agreed that algebraic eddy-viscosity or mixing-length models, of the type that form the basis of the present book, give good results provided separation has not occurred. Second, it cannot treat interaction problems in which separation does occur, for a basic assumption is that the important features occur in the outer part of the boundary layer.

Several attempts, notably by Inger and his associates [30], have been made to account for the sub-boundary layer in a rational way. The basis is the same as Lighthill's and in consequence is subject to the same sort of criticism. Thus the use of laminar equations to describe this layer means that Reynolds stresses are assumed to be relatively unimportant. In fact, Reynolds stresses are significant in the law-of-the-wall region, and the equation

$$\tau = \mu \frac{\partial u}{\partial y} - \rho \overline{u'v'}$$

must be considered in any explanation of the behavior of the wall shear stress τ_w. The implication is that even in the interactive region, the pressure gradient and the inertia terms lead to higher-order corrections for τ_w. At the present time, therefore, our understanding of the mechanisms that control extensive regions of separated flow is still largely incomplete, and this uncertainty is reflected in the numerical procedure adopted for the calculation of separating interactive flows of practical interest. The use of simple algebraic models for the eddy viscosity, of the type discussed in the present chapter and in Chapter 6, predicts the gross features of the flow, even when separation occurs, but not its detailed properties. More complex models

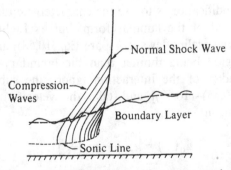

Figure 11.26. Incident normal shock wave with unseparated turbulent boundary layer [29].

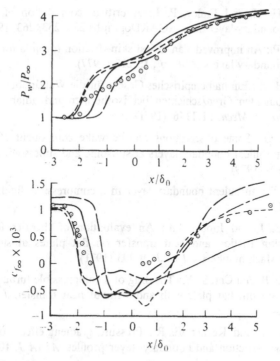

Figure 11.27. Comparison of computation and surface measurements for a supersonic compression corner with shock-wave/boundary-layer interaction. The ramp angle is 24°, the Reynolds number based on boundary-layer thickness is 1.33×10^6 and $M_\infty = 2.8$. The symbols denote experiment and (——) is obtained using an algebraic eddy-viscosity model, (— —) a one-equation turbulence model, (---) two-equation model A, and (— · —) two-equation Model B [30].

have been devised, and it might have been thought that these would permit more accurate predictions. In their review [23], Adamson and Messiter comment that "at the present time it appears that this promise has not been fulfilled, in the sense that it is not possible to choose a given model, with a given set of constants, that gives results more accurate than those given by a simple zero-equation model, for an arbitrary chosen interaction problem." A typical pair of graphs, due to Viegas and Horstmann [31] for pressure and skin friction, is shown in Fig. 11.27 comparing the predictions, using models of varying complexity, with experiment.

References

[1] Rotta, J. C.: Turbulent boundary layers with heat transfer in compressible flow. *AGARD Rept.* **281**, 1960.

[2] Van Driest, E. R.: Turbulent boundary layer in compressible fluids. *J. Aeronaut. Sci.* **18**:145 (1951).

[3] Fernholz, H. H. and Finley, P. J.: A critical compilation of compressible turbulent boundary-layer data. AGARDographs 223, 253, 263, 1977.

[4] Bradshaw, P.: An improved Van Driest skin-friction formula for compressible turbulent boundary layers. *AIAA J.*, **15**:212 (1977).

[5] Fernholz, H. H.: Ein halbempirisches Gesetz für die Wandreibung in kompressibilen turbulenten Grenzschichten bei isothermer und adiabater Wand. *Z. Angew. Math. u. Mech.* **51**:T146, (1971).

[6] Mabey, D. G.: Some observations on the wake component of the velocity profiles of turbulent boundary layers at subsonic and supersonic speeds. *Aero. Quart.* **30**:590 (1979).

[7] Coles, D.: The turbulent boundary layer in a compressible fluid. *Phys. Fluids* **7**:1403 (1964).

[8] Hopkins, E. J. and Inouye, M.: An evaluation of theories for predicting turbulent skin friction and heat transfer on flat plates at supersonic and hypersonic Mach numbers. *AIAA J.*, **9**:993 (1971).

[9] Spalding, D. B. and Chi, S. W.: The drag of a compressible turbulent boundary layer on a smooth flat plate with and without heat transfer. *J. Fluid Mech.* **18**:117 (1964).

[10] Hopkins, E. J. and Keener, E. R.: Pressure gradient effects on hypersonic turbulent skin–friction and boundary–layer profiles. *AIAA J.* **10**:1141 (1972).

[11] Cary, A. M. and Bertram, M. H.: Engineering prediction of turbulent skin friction and heat transfer in high-speed flow. NASA TN D-7507, 1974.

[12] Cary, A. M.: Summary of available information on Reynolds analogy for zero-pressure gradient, compressible turbulent-boundary-layer flow. NASA TN D-5560, 1970.

[13] Goddard, F. E., Jr.: Effect of uniformly distributed roughness on turbulent skin-friction drag at supersonic speeds. *J. Aero/Space Sci.*, **26**:1–15 (1959).

[14] Fenter, F. W.: The effect of heat transfer on the turbulent skin-friction of uniformly rough surfaces in compressible flow. The University of Texas, Defense Research Lab Rept. DLR-368, CM-839, April 1956.

[15] Moore, D. R. and Harkness, J.: Experimental investigation of the compressible turbulent boundary layer at very high Reynolds numbers, $M = 2.8$, Rept. No. 0.71000/4R-9, LTV Res. Center, 1964.

[16] Matting, F. W., Chapman, D. R., Nyholm, J. R. and Thomas, A. G.: Turbulent skin friction at high Mach numbers and Reynolds numbers in air and helium. NASA TR R-82, 1961.

[17] Cebeci, T., Smith, A. M. O. and Mosinskis, G. J.: Calculation of compressible adiabatic turbulent boundary layers. *AIAA J.* **8**:1973 (1970).

[18] Michel, R.: Etude de la transition sur les profiles d'aile; establissement d'un critere de determination de point de transition et calcul de la trainée de profile incompressible. ONERA Rept 1/578A, 1951.

[19] Pappas, C. S.: Measurement of heat transfer in the turbulent boundary layer on a flat plate in supersonic flow and comparison with skin-friction results. NACA Tech. Note No. 3222, 1954.

[20] Squire, L. C.: Further experimental investigations of compressible turbulent boundary layers with air injection. ARC R&M 3627, 1970.

[21] Pasiuk, L., Hastings, S. M., and Chatham, R.: Experimental Reynolds analogy factor for a compressible turbulent boundary layer with a pressure gradient. Naval Ordnance Rept. NOLTR 64-200, White Oak, Maryland, 1965.

[22] Lewis, J. E., Gran, R. L. and Kubota, T.: An experiment in the adiabatic compressible turbulent boundary layer in adverse and favorable pressure gradients. J. Fluid Mech., 51:657 (1972).

[23] Adamson, T. C. and Messiter, A. F.: Analysis of two-dimensional interactions between shock waves and boundary layers. Ann. Rev. Fluid Mech., 12:103–138 Annual Reviews, Palo Alto, 1980.

[24] Computation of Viscous-Inviscid Interactions. AGARD Conf. Proceedings No. 291, 1981.

[25] Melnik, R. E.: Turbulent interactions on airfoils at transonic speeds—Recent developments. AGARD CP 291, Paper 10, 1981.

[26] Law, C. H.: Supersonic turbulent boundary-layer separation. AIAA J. 12:1974.

[27] Roshko, A. and Thomke, G. J.: Supersonic turbulent boundary-layer interaction with a compression corner at very high Reynolds number. Proc. Symposium on Viscous Interaction Phenomena in Supersonic Hypersonic Flow. USAF Aerospace Research Labs., Wright-Patterson AFB, Ohio, Univ. of Dayton Press, May 1969.

[28] Hayakawa, K. and Squire, L. C.: The effect of the upstream boundary-layer state on the shock interaction at a compression corner. J. Fluid Mech., 122:369 (1982).

[29] Lighthill, M. J.: On boundary layers and upstream influence, II, Supersonic flow without separation. Proc. Royal Soc. A, 217:1953.

[30] Inger, G. R.: Nonasymptotic theory of unseparated turbulent boundary-layer-shock-wave interaction with application to transonic flows. In Numerical and Physical Aspects of Aerodynamic Flows (ed. T. Cebeci) p. 159, Springer-Verlag, New York 1982.

[31] Viegas, J. R. and Horstmann, C. C.: Comparison of multiequation turbulence models for several shock boundary-layer interaction flows. AIAA J. 17:811–820, 1979.

CHAPTER 12
Coupled Duct Flows

As in Chapters 5 and 7, the momentum and heat-transfer properties of coupled laminar and turbulent flows in two-dimensional and axisymmetric ducts can be studied by considering the case when both velocity and temperature profiles have developed and the case in which both profiles are developing. However, coupled duct flows usually continue to develop indefinitely as heat is added and the density decreases. Therefore, since the developing case is a general one and can be included in the approach discussed in Sections 5.3 and 7.3, here we only consider the calculation of momentum and heat transfer by the method described in these sections. The presence of compressibility requires slight modifications in the transformations and in the solution procedure, as we shall discuss below.

The governing equations for two-dimensional laminar and turbulent flows are given by Eqs. (11.37)–(11.39) and, with the concept of eddy viscosity and turbulent Prandtl number, by Eqs. (11.41) and (11.43). The latter equations can easily be generalized for axisymmetric flows and the resulting equations, including the continuity equation (3.48), can be written as

$$\frac{\partial}{\partial x}\left(r^K \rho u\right) + \frac{\partial}{\partial y}\left(r^K \overline{\rho v}\right) = 0 \tag{12.1}$$

$$\rho u \frac{\partial u}{\partial x} + \overline{\rho v}\frac{\partial u}{\partial y} = -\frac{dp}{dx} + \frac{1}{r^K}\frac{\partial}{\partial y}\left[r^K(\mu + \rho \varepsilon_m)\frac{\partial u}{\partial y}\right] \tag{12.2}$$

$$\rho u \frac{\partial H}{\partial x} + \overline{\rho v}\frac{\partial H}{\partial y} = \frac{1}{r^K}\frac{\partial}{\partial y}\left\{r^K\left(\frac{\mu}{\text{Pr}} + \rho \frac{\varepsilon_m}{\text{Pr}_t}\right)\frac{\partial H}{\partial y}\right.$$

$$\left. + r^K\left[\mu\left(1 - \frac{1}{\text{Pr}}\right) + \rho \varepsilon_m\left(1 - \frac{1}{\text{Pr}_t}\right)\right]u\frac{\partial u}{\partial y}\right\}. \tag{12.3}$$

As in uncoupled flows, we first seek the solution of the governing equations when they are expressed in transformed variables. When the shear layer thickness becomes comparable with the duct width, we revert to primitive variables.

To account for the compressibility, we modify the similarity variable η and the dimensionless stream function $f(x, \eta)$ given by Eqs. (5.45) as follows:

$$d\eta = \left(\frac{u_0}{v_0 x}\right)^{1/2} \frac{\rho}{\rho_0} \left(\frac{r}{L}\right)^K dy, \qquad \psi(x, y) = L^K (u_0 v_0 x)^{1/2} \rho_0 f(x, \eta).$$

$$(12.4)$$

Here and in the following discussion u_0, v_0, ρ_0 and H_0 denote reference values of velocity, kinematic viscosity, density and total enthalpy, respectively. Often these reference conditions are chosen, as in uncoupled flows, as the values at the entrance to the duct. As before, L is a reference length, which we will choose to be the half width of the duct for two-dimensional flows and the radius of the circular duct for axisymmetric flows.

With the use of these modified relations as given by Eq. (12.4), we can now follow the procedure that we used to derive the transformed momentum and energy equations and their boundary conditions for coupled boundary-layer flows in Section 11.4. This time, however, the definition of stream function that automatically satisfies the continuity equation (12.1) is

$$\rho u r^K = \frac{\partial \psi}{\partial y}, \qquad \overline{\rho v} r^K = -\frac{\partial \psi}{\partial x}, \qquad (12.5)$$

and also we denote the dimensionless total enthalpy H by $S = H/H_0$. The resulting equations for two-dimensional and axisymmetric flows can then be written, with a prime denoting differentiation with respect to η, in the following form:

$$(bf'')' + \tfrac{1}{2} ff'' = cx \frac{dp^*}{dx} + x\left(f' \frac{\partial f'}{\partial x} - f'' \frac{\partial f}{\partial x}\right), \qquad (12.6)$$

$$(eS' + df'f'')' + \tfrac{1}{2} fS' = x\left(f' \frac{\partial S}{\partial x} - S' \frac{\partial f}{\partial x}\right). \qquad (12.7)$$

Here, the dimensionless parameters b, c, p^*, e, and d are defined by

$$b = C(1-t)^{2K}(1+\varepsilon_m^+), \qquad c = \frac{\rho_0}{\rho}, \qquad p^* = \frac{p}{\rho_0 u_0^2},$$

$$e = \frac{C}{Pr}(1-t)^{2K}\left(1 + \frac{Pr}{Pr_t}\varepsilon_m^+\right), \qquad d = \frac{Cu_0^2}{H_0}\left[1 - \frac{1}{Pr} + \varepsilon_m^+\left(1 - \frac{1}{Pr_t}\right)\right](1-t)^{2K},$$

$$(12.8)$$

where

$$C = \frac{\rho\mu}{\rho_0\mu_0}, \qquad t = 1 - \left[1 - 2\left(\frac{\xi}{R_L}\right)\int_0^\eta c\, d\eta\right]^{1/2}, \qquad R_L = \frac{u_0 L}{\nu}. \qquad (12.9)$$

Before the shear layers merge, the boundary conditions for Eqs. (12.1)–(12.3) are identical to those given by Eq. (11.48). For the case of no mass transfer, and with the help of the Bernoulli equation, they can be written as

$$\eta = 0, \qquad f = f' = 0,$$

$$S = S_w(x) \quad \text{or} \quad S_w'(x) = -\frac{L}{k_0}\frac{c_{p_w}\dot{q}_w}{H_0}\left(\frac{k_0}{k_w}\right)\frac{c_w}{\sqrt{R_L}}\sqrt{\frac{x}{L}}. \qquad (12.10a)$$

$$\eta = \eta_e, \qquad \frac{d}{dx}\frac{(f')^2}{2} = -c\frac{dp^*}{dx}, \qquad S = 1. \qquad (12.10b)$$

To obtain the additional boundary condition due to the presence of the pressure-gradient term, we again use the principle of conservation of mass. For flow between parallel plates, we can write

$$\rho_0 u_0 L = \int_0^L \rho u \, dy, \qquad (12.11)$$

which in terms of transformed variables becomes

$$f(x, \eta_{sp}) = \sqrt{R_L\left(\frac{L}{x}\right)}, \qquad (12.12)$$

where η_{sp} is the value of η at the centerline $y = L$ and, with $\zeta = y/L$, is given by

$$\eta_{sp} = \sqrt{R_L\left(\frac{L}{x}\right)}\int_0^1 \frac{d\zeta}{c}. \qquad (12.13)$$

Since at a given x location, the dimensionless density ratio is not known but is to be calculated as the momentum and energy equations are solved, the dimensionless distance η_{sp} is also unknown, in contrast to the case of uncoupled flows; it must be found by iteration, but this does not introduce any severe computational difficulties.

For flow in a circular pipe, a similar procedure gives

$$f(x, \eta_{sp}) = \frac{1}{2}\sqrt{R_L\left(\frac{L}{x}\right)}, \qquad (12.14)$$

where now L denotes the pipe radius and

$$\eta_{sp} = \sqrt{R_L\left(\frac{L}{x}\right)}\int_0^1 \frac{1-\zeta}{c}\,d\zeta. \qquad (12.15)$$

Primitive Variables

When we use primitive variables, we define the dimensionless distance Y and the dimensionless stream function $F(x, Y)$ by

$$dY = \left(\frac{u_0}{\nu_0 L}\right)^{1/2}\frac{\rho}{\rho_0}\left(\frac{r}{L}\right)^K dy, \qquad \psi = L^K(u_0\nu_0 L)^{1/2}\rho_0 F(x, Y). \qquad (12.16)$$

With these new variables and with $\zeta = x/L$, and with a prime now denoting differentiation with respect to Y, the momentum and energy equations can be written as

$$(bF'')' = c\frac{dp^*}{dx} + F'\frac{\partial F'}{\partial \zeta} - F''\frac{\partial F}{\partial \zeta}, \tag{12.17}$$

$$(eS' + dF'F'')' = F'\frac{\partial S}{\partial \zeta} - S'\frac{\partial F}{\partial \zeta}. \tag{12.18}$$

Here and below the definitions of b, c, p^*, e, and d are the same as those defined in Eqs. (12.8) and (12.9) except that

$$t = 1 - \left(1 - \frac{2}{\sqrt{R_L}}\int_0^Y c\,dY\right)^{1/2}. \tag{12.19}$$

The wall boundary conditions for specified wall enthalpy or heat flux are

$$Y = 0, \qquad F = F' = 0,$$

$$S = S_w(\zeta) \quad \text{or} \quad S_w'(\zeta) = -\left(\frac{L}{k_0}\frac{c_{p_w}\dot{q}_w}{H_0}\right)\left(\frac{k_0}{k_w}\right)\frac{c_w}{\sqrt{R_L}}. \tag{12.20a}$$

The centerline boundary conditions for a two-dimensional duct flow

$$Y = Y_c, \qquad F'' = 0, \qquad S' = 0, \tag{12.20b}$$

and those for a circular pipe, with $L = r_0$, are

$$Y = Y_c, \qquad F'' = -\frac{1}{2}\sqrt{R_L}\,\frac{1}{Cc}\left[c\frac{dp^*}{d\xi} + \frac{1}{2}\frac{d}{d\xi}(F')^2\right],$$

$$S' = -\frac{1}{2}\sqrt{R_L}\,\frac{Pr}{Cc}\left[\frac{u_0^2}{H_0}(1 - Pr)F'F'' + F'\frac{\partial S}{\partial \zeta}\right]. \tag{12.20c}$$

From the conservation of mass we find that for flow between parallel plates,

$$F(\zeta, Y_c) = \sqrt{R_L}, \tag{12.21}$$

where the transformed centerline distance Y_c is given by

$$Y_c = \sqrt{R_L}\int_0^1 \frac{d\zeta}{c}. \tag{12.22}$$

Similarly, for flow in a circular pipe,

$$F(\xi, Y_c) = \tfrac{1}{2}\sqrt{R_L}, \tag{12.23}$$

where now

$$Y_c = \sqrt{R_L}\int_0^1 \frac{1-\zeta}{c}d\zeta. \tag{12.24}$$

12.1 Laminar Flow in a Tube with Uniform Heat Flux

As the first of two examples of the application of the above procedure, we consider the compressible laminar flow of helium in a round uniformly heated tube. The thermodynamic and transport properties are specified by the following relations

$$\frac{\mu}{\mu_0} = \frac{k}{k_0} = \left(\frac{T}{T_0}\right)^{0.68}, \qquad \frac{c_p}{c_{p_0}} = 1, \qquad \text{Pr} = \tfrac{2}{3}, \qquad \gamma = \tfrac{5}{3},$$

$$c_{p_0} = 5.231 \text{ kJ kg}^{-1}\text{K}^{-1}, \qquad R = 645.8 \text{ Nm kg}^{-1}\text{K}^{-1}. \qquad (12.25)$$

The density ratio c is also required and is obtained from the expression

$$c \equiv \frac{\rho_0}{\rho} = \frac{T}{T_0}\frac{p_0^*}{p^*}, \qquad (12.26)$$

which assumes that the pressure is constant across the layer and that T denotes the absolute static temperature. The fluid properties in the shear layer are calculated from the static-temperature distribution across the duct. For this purpose we make use of the total-enthalpy relation

$$T = \frac{H - u^2/2}{c_p}, \qquad (12.27)$$

and with the definitions of S and f' we rewrite it as

$$\frac{T}{T_0} = \frac{S - \left(u_0^2/2H_0\right)\left(f'\right)^2}{1 - u_0^2/2H_0}. \qquad (12.28)$$

Here the total entrance enthalpy H_0 can be computed from the expression

$$H_0 = c_{p_0}T_0\left(1 + \frac{\gamma - 1}{2}M_0^2\right). \qquad (12.29)$$

Calculations were carried out for an entrance Mach number of 0.06, Reynolds number $R_d = 2500$ and for values of the heat-flux parameter $Q^+[\equiv (\dot{q}_w/T_0 \cdot r_0/k_0)]$ of 0.135, 1.35, 13.5, and 27.0. Figure 12.1 shows the development of velocity profiles at several axial locations $x^+ = x/d\,R_m^{-1}$ for four values of Q^+. Here R_m denotes the Reynolds number, $u_m d/\nu_m$, where the fluid properties are evaluated at the bulk temperature T_m. Figure 12.1a shows the computed velocity profiles for $Q^+ = 0.135$ together with the velocity profile corresponding to the fully-developed incompressible (Poiseuille) profile. As expected, we see that the calculated profile at $x^+ \simeq 0.1$ agrees well with the Poiseuille profile, see Eq. (5.20). This orderly profile development begins to become distorted for flows of compressible flows with induced acceleration as shown in Figures 12.1b, c and d. In the entrance regions of the tube this distortion is observed in large fluid acceleration near the wall, resulting in higher heat inputs in velocity "overshoot", that is, velocities near the wall exceeding the centerline velocity. As observed by Presler [1], comparison of Figures 12.1a–d shows that the acceleration overshoot distortion changes from an observable contribution at the low $Q^+ = 0.135$ to a large contribution around $x^+ \sim 0.03$ for the high $Q^+ = 27.0$. The overshoot distortion is smoothed out in the downstream sections of the tube; but the effects of near-wall acceleration remain, and prevent a true fully-developed flow (parabolic profile) from being obtained (Fig. 12.1d).

Heating rate effects on velocities near the wall are shown in another way in Figure 12.2, where the centerline-to-bulk-velocity development with axial distance is plotted with Q^+ as parameter. As expected for the low Q^+ case

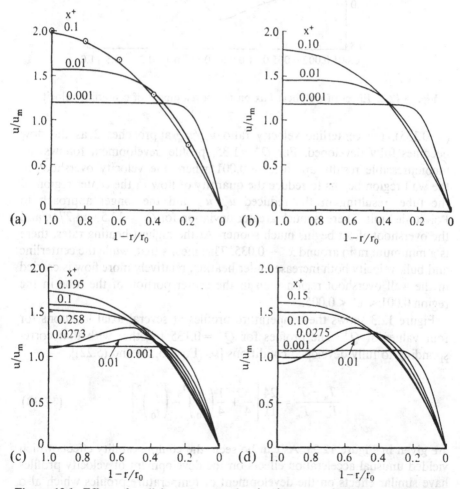

Figure 12.1. Effect of wall heat flux on the development of velocity profiles. (a) $Q^+ = 0.135$. Symbols denote the fully-developed Poiseuille profile, Eq. (5.20), (b) $Q^+ = 1.35$, (c) $Q^+ = 13.5$, (d) $Q^+ = 27.0$.

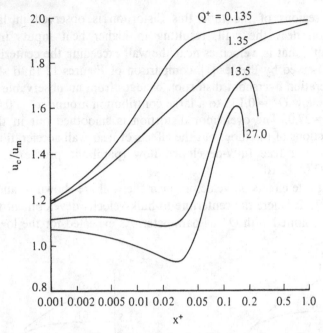

Figure 12.2. Effect of wall heat flux on the development of centerline velocity.

($\equiv 0.135$), the centerline velocity ratio u_c/u_m approaches 2 as the flow becomes fully developed. For $Q^+ = 1.35$, profile development follows the incompressible results up to $x^+ \sim 0.001$ where the velocity overshoot in the wall region begins to reduce the quantity of flow in the center region of the tube, resulting in the reduced u_c/u_m and the longer approach to Poiseuille limit. There is much greater deviation for $Q^+ = 13.5$ and 27.0, and the overshoot effect begins much sooner. At the highest heating rates, there is a minimum ratio around $x^+ \sim 0.035$. This means that, while the centerline and bulk velocity both increase under heating, relatively more flow is carried in the wall overshoot region than in the center portion of the tube in the region $0.001 < x^+ < 0.005$.

Figure 12.3 shows the temperature profiles at several axial locations for four values of Q^+. The profiles for $Q^+ = 0.135$ together with that corresponding to fully developed conditions [see Eqs. (5.21) and (5.22)],

$$\frac{T_w - T}{T_w - T_m} = \frac{24}{11}\left[\frac{3}{4} + \frac{1}{4}\left(\frac{r}{r_0}\right)^4 - \left(\frac{r}{r_0}\right)^2\right] \tag{12.30}$$

are given in Figure 12.3a. As can be seen, the compressibility effects which yielded unusual acceleration effects on the development of velocity profiles have similar effects on the development of temperature profiles which also suffer a similar distortion from the same causes. In the entrance region of the tube, the accelerating gas in the wall region convects most of the heat

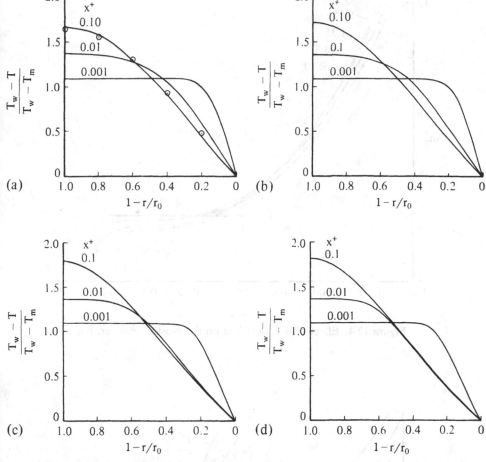

Figure 12.3. Effect of wall heat flux on the development of temperature profiles. (a) $Q^+ = 0.135$. Symbols denote the temperature profile corresponding to fully-developed conditions, Eq. (12.30), (b) $Q^+ = 1.35$, (c) $Q^+ = 13.5$, (d) $Q^+ = 27.0$.

flux from the wall; the central bulk of the flow remains at constant temperature up to axial lengths of $x^+ \sim 0.01$. Downstream of this location, heat flux penetrates to the centerline of the tube.

The variation of Nusselt number defined by

$$\text{Nu} = \frac{\dot{q}_w}{T_w - T_m} \frac{d}{k_m} \tag{12.31}$$

is shown in Figure 12.4. Since the initial Mach number is low, there are no choking or near choking effects in these data. An interesting result of these calculations is the insensitivity of the fully-developed Nusselt number to the level of wall heat flux. On the other hand, for values of $Q^+ = 13.5$ and 27.0, the Nusselt numbers begin to deviate from the classical uniform heat flux $\text{Nu} = 48/11 = 4.364$, see Eq. (5.25), as $x^+ \to \infty$.

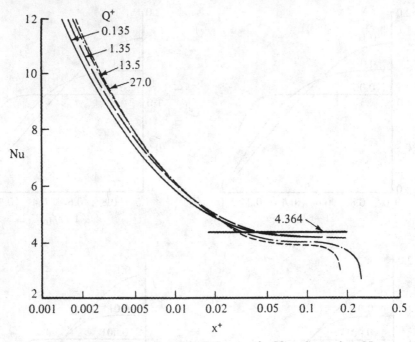

Figure 12.4. Effect of wall heat flux on the Nusselt number, Nu.

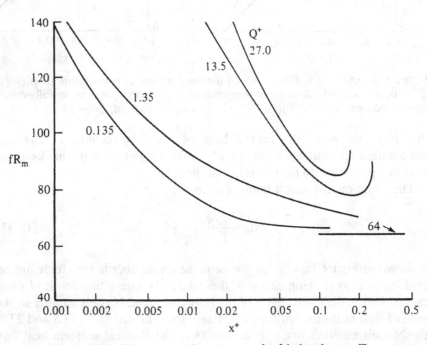

Figure 12.5. Effect of wall heat flux on the friction factor, fR_m.

Figure 12.5 shows the variation of the friction parameter fR_m with axial distance for different wall heat flux rates. Here the friction factor, f, with subscript denoting conditions evaluated at bulk temperature, T_m, is defined by

$$f = \frac{\tau_w}{\frac{1}{8}\rho_m u_m^2}. \tag{12.32}$$

We see from Fig. 12.5 that the friction factor is strongly influenced in the entrance region by the level of the heat flux parameter Q^+. This is a result of the fluid velocity overshoot near the wall, as indicated in Fig. 12.1. The steep velocity gradients in the entrance wall region naturally give rise to large wall shear. We also see from Figure 12.5 that the low Q^+ cases approach the fully-developed Poiseuille value of $fR_d = 64$, see Eq. (5.11). In contrast, higher Q^+ values approach a higher asymptotic value of around 85 before they begin to rise sharply due to high acceleration of the flow (Fig. 12.2).

12.2 Laminar, Transitional and Turbulent Flow in a Cooled Tube

As a second example, we extend the problem of Section 12.1 to include transitional and turbulent flow in a tube with specified wall temperature. The calculations were made with the procedure discussed in this chapter, a modified version of the computer program of Section 13.5 and the Cebeci–Smith eddy-viscosity formulation discussed in Chapters 6 and 11. They were started as laminar at $x = 0$ for given values of T_0, p_0, R_d, Pr and T_w/T_0 and continued as turbulent after the specified transition location x_0.

The variations of friction factor and Nusselt number are shown on Figures 12.6 and 12.7 for two values of the wall temperature. As expected,

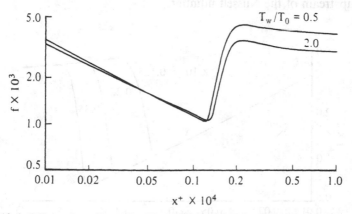

Figure 12.6. Effect of cooling ($T_w/T_0 = 0.5$) and heating ($T_w/T_0 = 2.0$) on friction factor with transition point fixed at $x_{tr}^+ = 0.125 \times 10^{-4}$.

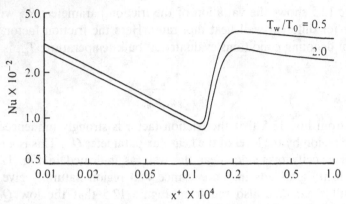

Figure 12.7. Effect of cooling ($T_w/T_0 = 0.5$) and heating ($T_w/T_0 = 2.0$) on Nusselt number with transition point fixed at $x_{tr}^+ = 0.125 \times 10^{-4}$.

the friction factor and Nusselt number behave in a similar manner with the wall temperature causing small differences in the laminar-flow region and larger differences in the turbulent-flow region. Large differences can, of course, be expected as the wall is further cooled or heated and the influence of the coupling between the momentum and energy equations increases and it seems probable that the effects will continue to be large for turbulent flow.

The results of Figures 12.6 and 12.7 confirm that the intermittency expression contained in the Cebeci–Smith turbulence model allows a plausible and smooth representation of the transition region. Figures 12.8 and 12.9 extend the calculations, and the use of this model, to show the effect of prescribing the onset of transition at different downstream locations, as might be achieved in practice by tripping the boundary layer with different wall roughness or with different duct-entry configurations. Again, it is evident that the model is satisfactory and that, for this particular case of wall cooling, the friction factor tends to reach the turbulent condition slightly upstream of the Nusselt number.

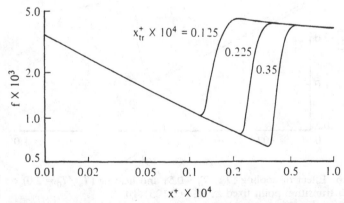

Figure 12.8. Effect of transition on friction factor for a cooled wall ($T_w/T_0 = 0.5$).

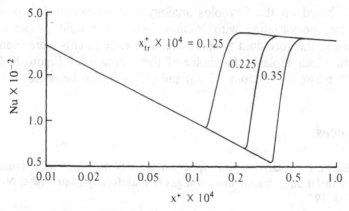

Figure 12.9. Effect of transition on Nusselt number for a cooled wall ($T_w/T_0 = 0.5$).

It is to be expected that the calculated results, based on conservation equations as they are, will correspond closely with physical observations. There are few experiments with which to support this contention, but those of Back et al. [2] do allow comparison between measured and calculated values of wall heat transfer for the extrance region of a pipe with cooled wall and a high temperature air flow. The results are presented on Figure 12.10 and correspond to a 12.7mm-diameter pipe, with an initial gas temperature of around $560\,°C$ and a uniform wall temperature of less than $100\,°C$. The measurements, represented by the dashed line, are typical of several reported by Back et al. and indicate trends similar to those of the calculations which assumed that transition occurred at the same location as observed experimentally. The measurements are some 15% lower than the calculations in the fully turbulent region and it is useful to note that the reference curve suggested by Back et al. for the turbulent region is 10% lower than the calculations. This reference curve is represented by the equation

$$St\,Pr^{0.6} = 0.03 R_x^{-0.2}, \tag{12.33}$$

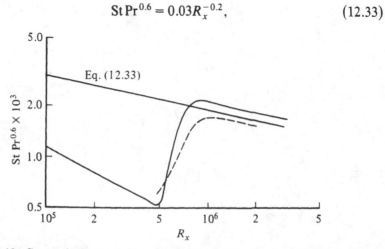

Figure 12.10. Comparison of calculated (solid line) and experimental (dashed line) results for the data of Back et al. [2]. Here $R_x = u_e x/\nu_e$.

which is based on the Reynolds analogy and an assumed $\frac{1}{5}$ power law profile for the variation with friction factor: the fluid properties were evaluated at the core fluid temperature. The experiments were reported in this form which explains the choice of the ordinate for Figure 12.10. The similarity between equations (12.33) and (6.70) should be noted.

References

[1] Presler, A. F.: Analytical and experimental study of compressible laminar-flow heat transfer and pressure drop of a gas in a uniformly heated tube. NASA TN D-6333, 1971.

[2] Back, L. H., Cuffel, R. F. and Massier, P. F.: Laminar, transitional and turbulent boundary-layer heat-transfer measurements with wall cooling in turbulent airflow through a tube. *J. Heat Transfer* **91**:477, 1969.

CHAPTER 13

Finite-Difference Solution of Boundary-Layer Equations

In this chapter we consider the finite-difference solution of the thin-shear-layer equations presented in previous chapters. In Section 13.1 we present a brief review of finite-difference techniques, discussing the relative advantages of implicit and explicit methods. As a result, the implicit Box scheme is preferred, and its use in internal and external flows is described in detail in Sections 13.2 to 13.5.

In Section 13.2 we discuss the solution of the energy equation alone, for developing thermal boundary layers in ducts with fully developed (uncoupled) velocity profile, and we show how the finite-difference equations can be solved by the so-called block-elimination method for a wide range of boundary conditions. In Section 13.3 we present a Fortran program that utilizes this numerical procedure and can be used to solve the energy equation for both laminar and turbulent flows in an axisymmetric duct. Although the program is written only for boundary conditions corresponding to uniform temperature, for the sake of simplicity, it can be used, with minor changes, to calculate flows with different thermal boundary conditions, or two-dimensional flows.

In Section 13.4 we describe the Box scheme used to obtain the solutions of the coupled mass, momentum and energy equations for two-dimensional laminar and turbulent boundary-layer flows, and in Section 13.5 we present the Fortran program employing this numerical procedure. It is easier to use the same program for uncoupled flows than to write a specially simplified program.

13.1 Review of Numerical Methods for Boundary-Layer Equations

Boundary-layer equations, which are parabolic in nature, can be solved by using several numerical methods such as finite-difference methods, the method of lines, the Galerkin method, and finite-element methods as discussed, for example, in Cebeci and Bradshaw [1]. Of these, finite-difference methods are at present the most common for the boundary-layer equations, and as a consequence, we shall consider only them.

To illustrate the application of finite-difference methods to boundary-layer equations, we first consider the energy equation for a two-dimensional constant-density flow in a symmetrical duct with fully developed velocity profile, that is,

$$u \frac{\partial T}{\partial x} = \frac{\nu}{\mathrm{Pr}} \frac{\partial^2 T}{\partial y^2}. \tag{13.1}$$

The solution of this equation, like any partial differential equation, requires *initial* and *boundary* conditions. Here let us assume the boundary conditions for Eq. (13.1) are given by

$$y = 0, \qquad T = T_w, \tag{13.2a}$$

$$y = \delta, \qquad T = T_e, \tag{13.2b}$$

and the initial conditions are written in the form

$$x = x_0, \qquad T = T(y), \quad \text{given.} \tag{13.3}$$

The solution of Eq. (13.1) by a finite-difference method requires that we approximate the derivatives by some difference approximations. Using Taylor's theorem and requiring that the function, say, w and its derivatives are single-valued, finite, and continuous functions of ζ, and with primes denoting differentiation with respect to ζ, we can write

$$w(\zeta + r) = w(\zeta) + rw'(\zeta) + \tfrac{1}{2}r^2 w''(\zeta) + \tfrac{1}{6}r^3 w'''(\zeta) + \cdots \tag{13.4a}$$

and

$$w(\zeta - r) = w(\zeta) - rw'(\zeta) + \tfrac{1}{2}r^2 w''(\zeta) - \tfrac{1}{6}r^3 w'''(\zeta) + \cdots. \tag{13.4b}$$

Adding Eqs. (13.4a) and (13.4b), we get

$$w(\zeta + r) + w(\zeta - r) = 2w(\zeta) + r^2 w''(\zeta)$$

provided the fourth- and higher-order terms are neglected. Thus,

$$w''(\zeta) = \frac{1}{r^2} [w(\zeta + r) - 2w(\zeta) + w(\zeta - r)] \tag{13.5}$$

with an error of order r^2.

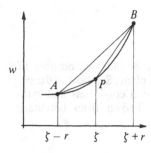

Figure 13.1. Notation for approximation of derivatives.

If Eq. (13.4b) is subtracted from Eq. (13.4a),

$$w'(\zeta) = \frac{1}{2r}[w(\zeta+r) - w(\zeta-r)].\qquad(13.6)$$

Equation (13.6) approximates the slope of the tangent at P by the slope of the chord AB (see Fig. 13.1) and is called a *central-difference* approximation. The slope of the tangent can also be approximated by either the slope of the chord PB, giving the *forward-difference* formula

$$w'(\zeta) = \frac{1}{r}[w(\zeta+r) - w(\zeta)],\qquad(13.7)$$

or the slope of the chord AP, giving the *backward-difference* formula

$$w'(\zeta) = \frac{1}{r}[w(\zeta) - w(\zeta-r)].\qquad(13.8)$$

Note that while Eq. (13.6) has an error of $O(r^2)$, both Eqs. (13.7) and (13.8) have errors of $O(r)$.

To demonstrate the finite-difference notation, Fig. 13.2 indicates a set of uniform net points on the xy plane; that is,

$$x = x_0, \qquad x_n = x_{n-1} + k, \qquad n = 1,2,\ldots,N,$$
$$y_0 = 0, \qquad y_j = y_{j-1} + h, \qquad j = 1,2,\ldots,J.\qquad(13.9)$$

With the compact notation

$$T(y_n, y_j) \equiv T_j^n\qquad(13.10)$$

the difference approximation of $\partial T/\partial x$ by central-difference, forward-difference, and backward-difference formulas follows from Eqs. (13.6)–(13.8)

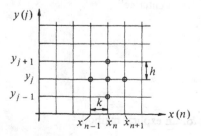

Figure 13.2. General finite-difference grid notation.

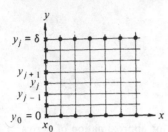

Figure 13.3. Initial and boundary conditions of Eq. (13.1) in the (x, y) plane, Symbols × denote values known from initial conditions and symbols • denote values known from boundary conditions.

and may be written, respectively, as

$$\frac{\partial T}{\partial x}(x_n, y_j) = \begin{cases} \dfrac{1}{2k}\left(T_j^{n+1} - T_j^{n-1}\right), & \text{(13.11a)} \\[2mm] \dfrac{1}{k}\left(T_j^{n+1} - T_j^{n}\right), & \text{(13.11b)} \\[2mm] \dfrac{1}{k}\left(T_j^{n} - T_j^{n-1}\right). & \text{(13.11c)} \end{cases}$$

Similarly, the difference approximation of $\partial^2 T/\partial y^2$ follows from Eq. (13.5) and may be written as

$$\frac{\partial^2 T}{\partial y^2}(x_n, y_j) = \frac{1}{h^2}\left(T_{j+1}^{n} - 2T_j^{n} + T_{j-1}^{n}\right). \tag{13.12}$$

For a given set of initial and boundary conditions (see Fig. 13.3), the solution of Eq. (13.1) may be obtained by using either an explicit or an implicit method. In an explicit method the value of T at the downstream station is expressed in terms of upstream quantities, and the corresponding equation can be solved explicitly. In an implicit method, T at the downstream station is expressed in terms of its downstream neighbors and the known quantities.

An explicit formulation may be obtained by representing $\partial T/\partial x$ by the forward-difference formula, Eq. (13.11b), and $\partial^2 T/\partial y^2$ by Eq. (13.12),

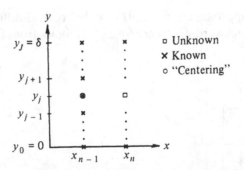

Figure 13.4. Finite-difference grid for an explicit method.

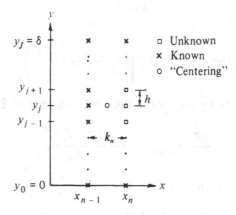

Figure 13.5. Finite-difference grid for the Crank–Nicolson method. Note that while h is uniform, k_n can be nonuniform.

locating both at the *net point* "upstream" (x_{n-1}, y_j) (see Fig. 13.4); that is,

$$u_j^{n-1} \frac{1}{k}\left(T_j^n - T_j^{n-1}\right) = \frac{\nu}{\Pr} \frac{1}{h^2}\left(T_{j+1}^{n-1} - 2T_j^{n-1} + T_{j-1}^{n-1}\right)$$

or

$$T_j^n = T_j^{n-1} + \frac{\nu}{\Pr} \frac{1}{u_j^{n-1}} \frac{k}{h^2}\left(T_{j+1}^{n-1} - 2T_j^{n-1} + T_{j-1}^{n-1}\right), \qquad j=1,2,\dots,J-1.$$

$$(13.13)$$

Note that u, which represents the velocity profile in Eq. (13.1), is independent of x and could be written as u_j without the superscript $n-1$. It is, however, written as u_j^{n-1} to remind the reader that u can be a function of x, in general.

We see from Eq. (13.13) that, by this explicit formulation, the value of T_j^n is expressed in terms of upstream values that are known, and the equation allows the value of T to be obtained for values of y corresponding to $j=1,2,\dots,J-1$. The values of T at $j=0$ and J are known from the boundary conditions. The numerical error inherent in this scheme can be shown to be of order $k + h^2$ and, as a result, the x step k must be maintained small to ensure acceptable accuracy. In addition, although explicit formulations are computationally simple, they can lead to instabilities unless the x step is also small; in this case, it is necessary to ensure that $k < \frac{1}{2}(\Pr/\nu)u^{n-1}h^2$.

In contrast, implicit methods are unconditionally stable, and significantly larger x steps, with corresponding economy, can be taken as long as accuracy is maintained. As an example, the Crank-Nicolson [2] method uses the finite-difference grid shown in Fig. 13.5, replacing $\partial^2 T/\partial y^2$ at $(x_{n-1/2}, y_j)$ by the average of the upstream and downstream values at $x_n - _1$

and x_n, respectively,

$$\frac{\partial^2 T}{\partial y^2}(x_{n-1/2}, y_j) = \frac{1}{2}\left[\left(\frac{\partial^2 T}{\partial y^2}\right)_j^n + \left(\frac{\partial^2 T}{\partial y^2}\right)_j^{n-1}\right],$$

and replacing $\partial T/\partial x$ by the central-difference formula of Eq. (13.11a),

$$\frac{\partial T}{\partial x}(x_{n-1/2}, y_j) = \frac{T_j^n - T_j^{n-1}}{k_n}.$$

Equation (13.1) can then be written as

$$u_j^{n-1/2}\frac{1}{k_n}\left(T_j^n - T_j^{n-1}\right) = \frac{\nu}{Pr}\frac{1}{2h^2}\left[\left(T_{j+1}^n - 2T_j^n + T_{j-1}^n\right)\right.$$
$$\left. + \left(T_{j+1}^{n-1} - 2T_j^{n-1} + T_{j-1}^{n-1}\right)\right]. \quad (13.14)$$

It can also be rewritten in the general form

$$a_j T_{j-1}^n + b_j T_j^n + c_j T_{j+1}^n = r_j, \qquad 1 \le j \le J-1, \qquad (13.15)$$

where

$$a_j = 1, \qquad b_j = -(2+\lambda_j), \qquad c_j = 1,$$
$$r_j = -\lambda_j T_j^{n-1} - \left(T_{j+1}^{n-1} - 2T_j^{n-1} + T_{j-1}^{n-1}\right), \qquad 1 \le j \le J-1,$$
$$\lambda_j = \frac{Pr}{\nu}\frac{2h^2}{k_n}u_j^{n-1/2}. \qquad (13.16)$$

Note that r contains only values of T^{n-1}, not T^n. At $j=0$ and J, the boundary conditions T_0^n and T_J^n are given. As a result, for $j=1$ and $J-1$, Eq. (13.15) can be written as

$$b_1 T_1^n + c_1 T_2^n = r_1 - a_1 T_0^n \equiv r_1^* \quad \text{for } j=1 \qquad (13.17a)$$

and

$$a_{J-1}T_{J-2}^n + b_{J-1}T_{J-1}^n = r_{J-1} - c_{J-1}T_J^n \equiv r_{J-1}^* \quad \text{for } j = J-1.$$
$$(13.17b)$$

The errors in this carefully centered Crank–Nicolson scheme are of order $h^2 + k_n^2$, but k_n need not be related to h for stability purposes. The scheme is unconditionally stable, and second-order accuracy may be achieved with uniform y spacing. On the other hand, the unknown value of T is expressed in terms of $J-1$ *other* values of T with two of them known; as a consequence, the computational arithmetic is more extensive than an explicit method.

However, the solution of Eqs. (13.15)–(13.17) can be obtained very easily. In vector notation, Eqs. (13.15) and (13.17) may be written as

$$AT = r, \tag{13.18}$$

where we have introduced the $(J-1)$-dimensional vectors

$$\mathbf{T} \equiv \begin{bmatrix} T_1 \\ T_2 \\ \vdots \\ T_{J-2} \\ T_{J-1} \end{bmatrix}, \quad \mathbf{r} \equiv \begin{bmatrix} r_1^* \\ r_2 \\ \vdots \\ r_{J-2} \\ r_{J-1}^* \end{bmatrix} \tag{13.19a}$$

and the $(J-1)$-order matrix with nonzero elements on only three diagonals (called *tridiagonal matrix*)

$$A = \begin{bmatrix} b_1 & c_1 & & & \\ a_2 & b_2 & c_2 & & \mathbf{0} \\ & \ddots & \ddots & \ddots & \\ \mathbf{0} & & a_{J-2} & b_{J-2} & c_{J-2} \\ & & & a_{J-1} & b_{J-1} \end{bmatrix}. \tag{13.19b}$$

Then the solution of Eq. (13.18) is obtained by two sweeps. In the so-called forward sweep, we compute

$$\beta_1 = b_1, \quad s_1 = r_1^*,$$

$$m_j = \frac{a_j}{\beta_{j-1}},$$

$$\beta_j = b_j - m_j c_{j-1}, \quad j = 2, 3, \ldots, J-1.$$

$$s_j = r_j - m_j s_{j-1}, \tag{13.20a}$$

In the backward sweep, we compute

$$T_{J-1} = \frac{s_{J-1}}{\beta_{J-1}}, \quad T_j = \frac{s_j - c_j T_{j+1}}{\beta_j}, \quad j = J-2, J-3, \ldots, 1. \tag{13.20b}$$

An alternative implicit method due to H. B. Keller [3] is now described and is referred to as the *Box method*. This method has several very desirable features that make it appropriate for the solution of all parabolic partial differential equations. The main features of this method are

1. Only slightly more arithmetic to solve than the Crank–Nicolson method.
2. Second-order accuracy with arbitrary (nonuniform) x and y spacings.

3. Allows very rapid x variations.
4. Allows easy programming of the solution of large numbers of coupled equations.

The solution of an equation by this method can be obtained by the following four steps:

1. Reduce the equation or equations to a first-order system.
2. Write difference equations using central differences.
3. Linearize the resulting algebraic equations (if they are nonlinear), and write them in matrix-vector form.
4. Solve the linear system by the block-tridiagonal-elimination method.

To solve Eq. (13.1) by this method, we first express it in terms of a system of two first-order equations by letting

$$T' = p \tag{13.21a}$$

and by writing Eq. (13.1) as

$$p' = \frac{\text{Pr}}{\nu} u \frac{\partial T}{\partial x}. \tag{13.21b}$$

Here the primes denote differentiation with respect to y. The finite-difference form of the *ordinary* differential equation (13.21a) is written for the midpoint $(x_n, y_{j-1/2})$ of the segment $P_1 P_2$ shown in Fig. 13.6, and the finite-difference form of the *partial* differential equation (13.21b) is written for the midpoint $(x_{n-1/2}, y_{j-1/2})$ of the rectangle $P_1 P_2 P_3 P_4$. This gives

$$\frac{T_j^n - T_{j-1}^n}{h_j} = \frac{p_j^n + p_{j-1}^n}{2} = p_{j-1/2}^n, \tag{13.22a}$$

$$\frac{1}{2} \left(\frac{p_j^n - p_{j-1}^n}{h_j} + \frac{p_j^{n-1} - p_{j-1}^{n-1}}{h_j} \right) = \frac{\text{Pr}}{\nu} u_{j-1/2}^{n-1/2} \frac{T_{j-1/2}^n - T_{j-1/2}^{n-1}}{k_n}. \tag{13.22b}$$

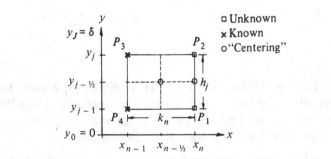

Figure 13.6. Finite-difference grid for the Box method. Note that both h and k can be nonuniform. Here $x_{n-\frac{1}{2}} \equiv 1/2(x_n + x_{n-1})$ and $y_{j-\frac{1}{2}} \equiv 1/2(y_j + y_{j-1})$.

Rearranging both expressions we can write them in the form

$$T_j^n - T_{j-1}^n - \frac{h_j}{2}\left(p_j^n + p_{j-1}^n \right) = 0, \tag{13.23a}$$

$$(s_1)_j p_j^n + (s_2)_j p_{j-1}^n + (s_3)_j \left(T_j^n + T_{j-1}^n \right) = R_{j-1/2}^{n-1}. \tag{13.23b}$$

Here

$$(s_1)_j = 1, \qquad (s_2)_j = -1, \qquad (s_3)_j = -\frac{\lambda_j}{2}, \tag{13.24a}$$

$$R_{j-1/2}^{n-1} = -\lambda_j T_{j-1/2}^{n-1} + p_{j-1}^{n-1} - p_j^{n-1}, \tag{13.24b}$$

$$\lambda_j = \frac{2\,\mathrm{Pr}}{\nu} u_{j-1/2}^{n-1/2} \frac{h_j}{k_n}. \tag{13.24c}$$

As before, the superscript on $u_{j-1/2}$ is not necessary but is included for generality.

Equations (13.23) are imposed for $j = 1, 2, \ldots, J-1$. At $j = 0$ and J, we have

$$T_0 = T_w, \qquad T_J = T_e, \tag{13.25}$$

respectively.

Since Eqs. (13.23) are linear, as are the corresponding boundary conditions (b.c.) given by Eq. (13.25), the system may be written at once in matrix-vector form as shown below, without the linearization needed in the case of the finite-difference equations for the velocity field.

$$
\begin{array}{c}
\text{b.c.} \\[4pt]
\text{Eq. (13.23a)} \\[8pt]
\text{Eq. (13.23b)} \\[16pt]
\text{b.c.}
\end{array}
\begin{bmatrix}
1 & 0 & 0 & 0 & & \\
-1 & \dfrac{-h_1}{2} & 1 & \dfrac{-h_1}{2} & & \\
(s_3)_j & (s_2)_j & (s_3)_j & (s_1)_j & 0 & 0 \\
0 & 0 & -1 & \dfrac{-h_{j+1}}{2} & 1 & \dfrac{-h_{j+1}}{2} \\
& & (s_3)_J & (s_2)_J & (s_3)_J & (s_1)_J \\
& & 0 & 0 & 1 & 0
\end{bmatrix}
$$

$$
\times
\begin{bmatrix}
\begin{pmatrix} T_0 \\ p_0 \end{pmatrix} \\
\begin{pmatrix} T_j \\ p_j \end{pmatrix} \\
\begin{pmatrix} T_J \\ p_J \end{pmatrix}
\end{bmatrix}
=
\begin{bmatrix}
\begin{pmatrix} (r_1)_0 \\ (r_2)_0 \end{pmatrix} \\
\begin{pmatrix} (r_1)_j \\ (r_2)_j \end{pmatrix} \\
\begin{pmatrix} (r_1)_J \\ (r_2)_J \end{pmatrix}
\end{bmatrix}
\tag{13.26}
$$

Here

$$(r_1)_0 = T_w, \qquad (r_1)_j = R_{j-1/2}^{n-1}, \qquad 1 \le j \le J,$$
$$(r_2)_j = 0, \qquad 0 \le j \le J-1, \qquad (r_2)_J = T_e. \tag{13.27}$$

The system of equations given by Eq. (13.26) can be rewritten as

$$\mathbf{A}\boldsymbol{\delta} = \mathbf{r}, \tag{13.28}$$

where

$$\mathbf{A} = \begin{bmatrix} A_0 & C_0 & & & & \\ B_1 & A_1 & C_1 & & & \\ & & \ddots & & & \\ & & B_j & A_j & C_j & \\ & & & B_{J-1} & A_{J-1} & C_{J-1} \\ & & & & B_J & A_J \end{bmatrix}, \quad \boldsymbol{\delta} = \begin{bmatrix} \delta_0 \\ \delta_1 \\ \vdots \\ \delta_j \\ \vdots \\ \delta_J \end{bmatrix}, \quad \mathbf{r} = \begin{bmatrix} r_0 \\ r_1 \\ \vdots \\ r_j \\ \vdots \\ r_J \end{bmatrix},$$

$$\tag{13.29}$$

$$\boldsymbol{\delta}_j = \begin{bmatrix} T_j \\ P_j \end{bmatrix}, \qquad \mathbf{r}_j = \begin{bmatrix} (r_1)_j \\ (r_2)_j \end{bmatrix} \tag{13.30a}$$

and $\mathbf{A}_j, \mathbf{B}_j, \mathbf{C}_j$ are 2×2 matrices defined as follows

$$\mathbf{A}_0 \equiv \begin{bmatrix} 1 & 0 \\ -1 & \dfrac{-h_1}{2} \end{bmatrix}, \qquad \mathbf{A}_j \equiv \begin{bmatrix} (s_3)_j & (s_1)_j \\ -1 & \dfrac{-h_{j+1}}{2} \end{bmatrix}, \qquad 1 \le j \le J-1,$$

$$\mathbf{A}_J \equiv \begin{bmatrix} (s_3)_J & (s_1)_J \\ 1 & 0 \end{bmatrix}, \qquad \mathbf{B}_j \equiv \begin{bmatrix} (s_3)_j & (s_2)_j \\ 0 & 0 \end{bmatrix}, \qquad 1 \le j \le J,$$

$$\mathbf{C}_j \equiv \begin{bmatrix} 0 & 0 \\ 1 & \dfrac{-h_{j+1}}{2} \end{bmatrix}, \qquad 0 \le j \le J-1. \tag{13.30b}$$

Note that, as in the Crank–Nicolson method, the implicit nature of the method has again generated a tridiagonal matrix, but the entries are 2×2 blocks rather than scalars.

The solution of Eq. (13.28) by the block-elimination method consists of two sweeps. In the *forward* sweep we compute Γ_j, Δ_j, and $\mathbf{w}_j$ from the recursion formulas given by

$$\Delta_0 = \mathbf{A}_0, \tag{13.31a}$$
$$\Gamma_j \Delta_{j-1} = \mathbf{B}_j, \tag{13.31b}$$
$$\Delta_j = \mathbf{A}_j - \Gamma_j \mathbf{C}_{j-1}, \qquad 1 \le j \le J \tag{13.31c}$$
$$\mathbf{w}_0 = \mathbf{r}_0, \tag{13.32a}$$
$$\mathbf{w}_j = \mathbf{r}_j - \Gamma_j \mathbf{w}_{j-1}, \qquad 1 \le j \le J. \tag{13.32b}$$

Here Γ_j has the same structure as B_j, that is,

$$\Gamma_j \equiv \begin{bmatrix} (\gamma_{11})_j & (\gamma_{12})_j \\ 0 & 0 \end{bmatrix},$$

and although the second row of Δ_j has the same structure as the second row of $\mathbf{A}_j$,

$$\Delta_j \equiv \begin{bmatrix} (\alpha_{11})_j & (\alpha_{12})_j \\ -1 & \dfrac{-h_{j+1}}{2} \end{bmatrix},$$

for generality we write it as

$$\Delta_j \equiv \begin{bmatrix} (\alpha_{11})_j & (\alpha_{12})_j \\ (\alpha_{21})_j & (\alpha_{22})_j \end{bmatrix}.$$

In the *backward* sweep, δ_j is computed from the following recursion formulas:

$$\Delta_J \delta_J = \mathbf{w}_J, \tag{13.33a}$$

$$\Delta_j \delta_j = \mathbf{w}_j - \mathbf{C}_j \delta_{j+1}, \qquad j = J-1, J-2, \dots, 0 \tag{13.33b}$$

13.2 Solution of the Energy Equation for Internal Flows with Fully Developed Velocity Profile

The energy equation for developing thermal boundary layers in ducts with fully developed velocity profile is given by Eq. (5.5). This equation, which is also given in dimensionless form by Eqs. (5.38) and (5.41), is a linear partial differential equation. For simple wall boundary conditions in laminar flows, its solution has been obtained by analytical methods. Although the solutions are in analytic form, their numerical evaluation is often quite tedious if they are in the form of slowly converging series. A more practical solution procedure for the energy equation is to use a finite-difference method. In this way we not only handle laminar flows for an arbitrarily wide range of boundary conditions, but we also handle turbulent flows with a prescribed variation of eddy viscosity and turbulent Prandtl number.

The numerical method that we shall use to solve this equation is the Box method, discussed in Section 13.1. Here we discuss the solution procedure in more detail and also consider more complicated boundary conditions.

Numerical Formulation

First we seek the solution of

$$(a_1 g')' + a_2 \frac{\eta}{2} g' - a_3 g = a_2 \hat{x} \frac{\partial g}{\partial \hat{x}} - a_3 \tag{5.38}$$

for the dimensionless temperature $g \equiv (T_w - T)/(T_w - T_e)$ subject to the boundary conditions

$$\eta = 0, \qquad g = 0 \quad \text{or} \quad g' = 1; \qquad \eta = \eta_e, \qquad g = 1. \tag{5.39}$$

For convenience we let $x = \hat{x}$, and analogous to the net points defined in the (xy) plane, we define new points in the $x\eta$ plane by

$$x_0 = 0, \qquad x_n = x_{n-1} + k_n, \qquad n = 1, 2, \dots, N,$$
$$\eta_0 = 0, \qquad \eta_j = \eta_{j-1} + h_j, \qquad j = 1, 2, \dots, J. \tag{13.34}$$

As before we express Eq. (5.38) in terms of two first-order equations by letting

$$g' = p \tag{13.35a}$$

and by writing Eq. (5.38) as

$$(a_1 p)' + a_2 \frac{\eta}{2} p - a_3 g = a_2 x \frac{\partial g}{\partial x} - a_3. \tag{13.35b}$$

The finite-difference form of Eq. (13.35a) is written for the midpoint $(x_n, \eta_{j-1/2})$ of the segment $P_1 P_2$ shown in Fig. 13.7, and the finite-difference form of (13.35b) is written for the midpoint $(x_{n-1/2}, \eta_{j-1/2})$ of the rectangle $P_1 P_2 P_3 P_4$. This gives

$$g_j^n - g_{j-1}^n - \frac{h_j}{2}\left(p_j^n + p_{j-1}^n\right) = 0, \tag{13.36a}$$

$$(s_1)_j p_j^n + (s_2)_j p_{j-1}^n + (s_3)_j \left(g_j^n + g_{j-1}^n\right) = R_{j-1/2}^{n-1}. \tag{13.36b}$$

Here

$$(s_1)_j = h_j^{-1}(a_1)_j^n + \tfrac{1}{4}(a_2)_{j-1/2}^n \eta_{j-1/2}, \tag{13.37a}$$

$$(s_2)_j = -h_j^{-1}(a_1)_{j-1}^n + \tfrac{1}{4}(a_2)_{j-1/2}^n \eta_{j-1/2}, \tag{13.37b}$$

$$(s_3)_j = -\tfrac{1}{2}(a_3)_{j-1/2}^n - (a_2)_{j-1/2}^n \alpha_n, \tag{13.37c}$$

$$R_{j-1/2}^{n-1} = -2(a_2)_{j-1/2}^{n-1/2} \alpha_n g_{j-1/2}^{n-1} - 2(a_3)_{j-1/2}^{n-1/2}$$
$$- \left\{ h_j^{-1}\left[(a_1 p)_j^{n-1} - (a_1 p)_{j-1}^{n-1}\right] \right\}$$
$$- \left[\tfrac{1}{2}(a_2)_{j-1/2}^{n-1} \eta_{j-1/2} p_{j-1/2}^{n-1} - (a_3)_{j-1/2}^{n-1} g_{j-1/2}^{n-1}\right], \tag{13.37d}$$

$$\alpha_n = \frac{x_{n-1/2}}{k_n}. \tag{13.37e}$$

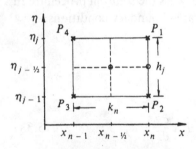

Figure 13.7. Net rectangle for difference approximations for the Box scheme.

Equations (13.36) are imposed for $j = 1, 2, \ldots, J - 1$. At $j = 0$ and $j = J$ we need boundary conditions. Since at the wall we may specify either the wall temperature or the heat flux, it is very convenient to write the wall boundary conditions, $j = 0$, as

$$\alpha_0 g_0 + \alpha_1 p_0 = \gamma_0. \tag{13.38a}$$

When the wall temperature is specified, we set $\alpha_0 = 1$, $\alpha_1 = 0$, and since $g_0 = 0$, $\gamma_0 = 0$. When the wall heat flux is specified, we set $\alpha_0 = 0$, $\alpha_1 = 1$, and γ_0 equals $p_0 \equiv 1$. The "edge" boundary conditions can also be formulated for a general case by using a similar procedure, and for $j = J$ they can be written as

$$\beta_0 g_J + \beta_1 p_J = \gamma_1. \tag{13.38b}$$

When the edge temperature is specified, we set $\beta_0 = 1$, $\beta_1 = 0$, and $\gamma_1 = g_J$. When the temperature gradient is specified, $\beta_0 = 0$, $\beta_1 = 1$, and $\gamma_1 = p_J$. Obviously to satisfy the edge boundary condition in Eq. (5.39), we set $\beta_0 = 1$, $\beta_1 = 0$, and take $\gamma_1 = 1$.

Block-Elimination Method

The linear system (13.36) and (13.38) can now be solved by the block-elimination method that was discussed in Section 13.1 for specified wall temperature or specified wall heat flux, as well as for specified edge temperature or specified centerline temperature gradient. As before, we write the linear system by Eq. (13.28) and redefine δ_j by

$$\delta_j = \begin{bmatrix} g_j \\ p_j \end{bmatrix} \tag{13.39}$$

Except for the $\mathbf{A}_0$ and $\mathbf{A}_J$ matrices, the rest of the matrices remain the same. The new definitions of $\mathbf{A}_0$ and $\mathbf{A}_J$, which result from the consideration of general boundary conditions, are

$$\mathbf{A}_0 = \begin{bmatrix} \alpha_0 & \alpha_1 \\ -1 & \dfrac{-h_1}{2} \end{bmatrix}, \qquad \mathbf{A}_J = \begin{bmatrix} (s_3)_J & (s_1)_J \\ \beta_0 & \beta_1 \end{bmatrix}. \tag{13.40}$$

Also

$$(r_1)_0 = \gamma_0, \qquad (r_2)_J = \gamma_1. \tag{13.41}$$

In the forward sweep, to compute Γ_j, Δ_j, and $\mathbf{w}_j$ from the recursion formulas given by Eq. (13.31), we first denote the elements of Γ_j by $(\gamma_{ik})_j$ and the elements of Δ_j by $(\alpha_{ik})_j (i, k = 1, 2)$. It then follows from Eq. (13.31a) that the values of $(\alpha_{ik})_0$ are

$$(\alpha_{11})_0 = \alpha_0, \qquad (\alpha_{12})_0 = \alpha_1,$$

$$(\alpha_{21})_0 = -1, \qquad (\alpha_{22})_0 = \dfrac{h_1}{-2}, \tag{13.42}$$

and the values of $(\alpha_{ik})_j$ for $j = 1, 2, \ldots, J-1$ are

$$(\alpha_{11})_j = (s_3)_j - (\gamma_{11})_j, \qquad (\alpha_{12})_j = (s_1)_j + \frac{h_j}{2}(\gamma_{12})_j,$$

$$(\alpha_{21})_j = -1, \qquad (\alpha_{22})_j = -\frac{h_{j+1}}{2}. \qquad (13.43)$$

The values of $(\gamma_{ik})_j$ for $1 \le j \le J$ are

$$(\gamma_{11})_j = \frac{(s_3)_j(\alpha_{22})_{j-1} - (s_2)_j(\alpha_{21})_{j-1}}{D_j},$$

$$(\gamma_{12})_j = \frac{(s_2)_j(\alpha_{11})_{j-1} - (s_3)_j(\alpha_{12})_{j-1}}{D_j}, \qquad (13.44)$$

where

$$D_j = (\alpha_{11})_{j-1}(\alpha_{22})_{j-1} - (\alpha_{21})_{j-1}(\alpha_{12})_{j-1}. \qquad (13.45)$$

We denote the components of the vectors $\mathbf{w}_j$ by

$$\mathbf{w}_j = \begin{bmatrix} (w_1)_j \\ (w_2)_j \end{bmatrix}, \qquad 0 \le j \le J.$$

It follows from Eq. (13.32) that for $j = 0$,

$$(w_1)_0 = \gamma_0, \qquad (w_2)_0 = (r_2)_0 \qquad (13.46a)$$

and for $1 \le j \le J$,

$$(w_1)_j = (r_1)_j - (\gamma_{11})_j(w_1)_{j-1} - (\gamma_{12})_j(w_2)_{j-1}, \qquad (w_2)_j = (r_2)_j. \qquad (13.46b)$$

In the backward sweep, the two components of δ_j follow from Eq. (13.33) and are given by

$$g_j = \frac{(w_1)_j(\alpha_{22})_j - e_j(\alpha_{12})_j}{D_{j+1}},$$

$$p_j = \frac{e_j(\alpha_{11})_j - (w_1)_j(\alpha_{21})_j}{D_{j+1}}, \qquad (13.47)$$

where

$$e_J = (w_2)_J, \qquad (\alpha_{21})_J = \beta_0, \qquad (\alpha_{22})_J = \beta_1 \quad \text{for} \quad j = J$$

and

$$e_j = (w_2)_j - g_{j+1} + \frac{h_{j+1}}{2}p_{j+1} \quad \text{for} \quad j < J.$$

13.3 Fortran Program for Internal Laminar and Turbulent Flows with Fully Developed Velocity Profile

In this section we present a Fortran program for solving the energy equation for internal flows in which the velocity profile is fully developed. To avoid the extra coding that would be needed to use the program for two-dimensional and axisymmetric flows with a wide range of boundary conditions, we consider only axisymmetric flows with *uniform* wall temperature. By using the eddy-viscosity and turbulent-Prandtl-number concepts to model the Reynolds-shear-stress and heat-flux terms, we express the energy equation in the form given by Eq. (7.1), that is

$$u\frac{\partial T}{\partial x} = \frac{\nu}{\text{Pr}}\frac{1}{r}\frac{\partial}{\partial y}\left[r\left(1+\frac{\text{Pr}}{\text{Pr}_t}\varepsilon_m^+\right)\frac{\partial T}{\partial y}\right], \tag{7.1}$$

and apply it to both laminar and turbulent flows. As discussed in Chapter 5, at the early stages when the thermal boundary-layer thickness is small, we use transformed variables to express Eq. (7.1) in the form given by Eq. (5.38), and later, before the thickness of the shear layer exceeds the radius of the duct, we switch back to physical variables and express Eq. (7.1) in dimensionless form as given by Eq. (5.41). The coefficients of these two equations (5.38) and (5.41), which are denoted by a_1, a_2, a_3 (with $a_3 \equiv 0$ for uniform wall temperature), can be used for both laminar and turbulent flows, with only minor modifications occurring in the definition of a_1.

$$a_1 = r\left(1+\frac{\text{Pr}}{\text{Pr}_t}\varepsilon_m^+\right). \tag{7.31}$$

Since the finite-difference equations given in Section 13.2 are general, they apply to both laminar and turbulent flows.

The program consists of a MAIN program and three subroutines: COEF, SOLV2, and OUTPUT. The following sections present a brief description of each routine.

The Fortran names for the symbols, unless otherwise noted, follow closely the notation used for symbols in the text.

MAIN

This routine contains the overall logic of computations, specifies the initial temperature profiles (g_j and p_j) at $x = x_0$, generates the grid normal to the flow, and accounts for the boundary-layer growth. Although a uniform grid across the shear layer is quite satisfactory for most laminar-flow calculations, it is not so for turbulent-flow calculations since close to the wall the velocity profiles change more rapidly than those for laminar flows. Consequently, it is necessary to use a nonuniform grid. Here we use one in which the ratio of lengths of any two adjacent intervals is a constant; that is, $\eta_j = \eta_{j-1} + h_j$, where $h_j = Kh_{j-1}$. The distance to the jth line is given by the following formula:

$$\eta_j = h_1\frac{K^j-1}{K-1}, \qquad j=1,2,3,\dots,J, \qquad K>1. \tag{13.48}$$

There are two parameters: h_1, the length of the first $\Delta\eta$ step, and K, the ratio of two successive steps. The total number of points, J, is calculated by the following formula:

$$J = \frac{\ln\left[1 + (K-1)(\eta_e/h_1)\right]}{\ln K}. \tag{13.49}$$

For laminar flows, we generally take K ($\equiv$ VGP) $= 1$, η_e ($\equiv$ ETAE) $= 8$ and choose h_1 [$\equiv$ DETA(1)] $= 0.2$ so that we have 41 grid points across the layer, which is sufficient for most laminar-flow calculations. For turbulent flows, we can take, say, $K = 1.15$, $\eta_e = 10$, and $h_1 = 0.025$; this gives initially about 30 grid points. Since the boundary-layer thickness grows for turbulent flows, at later x stations we may have around 40–50 grid points.

This routine also contains the input data that consist of

NXT: total number of x stations
ITURB: 0 for laminar and 1 for turbulent flow
IWBCOE: 1 for specified wall temperature and 0 for specified wall heat flux.
IEBCOE: 1 for specified edge temperature and 0 for zero edge-temperature gradient

and ETAE, DETA(1), VGP, PR, GWA, REY. In the present problem IWBCOE $= 1$, IEBCOE $= 1$, and GWA corresponds to the parameter needed in the smoothing function, Eq. (4.90), used for the wall temperature. We use this equation (see subroutine COEF) in order to avoid the difficulties associated with the jump in the wall temperature as was described in Section 4.3. The parameter REY is the Reynolds number based on radius, $u_0 r_0/\nu$ ($\equiv R_d/2$). Note that both PR and REY are needed only for turbulent flows, since for laminar flow, they do not appear in the dimensionless form of the equations.

```
      COMMON/BLCO/ NXT,IWBCOE,IEBCOE,ITURB,ICOORD,INDEX,N,NP,PR,VGP,
     1             GWA,REY,CEL,ETA(51),UP(51),DETA(51),A(51),YP(51),
     2             X(101),GW(101),PW(101),GE(101),G(51,2),P(51,2)
      COMMON/BLC1/ S1(51),S2(51),S3(51),R1(51),R2(51),A1(51,2),A2(51,2)
C - - - - - - - - - - - - - - - - - - - - - - - - - - - - - - - - - - -
C   READ IN $ PRINT OUT PARAMETERS
      READ(5 ,8000) NXT,IWBCOE,IEBCOE,ITURB,ETAE,DETA(1),VGP,PR,GWA,REY
      READ(5 ,8100) (X(I),I=1,NXT)
      WRITE(6,9000) NXT,IWBCOE,IEBCOE,ITURB,ETAE,DETA(1),VGP,PR,GWA,REY
      IF((VGP-1.0) .GT. 0.0001) GO TO 20
      NP     = ETAE/DETA(1)+1.0001
      GO TO 30
   20 NP     = ALOG((ETAE/DETA(1))*(VGP-1.0)+1.0)/ALOG(VGP)+1.0001
   30 IF(NP.LE.51) GOTO 40
      WRITE(6,9100)
      STOP
   40 ETA(1)= 0.0
      YP(1) = 0.
C   INITIAL TEMPERATURE PROFILE AT X = XO
      G(1,2)= 1.0
      P(1,2)= 0.0
      DO 50 J=2,51
      YP(J)  = 0.
      G(J,2)= 1.0
      P(J,2)= 0.0
      A1(J,2)= 1.0
      A2(J,2)= 0.0
C   GENERATION OF GRID
      DETA(J)=DETA(J-1)*VGP
      ETA(J)= ETA(J-1)+DETA(J-1)
```

```
   50 A(J)   = 0.5*DETA(J-1)
      N      = 1
      ICOORD = 1
      INDEX = 0
      GW(1) = 1.0
   60 WRITE(6,9200) N,X(N)
      IF(N.EQ.1) GOTO 80
      IF(ICOORD.EQ.2) CEL = 1.0/(X(N)-X(N-1))
      IF(ICOORD.EQ.1) CEL = 0.5*(X(N )+X(N-1))/(X(N)-X(N-1))
      IGROW = 0
   70 CALL COEF
      CALL SOLV2
C   CHECK FOR BOUNDARY LAYER GROWTH
      IF(ICOORD.EQ.2 .OR. NP.EQ.51) GOTO 80
      IF(IGROW.GE.3 .OR. ABS(P(NP,2)).LT.1.E-04) GOTO 80
      XSWTCH= 1.0/ETA(NP+1)**2
      IF(X(N).GE.XSWTCH) GOTO 80
      NP     = NP+1
      IGROW = IGROW+1
      G(NP,1)= G(NP-1,1)
      P(NP,1)= 0.0
      GOTO 70
   80 CALL OUTPUT
      GOTO 60
C - - - - - - - - - - - - - - - - - - - - - - - - - - -
 8000 FORMAT(4I5,6F10.0)
 8100 FORMAT(8F10.0)
 9000 FORMAT(1H0,7HNXT    =,I3,14X,7HIWBCOE=,I3,14X,7HIEBCOE=,I3,14X,
     1        7HITURB =,I3/1H ,7HETAE  =,E14.6,3X,7HDETA1 =,E14.6,3X,
     2        7HVGP   =,E14.6,3X,7HPR   =,E14.6/1H ,7HGWA   =,E14.6,3X,
     3        7HREY   =,E14.6)
 9100 FORMAT(1H0,34HNP EXCEEDED 51--PROGRAM TERMINATED)
 9200 FORMAT(/1H0,2HN=,I3,5X,3HX =,E14.6)
      END
```

Subroutine COEF

This subroutine defines the coefficients of the finite-difference energy equation given by Eqs. (13.37). Note that these definitions apply to the coefficients of the energy equation expressed either in transformed, SWITCH = 1.0 or physical, SWITCH = 0.0, variables. For turbulent flows the coefficients remain unchanged except a_1 becomes that given by Eq. (7.31).

This subroutine also contains expressions for calculating the velocity profiles for laminar and turbulent flows. In the former case the velocity profile is computed from [see Eq. (5.20)]

$$\hat{u} = 1 - (\hat{r})^2. \tag{13.50}$$

The velocity profile for a turbulent flow is computed from Eq. (7.13), with friction factor computed by Eq. (7.17).

The eddy-viscosity and turbulent-Prandtl-number calculations are done by using the expressions given by Eqs. (7.9), (7.10), and (6.26), respectively. Note that at $y = 0$ the latter becomes

$$\mathrm{Pr}_t = \frac{\kappa}{\kappa_h} \frac{B}{A}. \tag{13.51}$$

```
      SUBROUTINE COEF
      REAL K,KH
      COMMON/BLCO/ NXT,IWBCOE,IEBCOE,ITURB,ICOORD,INDEX,N,NP,PR,VGP,
     1            GWA,REY,CEL,ETA(51),UP(51),DETA(51),A(51),YP(51),
     2            X(101),GW(101),PW(101),GE(101),G(51,2),P(51,2)
      COMMON/BLC1/ S1(51),S2(51),S3(51),R1(51),R2(51),A1(51,2),A2(51,2)
```

```
        DIMENSION C(5),EDV(51),DUDY(51)
        DATA K,KH/0.4,0.44/
        DATA IGWALL,APLUS,C/0,26.0,34.96,28.79,33.95,6.3,-1.186/
C - - - - - - - - - - - - - - - - - - - - - - - - - - - - - - - - - - - - - -
        SWITCH = 0.0
        XP1 = 1.0
        JJ    = NP
        IF(ICOORD.EQ.2) GOTO 15
        SWITCH= 1.0
        XP1   = SQRT(X(N))
        GE(N) = 1.0
        DO 5 J=1,51
        YP(J) = XP1*ETA(J)
        IF(YP(J).GT.1.0) GOTO 10
    5 CONTINUE
        JJ    = 51
        GOTO 15
   10 JJ     = J-1
   15 IF(ITURB.EQ.1) GOTO 25
C   VELOCITY PROFILE $ COEFFS. OF THE ENERGY EQ. FOR LAMINAR FLOW
        DO 20 J=1,JJ
        YP(J) = XP1*ETA(J)
        RP    = 1.0-YP(J)
        UP(J) = 2.0*(1.0-RP**2)
        A1(J,2)= RP
   20 A2(J,2) = UP(J)*RP
        GOTO 45
C   VELOCITY PROFILE, EDDY VISCOSITY $ TURBULENT PRANDTL NUMBER FOR
C   TURBULENT FLOW
   25 CONTINUE
        F       = 0.3164/(2.*REY)**0.25
        ALOGPR= ALOG10(PR)
        SUM   = C(1)
        DO 30 I=2,5
   30 SUM    = SUM+C(I)*ALOGPR**(I-1)
        BPLUS = SUM/SQRT(PR)
        PRT   = K/KH*BPLUS/APLUS
        UP(1) = 0.0
        CYOA  = 0.5*REY/APLUS*SQRT(0.5*F)
        CYOB  = CYOA*APLUS/BPLUS
        DO 40 J=1,JJ
        YP(J) = XP1*ETA(J)
        RP    = 1.0-YP(J)
        YOA   = YP(J)*CYOA
        YOB   = YP(J)*CYOB
        EXPYOA= 0.0
        EXPYOB= 0.0
        IF(YOA.LT.50.0) EXPYOA = EXP(-YOA)
        IF(YOB.LT.50.0) EXPYOB = EXP(-YOB)
        CMIX  =(0.14-RP**2*(0.08+0.06*RP**2))*(1.-EXPYOA)
        DUDY(J)=0.25*REY*F*RP/(1.+SQRT(1.+0.5*F*RP*(REY*CMIX)**2))
        EDV(J) = CMIX**2*REY*DUDY(J)
        IF(J.EQ.1) GOTO 35
        UP(J) = UP(J-1)+0.5*(DUDY(J)+DUDY(J-1))*(YP(J)-YP(J-1))
        PRT    = K/KH*(1.-EXPYOA)/(1.-EXPYOB)
C   COEFFS OF THE ENERGY EQ. FOR TURBULENT FLOW
   35 A1(J,2)= RP*(1.0+PR/PRT*EDV(J))
        A2(J,2)= RP*UP(J)
   40 CONTINUE
   45 GW(N) = 0.0
        IF(IGWALL .EQ. 1) GO TO 50
        GW(N) = 0.5*(1.0+COS(3.14159*(X(N)-X(1))/GWA))
        IF(X(N).LT.(X(1)+GWA)) GOTO 50
        GW(N) = 0.0
        IGWALL= 1
C   COEFFICIENTS OF THE FINITE-DIFFERENCE EQUATIONS
   50 DO 55 J=2,NP
        ETAB   = 0.5*(ETA(J)+ETA(J-1))
        CGB    = 0.5*(G(J,1)+G(J-1,1))
        CPB    = 0.5*(P(J,1)+P(J-1,1))
        A2B    = 0.5*(A2(J,2)+A2(J-1,2))
        CA2B   = 0.5*(A2(J,1)+A2(J-1,1))
        DERA1P = (A1(J,1)*P(J,1)-A1(J-1,1)*P(J-1,1))/DETA(J-1)
        S1(J)  = A1(J,2)/DETA(J-1)+0.25*ETAB*A2B*SWITCH
        S2(J)  = -A1(J-1,2)/DETA(J-1)+0.25*ETAB*A2B*SWITCH
        S3(J)  = -0.5*(CA2B+A2B)*CEL
```

```
      R1(J) = 2.0*S3(J)*CGB-DERA1P-0.5*CA2B*ETAB*CPB*SWITCH
   55 R2(J-1)= 0.0
      RETURN
      END
```

Subroutine OUTPUT

This subroutine prints out the profiles g, p, u across the shear layer. It also computes the dimensionless mixed-mean temperature given by Eq. (5.4) and the Nusselt number given by Eq. (5.23), which in terms of dimensionless temperature and temperature gradient g'_w can be written as [see Eq. (P5.10)]

$$\text{Nu} = \frac{2g'_w}{g_m}. \tag{13.52}$$

In the region where the thermal shear layers have not yet merged, the dimensionless bulk temperature g_m is computed according to Eq. (P5.8), and g'_w is computed from $(\partial g / \partial y)_w$. When the thermal shear layers merge, they are computed from Eqs. (P5.9) and (P5.11), respectively. This subroutine also determines at what x (denoted by SW) we switch to physical variables. This is done as follows: since before the thermal layers merge we use the transformed variables η, whose edge value (η_e) is relatively constant, we can write

$$\eta_e = \frac{y_e}{\sqrt{\nu x / u_0}}.$$

From this expression it follows that

$$\frac{x_{sw}}{r_0} < \frac{1}{\eta_e^2} \left(\frac{u_0 r_0}{\nu} \right).$$

In addition, this subroutine shifts the profiles prior to the calculations at a new x station.

```
      SUBROUTINE OUTPUT
      COMMON/BLCO/ NXT,IWBCOE,IEBCOE,ITURB,ICOORD,INDEX,N,NP,PR,VGP,
     1             GWA,REY,CEL,ETA(51),UP(51),DETA(51),A(51),YP(51),
     2             X(101),GW(101),PW(101),GE(101),G(51,2),P(51,2)
      COMMON/BLC1/ S1(51),S2(51),S3(51),R1(51),R2(51),A1(51,2),A2(51,2)
C - - - - - - - - - - - - - - - - - - - - - - - - - - - - - - - - - - -
C    PRINT OUT DIMENSIONLESS TEMPERATURE PROFILE
      WRITE(6,9100) (J,ETA(J),G(J,2),P(J,2),UP(J),YP(J),J=1,NP)
      IF(N.EQ.1) GOTO 15
C    CALCULATE $ PRINT OUT DIMENSIONLESS MIX TEMPERATURE $ NUSSELT NUMBER
      CNUXO = 2.0*P(1,2)
      C     = 0.0
      IF(ICOORD.EQ.2) GOTO 5
      C     = 1.0
      CNUXO = CNUXO/SQRT(X(N))
    5 GMIX  = 0.0
      F2 = 0.
      DO 10 J=2,NP
      F1 = F2
      F2 = 2.*(1.-YP(J))*UP(J)*(G(J,2)-C)
      GMIX = GMIX+0.5*(F1+F2)*(YP(J)-YP(J-1))
   10 CONTINUE
      GMIX   = GMIX+C
      CNUXM = CNUXO/GMIX
      WRITE(6,9200) GMIX,CNUXM
   15 IF(N.EQ.NXT) STOP
```

```
      IF (INDEX.EQ.1) GOTO 25
      N = N+1
      IF (ICOORD.EQ.2) GOTO 35
      XSWTCH = 1./ETA(NP)**2
      IF (X(N).LT.XSWTCH) GOTO 35
      DO 20 II=N,NXT
      I    = NXT-II+N
      X(I+1)= X(I)
   20 CONTINUE
      X(N)  = XSWTCH
      NXT   = NXT+1
      INDEX = 1
      GOTO 35
C    SWITCH TO PRIMITIVE VARIABLES
   25 SQX   = SQRT(X(N))
      DO 30 J=1,51
      ETA(J)= ETA(J)*SQX
      DETA(J) = DETA(J)*SQX
      P(J,2)= P(J,2)/SQX
      IF (J.EQ.1) GOTO 30
      A(J) = 0.5*DETA(J-1)
   30 CONTINUE
      ICOORD= 2
      IEBCOE= 0
      WRITE(6,9300)
      WRITE(6,9100) (J,ETA(J),G(J,2),P(J,2),UP(J),YP(J),J=1,NP)
      CALL COEF
      N = N+1
      INDEX = 0
C    SHIFT PROFILES FOR NEXT STATION CALCULATION
   35 DO 40 J=1,51
      G(J,1)= G(J,2)
      P(J,1)= P(J,2)
      A1(J,1)=A1(J,2)
      A2(J,1)=A2(J,2)
   40 CONTINUE
      RETURN
C - - - - - - - - - - - - - - - - - - - - - - - - - - - - - -
 9100 FORMAT(1H0,2X,1HJ,3X,3HETA,10X,1HG,13X,1HP,12X,2HUP,12X,2HYP/
     1        (1H ,I3,F10.5,4E14.6))
 9200 FORMAT(1H0,6HGMIX =,E14.6,3X,6HNUXM =,E14.6)
 9300 FORMAT(1H0,31H***** PRIMITIVE VARIABLES *****)
      END
```

Subroutine SOLV2

The subroutine solves the linear system (13.36) and (13.38) by using the block-elimination method that was discussed in detail in Section 13.2. The notation follows closely that used in Section 13.2.

```
      SUBROUTINE SOLV2
      COMMON/BLCD/ NXT,IWBCOE,IEBCOE,ITURB,ICOORD,INDEX,N,NP,PR,VGP,
     1        GWA,REY,CEL,ETA(51),UP(51),DETA(51),A(51),YP(51),
     2        X(101),GW(101),PW(101),GE(101),G(51,2),P(51,2)
      COMMON/BLC1/ S1(51),S2(51),S3(51),R1(51),R2(51),A1(51,2),A2(51,2)
      DIMENSION G11(51),G12(51),A11(51),A12(51),A21(51),A22(51),W1(51),
     1          W2(51),DEN(51)
C - - - - - - - - - - - - - - - - - - - - - - - - - - - - -
      IF (IWBCOE.EQ.0) GOTO 10
C    SPECIFIED WALL TEMPERATURE
      ALFA0 = 1.0
      ALFA1 = 0.0
      G(1,2)= GW(N)
      P(1,2)= 0.0
      GOTO 20
C    SPECIFIED WALL HEAT FLUX
   10 ALFA0 = 0.0
      ALFA1 = 1.0
      G(1,2)= 0.0
      P(1,2)= PW(N)
```

```
   20 GAMMAO= ALFAO*G(1,2)+ALFA1*P(1,2)
      R1(1) = GAMMAO
      IF(IEBCOE.EQ.O) GOTO 30
C  SPECIFIED EDGE TEMPERATURE
      BETAO = 1.0
      BETA1 = 0.0
      G(NP,2)= GE(N)
      P(NP,2)= 0.0
      GOTO 40
C  SPECIFIED EDGE TEMPERATURE GRADIENT
   30 BETAO = 0.0
      BETA1 = 1.0
      G(NP,2)= 0.0
      P(NP,2)= 0.0
   40 GAMMA1= BETAO*G(NP,2)+BETA1*P(NP,2)
      R2(NP)= GAMMA1
C  W-ELEMENTS FOR J=1
      W1(1) = R1(1)
      W2(1) = R2(1)
C  ALFA ELEMENTS FOR J=1
      A11(1)= ALFAO
      A12(1)= ALFA1
      A21(1)= -1.0
      A22(1)= -0.5*DETA(1)
C  GAMMA ELEMENTS FOR J=2
      DET   = ALFA1-0.5*DETA(1)*ALFAO
      G11(2)= (S2(2)-0.5*DETA(1)*S3(2))/DET
      G12(2)= (ALFAO*S2(2)-ALFA1*S3(2))/DET
C  FORWARD SWEEP
      DO 60 J=2,NP
      DEN(J)= A11(J-1)*A22(J-1)-A21(J-1)*A12(J-1)
      IF(J.EQ.2) GOTO 50
      G11(J)= (S3(J)*A22(J-1)-S2(J)*A21(J-1))/DEN(J)
      G12(J)= (S2(J)*A11(J-1)-S3(J)*A12(J-1))/DEN(J)
   50 A11(J)= S3(J)-G12(J)
      A12(J)= S1(J)+A(J)*G12(J)
      A21(J)= -1.0
      A22(J)= -A(J+1)
      W1(J) = R1(J)-G11(J)*W1(J-1)-G12(J)*W2(J-1)
      W2(J) = R2(J)
   60 CONTINUE
C  BACKWARD SWEEP
      DENO  = A11(NP)*BETA1-A12(NP)*BETAO
      G(NP,2)= (W1(NP)*BETA1-W2(NP)*A12(NP))/DENO
      P(NP,2)= (W2(NP)*A11(NP)-BETAO*W1(NP))/DENO
      J     = NP
   70 J     = J-1
      E1    = W2(J)-G(J+1,2)+A(J+1)*P(J+1,2)
      G(J,2)= (W1(J)*A22(J)-E1*A12(J))/DEN(J+1)
      P(J,2)= (E1*A11(J)-W1(J)*A21(J))/DEN(J+1)
      IF(J.GT.1) GOTO 70
      RETURN
      END
```

Sample Calculations

To illustrate the use of the Fortran program, we now consider a laminar flow in a circular pipe with constant wall temperature. The results shown in Table 13.1 under "numerical method" were obtained by choosing $\eta_e = 8$, $h_1 = 0.20$, $K = 1$, and $a = 2 \times 10^{-6}$ [see Eq. (4.90)] for the definition of Nusselt number given by Eq. (5.23). The initial value of the dimensionless $\hat{x}$ location $\hat{x}_0$ was taken as 1×10^{-6}. In the region where g_w changes according to Eq. (4.90), 15 $\hat{x}$ stations were taken.

For this flow, there are analytic (series) solutions due to Sellars et al. [4]. According to [4], the local Nusselt number is given by the following

Table 13.1 Infinite-series solution functions for a circular pipe with constant wall temperature

n	λ_n	G_n
0	2.704	0.749
1	6.680	0.544
2	10.668	0.463

Note: For $n > 2$, $\lambda_n = 4n + \frac{8}{3}$, and $G_n = 1.01276 \lambda_n^{-1/3}$.

Table 13.2 Nu obtained by numerical and analytical methods for a laminar flow in a pipe with constant wall temperature

	Nu	
x^+/Pe	Numerical method	Analytical method
1×10^{-4}	28.3	28.2
5×10^{-4}	16.5	16.2
1×10^{-3}	12.9	12.8
4×10^{-3}	8.02	8.03
2×10^{-2}	4.92	4.92
6×10^{-2}	3.89	3.89
1×10^{-1}	3.71	3.71
3×10^{-1}	3.63	3.66
∞	3.63	3.66

formula:

$$\text{Nu} = \frac{\displaystyle\sum_{n=0}^{\infty} G_n \exp\left(-\lambda_n^2 \frac{x^+}{\text{Pe}}\right)}{2 \displaystyle\sum_{n=0}^{\infty} \left(G_n/\lambda_n^2\right)\exp\left(-\lambda_n^2 \frac{x^+}{\text{Pe}}\right)}, \qquad (13.53)$$

where $x^+ = x/r_0$. The constants G_n and eigenvalues λ_n in Eq. (13.53) are given in Table 13.1.

Table 13.2 shows a comparison between the numerical and the analytical results at various values of x^+/Pe. We see that with 41 grid points across the layer, very accurate results can be obtained with this code.

13.4 Solution of Mass, Momentum, and Energy Equations for Boundary-Layer Flows

In this section we use the Box method to obtain the solutions of the coupled mass, momentum, and energy equations for external flows. In Section 13.5 we present a Fortran program that solves these equations. The numerical method and the computer program are presented in such a way that the reader can easily apply them to other problems, including internal flows. Examples of such extensions to the program are given in Chapter 14, assuming that in the case of turbulent flows a suitable turbulence model (for example, an eddy-viscosity correlation) is available.

Numerical Formulation

To describe the numerical solution of the boundary-layer equations for two-dimensional coupled laminar and turbulent *external* flows, we start with the transformed equations and boundary conditions given by Eqs. (11.45), (11.46), and (11.49), which use the concepts of eddy viscosity and turbulent Prandtl number for turbulent flows:

$$(bf'')' + m_1 ff'' + m_2\left[c - (f')^2\right] = x\left(f'\frac{\partial f'}{\partial x} - f''\frac{\partial f}{\partial x}\right), \quad (11.45)$$

$$(eS' + df'f'')' + m_1 fS' = x\left(f'\frac{\partial S}{\partial x} - S'\frac{\partial f}{\partial x}\right), \quad (11.46)$$

$$\eta = 0, \quad f' = 0, \quad f = f_w(x), \quad S = S_w(x) \quad \text{or} \quad S' = S_w'(x),$$
$$(11.49a)$$

$$\eta = \eta_e, \quad f' = 1, \quad S = 1. \quad (11.49b)$$

Although different turbulence models employing algebraic eddy-viscosity (or mixing-length) and turbulent-Prandtl-number formulations can be used in these equations, we shall use the Cebeci–Smith model that has been used throughout this book.

First we write Eqs. (11.45) and (11.46) and their boundary conditions in terms of a first-order system. For this purpose we introduce new dependent variables $u(x, \eta)$, $v(x, \eta)$, and $p(x, \eta)$ so that the transformed momentum and energy equations can be written as

$$f' = u, \quad (13.54a)$$

$$u' = v, \quad (13.54b)$$

$$g' = p, \quad (13.54c)$$

$$(bv)' + m_1 fv + m_2(c - u^2) = x\left(u\frac{\partial u}{\partial x} - v\frac{\partial f}{\partial x}\right), \quad (13.54d)$$

$$(ep + duv)' + m_1 fp = x\left(u\frac{\partial g}{\partial x} - p\frac{\partial f}{\partial x}\right), \quad (13.54e)$$

where $g \equiv H/H_e$ is the total-enthalpy ratio denoted by S in Eq. (11.46). In terms of the new dependent variables, the boundary conditions given by Eqs. (11.49) become

$$f(x,0) = f_w(x), \quad u(x,0) = 0,$$

$$g(x,0) = g_w(x) \quad \text{or} \quad p(x,0) = p_w'(x) \quad (13.55a)$$

$$u(x, \eta_e) = 1, \quad g(x, \eta_e) = 1 \quad (13.55b)$$

We now consider the same net rectangle in the $x\eta$ plane shown in Fig. 13.7 and the net points defined by Eq. (13.34). As in Sections 13.1 and 13.2, where we used the Box method, the finite-difference approximations to the three first-order ordinary differential equations (13.54a)–(13.54c) are written for the midpoint $(x_n, \eta_{j-1/2})$ of the segment $P_1 P_2$ shown in Fig. 13.7, and, as in Eqs. (13.21b) or (13.35b), the finite-difference approximations to the two first-order partial differential equations (13.54d) and (13.54e) are written for the midpoint $(x_{n-1/2}, \eta_{j-1/2})$ of the rectangle $P_1 P_2 P_3 P_4$. This procedure gives

$$h_j^{-1}\left(f_j^n - f_{j-1}^n\right) = u_{j-1/2}^n, \tag{13.56a}$$

$$h_j^{-1}\left(u_j^n - u_{j-1}^n\right) = v_{j-1/2}^n, \tag{13.56b}$$

$$h_j^{-1}\left(g_j^n - g_{j-1}^n\right) = p_{j-1/2}^n, \tag{13.56c}$$

$$h_j^{-1}\left(b_j^n v_j^n - b_{j-1}^n v_{j-1}^n\right) + \left(m_1^n + \alpha_n\right)(fv)_{j-1/2}^n - \left(m_2^n + \alpha_n\right)(u^2)_{j-1/2}^n$$
$$+ \alpha_n\left(v_{j-1/2}^{n-1} f_{j-1/2}^n - f_{j-1/2}^{n-1} v_{j-1/2}^n\right) = R_{j-1/2}^{n-1}, \tag{13.56d}$$

$$h_j^{-1}\left(e_j^n p_j^n - e_{j-1}^n p_{j-1}^n\right) + h_j^{-1}\left(d_j^n u_j^n v_j^n - d_{j-1}^n u_{j-1}^n v_{j-1}^n\right)$$
$$+ \left(m_1^n + \alpha_n\right)(fp)_{j-1/2}^n$$
$$- \alpha_n\left[(ug)_{j-1/2}^n + u_{j-1/2}^{n-1} g_{j-1/2}^n - g_{j-1/2}^{n-1} u_{j-1/2}^n\right.$$
$$\left. + f_{j-1/2}^{n-1} p_{j-1/2}^n - p_{j-1/2}^{n-1} f_{j-1/2}^n\right] = T_{j-1/2}^{n-1}, \tag{13.56e}$$

where

$$\alpha_n = \frac{x^{n-1/2}}{k_n}, \tag{13.57}$$

$$R_{j-1/2}^{n-1} = - L_{j-1/2}^{n-1} + \alpha_n\left[(fv)_{j-1/2}^{n-1} - (u^2)_{j-1/2}^{n-1}\right] - m_2^n c_{j-1/2}^n, \tag{13.58a}$$

$$L_{j-1/2}^{n-1} = \left\{h_j^{-1}\left(b_j v_j - b_{j-1} v_{j-1}\right) + m_1(fv)_{j-1/2}\right.$$
$$\left. + m_2\left[c_{j-1/2} - (u^2)_{j-1/2}\right]\right\}^{n-1}, \tag{13.58b}$$

$$T_{j-1/2}^{n-1} = - M_{j-1/2}^{n-1} + \alpha_n\left[(fp)_{j-1/2}^{n-1} - (ug)_{j-1/2}^{n-1}\right], \tag{13.59a}$$

$$M_{j-1/2}^{n-1} = \left[h_j^{-1}\left(e_j p_j - e_{j-1} p_{j-1}\right) + h_j^{-1}\left(d_j u_j v_j - d_{j-1} u_{j-1} v_{j-1}\right)\right.$$
$$\left. + m_1(fp)_{j-1/2}\right]^{n-1}. \tag{13.59b}$$

As in Section 13.2, in order to account for the mixed wall boundary conditions for the energy equation, we use Eq. (13.38a) and write the wall boundary conditions for the momentum and energy equations as

$$f_0^n = 0, \qquad u_0^n = 0, \qquad \alpha_0 g_0^n + \alpha_1 p_0^n = \gamma_0^n \tag{13.60a}$$

and the edge boundary conditions as

$$u_j^n = 1, \qquad g_j^n = 1. \tag{13.60b}$$

As before, when the wall temperature is specified, we set $\alpha_0 = 1$, $\alpha_1 = 0$, and since the enthalpy ratio at the surface, $g_0(x)$, is given, $\gamma_0 = g_0(x)$. When the wall heat flux is specified, we set $\alpha_0 = 0$, $\alpha_1 = 1$, and γ_0 equals the dimensionless total-enthalpy gradient at the wall.

Newton's Method

If we assume f_j^{n-1}, u_j^{n-1}, v_j^{n-1}, g_j^{n-1}, and p_j^{n-1} to be known for $0 \le j \le J$, then Eqs. (13.56) and (13.60), unlike the previous examples of Sections 13.1 and 13.2, form a system of $5J+5$ nonlinear equations for the solution of $5J+5$ unknowns $(f_j^n, u_j^n, v_j^n, g_j^n, p_j^n)$, $j = 0, 1, \ldots, J$. To solve this nonlinear system, we use Newton's method; we introduce the iterates $[f_j^{(i)}, u_j^{(i)}, v_j^{(i)}, g_j^{(i)}, p_j^{(i)}]$, $i = 0, 1, 2, \ldots$, with initial values equal to those at the previous x station (which is usually the best initial guess available). For the higher iterates we set

$$f_j^{(i+1)} = f_j^{(i)} + \delta f_j^{(i)}, \qquad u_j^{(i+1)} = u_j^{(i)} + \delta u_j^{(i)},$$

$$v_j^{(i+1)} = v_j^{(i)} + \delta v_j^{(i)}, \qquad g_j^{(i+1)} = g_j^{(i)} + \delta g_j^{(i)},$$

$$p_j^{(i+1)} = p_j^{(i)} + \delta p_j^{(i)}. \tag{13.61}$$

We then insert the right-hand sides of these expressions in place of f_j, u_j, v_j, g_j, and p_j in Eqs. (13.56) and (13.60) and drop the terms that are quadratic in $\delta f_j^{(i)}$, $\delta u_j^{(i)}$, $\delta v_j^{(i)}$, $\delta g_j^{(i)}$, and $\delta p_j^{(i)}$. This procedure yields the following linear system (the superscript i in δ quantities is dropped for simplicity):

$$\delta f_j - \delta f_{j-1} - \frac{h_j}{2}(\delta u_j + \delta u_{j-1}) = (r_1)_j, \tag{13.62a}$$

$$\delta u_j - \delta u_{j-1} - \frac{h_j}{2}(\delta v_j + \delta v_{j-1}) = (r_4)_{j-1}, \tag{13.62b}$$

$$\delta g_j - \delta g_{j-1} - \frac{h_j}{2}(\delta p_j + \delta p_{j-1}) = (r_5)_{j-1}, \tag{13.62c}$$

$$(s_1)_j \delta v_j + (s_2)_j \delta v_{j-1} + (s_3)_j \delta f_j + (s_4)_j \delta f_{j-1} + (s_5)_j \delta u_j$$
$$+ (s_6)_j \delta u_{j-1} + (s_7)_j \delta g_j + (s_8)_j \delta g_{j-1} = (r_2)_j, \tag{13.62d}$$

$$(\beta_1)_j \delta p_j + (\beta_2)_j \delta p_{j-1} + (\beta_3)_j \delta f_j + (\beta_4)_j \delta f_{j-1} + (\beta_5)_j \delta u_j$$
$$+ (\beta_6)_j \delta u_{j-1} + (\beta_7)_j \delta g_j + (\beta_8)_j \delta g_{j-1} + (\beta_9)_j \delta v_j$$
$$+ (\beta_{10})_j \delta v_{j-1} = (r_3)_j. \tag{13.62e}$$

where

$$(r_1)_j = f_{j-1}^{(i)} - f_j^{(i)} + h_j u_{j-1/2}^{(i)}, \tag{13.63a}$$

$$(r_4)_{j-1} = u_{j-1}^{(i)} - u_j^{(i)} + h_j v_{j-1/2}^{(i)}, \tag{13.63b}$$

$$(r_5)_{j-1} = g_{j-1}^{(i)} - g_j^{(i)} + h_j p_{j-1/2}^{(i)}, \tag{13.63c}$$

$$(r_2)_j = R_{j-1/2}^{n-1} - \left[h_j^{-1} \left(b_j^{(i)} v_j^{(i)} - b_{j-1}^{(i)} v_{j-1}^{(i)} \right) + (m_1^n + \alpha_n)(fv)_{j-1/2}^{(i)} \right.$$
$$\left. - (m_2^n + \alpha_n)(u^2)_{j-1/2}^{(i)} + \alpha_n \left(v_{j-1/2}^{n-1} f_{j-1/2}^{(i)} - f_{j-1/2}^{n-1} v_{j-1/2}^{(i)} \right) \right], \tag{13.63d}$$

$$(r_3)_j = T_{j-1/2}^{n-1} - \left\{ h_j^{-1} \left(e_j^{(i)} p_j^{(i)} - e_{j-1}^{(i)} p_{j-1}^{(i)} \right) \right.$$
$$+ h_j^{-1} \left(d_j^{(i)} u_j^{(i)} v_j^{(i)} - d_{j-1}^{(i)} u_{j-1}^{(i)} v_{j-1}^{(i)} \right)$$
$$+ (m_1^n + \alpha_n)(fp)_{j-1/2}^{(i)}$$
$$- \alpha_n \left[(ug)_{j-1/2}^{(i)} + u_{j-1/2}^{n-1} g_{j-1/2}^{(i)} - g_{j-1/2}^{n-1} u_{j-1/2}^{(i)} \right.$$
$$\left. \left. + f_{j-1/2}^{n-1} p_{j-1/2}^{(i)} - p_{j-1/2}^{n-1} f_{j-1/2}^{(i)} \right] \right\}. \tag{13.63e}$$

In writing the system given by Eqs. (13.62) and (13.63), we have used a certain order for them. The reason for this order is to ensure that the $\mathbf{A}_0$ matrix in Eq. (13.29) is not singular. The present order satisfies this requirement. Of course other orderings can also be used as long as the $\mathbf{A}_0$ matrix is not singular.

The coefficients of the momentum equation are

$$(s_1)_j = h_j^{-1} b_j^{(i)} + \frac{m_1^n + \alpha_n}{2} f_j^{(i)} - \frac{\alpha_n}{2} f_{j-1/2}^{n-1}, \tag{13.64a}$$

$$(s_2)_j = -h_j^{-1} b_{j-1}^{(i)} + \frac{m_1^n + \alpha_n}{2} f_{j-1}^{(i)} - \frac{\alpha_n}{2} f_{j-1/2}^{n-1}, \tag{13.64b}$$

$$(s_3)_j = \frac{m_1^n + \alpha_n}{2} v_j^{(i)} + \frac{\alpha_n}{2} v_{j-1/2}^{n-1}, \tag{13.64c}$$

$$(s_4)_j = \frac{m_1^n + \alpha_n}{2} v_{j-1}^{(i)} + \frac{\alpha_n}{2} v_{j-1/2}^{n-1}, \tag{13.64d}$$

$$(s_5)_j = -(m_2^n + \alpha_n) u_j^{(i)}, \tag{13.64e}$$

$$(s_6)_j = -(m_2^n + \alpha_n) u_{j-1}^{(i)}, \tag{13.64f}$$

$$(s_7)_j = 0, \tag{13.64g}$$

$$(s_8)_j = 0. \tag{13.64h}$$

Note that the coefficients $(s_7)_j$ and $(s_8)_j$, which are zero in this case, are included here for the generality needed later (see Section 14.1). The coefficients of the energy equation are

$$(\beta_1)_j = h_j^{-1}e_j^{(i)} + \frac{m_1^n + \alpha_n}{2}f_j^{(i)} - \frac{\alpha_n}{2}f_{j-1/2}^{n-1},\qquad(13.65a)$$

$$(\beta_2)_j = -h_j^{-1}e_{j-1}^{(i)} + \frac{m_1^n + \alpha_n}{2}f_{j-1}^{(i)} - \frac{\alpha_n}{2}f_{j-1/2}^{n-1},\qquad(13.65b)$$

$$(\beta_3)_j = \frac{m_1^n + \alpha_n}{2}p_j^{(i)} + \frac{\alpha_n}{2}p_{j-1/2}^{n-1},\qquad(13.65c)$$

$$(\beta_4)_j = \frac{m_1^n + \alpha_n}{2}p_{j-1}^{(i)} + \frac{\alpha_n}{2}p_{j-1/2}^{n-1},\qquad(13.65d)$$

$$(\beta_5)_j = h_j^{-1}d_j^{(i)}v_j^{(i)} - \frac{\alpha_n}{2}\left(g_j^{(i)} - g_{j-1/2}^{n-1}\right),\qquad(13.65e)$$

$$(\beta_6)_j = -h_j^{-1}d_{j-1}^{(i)}v_{j-1}^{(i)} - \frac{\alpha_n}{2}\left(g_{j-1}^{(i)} - g_{j-1/2}^{n-1}\right),\qquad(13.65f)$$

$$(\beta_7)_j = -\frac{\alpha_n}{2}\left(u_j^{(i)} + u_{j-1/2}^{n-1}\right),\qquad(13.65g)$$

$$(\beta_8)_j = -\frac{\alpha_n}{2}\left(u_{j-1}^{(i)} + u_{j-1/2}^{n-1}\right),\qquad(13.65h)$$

$$(\beta_9)_j = h_j^{-1}d_j^{(i)}u_j^{(i)},\qquad(13.65i)$$

$$(\beta_{10})_j = -h_j^{-1}d_{j-1}^{(i)}u_{j-1}^{(i)}.\qquad(13.65j)$$

The boundary conditions, Eq. (13.60), become

$$\delta f_0 = 0,\qquad \delta u_0 = 0,\qquad \alpha_0\delta g_0 + \alpha_1\delta p_0 = 0,\qquad(13.66a)$$

$$\delta u_J = 0,\qquad \delta g_J = 0,\qquad(13.66b)$$

which just express the requirement for the boundary conditions to remain constant during the iteration process.

Block-Elimination Method

We next write the resulting linear system, Eqs. (13.62) and (13.66), in the matrix-vector form given by Eq. (13.28) and use the block-elimination method discussed in Sections 13.1 and 13.2 for two first-order equations. The block-elimination method is a general one and can be used for any number of first-order equations. The amount of algebra involved in writing the recursion formulas, however, gets more tedious as the order of the system goes up. For a system consisting of five first-order equations, as in our case here, the amount of algebra is reasonable.

Once the algorithm to solve the linear system given by Eqs. (13.62) and (13.66) is written by this method, the resulting algorithm can easily be modified to solve other linear systems consisting of five first-order equations with different coefficients and boundary conditions, as we shall see in Chapter 14.

We now define the vectors $\boldsymbol{\delta}_j$ and $\mathbf{r}_j$ in Eq. (13.28), for each value of j, by

$$\boldsymbol{\delta}_j \equiv \begin{bmatrix} \delta f_j \\ \delta u_j \\ \delta v_j \\ \delta g_j \\ \delta p_j \end{bmatrix}, \quad 0 \le j \le J, \quad \mathbf{r}_0 = \begin{bmatrix} 0 \\ 0 \\ 0 \\ (r_4)_1 \\ (r_5)_1 \end{bmatrix}, \quad (13.67a)$$

$$\mathbf{r}_j \equiv \begin{bmatrix} (r_1)_j \\ (r_2)_j \\ (r_3)_j \\ (r_4)_j \\ (r_5)_j \end{bmatrix}, \quad 1 \le j \le J-1, \quad \mathbf{r}_J \equiv \begin{bmatrix} (r_1)_J \\ (r_2)_J \\ (r_3)_J \\ 0 \\ 0 \end{bmatrix} \quad (13.67b)$$

and the 5×5 matrices $\mathbf{A}_j, \mathbf{B}_j, \mathbf{C}_j$ by

$$\mathbf{A}_0 \equiv \begin{bmatrix} 1 & 0 & 0 & 0 & 0 \\ 0 & 1 & 0 & 0 & 0 \\ 0 & 0 & 0 & \alpha_0 & \alpha_1 \\ 0 & -1 & -\dfrac{h_1}{2} & 0 & 0 \\ 0 & 0 & 0 & -1 & -\dfrac{h_1}{2} \end{bmatrix},$$

$$\mathbf{A}_J \equiv \begin{bmatrix} 1 & -\dfrac{h_J}{2} & 0 & 0 & 0 \\ (s_3)_J & (s_5)_J & (s_1)_J & (s_7)_J & 0 \\ (\beta_3)_J & (\beta_5)_J & (\beta_9)_J & (\beta_7)_J & (\beta_1)_J \\ 1 & 1 & 0 & 0 & 0 \\ 0 & 0 & 0 & 1 & 0 \end{bmatrix}, \quad (13.68a)$$

$$\mathbf{A}_j \equiv \begin{bmatrix} 1 & -\dfrac{h_j}{2} & 0 & 0 & 0 \\ (s_3)_j & (s_5)_j & (s_1)_j & (s_7)_j & 0 \\ (\beta_3)_j & (\beta_5)_j & (\beta_9)_j & (\beta_7)_j & (\beta_1)_j \\ 0 & -1 & -\dfrac{h_{j+1}}{2} & 0 & 0 \\ 0 & 0 & 0 & -1 & -\dfrac{h_{j+1}}{2} \end{bmatrix}, \quad 1 \le j \le J-1,$$

$$(13.68b)$$

$$
\mathbf{B}_j \equiv
\begin{bmatrix}
-1 & -\dfrac{h_j}{2} & 0 & 0 & 0 \\[2mm]
(s_4)_j & (s_6)_j & (s_2)_j & (s_8)_j & 0 \\[1mm]
(\beta_4)_j & (\beta_6)_j & (\beta_{10})_j & (\beta_8)_j & (\beta_2)_j \\[1mm]
0 & 0 & 0 & 0 & 0 \\[1mm]
0 & 0 & 0 & 0 & 0
\end{bmatrix},
\qquad 1 \le j \le J,
$$

$$(13.68c)$$

$$
\mathbf{C}_j \equiv
\begin{bmatrix}
0 & 0 & 0 & 0 & 0 \\[1mm]
0 & 0 & 0 & 0 & 0 \\[1mm]
0 & 0 & 0 & 0 & 0 \\[1mm]
0 & 1 & -\dfrac{h_{j+1}}{2} & 0 & 0 \\[2mm]
0 & 0 & 0 & 1 & -\dfrac{h_{j+1}}{2}
\end{bmatrix},
\qquad 0 \le j \le J-1.
$$

$$(13.68d)$$

In terms of these definitions, the solution of Eq. (13.28) can be obtained by the two sweeps given by Eqs. (13.31)–(13.33). Except for the algebra, the procedure is straightforward; the resulting algorithm, called SOLV5, is given below.

```
      SUBROUTINE SOLV5
      COMMON /INPT1/ WW(60),ALFA0,ALFA1
      COMMON /BLCO/ NP,NPT,NX,NXT,NTR,IT
      COMMON /GRD/ X(60),ETA(61),DETA(61),A(61)
      COMMON /BLC1/ F(61,2),U(61,2),V(61,2),B(61,2),G(61,2),P(61,2),
     1             C(61,2),D(61,2),E(61,2),RMU(61),BC(61)
      COMMON /BLC3/ DELF(61),DELU(61),DELV(61),DELG(61),DELP(61)
      COMMON/BLC6/ S1(61),S2(61),S3(61),S4(61),S5(61),S6(61),S7(61),
     1             S8(61),B1(61),B2(61),B3(61),B4(61),B5(61),B6(61),
     2             B7(61),B8(61),B9(61),B10(61),R(5,61)
      DIMENSION    A11(61),A12(61),A13(61),A14(61),A15(61),A21(61),
     1             A22(61),A23(61),A24(61),A25(61),A31(61),A32(61),
     2             A33(61),A34(61),A35(61),G11(61),G12(61),G13(61),
     3             G14(61),G15(61),G21(61),G22(61),G23(61),G24(61),
     4             G25(61),G31(61),G32(61),G33(61),G34(61),G35(61),
     5             W1(61),W2(61),W3(61),W4(61),W5(61)
C - - - - - - - - - - - - - - - - - - - - - - - - - - - - - - - - - - -
C   ELEMENTS OF TRIANGLE MATRIX, ACCORDING TO EQ.(13.68A)
      A11(1)=1.0
      A12(1)=0.0
      A13(1)=0.0
      A14(1)=0.0
      A15(1)=0.0
      A21(1)=0.0
      A22(1)=1.0
      A23(1)=0.0
      A24(1)=0.0
      A25(1)=0.0
      A31(1)=0.0
      A32(1)=0.0
      A33(1)=0.0
      A34(1)=ALFA0
      A35(1)=ALFA1
C   ELEMENTS O  W-VECTOR, ACCORDING TO EQ.(13.32A)
      W1 (1)=R(1,1)
      W2 (1)=R(2,1)
```

```
          W3  (1) =R (3,1)
          W4  (1) =R (4,1)
          W5  (1) =R (5,1)
C - FORWARD SWEEP
C   DEFINITIONS
          DO 30 J=2,NP
          AA1=A(J)*A24(J-1)-A25(J-1)
          AA2=A(J)*A34(J-1)-A35(J-1)
          AA3=A(J)*A12(J-1)-A13(J-1)
          AA4=A(J)*A22(J-1)-A23(J-1)
          AA5=A(J)*A32(J-1)-A33(J-1)
          AA6=A(J)*A14(J-1)-A15(J-1)
          AA7=A(J)*S6(J)-S2(J)
          AA8=S8(J)*A(J)
          AA9=A(J)*B6(J)-B10(J)
          AA10=A(J)*B8(J)-B2(J)
C   ELEMENTS OF TRIANGLE MATRIX, ACCORDING TO EQ.(13.31B)
          DET=A11(J-1)*(AA4*AA2-AA1*AA5)-A21(J-1)*(AA3*AA2-AA5*AA6)+
         1      A31(J-1)*(AA3*AA1-AA4*AA6)
          G11(J)=(-(AA4*AA2-AA5*AA1)+A(J)**2*(A21(J-1)*AA2-A31(J-1)*AA1))/
         1      DET
          G12(J)=((AA3*AA2-AA5*AA6)-A(J)**2*(A11(J-1)*AA2-A31(J-1)*AA6))/DET
          G13(J)=(-(AA3*AA1-AA4*AA6)+A(J)**2*(A11(J-1)*AA1-A21(J-1)*AA6))/
         1      DET
          G14(J)=G11(J)*A12(J-1)+G12(J)*A22(J-1)+G13(J)*A32(J-1)+A(J)
          G15(J)=G11(J)*A14(J-1)+G12(J)*A24(J-1)+G13(J)*A34(J-1)
          G21(J)=(S4(J)*(AA2*AA4-AA1*AA5)+A31(J-1)*(AA1*AA7-AA4*AA8)+
         1      A21(J-1)*(AA5*AA8-AA7*AA2))/DET
          G22(J)=(A11(J-1)*(AA2*AA7-AA5*AA8)+A31(J-1)*(AA3*AA8-AA6*AA7)+
         1      S4(J)*(AA5*AA6-AA2*AA3))/DET
          G23(J)=(A11(J-1)*(AA4*AA8-AA1*AA7)+S4(J)*(AA3*AA1-AA4*AA6)+
         1      A21(J-1)*(AA7*AA6-AA3*AA8))/DET
          G24(J)=G21(J)*A12(J-1)+G22(J)*A22(J-1)+G23(J)*A32(J-1)-S6(J)
          G25(J)=G21(J)*A14(J-1)+G22(J)*A24(J-1)+G23(J)*A34(J-1)-S8(J)
          G31(J)=(B4(J)*(AA4*AA2-AA5*AA1)-AA9*(A21(J-1)*AA2-A31(J-1)*AA1)+
         1      AA10*(A21(J-1)*AA5-A31(J-1)*AA4))/DET
          G32(J)=(-B4(J)*(AA3*AA2-AA5*AA6)+AA9*(A11(J-1)*AA2-A31(J-1)*AA6)-
         1      AA10*(A11(J-1)*AA5-A31(J-1)*AA3))/DET
          G33(J)=(B4(J)*(AA3*AA1-AA4*AA6)-AA9*(A11(J-1)*AA1-A21(J-1)*AA6)+
         1      AA10*(A11(J-1)*AA4-A21(J-1)*AA3))/DET
          G34(J)=G31(J)*A12(J-1)+G32(J)*A22(J-1)+G33(J)*A32(J-1)-B6(J)
          G35(J)=G31(J)*A14(J-1)+G32(J)*A24(J-1)+G33(J)*A34(J-1)-B8(J)
C   ELEMENTS OF TRIANGLE MATRIX, ACCORDING TO EQ.(13.31C)
          A11(J)=1.0
          A12(J)=-A(J)-G14(J)
          A13(J)=A(J)*G14(J)
          A14(J)=-G15(J)
          A15(J)=A(J)*G15(J)
          A21(J)=S3(J)
          A22(J)=S5(J)-G24(J)
          A23(J)=S1(J)+A(J)*G24(J)
          A24(J)=-G25(J)+S7(J)
          A25(J)=A(J)*G25(J)
          A31(J)=B3(J)
          A32(J)=B5(J)-G34(J)
          A33(J)=B9(J)+A(J)*G34(J)
          A34(J)=B7(J)-G35(J)
          A35(J)=B1(J)+A(J)*G35(J)
C   ELEMENTS OF W-VECTOR, ACCORDING TO EQ.(13.32B)
          W1(J) =R(1,J)-G11(J)*W1(J-1)-G12(J)*W2(J-1)-G13(J)*W3(J-1)-
         1      G14(J)*W4(J-1)-G15(J)*W5(J-1)
          W2(J) =R(2,J)-G21(J)*W1(J-1)-G22(J)*W2(J-1)-G23(J)*W3(J-1)-
         1      G24(J)*W4(J-1)-G25(J)*W5(J-1)
          W3(J) =R(3,J)-G31(J)*W1(J-1)-G32(J)*W2(J-1)-G33(J)*W3(J-1)-
         1      G34(J)*W4(J-1)-G35(J)*W5(J-1)
          W4(J) =R(4,J)
          W5(J) =R(5,J)
       30 CONTINUE
C - BACKWARD SWEEP
          J      =NP
C   DEFINITIONS
          DP =      -(A31(J)*(A13(J)*W2(J)-W1(J)*A23(J))-A32(J)*(A11(J)*
         1      W2(J)-W1(J)*A21(J))  +  W3(J)*(A11(J)*A23(J)-A13(J)*A21(J)))
          DV =      -(A31(J)*(W1(J)*A25(J)-W2(J)*A15(J))-W3(J)*(A11(J)*A25(J)
         1      -A15(J)*A21(J))+A35(J)*(A11(J)*W2(J)-W1(J)*A21(J)))
          DF =      -(W3(J)*(A13(J)*A25(J)-A23(J)*A15(J))-A33(J)*(W1(J)*A25(J)
```

```
     1     -A15(J)*W2(J)) + A35(J)*(W1(J)*A23(J)-A13(J)*W2(J)))
       D1 =    -(A31(J)*(A13(J)*A25(J)-A23(J)*A15(J))-A33(J)*(A11(J)*
     1     A25(J)-A21(J)*A15(J))+A35(J)*(A11(J)*A23(J)-A21(J)*A13(J)))
C  ELEMENTS OF DELTA-VECTOR FOR J=NP, ACCORDING TO EQ.(13.33A)
       DELP(J) = DP/D1
       DELV(J) = DV/D1
       DELF(J) = DF/D1
       DELG(J) = 0.0
       DELU(J) = 0.0
    40 J = J-1
C  DEFINITIONS
       BB1=DELU(J+1)-A(J+1)*DELV(J+1)-W4(J)
       BB2=DELG(J+1)-A(J+1)*DELP(J+1)-W5(J)
       CC1=W1(J)-A12(J)*BB1-A14(J)*BB2
       CC2=W2(J)-A22(J)*BB1-A24(J)*BB2
       CC3=W3(J)-A32(J)*BB1-A34(J)*BB2
       DD1=A13(J)-A12(J)*A(J+1)
       DD2=A23(J)-A22(J)*A(J+1)
       DD3=A33(J)-A32(J)*A(J+1)
       EE1=A15(J)-A14(J)*A(J+1)
       EE2=A25(J)-A24(J)*A(J+1)
       EE3=A35(J)-A34(J)*A(J+1)
       DETT=A11(J)*DD2*EE3+A21(J)*DD3*EE1+A31(J)*DD1*EE2
     1     -A31(J)*DD2*EE1-A21(J)*DD1*EE3-A11(J)*DD3*EE2
C  ELEMENTS OF DELTA-VECTOR, ACCORDING TO EQ.(13.33B)
       DELF(J)=(CC1*DD2*EE3+CC2*DD3*EE1+CC3*DD1*EE2-CC3*DD2*EE1
     1     -CC2*DD1*EE3-CC1*DD3*EE2)/DETT
       DELV(J)=(A11(J)*CC2*EE3+A21(J)*CC3*EE1+A31(J)*CC1*EE2-
     1     A31(J)*CC2*EE1-A21(J)*CC1*EE3-A11(J)*CC3*EE2)/DETT
       DELP(J)=(A11(J)*CC3*DD2+A21(J)*CC1*DD3+A31(J)*CC2*DD1-
     1     A31(J)*CC1*DD2-A21(J)*CC3*DD1-A11(J)*CC2*DD3)/DETT
       DELU(J)=BB1-A(J+1)*DELV(J)
       DELG(J)= BB2-A(J+1)*DELP(J)
       IF(J .GT. 1) GO TO 40
C  NEW VALUES OF F,U,V,G,P, ACCORDING TO EQ.(13.61)
       DO 50 J=1,NP
       F(J,2)=F(J,2)+DELF(J)
       U(J,2)=U(J,2)+DELU(J)
       V(J,2)=V(J,2)+DELV(J)
       G(J,2)=G(J,2)+DELG(J)
       P(J,2)=P(J,2)+DELP(J)
    50 CONTINUE
       U(1,2) = 0.0
       RETURN
       END
```

13.5 Fortran Program for Coupled Boundary-Layer Flows

In this section we present a Fortran program for solving the coupled mass, momentum, and energy equations for two-dimensional laminar and turbulent boundary layers by the numerical procedure described in the previous section. This program, coded in Fortran IV for the IBM 370/165 but containing no machine-dependent features such as "A" formats, can also be used to solve other fifth-order ordinary and partial differential equations, as will be described later in Chapter 14.

The program consists of a MAIN routine, which contains the overall logic of the computations, and also the subroutines INPUT, IVPL, EDDY, COEF, OUTPUT, and SOLV5. The following sections present a brief description of each routine except for SOLV5, which was presented in Section 13.4.

The Fortran names for the symbols, unless otherwise noted, follow closely the notation used for symbols in the text. All quantities are dimensionless unless noted. A flow diagram is shown in Fig. 13.8.

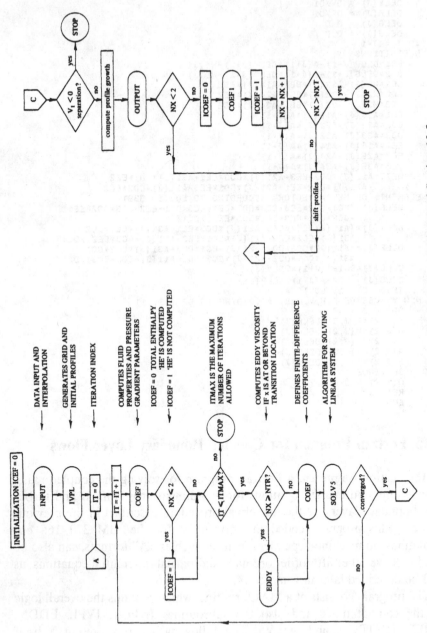

Figure 13.8. Flow diagram for the computer program of Section 13.5.

MAIN

This routine contains the overall logic of the computations, generates the grid normal to the flow, calculates the fluid properties, and accounts for the boundary-layer growth. It also checks the convergence of the iterations by using the wall shear parameter v_0 as the convergence criterion. For laminar flows, calculations are stopped when

$$|\delta v_0^{(i)}| < 10^{-5}, \tag{13.69}$$

which gives about four-figure accuracy for most predicted quantities. If v_0 becomes negative during any iteration, the calculations are stopped. For turbulent flows, Eq. (13.69) is expressed on a percentage basis and is written as

$$\left| \frac{\delta v_0^{(i)}}{v_0 + \delta v_0^{(i)}/2} \right| < 0.02. \tag{13.70}$$

The grid generated across the layer is identical to the one described in Section 13.3, with typical values of h_1 and K being equal to 0.2 and 1, respectively, for laminar flows and equal to 0.01 and 1.14, respectively, for turbulent flows. For a flow consisting of both laminar and turbulent flows, one can take them as 0.01 and 1.14.

The parameters μ, c, C, b, e, and d appearing in the momentum and energy equations are functions of static temperature. In this routine c, C, b, e, and d are defined by Eq. (10.7) for laminar flows and by Eq. (11.47) for turbulent flows. The dynamic viscosity μ is computed from Sutherland's formula, Eq. (1.4a), once the total-enthalpy ratio g is computed from the energy equation.

For most laminar boundary-layer flows the transformed boundary-layer thickness $\eta_e(x)$ is almost constant. A value of $\eta_e = 8$ is sufficient for the velocity to attain 0.9999 of its external-stream value. However, for turbulent boundary layers, $\eta_e(x)$ generally increases with increasing x. To determine an estimate of $\eta_e(x)$ we always require that $\eta_e(x_n) \geq \eta_e(x_{n-1})$. Then in routine MAIN, when the computations on $x = x_n$ (for any $n \geq 1$) have been completed, we test to see if $|v_J^n| \leq \varepsilon_v$, where $\eta_J = \eta_e(x_n)$ and, say, $\varepsilon_v = 10^{-3}$. If this test is satisfied, we set $\eta_e(x_{n+1}) = \eta_e(x_n)$. Otherwise, we set $J_{new} = J_{old} + t$, where t is a number of added points, say, $t = 3$. In this case we also specify values of f_j^n, u_j^n, v_j^n, g_j^n, p_j^n, b_j^n, and e_j^n for the new η_j points. We take the values of $u_J = 1$, $g_J = 1$, $v_j^n = v_J^n$, $p_j^n = p_J^n$, $c_j^n = c_J^n$, $d_j^n = d_J^n$, $e_j^n = e_J^n$, $\mu_j^n = \mu_J^n$, $C_j^n = C_J^n$, $f_j^n = (\eta_j - \eta_e)u_J^n + f_J^n$. This is also done for the same parameters of the previous station $(n-1)$.

Fortran name	Symbol
ITMAX	Iteration count
CEL, P1P, P2P	α_n, $m_1^n + \alpha_n$, $m_2^n + \alpha_n$, respectively
DELV(1)	δv_0
NP	J

```
C      MAIN
       COMMON /INPT2/ ETAE,VGP
       COMMON /BLC0/ NP,NPT,NX,NXT,NTR,IT
       COMMON /AK1/ RMUI,TI,RMI,UI,PR,HE
       COMMON /GRD/ X(60),ETA(61),DETA(61),A(61)
       COMMON /BLC3/ DELF(61),DELU(61),DELV(61),DELG(61),DELP(61)
       COMMON /BLC1/ F(61,2),U(61,2),V(61,2),B(61,2),G(61,2),P(61,2),
      1              C(61,2),D(61,2),E(61,2),RMU(61),BC(61)
       COMMON /EDGE/ UE(60),TE(60),RHOE(60),RMUE(60),PE(60),P1(60),P2(60)
C
C - - - - - - - - - - - - - - - - - - - - - - - - - - - - - - - - - - -
C
       NPT = 61
       ITMAX = 6
       NX    = 1
       CALL INPUT
C
C  GRID GENERATION
       IF((VGP-1.0) .LE. 0.001) GO TO 5
       NP    = ALOG((ETAE/DETA(1))*(VGP-1.0)+1.0)/ALOG(VGP) + 1.0001
       GO TO 10
    5  NP    = ETAE/DETA(1) + 1.0001
   10  IF(NP .LE. 61) GO TO 15
       WRITE(6,9000)
       STOP
   15  ETA(1)= 0.0
       DO 20 J=2,NPT
       DETA(J)=VGP*DETA(J-1)
       A(J)   = 0.5*DETA(J-1)
   20  ETA(J)= ETA(J-1)+DETA(J-1)
       CALL IVPL
C
   30  WRITE(6,9100) NX,X(NX)
       IT    = 0
   40  IT    = IT+1
C
C  FLUID PROPERTIES
       DO 50 J=1,NP
       T = (HE*G(J,2) - 0.5*(UE(NX)*U(J,2))**2)/1004.3
       RMU(J) = 1.45E-6*(T**1.5)/(T + 110.33)
       C(J,2) = T/TE(NX)
       BC(J) = RMU(J)/RMUE(NX)/C(J,2)
       D(J,2) = BC(J)*(UE(NX)**2)*(1.0 - 1.0/PR)/HE
C
       IF(NX .GE. NTR) GO TO 50
       B(J,2) = BC(J)
       E(J,2) = BC(J)/PR
   50  CONTINUE
       IF(IT .LE. ITMAX) GO TO 60
       WRITE(6,2500)
       STOP
C
   60  IF(NX .GE. NTR) CALL EDDY
       CALL COEF
       CALL SOLV5
       WRITE (6,3000) V(1,2), DELV(1)
C
C  CHECK FOR CONVERGENCE
       IF(NX .GE. NTR) GO TO 80
C   LAMINAR FLOW
       IF(ABS(DELV(1)) .GT. 1.0E-05) GO TO 40
       GO TO 100
C   TURBULENT FLOW
   80  IF(ABS(DELV(1)/(V(1,2)+0.5*DELV(1))) .GT. 0.02) GO TO 40
C
C  CHECK FOR GROWTH
  100  IF(V(1,2) .LT. 0.0) STOP
       IF(NP .EQ. NPT) GO TO 120
       IF(ABS(V(NP,2)) .LE. 1.0E-03) GO TO 120
       NPO   = NP
       NP1   = NP+1
       NP    = NP+1
       IF(NP .GT. NPT) NP = NPT
C
C  DEFINITION OF PROFILES FOR NEW NP
       DO 110 L=1,2
       DO 110 J=NP1,NP
```

```
      F(J,L) = F(NPO,L)  + ETA(J)  - ETA(NPO)
      U(J,L) = 1.0
      V(J,L) = V(NPO,L)
      B(J,L)= B(NPO,L)
      G(J,L)= G(NPO,L)
      P(J,L)= P(NPO,L)
      C(J,L) = C(NPO,L)
      D(J,L) = D(NPO,L)
      E(J,L) = E(NPO,L)
  110 CONTINUE
      IT    = 0
      GO TO 40
  120 CALL OUTPUT
      GO TO 30
C - - - - - - - - - - - - - - - - - - - - - - - - - - - - - - - - - -
 2500 FORMAT(1HO,16X,25HITERATIONS EXCEEDED ITMAX)
 3000 FORMAT(1H ,8HV(WALL)=,E13.6,1X,6HDELV =,E13.6)
 9000 FORMAT(1HO,'  NP EXCEEDED NPT -- PROGRAM TERMINATED')
 9100 FORMAT(1HO,4HNX =,I3,5X,3HX =,F10.3)
      END
```

Subroutine INPUT

This subroutine specifies the wall boundary conditions for the energy equation, the total number of x stations (NXT), the transition location (NTR), the dimensionless pressure gradient m_2 at the first x station P2(1), and the variable grid parameters h_1 and K. In addition, we specify the freestream parameters $M_\infty, T_\infty, p_\infty$ (RMI, TI, PI, respectively) and the molecular Prandtl number Pr. We also specify u_e/u_∞ as a function of surface distance x. The dimensionless pressure-gradient parameters m_1 and m_2 are computed numerically at x stations other than the first. The derivative du_e/dx is obtained by using three-point Lagrange interpolation formulas given for all stations except the first and last by

$$\left(\frac{du_e}{dx}\right)_n = -\frac{u_e^{n-1}}{A_1}(x_{n+1}-x_n)$$

$$+\frac{u_e^n}{A_2}(x_{n+1}-2x_n+x_{n-1})+\frac{u_e^{n+1}}{A_3}(x_n-x_{n-1}). \quad (13.71)$$

Here

$$A_1 = (x_n - x_{n-1})(x_{n+1}-x_{n-1}),$$

$$A_2 = (x_n - x_{n-1})(x_{n+1}-x_n),$$

$$A_3 = (x_{n+1}-x_n)(x_{n+1}-x_{n-1}).$$

The derivative du_e/dx at the endpoint $n = N$ is given by

$$\left(\frac{du_e}{dx}\right)_N = \frac{u_e^{N-2}}{A_1}(x_N-x_{N-1})-\frac{u_e^{N-1}}{A_2}(x_N-x_{N-2})$$

$$+\frac{u_e^N}{A_3}(2x_N-x_{N-2}-x_{N-1}), \quad (13.72)$$

where now

$$A_1 = (x_{N-1} - x_{N-2})(x_N - x_{N-2}),$$
$$A_2 = (x_{N-1} - x_{N-2})(x_N - x_{N-1}),$$
$$A_3 = (x_N - x_{N-1})(x_N - x_{N-2}).$$

The freestream values μ_∞, u_∞, ρ_∞, and H_e are calculated for air from the following formulas:

$$\mu_\infty = 1.45 \times 10^{-6} \frac{T_\infty^{3/2}}{T_\infty + 110} \qquad \text{kg m}^{-1}\text{s}^{-1}, \tag{13.73a}$$

$$u_\infty = 20.04 M_\infty \sqrt{T_\infty} \qquad \text{m s}^{-1}, \tag{13.73b}$$

$$\rho_\infty = \frac{p_\infty}{287 T_\infty} \qquad \text{kg m}^{-3}, \tag{13.73c}$$

$$H_e = 1004 T_\infty + \tfrac{1}{2} u_\infty^2 \qquad \text{m}^2 \text{ s}^{-2}, \tag{13.73d}$$

where the temperature is in Kelvin (K).

To compute the edge values of T_e and p_e, we use the following expressions:

$$\frac{T_e}{T_\infty} = 1 - \frac{\gamma - 1}{2} M_\infty^2 \left[\left(\frac{u_e}{u_\infty} \right)^2 - 1 \right], \tag{13.74a}$$

$$\frac{p_e}{p_\infty} = \left(\frac{T_e}{T_\infty} \right)^{\gamma/\gamma - 1} \tag{13.74b}$$

The edge values of μ and ρ are computed by using formulas identical to those given by Eq. (13.73) except that freestream values of temperature and pressure are replaced by their edge values.

The solution of the energy equation can be obtained either for specified wall temperature by reading $\alpha_0 = \text{ALFA0} = 1.0$, $\alpha_1 = \text{ALFA1} = 0.0$ or for specified heat flux by reading $\alpha_0 = 0.0$, $\alpha_1 = 1.0$. In addition, we read in the dimensionless wall temperature g_w or the dimensionless temperature gradient p_w through WW(I). Note that the read-in values of UE(I) are in dimensional form (m s^{-1}).

```
      SUBROUTINE INPUT
      COMMON /INPT2/ ETAE,VGP
      COMMON /AK1/ RMUI,TI,RMI,UI,PR,HE
      COMMON /INPT1/ WW(60),ALFA0,ALFA1
      COMMON /BLC0/ NP,NPT,NX,NXT,NTR,IT
      COMMON /GRD/ X(60),ETA(61),DETA(61),A(61)
      COMMON /EDGE/ UE(60),TE(60),RHOE(60),RMUE(60),PE(60),P1(60),P2(60)
C - - - - - - - - - - - - - - - - - - - - - - - - - - - - - - - - - -
C     ALFA0 = 1.0 ,   ALFA1 = 0.0    SPECIFIED WALL TEMP
C     ALFA0 = 0.0 ,   ALFA1 = 1.0    SPECIFIED HEAT FLUX
C     WW  EITHER G(1,2) FOR WALL TEMP. SPECIFIED
```

```
C           OR      P(1,2) FOR HEAT FLUX SPECIFIED
        ETAE = 8.0
        READ(5,8000) NXT,NTR,P2(1),DETA(1),VGP
        READ(5,8100) RMI,TI,PI,PR,ALFAO,ALFA1
        READ(5,8200) (X(I),UE(I),WW(I),I=1,NXT)
        WRITE(6,9010)
C    FREE-STREAM CONDITIONS
        RMUI = 1.45E-6*(TI**1.5)/(TI+110.33)
        UI = 20.04*RMI*SORT(TI)
        RHOI = PI/TI/287.0
        HE = 1004.3*TI + 0.5*UI**2
        WRITE(6,9100) RMI,TI,PI,UI,RHOI,HE
C    EDGE CONDITIONS
        DO 20 I=1,NXT
        UE(I) = UE(I)/UI
        TE(I) = TI*(1.0 - 0.2*(RMI**2)*(UE(I)**2 - 1.0))
        RMUE(I) = 1.45E-6*(TE(I)**1.5)/(TE(I)+110.33)
        PE(I) = PI*(TE(I)/TI)**3.5
        RHOE(I) = PE(I)/(287.0*TE(I))
        UE(I) = UI*UE(I)
20      CONTINUE
C    CALCULATION OF PRESSURE-GRADIENT PARAMETER P=P2
        P1(1) = 0.5*(P2(1) + 1.0)
        DO 80 I=2,NXT
        IF(I .EQ. NXT) GO TO 60
        A1    = (X(I)-X(I-1))*(X(I+1)-X(I-1))
        A2    = (X(I)-X(I-1))*(X(I+1)-X(I))
        A3    = (X(I+1)-X(I))*(X(I+1)-X(I-1))
        DUDS  = -(X(I+1)-X(I))/A1*UE(I-1) + (X(I+1)-2.0*X(I)+X(I-1))/
       1        A2*UE(I) + (X(I)-X(I-1))/A3*UE(I+1)
        GO TO 70
60      A1    = (X(I-1)-X(I-2))*(X(I)-X(I-2))
        A2    = (X(I-1)-X(I-2))*(X(I)-X(I-1))
        A3    = (X(I)-X(I-1))*(X(I)-X(I-2))
        DUDS  = (X(I)-X(I-1))/A1*UE(I-2) - (X(I)-X(I-2))/A2*UE(I-1) +
       1        (2.0*X(I)-X(I-2)-X(I-1))/A3*UE(I)
70      P2(I) = X(I)/UE(I)*DUDS
        P1(I) = 0.5*(1.0 + P2(I) + X(I)*(RHOE(I)*RMUE(I) - RHOE(I-1)*
       1        RMUE(I-1))/(X(I)-X(I-1))/RHOE(I)/RMUE(I))
80      CONTINUE
        WRITE(6,9200) (I,UE(I),TE(I),RHOE(I),RMUE(I),P1(I),P2(I),I=1,NXT)
        WRITE(6,9300)
        RETURN
C - - - - - - - - - - - - - - - - - - - - - - - - - - C - - - - - -
8000    FORMAT(2I3,3F10.0)
8100    FORMAT(8F10.0)
8200    FORMAT(3F10.0)
9000    FORMAT(1H0,6HNXT  =,I3,14X,6HNTR  =,I3/1H ,6HDETA1=,E14.6,3X,
       1        6HVGP  =,E14.6,3X,3HPR=,F7.3)
9010    FORMAT(1H0,30X,'INPUT DATA'///)
9100    FORMAT(1H0,'MACH NO =',E12.5,3X,'T INF. =',E12.5,3X,'PI =',
       1        E12.5,/1H0,'UI =',E12.5,3X,'RHOI =',E12.5,3X,'HE =',
       2        E12.5)
9300    FORMAT(1H0,30X,'OUTPUT DATA'///)
9200    FORMAT(1H0,2HNX,8X,2HUE,13X,2HTE,12X,4HRHOE,11X,4HRMUE,12X,2HP1,
       1        13X,2HP2/(1H ,I2,6E15.5))
        END
```

Subroutine IVPL

This subroutine is used to generate the initial velocity profiles for a similar compressible laminar flow. At first, however, the calculations are done for a similar incompressible flow in order to generate better profiles for the compressible flow by iteration. The total dimensionless enthalpy ratio g_j is taken as unity, and its derivative p_j is set equal to zero. To compute the initial profiles of f_j, u_j, and v_j, we use the velocity profile given by Eq. (4.59) and write

$$f_j = \frac{\eta_e}{4} \left(\frac{\eta_j}{\eta_e} \right)^2 \left[3 - \frac{1}{2} \left(\frac{\eta_j}{\eta_e} \right)^2 \right],$$

$$u_j = \frac{3}{2} \frac{\eta_j}{\eta_e} - \frac{1}{2} \left(\frac{\eta_j}{\eta_e} \right)^3,$$

$$v_j = \frac{3}{2} \frac{1}{\eta_e} \left[1 - \left(\frac{\eta_j}{\eta_e} \right)^2 \right].$$

The calculations for turbulent flow can also be started by writing a separate subroutine for the initial velocity and temperature profiles. For details see Bradshaw et al. [5].

```
      SUBROUTINE IVPL
      COMMON /AK1/ RMUI,TI,RMI,UI,PR,HE
      COMMON /BLCO/ NP,NPT,NX,NXT,NTR,IT
      COMMON /GRD/ X(60),ETA(61),DETA(61),A(61)
      COMMON /BLC1/ F(61,2),U(61,2),V(61,2),B(61,2),G(61,2),P(61,2),
     1              C(61,2),D(61,2),E(61,2),RMU(61),BC(61)
      COMMON /BLC3/ DELF(61),DELU(61),DELV(61),DELG(61),DELP(61)
C - - - - - - - - - - - - - - - - - - - - - - - - - - - - - - - - - -
C     GENERATE INITIAL PROFILE BY SOLVING THE INCOMPRESSIBLE FLOW
      ETANPQ= 0.25*ETA(NP)
      ETAU15= 1.5/ETA(NP)
      DO 30 J=1,NP
      ETAB  = ETA(J)/ETA(NP)
      ETAB2 = ETAB**2
      F(J,2)= ETANPQ*ETAB2*(3.0-0.5*ETAB2)
      U(J,2)= 0.5*ETAB*(3.0-ETAB2)
      V(J,2)= ETAU15*(1.0-ETAB2)
      G(J,2) = ETAB
      P(J,2) = 1.0/ETA(NP)
      B(J,2) = 1.0
      C(J,2) = 1.0
      E(J,2) = 1.0/PR
      D(J,2) = 0.0
   30 CONTINUE
      IT = 0
   50 IT = IT + 1
      IF(IT .LE. 8) GO TO 70
      WRITE(6,9900)
      STOP
   70 CONTINUE
      CALL COEF
      CALL SOLV5
      IF(ABS(DELV(1)) .GT. 1.0E-05) GO TO 50
      RETURN
C - - - - - - - - - - - - - - - - - - - - - - - - - - - - - - - - - -
 9900 FORMAT(1H0,'INCOMPRESSIBLE DID NOT CONVERGE')
      END
```

Subroutine EDDY

This subroutine contains the formulas used in the Cebeci–Smith eddy-viscosity model. For simplicity we do not include the low-Reynolds-number and the mass-transfer effects in these formulas. These capabilities, if desired, can easily be incorporated. With subscripts i and o denoting inner and outer regions, the formulas for eddy-viscosity expressions are given by

$$(\varepsilon_m)_i = L^2 \frac{\partial u}{\partial y} \gamma_{tr}\gamma, \qquad (\varepsilon_m)_i \leq (\varepsilon_m)_o, \qquad (13.75a)$$

$$(\varepsilon_m)_o = 0.0168 \left| \int_0^\infty (u_e - u) \, dy \right| \gamma_{tr}\gamma, \qquad (\varepsilon_m)_o \geq (\varepsilon_m)_i, \quad (13.75b)$$

where

$$L = 0.4y\left[1 - \exp\left(-\frac{y}{A}\right)\right], \qquad = 26\left(\frac{\rho}{\rho_w}\right)^{1/2} \frac{\nu_e}{N} u_\tau^{-1},$$

$$N^2 = 1 - 11.8 \frac{\mu_w}{\mu_e}\left(\frac{\rho_e}{\rho_w}\right)^2 p^+, \qquad u_\tau = \left(\frac{\tau_w}{\rho_w}\right)^{1/2}, \qquad p^+ = \frac{\nu_e u_e}{u_\tau^3}\frac{du_e}{dx} \qquad (13.76)$$

$$\gamma_{tr} = 1 - \exp\left[-G_{tr}(x - x_{tr})\int_{x_{tr}}^x \frac{dx}{u_e}\right], \qquad \gamma = \left[1 + 5.5\left(\frac{y}{\delta}\right)^6\right]^{-1},$$

$$G_{tr} = 8.33 \times 10^{-4} \frac{u_e^3}{\nu_e^2}(R_x)^{-1.34}, \qquad R_x = \frac{u_e x}{\nu_e}.$$

In terms of transformed variables these formulas become (noting that $v = f''$)

$$(\varepsilon_m^+)_i = \frac{0.16}{c^2}\frac{\mu_e}{\mu}\sqrt{R_x}\, I^2 v^2\left[1 - \exp\left(-\frac{y}{A}\right)\right]^2 \gamma_{tr}\gamma, \qquad (13.77a)$$

$$(\varepsilon_m^+)_o = \frac{0.0168}{c}\frac{\mu_e}{\mu}\sqrt{R_x}\left[\int_0^{\eta_e} c(1 - u)\, d\eta\right]\gamma_{tr}\gamma, \qquad (13.77b)$$

where with $\varepsilon_m^+ = \varepsilon_m/\nu$,

$$I = \int_0^\eta c\, d\eta, \qquad \frac{y}{A} = \frac{N}{26} c^{-3/2}\frac{C_w}{C} R_x^{1/4} I v_w^{1/2},$$

$$N^2 = 1 - 11.8\frac{\mu_w}{\mu_e}C_w^2 p^+, \qquad p^+ = \frac{m_2}{R_x^{1/4}}\left(\frac{\mu_e}{\mu_w}\right)^{3/2}\frac{1}{v_w^{3/2}}.$$

```
      SUBROUTINE EDDY
      COMMON /OTPT1/ RX,CNUE
      COMMON /AK1/ RMUI,TI,RMI,UI,PR,HE
      COMMON /BLCO/ NP,NPT,NX,NXT,NTR,IT
      COMMON /GRD/ X(60),ETA(61),DETA(61),A(61)
      COMMON /BLC1/ F(61,2),U(61,2),V(61,2),B(61,2),G(61,2),P(61,2),
     1              C(61,2),D(61,2),E(61,2),RMU(61),BC(61)
      COMMON /EDGE/ UE(60),TE(60),RHOE(60),RMUE(60),PE(60),P1(60),P2(60)
      DATA ITRN,PRT/0,0.9/
C     ----------------
      IED = 0
      IF(NX .NE. NTR)   GO TO 5
      RXNTR = UE(NTR-1)*X(NTR-1)*RHOE(NTR-1)/RMUE(NTR-1)
      GG = 8.33E-4*UE(NTR-1)**3/(CNUE**2*(RXNTR**1.34))
    5 IF(IT .GT. 1) GO TO 30
      IF(ITRN .EQ. 1) GO TO 30
C     CALCULATE GAMMA TRANSITION
      UEING = 0.0
      U1 = 1.0/UE(NTR-1)
      DO 10 I=NTR,NX
      U2 = 1.0/UE(I)
      UEING = UEING + 0.5*(U1+U2)*(X(I) - X(I-1))
      U1 = U2
   10 CONTINUE
      ROING = X(NX) - X(NTR-1)
```

```
      EXPT = GG*ROING*UEING
      IF(EXPT .LE. 10.0) GO TO 20
      ITRN = 1
      GAMTR = 1.0
      GO TO 30
  20  GAMTR = 1.0 - EXP(-EXPT)
  30  CONTINUE
      TERMP = C(1,2)
      SUM = 0.0
      CII = 0.0
      DO 40 J=2,NP
      CII = CII + A(J)*(C(J-1,2) + C(J,2))
      TERM = C(J,2)*(1.0 - U(J,2))
      SUM = SUM + A(J)*(TERM+TERMP)
      TERMP = TERM
  40  CONTINUE
      CI = 0.0
      PPLUS = P2(NX)/(RX**0.25)*(RMU(NP)/(RMU(1)*V(1,2)))**1.5
      CN = SQRT(1.0 - 11.8*PPLUS*RMU(1)/RMU(NP)*C(1,2)**2)
      CYA = CN*(RX**0.25)*SQRT(BC(1)*V(1,2))/26.0
      DO 100 J=2,NP
      CI = CI + A(J)*(C(J-1,2) + C(J,2))
      EDVO = 0.0168*SUM*(RX**0.5)/BC(J)/C(J,2)**2
      GAMINT = 1.0/(1.0 + 5.5*(CI/CII)**6)
      EDVO = EDVO*GAMTR*GAMINT
      IF(IED .EQ. 1) GO TO 50
      YOA = CYA*CI/BC(J)/C(J,2)**1.5
      EL = 1.0
      IF(YOA .LE. 10.0) EL = 1.0 - EXP(-YOA)
      EDVI = 0.16*SQRT(RX)*GAMTR*V(J,2)*(CI*EL)**2/BC(J)/C(J,2)**3
      IF(EDVI .LT. EDVO) GO TO 70
      IED = 1
  50  EDV = EDVO
      GO TO 90
  70  EDV = EDVI
  90  B(J,2) = (1.0 + EDV)*BC(J)
      E(J,2) = BC(J)*((1.0 + EDV*PR/PRT)/PR)
 100  CONTINUE
      B(1,2) = BC(1)
      E(1,2) = BC(1)/PR
      RETURN
      END
```

Subroutine COEF

This subroutine contains the coefficients of the linearized momentum and energy equations written in the form given by Eq. (13.62). The Fortran notation for some of the typical parameters are given in the accompanying table.

Fortran name	Symbol
FB, UB, GB, CB	$f^n_{j-1/2}, u^n_{j-1/2}, g^n_{j-1/2}, c^n_{j-1/2}$
USB, FVB, FPB	$(u^2)^n_{j-1/2}, (fv)^n_{j-1/2}, (fp)^n_{j-1/2}$
DERBV	$[(bv)^n_j - (bv)^n_{j-1}]h^{-1}_j$
CFB, CUB, CGB, CCB	$f^{n-1}_{j-1/2}, u^{n-1}_{j-1/2}, g^{n-1}_{j-1/2}, c^{n-1}_{j-1/2}$
CDEREP	$h^{-1}_j[(ep)^{n-1}_j - (ep)^{n-1}_{j-1}]$
CRB, CLB	$R^{n-1}_{j-1/2}, L^{n-1}_{j-1/2}$ [see Eqs. (13.58a) and (13.58b)]
CTB, CMB	$T^{n-1}_{j-1/2}, M^{n-1}_{j-1/2}$ [see Eqs. (13.59a) and (13.59)]
S1(J) to S8(J)	$(s_1)_j$ to $(s_8)_j$ [see Eq. (13.64)]
B1(J) to B10(J)	$(\beta_1)_j$ to $(\beta_{10})_j$ [see Eq. (13.65)]
R(1,J) to R(5,J)	$(r_1)_j$ to $(r_5)_j$ [see Eq. (13.63)]

```
      SUBROUTINE COEF
      COMMON /INPT1/ WW(60),ALFA0,ALFA1
      COMMON /BLCO/ NP,NPT,NX,NXT,NTR,IT
      COMMON /BLC1/ F(61,2),U(61,2),V(61,2),B(61,2),G(61,2),P(61,2),
     1              C(61,2),D(61,2),E(61,2),RMU(61),BC(61)
      COMMON /GRD/ X(60),ETA(61),DETA(61),A(61)
      COMMON /EDGE/ UE(60),TE(60),RHOE(60),RMUE(60),PE(60),P1(60),P2(60)
      COMMON/BLC6/ S1(61),S2(61),S3(61),S4(61),S5(61),S6(61),S7(61),
     1             S8(61),B1(61),B2(61),B3(61),B4(61),B5(61),B6(61),
     2             B7(61),B8(61),B9(61),B10(61),R(5,61)
C - - - - - - - - - - - - - - - - - - - - - - - - - - - - - - - - - - -
      IF(IT .GT. 1) GO TO 5
      CEL = 0.0
      IF(NX .GT. 1) CEL = 0.5*(X(NX)+X(NX-1))/(X(NX)-X(NX-1))
      P1P = P1(NX) + CEL
      P2P = P2(NX) + CEL
      IF(ALFA0 .GT. 0.1) G(1,2) = WW(NX)
      IF(ALFA1 .GT. 0.1) P(1,2) = WW(NX)
    5 CONTINUE
      DO 100 J=2,NP
C  PRESENT STATION
      USB   = 0.5*(U(J,2)**2+U(J-1,2)**2)
      FVB   = 0.5*(F(J,2)*V(J,2)+F(J-1,2)*V(J-1,2))
      FPB   = 0.5*(F(J,2)*P(J,2)+F(J-1,2)*P(J-1,2))
      UGB   = 0.5*(U(J,2)*G(J,2)+U(J-1,2)*G(J-1,2))
      UB    = 0.5*(U(J,2)+U(J-1,2))
      VB    = 0.5*(V(J,2)+V(J-1,2))
      FB    = 0.5*(F(J,2)+F(J-1,2))
      GB    = 0.5*(G(J,2)+G(J-1,2))
      PB    = 0.5*(P(J,2)+P(J-1,2))
      CB = 0.5*(C(J,2) + C(J-1,2))
      DERBV = (B(J,2)*V(J,2)-B(J-1,2)*V(J-1,2))/DETA(J-1)
      DEREP = (E(J,2)*P(J,2)-E(J-1,2)*P(J-1,2))/DETA(J-1)
      DRDUV = (D(J,2)*U(J,2)*V(J,2) - D(J-1,2)*U(J-1,2)*V(J-1,2))/
     1          DETA(J-1)
      IF(NX .GT. 1) GO TO 10
C  PREVIOUS STATION
      CFB    = 0.0
      CVB    = 0.0
      CPB    = 0.0
      CUB    = 0.0
      CGB    = 0.0
      CUGB   = 0.0
      CFPB   = 0.0
      CFVB   = 0.0
      CUSB   = 0.0
      CDERBV = 0.0
      CDEREP = 0.0
      CRB = -P2(NX)*CB
      CTB = 0.0
      GO TO 20
   10 CFB    = 0.5*(F(J,1)+F(J-1,1))
      CVB    = 0.5*(V(J,1)+V(J-1,1))
      CPB    = 0.5*(P(J,1)+P(J-1,1))
      CUB    = 0.5*(U(J,1)+U(J-1,1))
      CGB    = 0.5*(G(J,1)+G(J-1,1))
      CFVB   = 0.5*(F(J,1)*V(J,1)+F(J-1,1)*V(J-1,1))
      CFPB   = 0.5*(F(J,1)*P(J,1)+F(J-1,1)*P(J-1,1))
      CUGB   = 0.5*(U(J,1)*G(J,1)+U(J-1,1)*G(J-1,1))
      CUSB   = 0.5*(U(J,1)**2+U(J-1,1)**2)
      CCB = 0.5*(C(J,1) + C(J-1,1))
      CDERBV= (B(J,1)*V(J,1)-B(J-1,1)*V(J-1,1))/DETA(J-1)
      CDEREP= (E(J,1)*P(J,1)-E(J-1,1)*P(J-1,1))/DETA(J-1)
      CDRDUV= (D(J,1)*U(J,1)*V(J,1) - D(J-1,1)*U(J-1,1)*V(J-1,1))/
     1          DETA(J-1)
      CLB = CDERBV + P1(NX-1)*CFVB + P2(NX-1)*(CCB-CUSB)
      CRB = -CLB - P2(NX)*CB - CEL*CUSB + CEL*CFVB
      CMB = CDEREP + CDRDUV + P1(NX-1)*CFPB
      CTB = -CMB + CEL*(CFPB-CUGB)
C  COEFFICIENTS OF THE DIFFERENCED MOMENTUM EQ.
   20 CONTINUE
      S1(J) = B(J,2)/DETA(J-1) + 0.5*P1P*F(J,2) - 0.5*CEL*CFB
      S2(J) = -B(J-1,2)/DETA(J-1) + 0.5*P1P*F(J-1,2) - 0.5*CEL*CFB
      S3(J) = 0.5*(P1P*V(J,2) + CEL*CVB)
      S4(J) = 0.5*(P1P*V(J-1,2) + CEL*CVB)
      S5(J) = -P2P*U(J,2)
```

```
       S6(J) = -P2P*U(J-1,2)
       S7(J) = 0.0
       S8(J) = 0.0
       R(2,J) = CRB - (DERBV + P1P*FVB - P2P*USB + CEL*(FB*CVB-VB*CFB))
C   COEFFICIENTS OF DIFFERENCED ENERGY EQ.
       B1(J) = E(J,2)/DETA(J-1) + 0.5*P1P*F(J,2) - 0.5*CEL*CFB
       B2(J) = -E(J-1,2)/DETA(J-1) + 0.5*P1P*F(J-1,2) - 0.5*CEL*CFB
       B3(J) = 0.5*(P1P*P(J,2) + CEL*CPB)
       B4(J) = 0.5*(P1P*P(J-1,2) + CEL*CPB)
       B5(J) = D(J,2)*V(J,2)/DETA(J-1) - 0.5*CEL*(G(J,2)-CGB)
       B6(J) = -D(J-1,2)*V(J-1,2)/DETA(J-1) - 0.5*CEL*(G(J-1,2)-CGB)
       B7(J) = -0.5*CEL*(U(J,2)+CUB)
       B8(J) = -0.5*CEL*(U(J-1,2)+CUB)
       B9(J) = D(J,2)*U(J,2)/DETA(J-1)
       B10(J) = -D(J-1,2)*U(J-1,2)/DETA(J-1)
       R(3,J) = CTB - (DEREP + DRDUV + P1P*FPB - CEL*(UGB-CGB*UB+CUB*GB) +
      1             CEL*(CPB*FB-CFB*PB))
C   DEFINITIONS OF RJ
       R(1,J) = F(J-1,2)-F(J,2)+DETA(J-1)*UB
       R(4,J-1) = U(J-1,2)-U(J,2)+DETA(J-1)*VB
       R(5,J-1) = G(J-1,2)-G(J,2)+DETA(J-1)*PB
  100 CONTINUE
       R(1,1) = 0.0
       R(2,1) = 0.0
       R(3,1) = 0.0
       R(4,NP)=0.0
       R(5,NP)=0.0
       RETURN
       END
```

Subroutine OUTPUT

This subroutine prints out the desired profiles such as f_j, u_j, v_j, g_j, p_j, and b_j as functions of η_j. It also computes the boundary-layer parameters θ, δ^*, H, c_f, Nu_x, St_x, R_θ, R_{δ^*}, and R_x. The definitions of these boundary-layer parameters are summarized below for completeness.

$$\theta = \int_0^\infty \frac{\rho u}{\rho_e u_e}\left(1 - \frac{u}{u_e}\right) dy = \frac{x}{\sqrt{R_x}} \int_0^\infty f'(1 - f') \, d\eta. \quad (13.78a)$$

$$\delta^* = \int_0^\infty \left(1 - \frac{\rho u}{\rho_e u_e}\right) dy = \frac{x}{\sqrt{R_x}} \int_0^\infty (c - f') \, d\eta. \quad (13.78b)$$

$$H = \frac{\delta^*}{\theta}. \quad (13.78c)$$

$$c_f = \frac{\tau_w}{\frac{1}{2}\rho_e u_e^2} = \frac{2C_w}{\sqrt{R_x}} f_w''. \quad (13.78d)$$

$$\mathrm{Nu}_x = \frac{\dot{q}_w x}{(T_w - T_e)k} = \frac{C_w g_w' \sqrt{R_x}}{1 - g_w} \quad (13.78e)$$

$$\mathrm{St}_x = \frac{\dot{q}_w}{\rho_e u_e (H_w - H_e)} = \frac{C_w g_w'}{\mathrm{Pr}\sqrt{R_x}(1 - g_w)}. \quad (13.78f)$$

$$R_\theta = \frac{u_e \theta}{v_e}, \qquad R_{\delta^*} = \frac{u_e \delta^*}{v_e}, \qquad R_x = \frac{u_e x}{v_e}. \quad (13.78g)$$

```
      SUBROUTINE OUTPUT
      COMMON /OTPT1/ RX,CNUE
      COMMON /AK1/ RMUI,TI,RMI,UI,PR,HE
      COMMON /BLCO/ NP,NPT,NX,NXT,NTR,IT
      COMMON /GRD/ X(60),ETA(61),DETA(61),A(61)
      COMMON /BLC1/ F(61,2),U(61,2),V(61,2),B(61,2),G(61,2),P(61,2),
     1             C(61,2),D(61,2),E(61,2),RMU(61),BC(61)
      COMMON /EDGE/ UE(60),TE(60),RHOE(60),RMUE(60),PE(60),P1(60),P2(60)
C - - - - - - - - - - - - - - - - - - - - - - - - - - - - - - - - - -
      WRITE(6,4400)
      WRITE(6,4500) (J,ETA(J),F(J,2),U(J,2),V(J,2),G(J,2),P(J,2),
     1             B(J,2),J=1,NP,5)
      IF(J .NE. (NP+5)) WRITE(6,4500) NP,ETA(NP),F(NP,2),U(NP,2),
     1       V(NP,2),G(NP,2),P(NP,2),B(NP,2)
      IF(NX .EQ. 1) GO TO 210
      TERMP = 0.0
      SUM = 0.0
      SUMC = 0.0
      DO 20 J=2,NP
      SUMC = SUMC + A(J)*(C(J,2) + C(J-1,2))
      TERM = U(J,2)*(1.0 - U(J,2))
      SUM = SUM + A(J)*(TERM + TERMP)
      TERMP = TERM
   20 CONTINUE
      DELS = X(NX)/SQRT(RX)*(SUMC - F(NP,2))
      RDELS = SQRT(RX)*(SUMC - F(NP,2))
      THETA = X(NX)/SQRT(RX)*SUM
      RTHETA = SQRT(RX)*SUM
      H = DELS/THETA
      CF2 = BC(1)*V(1,2)/SQRT(RX)
      IF(ABS(G(1,2)-1.0) .LT. 0.00001) GO TO 200
      ANUX = P(1,2)*BC(1)*SQRT(RX)/(1.0 - G(1,2))
      STX = ANUX/RX/PR
      WRITE(6,9000) RX,DELS,THETA,ANUX,H,CF2,RDELS,RTHETA,STX
      GO TO 210
  200 CONTINUE
      WRITE(6,9100) RX,DELS,THETA,H,CF2,RDELS,RTHETA
  210 NX    = NX+1
      IF(NX .GT. NXT) STOP
      CNUE = RMUE(NX)/RHOE(NX)
      RX = UE(NX)*X(NX)/CNUE
C  SHIFT PROFILES
      DO 250 J=1,NP
      F(J,1)= F(J,2)
      U(J,1)= U(J,2)
      V(J,1)= V(J,2)
      G(J,1)= G(J,2)
      P(J,1)= P(J,2)
      E(J,1)= E(J,2)
      C(J,1) = C(J,2)
      D(J,1) = D(J,2)
  250 B(J,1)= B(J,2)
      RETURN
C - - - - - - - - - - - - - - - - - - - - - - - - - - - - - - - - - -
 4400 FORMAT(1H0,2X,1HJ,4X,3HETA,9X,1HF,13X,1HU,13X,1HV,13X,1HG,13X,1HP,
     1       13X,1HB)
 4500 FORMAT(1H ,I3,F10.3,6E14.6)
 9000 FORMAT(1H0,5HRX  =,E12.5,3X,6HDELS =,E12.5,3X,7HTHETA =,E12.5,
     1       3X,'NUSSELT NO=',E12.5,3X,2HH=,E12.5/1H0,5HCF/2=,E12.5,3X,
     2       6HRDELS=,E12.5,3X,7HRTHETA=,E12.5,3X,'STANTON NO =',E12.5/
     3       1H0,'  ---------------------------------------'/)
 9100 FORMAT(1H0,5HRX  =,E12.5,3X,6HDELS =,E12.5,3X,7HTHETA =,E12.5,
     1       3X,2HH=,E12.5/1H0,5HCF/2=,E12.5,3X,6HRDELS=,E12.5,3X,
     2       7HRTHETA=,E12.5/1H0,'----------------------------------'/)
      END
```

References

[1] Cebeci, T. and Bradshaw, P.: *Momentum Transfer in Boundary Layers*. Hemisphere, Washington, DC, 1977.

[2] Crank, J. and Nicolson, P.: A practical method for numerical evaluation of solutions of partial-differential equations of the heat-conduction type. *Proc. Camb. Phil. Soc.* **43**:50 (1947).

[3] Keller, H. B.: A new difference scheme for parabolic problems, in *Numerical Solution of Partial-Differential Equations* (J. Bramble, ed.), Vol. II. Academic, New York, 1970.

[4] Sellars, J. R., Tribus, M., and Klein, J. S.: Heat transfer to laminar flow in a round tube or flat conduit—The Graetz Problem extended. *Trans ASME* **78**:441 (1956).

[5] Bradshaw, P., Cebeci, T., and Whitelaw, J. H.: *Engineering Calculation Methods for Turbulent Flows*. Academic, London, 1981.

CHAPTER 14

Applications of a Computer Program to Heat-Transfer Problems

Sections 13.4 and 13.5 described the numerical method and computer program used to obtain the boundary-layer flow results of Chapters 4 to 12. The following four sections of this chapter describe the steps required to obtain results for four new problems, in sufficient detail that a reader may reproduce earlier results for him- or herself and make use of the program to solve new problems. It should be remembered that the program is limited to steady two-dimensional boundary-layer equations; it can, however, be extended to steady axisymmetric flows, and also to unsteady flows using the procedures described, for example, by Bradshaw et al. [1]. The accuracy of solutions can only be as good as the numerical approximations and physical assumptions; the Box scheme, with a reasonable choice of net spacing, guarantees second-order numerical accuracy, and the well-tested Cebeci–Smith model of turbulence has been shown to be adequate for a wide range of turbulent boundary-layer flows [2], though in difficult cases (for example, with turbulent buoyant flows) it may only give qualitatively correct results.

The physical situations of the following four sections correspond to (1) an internal laminar flow with natural and forced convection, (2) a vertical wall jet, again with laminar flow and with natural and forced convection, (3) a heated turbulent jet, and (4) a laminar mixing layer between two uniform streams at different temperatures. Thus, laminar and turbulent flows are represented with wall, free, and symmetry boundary conditions covering most of the problems likely to be encountered in more complicated flows.

14.1 Forced and Free Convection between Two Vertical Parallel Plates

To illustrate the applications of the computer program of Section 13.5, we first consider the forced- and free-convection flow between two vertical parallel plates discussed in Section 9.4. To keep the coding simple, we assume that both vertical surfaces are heated to the same uniform wall temperature, that the flow is laminar, and that neither the velocity nor the temperature boundary layers have merged. It is not difficult to obtain solutions for different boundary conditions, including nonuniform temperature distributions or specified heat flux, and for velocity and temperature boundary layers that have merged. The method can also be easily extended to transitional and turbulent flows with a suitable turbulence model.

The differential equations and the boundary conditions that represent the flow in terms of transformed coordinates correspond to Eqs. (9.79)–(9.82); that is,

$$f''' + \frac{1}{2}ff'' = \xi\left(\frac{dw}{d\xi} + f'\frac{\partial f'}{\partial \xi} - f''\frac{\partial f}{\partial \xi}\right) + \xi\theta, \qquad (9.79)$$

$$\frac{\theta''}{\mathrm{Pr}} + \frac{1}{2}f\theta' = \xi\left(f'\frac{\partial \theta}{\partial \xi} - \theta'\frac{\partial f}{\partial \xi}\right), \qquad (9.80)$$

$$\eta = 0, \qquad f = f' = 0, \qquad \theta = 1, \qquad (9.81a)$$

$$\eta = \eta_e, \qquad f' = \bar{u}_e, \qquad \theta = 0, \qquad (9.81b)$$

$$\bar{u}_e(\eta_{sp} - \eta_e) + f_e = \eta_{sp} = \sqrt{\frac{\mathrm{Gr}_L}{R_L}\frac{1}{\xi}}, \qquad (9.82b)$$

where $w \equiv p^*$ and the relation between $\bar{u}_e$ and w follows from Bernoulli's equation written as

$$\frac{d}{d\xi}\left(\frac{\bar{u}_e^2}{2}\right) \equiv -\frac{dw}{d\xi}. \qquad (14.1)$$

The above system of equations can be solved by either the nonlinear eigenvalue method or by the Mechul method discussed by Bradshaw et al. [1]. The former method requires fewer changes in the computer program of Section 13.5 and is described below. It should be noted that with wall cooling in upward flow (as opposed to the wall heating implied in the present problem), flow separation can occur, in which case the Mechul method must be used, as discussed, for example, by Cebeci et al. [3].

Numerical Formulation

As in Section 13.4, the transformed equations are written in terms of a first-order system. With the definitions of Eqs. (13.54a)–(13.54c) and Eq.

(14.1) and g replacing θ, Eqs. (9.79) to (8.81) become

$$v' + \frac{1}{2}fv = \xi\left(\frac{dw}{d\xi} + u\frac{\partial u}{\partial \xi} - v\frac{\partial f}{\partial \xi}\right) + \xi g, \tag{14.2}$$

$$\frac{p'}{\text{Pr}} + \frac{1}{2}fp = \xi\left(u\frac{\partial g}{\partial \xi} - p\frac{\partial f}{\partial \xi}\right), \tag{14.3}$$

$$\eta = 0, \qquad f = u = 0, \qquad g = 1, \tag{14.4a}$$

$$\eta = \eta_e, \qquad u = \bar{u}_e, \qquad g = 0, \tag{14.4b}$$

where the relation between $\bar{u}_e$ and w is given by Eq. (14.1).

The finite-difference approximations for Eqs. (14.2) and (14.3) are written in the same manner as Eqs. (13.56d) and (13.56e); that is,

$$h_j^{-1}(v_j^n - v_{j-1}^n) + \left(\frac{1}{2} + \alpha_n\right)(fv)_{j-1/2}^n - \alpha_n(u^2)_{j-1/2}^n$$

$$+ \alpha_n\left(v_{j-1/2}^{n-1}f_{j-1/2}^n - f_{j-1/2}^{n-1}v_{j-1/2}^n\right) - \lambda\xi_{n-1/2}g_{j-1/2}^n = R_{j-1/2}^{n-1}, \tag{14.5}$$

$$h_j^{-1}(p_j^n - p_{j-1}^n) + \left(\frac{1}{2} + \alpha_n\right)(fp)_{j-1/2}^n - \alpha_n\left[(ug)_{j-1/2}^n + u_{j-1/2}^{n-1}g_{j-1/2}^n\right.$$

$$\left. - g_{j-1/2}^{n-1}u_{j-1/2}^n + f_{j-1/2}^{n-1}p_{j-1/2}^n - p_{j-1/2}^{n-1}f_{j-1/2}^n\right] = T_{j-1/2}^{n-1}, \tag{14.6}$$

where α_n (with x replaced by ξ) is given by Eq. (13.57), $\beta = w^{n-1} - w^n$, $T_{j-1/2}^{n-1}$ is given by Eq. (13.59a), and $R_{j-1/2}^{n-1}$, $L_{j-1/2}^{n-1}$, and $M_{j-1/2}^{n-1}$ are given by

$$R_{j-1/2}^{n-1} = -L_{j-1/2}^{n-1} + \lambda\xi_{n-1/2}g_{j-1/2}^{n-1} + \alpha_n\left[(fv)_{j-1/2}^{n-1} - (u^2)_{j-1/2}^{n-1} - 2\beta\right], \tag{14.7a}$$

$$L_{j-1/2}^{n-1} = \left[h_j^{-1}(v_j - v_j) + \frac{1}{2}(fv)_{j-1/2}\right]^{n-1}, \tag{14.7b}$$

$$M_{j-1/2}^{n-1} = \left[h_j^{-1}(p_j - p_{j-1}) + \frac{1}{2}(fp)_{j-1/2}\right]^{n-1}. \tag{14.7c}$$

Similarly, the wall boundary conditions with $\alpha_0 = 1$ and $\alpha_1 = 0$ [see Eq. (13.60a)] may be written as

$$f_0^n = u_0^n = 0, \qquad g_0^n = 1. \tag{14.8a}$$

The edge boundary conditions follow from Eqs. (14.1) and (14.4b),

$$u_J^n = \sqrt{(u^2)_J^{n-1} + 2\beta}, \qquad g_J^n = 0. \tag{14.8b}$$

The above edge boundary condition for the momentum equation now replaces the previous edge boundary condition $u_J^n = 1$ given by Eq. (13.60b).

As in Section 13.4, we now linearize Eqs. (14.5) and (14.6) and their boundary conditions, Eq. (14.8), by using Newton's method. The resulting equations expressed in the form given by Eqs. (13.62d) and (13.62e) have coefficients that show great similarities to those given by Eqs. (13.63d), (13.63e), (13.64), and (13.65). The new coefficients of the momentum equation $(s_1)_j$ to $(s_6)_j$ are identical to those given by Eqs. (13.64a)–(13.64h), provided that we take $b_j = 1$, $m_1 = \frac{1}{2}$ and set $m_2 = 0$. The coefficients $(s_7)_j$ and $(s_8)_j$ are not zero and are given by

$$(s_7)_j = (s_8)_j = -\frac{1}{2}\xi_{n-1/2}.$$

Similarly the new coefficients of the energy equation are identical to those given by Eqs. (13.65a)–(13.65j), provided that we take $e_j = 1/\text{Pr}$, $m_1 = \frac{1}{2}$ and set $d_j = 0$.

With $R_{j-1/2}^{n-1}$ given by Eq. (14.7a) and $M_{j-1/2}^{n-1}$ in Eq. (13.59a) given by Eq. (14.7c), the new definitions of $(r_2)_j$ and $(r_3)_j$ are identical to those given by Eqs. (13.63d) and (13.63e), respectively, provided, of course, we take $b_j = 1$, $m_1 = \frac{1}{2}$, $e_j = 1/\text{Pr}$ and set $m_2 = d_j = 0$.

The boundary conditions given by Eqs. (14.8) with $\alpha_0 = 1$ and $\alpha_1 = 0$ are identical to those given by Eq. (13.66).

Newton's Method

As discussed in Section 5.3, the main difference between the calculation of boundary-layer and internal flows is due to the pressure-gradient term. Although it is given for boundary-layer flows, it must be computed for internal flows where the duct area is specified instead. For the latter case, we can use two different computational techniques: the Mechul method and the nonlinear eigenvalue method. Here we use the latter method in which, with the coefficients and boundary conditions of the momentum and energy equations modified as discussed above, we solve the system of equations (13.62) for an assumed value of the pressure-gradient term β^n by using the block-elimination method discussed in Section 13.4. For brevity we shall call it the *standard* problem. The resulting solution obtained from the standard problem can then be used to check whether the relation given by Eq. (9.82b) is satisfied. If not, then a new value of w^n is obtained by Newton's method as described below.

Noting that $\bar{u}_e = u_J^n$, $\eta_e = \eta_J$, and $f_e = f_J^n$, we write Eq. (9.82b) as

$$\phi(\beta_n^\nu) = \frac{1}{\eta_{sp}}\left[u_J^n(\eta_{sp} - \eta_J) + f_J^n\right] - 1, \tag{14.9}$$

and then by using Newton's method,

$$\beta_n^{\nu+1} = \beta_n^\nu - \frac{\phi(\beta_n^\nu)}{\partial\phi(\beta_n^\nu)/\partial\beta}. \tag{14.10a}$$

The derivative of ϕ with respect to β_n is obtained from Eq. (14.9):

$$\frac{\partial\phi(\beta_n^\nu)}{\partial\beta} = \frac{1}{\eta_{sp}}\left[\frac{1}{u_J^n}(\eta_{sp} - \eta_J) + \left(\frac{\partial f}{\partial\beta}\right)_J^n\right]. \tag{14.10b}$$

To compute the derivative of f with respect to β, we first take the derivative of the finite-difference approximations of Eqs. (14.2) and (14.3) and their boundary conditions, Eq. (14.4), with respect to β. This essentially amounts to differentiating Eqs. (13.56a)–(13.56c) and modified versions of Eqs. (13.56d) and (13.56e) [namely, Eqs. (14.5) and (14.6)] together with the boundary conditions given by Eqs. (14.8). This procedure leads to a aystem of *linear* difference equations known as the *variational equations*. With f_1, u_1, v_1, g_1, and p_1 defined by

$$f_1 \equiv \frac{\partial f}{\partial \beta}, \quad u_1 \equiv \frac{\partial u}{\partial \beta}, \quad v_1 \equiv \frac{\partial v}{\partial \beta}, \quad g_1 \equiv \frac{\partial g}{\partial \beta}, \quad p_1 \equiv \frac{\partial p}{\partial \beta},$$

$$(14.11)$$

they can be written in a form analogous to those given by Eqs. (13.62):

$$(f_1)_j - (f_1)_{j-1} - \frac{h_j}{2}\left[(u_1)_j + (u_1)_{j-1}\right] = 0, \qquad (14.12a)$$

$$(u_1)_j - (u_1)_{j-1} - \frac{h_j}{2}\left[(v_1)_j + (v_1)_{j-1}\right] = 0, \qquad (14.12b)$$

$$(g_1)_j - (g_1)_{j-1} - \frac{h_j}{2}\left[(p_1)_j + (p_1)_{j-1}\right] = 0, \qquad (14.12c)$$

$$(s_1)_j(v_1)_j + (s_2)_j(v_1)_j + (s_3)_j(f_1)_j + (s_4)_j(f_1)_{j-1} + (s_5)_j(u_1)_j$$
$$(s_6)_j(u_1)_{j-1} + (s_7)_j(g_1)_j + (s_8)_j(g_1)_{j-1} = -2\alpha_n, \qquad (14.12d)$$

$$(\beta_1)_j(p_1)_j + (\beta_2)_j(p_1)_{j-1} + (\beta_3)_j(f_1)_j + (\beta_4)_j(f_1)_{j-1} + (\beta_5)_j(u_1)_j$$
$$+ (\beta_6)_j(u_1)_{j-1} + (\beta_7)_j(g_1)_j + (\beta_8)_j(g_1)_{j-1} + (\beta_9)_j(v_1)_j$$
$$+ (\beta_{10})_j(v_1)_{j-1} = 0. \qquad (14.12e)$$

Here the coefficients $(s_k)_j$ and $(\beta_l)_j$ $(k = 1,\ldots,8, l = 1,\ldots,10)$ are identical to those given by Eqs. (13.64) and (13.65) with the restrictions discussed in the numerical formulation section. The coefficients $(r_1)_j, (r_4)_{j-1}, (r_5)_{j-1}, (r_2)_j,$ and $(r_3)_j$ that correspond to the right-hand sides of Eqs. (14.12a)–(14.12e), respectively, are all *zero* except for $(r_2)_j$, which is equal to

$$(r_2)_j = -2\alpha_n. \qquad (14.13)$$

Similarly, differentiating the boundary conditions, Eq. (14.8), with respect to β, we get

$$(f_1)_0^n = 0, \quad (u_1)_0^n = 0, \quad (g_1)_0^n = 0, \qquad (14.14a)$$

$$(u_1)_J^n = \frac{1}{u_J^n}, \quad (g_1)_J^n = 0. \qquad (14.14b)$$

The linear system given by Eqs. (14.12) and (14.14) can be written in the form given by Eq. (13.28) and can be solved by the block-elimination method. Since the coefficients $(s_k)_j$ and $(\beta_l)_j$ of the variational equations are identical to those of the standard problem, the **A** matrix in Eq. (13.28) does not need to be computed for the *variational* problem. Since the boundary conditions of the variational problem are also identical to those of the standard problem, SOLV5 algorithm requires no changes. As a result, when the equations for the variational problem are being solved, all we need to do is to set all $(r_k)_j (k = 1, \ldots, 5)$ except $(r_2)_j$ to zero and, with all the coefficients $(s_k)_j$ and $(\beta_l)_j$ known from the solution of the standard problem, compute δ_j, which is now defined by

$$\delta_j \equiv \begin{bmatrix} (f_1)_j \\ (u_1)_j \\ (v_1)_j \\ (g_1)_j \\ (p_1)_j \end{bmatrix}. \tag{14.15}$$

In contrast to the standard problem, however, we do not need to compute δ for all j. Instead we only need to compute δ_J since we are only interested in the value of $(f_1)_J$.

Once the value of $(f_1)_J^n$ is obtained from the solution of the variational problem, we solve the standard problem with the new value of $\beta_n^{\nu+1}$ computed from Eq. (14.10). Then we test to see whether

$$\left| \phi\left(\beta_n^{\nu+1} \right) \right| < \varepsilon \tag{14.16}$$

with ε equal to 10^{-4}, say. If not, we solve the variational equations to compute $(f_1)_J^n$ again so that we can calculate a new value of β_n from Eq. (14.10) to be used in the subsequent solution of the standard problem. This procedure is repeated until the convergence criterion set by Eq. (14.16) is satisfied.

Fortran Program

To solve our problem with the computer program of Section 13.5 and to ensure the minimum number of modifications, we use a revised version of MAIN and the three subroutines IVPL, COEF, and SOLV5. If we define a flag called IPROB that specifies whether the equations are solved for the standard problem (IPROB = 1) or for the variational problem (IPROB = 2), we can use the SOLV5 subroutine without any changes except with the statement

IF(IPROB.EQ.2)RETURN   (14.17)

inserted immediately ahead of statement number 40 in that routine.

We use the IVPL subroutine of Section 13.5 to define the initial velocity profiles without any modifications. We set $b_j = 1$, $d_j = 0$, $e_j = 1/\text{Pr}$, and $c_j = 1$. We also use this subroutine to calculate the initial temperature profile, with $g_w = 1$, and its derivative, from

$$g = 1 - \frac{\eta}{\eta_e} \qquad p = \frac{-1}{\eta_e}. \tag{14.18}$$

These modifications are minor, and the revised listing for this subroutine is not given here.

We can make use of most of the existing lines of subroutine COEF of Section 13.5 to prepare the revised COEF subroutine. For this purpose we set $P2(NX) = 0$ and $P1(NX) = 0.5$ in MAIN. The coefficients $(s_k)_j (k = 1,...,6)$ and $(\beta_l)_j$ remain the same. The coefficients $(s_7)_j$ and $(s_8)_j$ are defined by new expressions. The $(r_k)_j$ are also modified for standard and variational problems as indicated in the revised COEF subroutine shown below.

```
      SUBROUTINE COEF
      COMMON /INPT1/ WW(60),ALFAO,ALFA1
      COMMON /NCV1/ IPROB,RHG,BETA
      COMMON /BLCO/ NP,NPT,NX,NXT,NTR,IT
      COMMON /BLC1/ F(61,2),U(61,2),V(61,2),B(61,2),G(61,2),P(61,2),
     1              C(61,2),D(61,2),E(61,2),RMU(61),BC(61)
      COMMON /GRD/ X(60),ETA(61),DETA(61),A(61)
      COMMON /EDGE/ UE(60),TE(60),RHOE(60),RMUE(60),PE(60),P1(60),P2(60)
      COMMON/BLC6/ S1(61),S2(61),S3(61),S4(61),S5(61),S6(61),S7(61),
     1             S8(61),B1(61),B2(61),B3(61),B4(61),B5(61),B6(61),
     2             B7(61),B8(61),B9(61),B10(61),R(5,61)
C - - - - - - - - - - - - - - - - - - - - - - - - - - - - - - - - - -
      IF(IT .GT. 1) GO TO 5
      IF(NX .GT. 1) GO TO 3
      CEL = 0.0
      P1P = 0.5
      P2P = 0.0
      XB  = 0.0
      GO TO 5
    3 CONTINUE
      XB = 0.5*(X(NX) + X(NX-1))
      CEL = 0.5*(X(NX) + X(NX-1))/(X(NX) - X(NX-1))
      U(NP,2) = SORT(2.0*BETA + U(NP,1)**2)
      P1P = P1(NX) + CEL
      P2P = P2(NX) + CEL
    5 CONTINUE
      DO 100 J=2,NP
      IF(IPROB .EQ. 2) GO TO 80
C   PRESENT STATION
      USB  = 0.5*(U(J,2)**2+U(J-1,2)**2)
      FVB  = 0.5*(F(J,2)*V(J,2)+F(J-1,2)*V(J-1,2))
      FPB  = 0.5*(F(J,2)*P(J,2)+F(J-1,2)*P(J-1,2))
      UGB  = 0.5*(U(J,2)*G(J,2)+U(J-1,2)*G(J-1,2))
      UB   = 0.5*(U(J,2)+U(J-1,2))
      VB   = 0.5*(V(J,2)+V(J-1,2))
      FB   = 0.5*(F(J,2)+F(J-1,2))
      GB   = 0.5*(G(J,2)+G(J-1,2))
      PB   = 0.5*(P(J,2)+P(J-1,2))
      CB   = 0.5*(C(J,2) + C(J-1,2))
      DERBV = (B(J,2)*V(J,2)-B(J-1,2)*V(J-1,2))/DETA(J-1)
      DEREP = (E(J,2)*P(J,2)-E(J-1,2)*P(J-1,2))/DETA(J-1)
      DRDUV = (D(J,2)*U(J,2)*V(J,2)  - D(J-1,2)*U(J-1,2)*V(J-1,2))/
     1              DETA(J-1)
      IF(NX .GT. 1) GO TO 10
C   PREVIOUS STATION
      CFB  = 0.0
      CVB  = 0.0
      CPB  = 0.0
      CUB  = 0.0
      CGB  = 0.0
      CUGB = 0.0
      CFPB = 0.0
      CFVB = 0.0
      CUSB = 0.0
      CDERBV = 0.0
      CDEREP = 0.0
      CRB = -P2(NX)*CB
      CTB = 0.0
      GO TO 20
```

```
   10 CFB   = 0.5*(F(J,1)+F(J-1,1))
      CVB   = 0.5*(V(J,1)+V(J-1,1))
      CPB   = 0.5*(P(J,1)+P(J-1,1))
      CUB   = 0.5*(U(J,1)+U(J-1,1))
      CGB   = 0.5*(G(J,1)+G(J-1,1))
      CFVB  = 0.5*(F(J,1)*V(J,1)+F(J-1,1)*V(J-1,1))
      CFPB  = 0.5*(F(J,1)*P(J,1)+F(J-1,1)*P(J-1,1))
      CUGB  = 0.5*(U(J,1)*G(J,1)+U(J-1,1)*G(J-1,1))
      CUSB  = 0.5*(U(J,1)**2+U(J-1,1)**2)
      CCB = 0.5*(C(J,1) + C(J-1,1))
      CDERBV= (B(J,1)*V(J,1)-B(J-1,1)*V(J-1,1))/DETA(J-1)
      CDEREP= (E(J,1)*P(J,1)-E(J-1,1)*P(J-1,1))/DETA(J-1)
      CDRDUV = (D(J,1)*U(J,1)*V(J,1) - D(J-1,1)*U(J-1,1)*V(J-1,1))/
     1             DETA(J-1)
      CLB = CDERBV + P1(NX-1)*CFVB + P2(NX-1)*(CCB-CUSB)
      CRB = -CLB - P2(NX)*CB - CEL*CUSB + CEL*CFVB - XB*CGB - 2.0*CEL
     1        *BETA
      CMB = CDEREP + CDRDUV + P1(NX-1)*CFPB
      CTB = -CMB + CEL*(CFPB-CUGB)
C  COEFFICIENTS OF THE DIFFERENCED MOMENTUM EQ.
   20 CONTINUE
      S1(J) = B(J,2)/DETA(J-1) + 0.5*P1P*F(J,2) - 0.5*CEL*CFB
      S2(J) = -B(J-1,2)/DETA(J-1) + 0.5*P1P*F(J-1,2) - 0.5*CEL*CFB
      S3(J) = 0.5*(P1P*V(J,2) + CEL*CVB)
      S4(J) = 0.5*(P1P*V(J-1,2) + CEL*CVB)
      S5(J) = -P2P*U(J,2)
      S6(J) = -P2P*U(J-1,2)
      S7(J) = 0.5*XB
      S8(J) = S7(J)
      R(2,J) = CRB - (DERBV + P1P*FVB - P2P*USB + CEL*(FB*CVB-VB*CFB) +
     1             XB*GB)
C  COEFFICIENTS OF DIFFERENCED ENERGY EQ.
      B1(J) = E(J,2)/DETA(J-1) + 0.5*P1P*F(J,2) - 0.5*CEL*CFB
      B2(J) = -E(J-1,2)/DETA(J-1) + 0.5*P1P*F(J-1,2) - 0.5*CEL*CFB
      B3(J) = 0.5*(P1P*P(J,2) + CEL*CPB)
      B4(J) = 0.5*(P1P*P(J-1,2) + CEL*CPB)
      B5(J) = D(J,2)*V(J,2)/DETA(J-1) - 0.5*CEL*(G(J,2)-CGB)
      B6(J) = -D(J-1,2)*V(J-1,2)/DETA(J-1) - 0.5*CEL*(G(J-1,2)-CGB)
      B7(J) = -0.5*CEL*(U(J,2)+CUB)
      B8(J) = -0.5*CEL*(U(J-1,2)+CUB)
      B9(J) = D(J,2)*U(J,2)/DETA(J-1)
      B10(J) = -D(J-1,2)*U(J-1,2)/DETA(J-1)
      R(3,J) = CTB - (DEREP + DRDUV + P1P*FPB - CEL*(UGB-CGB*UB+CUB*GB)+
     1             CEL*(CPB*FB-CFB*PB))
C  DEFINITIONS OF RJ FOR STANDARD PROBLEM
      R(1,J)= F(J-1,2)-F(J,2)+DETA(J-1)*UB
      R(4,J-1) = U(J-1,2)-U(J,2)+DETA(J-1)*VB
      R(5,J-1) = G(J-1,2)-G(J,2)+DETA(J-1)*PB
      GO TO 100
   80 CONTINUE
C  DEFINITIONS OF RJ FOR VARIATIONAL PROBLEM
      R(1,J) = 0.0
      R(2,J) = -2.0*CEL
      R(3,J) = 0.0
      R(4,J-1) = 0.0
      R(5,J-1) = 0.0
  100 CONTINUE
C  DEFINITIONS OF RJ FOR BOUNDARY CONDITIONS
      R(1,1) = 0.0
      R(2,1) = 0.0
      R(3,1) = 0.0
      R(4,NP)=0.0
      R(5,NP)=0.0
      IF(IPROB .EQ. 2) R(4,NP) = 1.0/U(NP,2)
      RETURN
      END
```

The revised MAIN routine contains the INPUT, GRID, and OUTPUT subroutines and the logic of the calculations. It also contains Eqs. (14.9) and (14.10b). Note also that g_0 ($\equiv$ GW) $= 1.0$ and RHG $= R_L/\mathrm{Gr}_L$. This routine is shown below.

```
C      MAIN
       COMMON /INPT1/ WW(60),ALFA0,ALFA1
       COMMON /BLC0/ NP,NPT,NX,NXT,NTR,IT
       COMMON /BLC1/ F(61,2),U(61,2),V(61,2),B(61,2),G(61,2),P(61,2),
      1              C(61,2),D(61,2),E(61,2),RMU(61),BC(61)
       COMMON /GRD/ X(60),ETA(61),DETA(61),A(61)
       COMMON /EDGE/ UE(60),TE(60),RHOE(60),RMUE(60),PE(60),P1(60),P2(60)
       COMMON /BLC3/ DELF(61),DELU(61),DELV(61),DELG(61),DELP(61)
       COMMON /NCV1/ IPROB,RHG,BETA
       COMMON /AK1/GW,PR
C - - - - - - - - - - - - - - - - - - - - - - - - - - - - - - - - - - - -
       ITMAX = 6
       NX    = 1
       NPT = 61
       ETAE = 8.0
C  INPUT DATA
       READ(5,8000) NXT
       READ(5,8100) DETA(1),VGP,ALFA0,ALFA1
       READ(5,8100) GW,PR,RHG
       READ(5,8100) (X(I),I=1,NXT)
       WRITE(6,7000) PR,RHG,NXT,ETAE,DETA(1),VGP
C  GRID GENERATION
       IF (VGP-1.0) .LE. 0.001) GO TO 5
       NP   = ALOG((ETAE/DETA(1))*(VGP-1.0)+1.0)/ALOG(VGP) + 1.0001
       GO TO 10
     5 NP   = ETAE/DETA(1) + 1.0001
    10 IF(NP .LE. 61) GO TO 15
       WRITE(6,9000)
       STOP
    15 ETA(1) = 0.0
       DO 20 J=2,NPT
       DETA(J) =VGP*DETA(J-1)
       A(J)   = 0.5*DETA(J-1)
    20 ETA(J) = ETA(J-1)+DETA(J-1)
       CALL IVPL
       BETA = 0.0
    30 WRITE(6,9100) NX,X(NX)
       P1(NX) = 0.5
       P2(NX) = 0.0
       IT     = 0
       IPHI = 0
       WRITE(6,9410)
    40 IT    = IT+1
       IPROB = 1
       IF(IT .LE. ITMAX) GO TO 60
       WRITE(6,2500)
       GO TO 80
    60 CONTINUE
       CALL COEF
       CALL SOLV5
       WRITE(6,3000) V(1,2),DELV(1)
C  CHECK FOR CONVERGENCE
       IF(ABS(DELV(1)) .GT. 1.0E-05) GO TO 40
       IF(V(1,2) .LT. 0.0) STOP
       IF(NX .EQ. 1) GO TO 80
C  SOLVE VARIATIONAL EQNS
       IF(IPHI .GT. 6) GO TO 80
       IPROB = 2
       CALL COEF
       CALL SOLV5
C  CHECK FOR Q/(DQ/DB)
       IPHI = IPHI + 1
       ETASP = 1.0/SQRT(X(NX)*RHG)
       PHI = (F(NP,2) + U(NP,2)*(ETASP - ETA(NP)))/ETASP - 1.0
       DPHI = (DELF(NP) + (ETASP - ETA(NP))/U(NP,2))/ETASP
       DBETA = PHI/DPHI
       WRITE(6,9400) BETA,DBETA
       BETA = BETA - DBETA
       IT = 0
       IF(ABS(DBETA) .GT. 0.0001) GO TO 40
    80 CONTINUE
C  OUTPUT DATA
       WRITE(6,4400)
       WRITE(6,4500) (J,ETA(J),F(J,2),U(J,2),V(J,2),G(J,2),P(J,2),
      1               B(J,2),J=1,NP,5)
       WRITE(6,4500) NP,ETA(NP),F(NP,2),U(NP,2),V(NP,2),G(NP,2),P(NP,2),
```

```
      1         B(NP,2)
      NX     = NX+1
      IF(NX .GT. NXT) STOP
C  SHIFT PROFILES
      DO 110 J=1,NP
      F(J,1)= F(J,2)
      U(J,1)= U(J,2)
      V(J,1)= V(J,2)
      G(J,1)= G(J,2)
      P(J,1)= P(J,2)
      E(J,1)= E(J,2)
  110 B(J,1)= B(J,2)
      GO TO 30
C - - - - - - - - - - - - - - - - - - - - - - - - - - - - - - - - -
 8000 FORMAT(8I3)
 8100 FORMAT(8F10.0)
 3000 FORMAT(1H ,8HV(WALL)=,E13.6,2X,6HDELV =,E13.6)
 4400 FORMAT(1H0,2X,1HJ,4X,3HETA,9X,1HF,13X,1HU,13X,1HV,13X,1HG,13X,1HP,
     1        13X,1HB)
 4500 FORMAT(1H ,I3,F10.3,6E14.6)
 7000 FORMAT(1H0,4HPR=,E12.5,5X,'RH/GR =',E12.5/1H0,5HNXT =,I3,
     1        5X,6HETAE =,E12.5,3X,6HDETA1=,E12.5,3X,6HVGP  =,E12.5)
 2500 FORMAT(1H0,16X,25HITERATIONS EXCEEDED ITMAX)
 9000 FORMAT(1H0,'NP EXCEEDED NPT -- PROGRAM TERMINATED')
 9100 FORMAT(1H0,4HNX =,I3,5X,3HX =,F10.3/)
 9400 FORMAT(1H ,79X,2E14.6)
 9410 FORMAT(1H0,84X,4HBETA,9X,5HDBETA)
      END
```

Sample Calculations

Table 14.1 shows the computed wall shear and heat-transfer parameters for uniform wall temperature with Pr = 0.72, $R_L/\mathrm{Gr}_L = 1$, $h_1 = 0.16$, $K = 1$ for a total of six ξ stations. Note that at $\xi = 0$, the flow is identical with unheated flow in a constant-pressure boundary layer, and the solution procedure obviously does not make use of the eigenvalue approach at this station.

We should remember that near the entrance of the duct the effect of buoyancy is small; specifically, the Richardson number defined by Eq. (9.66) is small due to small shear-layer thickness. As ξ becomes bigger, the buoyancy effects become important. The calculations shown in Table 14.1 are limited to small values of ξ because for simplicity we have restricted the calculations to the situation in which the shear layers have not merged. To

Table 14.1 Computed wall shear and heat-transfer parameters for uniform wall temperature, Pr = 0.72, and $R_L/\mathrm{Gr}_L = 1$

$\xi \times 10^3$	f_w''	g_w'
0	0.3320	-0.2956
0.0025	0.4881	-0.3327
0.005	0.4802	-0.3305
0.0075	0.5366	-0.3419
0.01	0.5490	-0.3444
0.015	0.6088	-0.3556

observe greater effects of buoyancy, as discussed in Section 9.4, it is necessary to go to higher values of ξ and include the capability of calculating the shear layers when they do merge.

14.2 Wall Jet and Film Heating

We now consider the calculation of a wall-jet flow at initial temperature T_j, blowing up a vertical wall at constant temperature T_w and below a vertical boundary-layer flow upward of external-stream speed u_e, as described in Section 9.3 with the exception that the adiabatic-wall boundary condition is replaced by uniform wall temperature. The differential equations and the boundary conditions that represent this flow in terms of transformed coordinates for small values of ξ correspond to Eq. (9.76). As discussed in Section 9.3, we also need initial conditions at $\xi = \xi_0$. Here we specify the initial velocity profiles (see Sec. 4.5) in terms of transformed variables by the following expressions:

$$
u = \begin{cases}
6u_c\left(\dfrac{\eta}{\eta_c}\right)\left(1 - \dfrac{\eta}{\eta_c}\right) & 0 \leq \eta \leq \eta_c, \quad (14.19a) \\[2ex]
u_e\sin\left[\dfrac{\pi}{2}\dfrac{\eta - \eta_c}{\eta_e - \eta_c}\right] & \eta_c \leq \eta \leq \eta_e. \quad (14.19b)
\end{cases}
$$

Here u_c denotes the average velocity of the jet. As can be seen from Fig. 4.18, there is a discontinuity in the velocity profile at η_c, which corresponds to transformed slot height y_c. Hermite or oscillatory interpolation (see Isaacson and Keller [4]) is used to avoid the related problems, resulting in the blending velocity profile shown in Fig. 4.19, and is defined as

$$
u(\eta) = u(\eta_1)\psi_1(\eta) + u(\eta_2)\psi_2(\eta) + u'(\eta_1)\bar{\psi}_1(\eta) + u'(\eta_2)\bar{\psi}_2(\eta) \quad (14.20)
$$

Here the prime denotes differentiation with respect to η, and

$$
\psi_1(\eta) = \left(1 - 2\frac{\eta - \eta_1}{\eta_1 - \eta_2}\right)\left(\frac{\eta - \eta_2}{\eta_1 - \eta_2}\right)^2,
$$

$$
\bar{\psi}_1(\eta) = (\eta - \eta_1)\left(\frac{\eta - \eta_2}{\eta_1 - \eta_2}\right)^2,
$$

$$
\psi_2(\eta) = \left(1 + 2\frac{\eta - \eta_2}{\eta_1 - \eta_2}\right)\left(\frac{\eta_1 - \eta}{\eta_1 - \eta_2}\right)^2,
$$

$$
\bar{\psi}_2(\eta) = (\eta - \eta_2)\left(\frac{\eta_1 - \eta}{\eta_1 - \eta_2}\right)^2. \quad (14.21)
$$

The initial temperature distribution is presumed here to result from a uniform distribution in the exit plane of the slot, which is higher than the uniform temperature of the freestream and less than the uniform wall temperature. To avoid the difficulties associated with the discontinuities at the wall and at the outer edge of the slot (see Section 4.5), simple smoothing functions are again introduced. The initial temperature profile may be expressed by Eq. (8.37), which is now written as

$$g = \tfrac{1}{2}\left[1 - \tanh \beta(\eta - \eta_c)\right]. \tag{14.22}$$

Numerical Formulation

Rewritten as a first-order sytem, the governing equations correspond to slightly modified versions of Eqs. (14.2) and (14.3), that is

$$v' + \frac{1}{2}fv = \text{Ri } zg + z\left(u\frac{\partial u}{\partial z} - v\frac{\partial f}{\partial z}\right), \tag{14.23}$$

$$\frac{p'}{\text{Pr}} + \frac{1}{2}fp = z\left(u\frac{\partial g}{\partial z} - p\frac{\partial f}{\partial z}\right), \tag{14.24}$$

$$\eta = 0, \quad f = u = 0, \quad g = 1; \quad \eta = \eta_e, \quad u = 1, \quad g = 0 \tag{14.25}$$

and, in finite-difference form, to Eqs. (14.5)–(14.7) with obvious changes resulting from the difference between Eqs. (14.2) and (14.3) and Eqs. (14.23) and (14.24). The boundary conditions given by Eq. (14.25) now become

$$f_0 = u_0 = 0, \quad g_0 = 1; \quad u_J = 1, \quad g_J = 0. \tag{14.26}$$

Note that Ri contains the gravitational acceleration; g is the dimensionless temperature difference.

Fortran Program

The computer program of Section 14.1 can now be used for this problem. Since no variational equations are solved, the SOLV5 subroutine contains no extra statement and is the same as that in Section 13.4. The subroutine COEF now contains only the definitions pertinent to the *standard problem*, and there is no need to include the statements used for the variational problem.

The only major change occurs in the subroutine IVPL where we define the initial velocity and temperature profiles according to Eqs. (14.19) and (14.22). To obtain f'' and g', we differentiate Eqs. (14.19) and (14.22). To obtain the blending velocity profile according to Eqs. (14.20) and (14.21), we use another subroutine called subroutine HER. Here $\psi_1(\eta) = \text{SY1}$, $\psi_2(\eta) = \text{SY2}$, $\bar{\psi}_1(\eta) = \text{CSY1}$, and $\bar{\psi}_2(\eta) = \text{CSY2}$. Both subroutines together with MAIN are given below, following sample calculations.

Table 14.2 Computed wall shear and heat-transfer parameters for a laminar wall jet on a vertical flat plate: $Pr = 0.72$, $Ri = 0.10$, $u_c/u_e = 0.50$

z	θ'_w	f''_w
1	0	1.0
1.005	-0.00005	0.874
1.01	-0.00009	0.813
1.02	-0.00012	0.793
1.05	-0.00036	0.713
$\vdots$	$\vdots$	$\vdots$
1.35	-0.040	0.503
1.40	-0.050	0.473
1.45	-0.061	0.482
1.50	-0.071	0.458

Sample Calculations

Table 14.2 shows the results computed with the revised computer program for $u_c/u_e = 0.50$. The calculations were started at $z = 1$, $Ri = 0.10$ by entering the initial and temperature profiles as described before. Because there are rapid changes in the velocity and temperature profiles, especially in the blending region, it is essential that we use a variable grid across the layer. (It is for this reason that it is convenient to use analytic initial profiles rather than tables of values that would have to be interpolated.) In our calculation we set $\eta_c = 3$, $\eta_1 = 0.95\eta_c$, $\eta_2 = 1.05\eta_c$, and $\eta_e = 8.85$. In the regions defined by $0 \leq \eta \leq \eta_1$ and $\eta_2 \leq \eta \leq 8.85$ we set $\Delta\eta = 0.15$ and $K = 1$, and in the blending region we again choose uniform spacing in $\Delta\eta$, this time with $\Delta\eta$ equal to 0.075. It is quite possible that in some cases one may need to take a smaller spacing than 0.075 in the blending region.

```
C          DATA SET AKNWMAIN    AT LEVEL 004 AS OF 04/01/83
C      MAIN
       COMMON /AK1/ RL,PR,UC,FR
       COMMON /EDGE/ P1(60),P2(60)
       COMMON /INPT1/ FN,ALFA0,ALFA1
       COMMON /BLC0/ NP,NPT,NX,NXT,NTR,IT,NXS
       COMMON /BLC1/ F(99,2),U(99,2),V(99,2),B(99,2),G(99,2),P(99,2),
      1            E(99,2)
       COMMON /GRD/ X(60),ETA(99),DETA(99),A(99)
       COMMON /BLC3/ DELF(99),DELU(99),DELV(99),DELG(99),DELP(99)
       COMMON /GRD2/ DETA1,DETA2,DETA3,DETA4,XETA1,XETA2,ETAC,ETAE
       COMMON /STORE/ FS(99),US(99),VS(99),GS(99),PS(99)
C - - - - - - - - - - - - - - - - - - - - - - - - - - - - - - - - -
C  FN = -1.0  COOLING I.E. BUOYANCY FORCE AGAINST THE DN OF THE FLOW
C  FN =  1.0  HEATING I.E. BUOYANCY FORCE IN THE DN OF THE FLOW
C  ALFA0 = 1.0 ,  ALFA1 = 0.0  SPECIFIED WALL TEMP  (G(1,2) = 1.0)
C  ALFA0 = 0.0 ,  ALFA1 = 1.0  SPECIFIED HEAT FLUX  (P(1,2) IS READ)
C  ******************** INPUT DATA   ********************
       ITMAX = 6
       NX    = 1
       NPT = 99
```

```
      READ(5,8000) NXT,NXS,NXSM
      READ(5,8100) DETA(1),VGP,ETAE,ETAC,XETA1,XETA2,ALFA0,ALFA1
      READ(5,8100) PR,FN,RL,UAV,FR,DETA2,DETA3
      READ(5,8100) (X(I),I=1,NXT)
      UC = 6.0*UAV
      WRITE(6,7000) PR,FN,UAV,FR
      WRITE(6,7100) NXT,NTR,NXS,NXSM,ETAE,DETA(1),DETA2,DETA3,VGP
      DO 20 I=1,NXT
   20 X(I) = X(I)/FR
C  GRID IS GENERATED IN IVPL
      CALL IVPL
   30 IT      = 0
      DX = (X(NX) - X(1))*FR
      WRITE(6,9100) NX,X(NX),DX
      P1(NX) = 0.0
      P2(NX) = 0.0
      IF(NX .GT. NXS) P1(NX) = 0.5
      IF(NX .EQ. 1) GO TO 100
   40 IT      = IT+1
      IF(IT .LE. ITMAX) GO TO 60
      WRITE(6,2500)
      GO TO 80
   60 CONTINUE
      CALL COEF
      CALL SOLV5
      WRITE(6,3000) V(1,2),DELV(1)
C  CHECK FOR CONVERGENCE
      IF(ABS(DELV(1)) .GT. 1.0E-05) GO TO 40
      IF(V(1,2) .LT. 0.0) STOP
   80 CONTINUE
      IF(NP .GE. NPT) GO TO 100
      IF(ABS(V(NP,2)) .LT. 0.0005) GO TO 100
      NP = NP + 1
      DO 90 L=1,2
      DO 85 J=NP,NP
      U(J,L) = U(J-1,L)
      V(J,L) = 0.0
      G(J,L) = 0.0
      P(J,L) = 0.0
      F(J,L) = F(J-1,L) + DETA(J-1)*U(J,L)
      B(J,L) = B(J-1,L)
      E(J,L) = E(J-1,L)
   85 CONTINUE
   90 CONTINUE
      IT = 1
      WRITE(6,4600)
      GO TO 40
  100 CONTINUE
C  OUTPUT DATA
      WRITE(6,4400)
      WRITE(6,4500) (J,ETA(J),F(J,2),U(J,2),V(J,2),G(J,2),P(J,2),
     1              B(J,2),J=1,NP)
      IF(NX .NE. NXSM) GO TO 120
      NXSM = 1000
      X1 = X(NX) - X(NX-1)
      X2 = X(NX-1) - X(NX-2)
      DO 110 J=1,NP
      F(J,1) = 0.5*(X2*(F(J,2)+F(J,1)) + X1*(F(J,1)+FS(J)))/(X1+X2)
      U(J,1) = 0.5*(X2*(U(J,2)+U(J,1)) + X1*(U(J,1)+US(J)))/(X1+X2)
      V(J,1) = 0.5*(X2*(V(J,2)+V(J,1)) + X1*(V(J,1)+VS(J)))/(X1+X2)
      G(J,1) = 0.5*(X2*(G(J,2)+G(J,1)) + X1*(G(J,1)+GS(J)))/(X1+X2)
      P(J,1) = 0.5*(X2*(P(J,2)+P(J,1)) + X1*(P(J,1)+PS(J)))/(X1+X2)
  110 CONTINUE
      IT = 1
      GO TO 40
  120 CONTINUE
      IF(NX .EQ. NXS) P1(NX) = 0.5
      NX      = NX+1
      IF(NX .GT. NXT) STOP
C  SHIFT PROFILES
      DO 140 J=1,NP
      F(J,1)= F(J,2)
      U(J,1)= U(J,2)
      V(J,1)= V(J,2)
      G(J,1)= G(J,2)
      P(J,1)= P(J,2)
```

```
       E(J,1)= E(J,2)
       B(J,1)= B(J,2)
  140 CONTINUE
       IF(NX .NE. (NXSM-2)) GO TO 30
       DO 160 J=1,NPT
       US(J)= U(J,2)
       VS(J)= V(J,2)
       GS(J)= G(J,2)
       PS(J)= P(J,2)
       FS(J)= F(J,2)
       IF(J .LE. NP) GO TO 160
       US(J)  = US(J-1)
       VS(J)  = 0.0
       GS(J)  = 0.0
       PS(J)  = 0.0
       FS(J)  = FS(J-1) + DETA(J-1)
  160 CONTINUE
       GO TO 30
C - - - - - - - - - - - - - - - - - - - - - - - - - - - - - - - - -
 8000 FORMAT(8I3)
 8100 FORMAT(8F10.0)
 3000 FORMAT(1H ,8HV(WALL)=,E13.6,2X,6HDELV =,E13.6)
 4400 FORMAT(1H0,2X,1HJ,4X,3HETA,9X,1HF,13X,1HU,13X,1HV,13X,1HG,13X,1HP,
     1          13X,1HB)
 4500 FORMAT(1H ,I3,F10.3,6E14.6)
 4600 FORMAT(1H0)
 7000 FORMAT(1H0,'PR =',F6.2,3X,'FN =',F5.2,3X,'UAV =',F6.2,3X,
     1          'FR NO =',F6.2/)
 7100 FORMAT(1H0,'NXT=',I3,2X,'NTR=',I3,2X,'NXS=',I3,2X,'NXSM=',I2,
     1      3X,'ETAE=',F5.2,2X,'DETA1=',F5.3,2X,'DETA2=',F5.3,2X,
     2      'DETA3=',F5.3,2X,'VGP=',F5.3/)
 2500 FORMAT(1H0,16X,25HITERATIONS EXCEEDED ITMAX)
 9100 FORMAT(1H0,4HNX =,I3,5X,3HX =,F10.3,3X,'X-XO =',E12.5/)
       END



       SUBROUTINE IVPL
       COMMON /BLC0/ NP,NPT,NX,NXT,NTR,IT,NXS
       COMMON /BLC1/ F(99,2),U(99,2),V(99,2),B(99,2),G(99,2),P(99,2),
     1          E(99,2)
       COMMON /GRD2/ DETA1,DETA2,DETA3,DETA4,XETA1,XETA2,ETAC,ETAE
       COMMON /GRD/ X(60),ETA(99),DETA(99),A(99)
       COMMON /AK1/ RL,PR,UC,FR
C - - - - - - - - - - - - - - - - - - - - - - - - - - - - - - - - -
C      NC LAST POINT FOR JET
C      NMJ LAST POINT FOR HER.POLY
C      DETA1 SPACING FOR JET REGION
C      DETA2 SPACING FOR HER.POLY REGION
C      DETA3 SPACING FOR BL REGION
C-----------------
       PI = 3.14159
       UO = 1.0
       DETA1 = DETA(1)
       ETA1 = XETA1*ETAC
       ETA2 = XETA2*ETAC
       NC = ETA1/DETA1 + 1.0001
       NMJ= (ETA2-ETA1)/DETA2 + NC
       NC1= NC+1
       NP = (ETAE-ETA2)/DETA3 + NMJ
C  CREATION OF THE GRID
       ETA(1) = 0.0
       DO 10 J=2,NC
   10 ETA(J) = ETA(J-1) + DETA1
       DO 20 J=NC1,NMJ
   20 ETA(J) = ETA(J-1) + DETA2
       NMJ1 = NMJ +1
       DO 30 J=NMJ1,NPT
   30 ETA(J) = ETA(J-1) + DETA3
C  ****************    COMPUTATION OF U,V PROFILES  **********
C  ****************        LOWER PART (JET)
       DO 50 J=1,NC1
       EBETAC = ETA(J)/ETAC
       U(J,2) = UC*EBETAC*(1.0-EBETAC)
       V(J,2) = UC/ETAC*(1.0-2.0*EBETAC)
   50 CONTINUE
C  *******************    UPPER PART    ********
```

```
      DO 60 J=NMJ,NP
      U(J,2) = UO*SIN(0.5*PI*(ETA(J)-ETAC)/(ETA(NP)-ETAC))
      V(J,2) = 0.5*PI*UO*COS(0.5*PI*(ETA(J)-ETAC)/(ETA(NP)-ETAC))/
     1         (ETA(NP)-ETAC)
   60 CONTINUE
      DO 70 J=NC1,NMJ
   70 CALL HER(NC1,NMJ,J,ETAC,ETA)
      DO 140 J=1,NP
      G(J,2) = 0.5*(1.0 - TANH(2.0*(ETA(J) - ETAC)))
      P(J,2) = -1.0/COSH(2.0*(ETA(J) - ETAC))**2
      B(J,2) = 1.0
      E(J,2) = B(J,2)/PR
  140 CONTINUE
      P(NP,2) = 0.0
      P(1,2) = 0.0
      SQFR = 1.0
      DO 150 J=2,NPT
      ETA(J) = ETA(J)/SQFR
      DETA(J-1) = ETA(J) - ETA(J-1)
      A(J)      = 0.5*DETA(J-1)
  150 CONTINUE
      DO 160 J=1,NP
      V(J,2) = V(J,2)*SQFR
      P(J,2) = P(J,2)*SQFR
      IF(J .EQ. 1) GO TO 160
      F(J,2) = F(J-1,2) + A(J)*(U(J,2) + U(J-1,2))
  160 CONTINUE
      WRITE(6,9000) ETAC,XETA1,XETA2,ETA(NC),ETA(NMJ)
      WRITE(6,9500)(J,ETA(J),F(J,2),U(J,2),V(J,2),G(J,2),P(J,2),J=1,NP)
      RETURN
C - - - - - - - - - - - - - - - - - - - - - - - - - - - - - - -
 9000 FORMAT(1H0,5HETAC=,F10.6,5X,6HXETA1=,F10.6,5X,6HXETA2=,F10.6,5X,
     1           5HETA1=,F10.6,5X,5HETA2=,F10.6)
 9500 FORMAT(1H0,2X,1HJ,7X,3HETA,14X,1HF,17X,1HU,17X,1HV,17X,1HG,17X,1HP
     1           /(1H ,I4,F12.5,5E18.6))
      END
      SUBROUTINE HER (NC,NMJ,J,YC,Y)
      COMMON /BLC1/ F(99,2),U(99,2),V(99,2),B(99,2),G(99,2),P(99,2),
     1      E(99,2)
      DIMENSION Y(1)
C - - - - - - - - -
      Y12 = Y(NC) - Y(NMJ)
      Y12SQ = Y12*Y12
      SY1 = (1.0-2.0*(Y(J)-Y(NC))/Y12)*((Y(J)-Y(NMJ))/Y12)**2
      SY2 = (1.0+2.0*(Y(J)-Y(NMJ))/Y12)*((Y(NC)-Y(J))/Y12)**2
      CSY1= (Y(J)-Y(NC))*((Y(J)-Y(NMJ))/Y12)**2
      CSY2= (Y(J)-Y(NMJ))*((Y(NC)-Y(J))/Y12)**2
      DSY1 = (2.0*(Y(J)-Y(NMJ))*(1.0-2.0*(Y(J)-Y(NC))/Y12)-2.0/Y12*
     1         (Y(J)-Y(NMJ))**2)/Y12SQ
      DSY2 = (-2.0*(Y(NC)-Y(J))*(1.0+2.0*(Y(J)-Y(NMJ))/Y12)+2.0/Y12*
     1         (Y(NC)-Y(J))**2)/Y12SQ
      DCSY1 = (2.0*(Y(J)-Y(NC))*(Y(J)-Y(NMJ)) + (Y(J)-Y(NMJ))**2)/Y12SQ
      DCSY2 = (-2.0*(Y(NC)-Y(J))*(Y(J)-Y(NMJ)) + (Y(NC)-Y(J))**2)/Y12SQ
      PY1 =((1.0-2.0*(Y(J)-Y(NC))/Y12)*(Y(J)-Y(NMJ))**3/3.0+(Y(J)-
     1       Y(NMJ))**4/(6.0*Y12))/Y12SQ
      CPY1 = ((Y(J)-Y(NC))*(Y(J)-Y(NMJ))**3/3.0 -(Y(J)-Y(NMJ))**4/12.0)/
     1       Y12SQ
      PY2 = ((1.0+2.0*(Y(J)-Y(NMJ))/Y12)*(Y(J)-Y(NC))**3/3.0 - (Y(J)-
     1       Y(NC))**4/(6.0*Y12))/Y12SQ
      CPY2 = ((Y(J)-Y(NMJ))*(Y(J)-Y(NC))**3/3.0-(Y(J)-Y(NC))**4/12.0)/
     1       Y12SQ
      U(J,2) = U(NC,2)*SY1+U(NMJ,2)*SY2+V(NC,2)*CSY1+V(NMJ,2)*CSY2
      V(J,2) = U(NC,2)*DSY1+U(NMJ,2)*DSY2+V(NC,2)*DCSY1+V(NMJ,2)*DCSY2
      RETURN
      END
```

14.3 Turbulent Free Jet

In Sections 8.1 and 8.3 we discussed the solutions of a laminar and a turbulent jet issuing into a still atmosphere in the context of uncoupled flows. Here we discuss the use of the computer program of Section 13.5 to calculate the properties of a two-dimensional, nonsimilar, heated turbulent jet.

The transformed boundary-layer equations for a two-dimensional laminar heated jet are given by Eqs. (8.35) and (8.36). With the concepts of eddy viscosity and turbulent Prandtl number, these equations can be written as

$$(bf'')' + (f')^2 + ff'' = 3\xi\left(f'\frac{\partial f'}{\partial \xi} - f''\frac{\partial f}{\partial \xi}\right), \tag{14.27}$$

$$(e\theta')' + f\theta' = 3\xi\left(f'\frac{\partial \theta}{\partial \xi} - \theta'\frac{\partial f}{\partial \xi}\right). \tag{14.28}$$

Here θ is the dimensionless temperature and

$$b = 1 + \varepsilon_m^+, \qquad e = \frac{1}{\text{Pr}} + \frac{\varepsilon_m^+}{\text{Pr}_t}. \tag{14.29}$$

The boundary conditions for Eqs. (14.27) and (14.28) follow from Eq. (8.18) and can be rewritten as

$$\eta = 0; \quad f = f'' = 0, \quad \theta' = 0; \quad \eta = \eta_e, \quad f' = 0, \quad \theta = 0. \tag{14.30}$$

The system given by Eqs. (14.27)–(14.30) requires initial conditions. Here for simplicity we assume uniform velocity and temperature profiles and define them in dimensionless form by the expression given in Eq. (8.37). In terms of the transformation defined by Eq. (8.34), we can express Eq. (8.37) in the following form:

$$\frac{f'}{3\xi_0^{1/3}} = \theta = \frac{1}{2}\left\{ 1 - \tanh\beta\left[\frac{3\xi_0^{2/3}}{\sqrt{R_L}}(\eta - \eta_c)\right]\right\}. \tag{14.31}$$

Here R_L is a dimensionless Reynolds number, $u_0 L/\nu$, and ξ_0 is the ξ location at which the initial profiles are specified. L is a reference length taken to be equal to the half-width of the duct.

We use the eddy-viscosity model given by Eq. (8.59) as the turbulence model, and we take Pr_t to be constant and equal to 0.5. Since ε_m^+ is much larger than 1, we set $b = \varepsilon_m^+$ and in terms of transformed variables write it as

$$\varepsilon_m^+ = 0.037\sqrt{R_L}\,\xi^{1/3}\eta_{1/2}f_c'. \tag{14.32}$$

Here $\eta_{1/2}$ is the transformed η distance where $u = \frac{1}{2}u_c$, and f_c' is the dimensionless centerline velocity.

Table 14.3 Variation of dimensionless centerline velocity and temperature for a turbulent heated jet: $Pr = 0.72$, $R_L = 5300$

ξ	f_c'	g_c
1	3.0	0.999997
1.005	3.005	0.999997
1.01	3.010	0.999997
1.02	3.020	0.999996
1.05	3.049	0.999995
$\vdots$	$\vdots$	$\vdots$
1.50	3.436	0.999388
1.55	3.473	0.999104
1.60	3.509	0.99873
$\vdots$	$\vdots$	$\vdots$
10.0	4.5925	0.692689

Numerical Formulation

As in Section 13.4, the transformed equations are written in terms of a first-order system. Using the definitions of Eqs. (13.54a)–(13.54c) and again replacing θ with g, Eqs. (14.27), (14.28), and (14.30) become

$$(bv)' + u^2 + fv = 3\xi\left(u\frac{\partial u}{\partial \xi} - v\frac{\partial f}{\partial \xi}\right), \tag{14.33}$$

$$(ep)' + fp = 3\xi\left(u\frac{\partial g}{\partial \xi} - p\frac{\partial f}{\partial \xi}\right), \tag{14.34}$$

$$\eta = 0, \quad f = v = p = 0; \quad \eta = \eta_e, \quad u = g = 0. \tag{14.35}$$

The finite-difference approximations for Eqs. (14.33) and (14.34) are written in the same manner as Eqs. (13.56d) and (13.56e); that is,

$$h_j^{-1}\left(b_j^n v_j^n - b_{j-1}^n v_{j-1}^n\right) + (1 + \alpha_n)(fv)_{j-1/2}^n + (1 - \alpha_n)(u^2)_{j-1/2}^n$$
$$+ \alpha_n\left(v_{j-1/2}^{n-1} f_{j-1/2}^n - f_{j-1/2}^{n-1} v_{j-1/2}^n\right) = R_{j-1/2}^{n-1}, \tag{14.36a}$$

$$h_j^{-1}\left(e_j^n p_j^n - e_{j-1}^n p_{j-1}^n\right) + (fp)_{j-1/2}^n - \alpha_n\left[(ug)_{j-1/2}^n + u_{j-1/2}^{n-1} g_{j-1/2}^n\right.$$
$$\left. - g_{j-1/2}^{n-1} u_{j-1/2}^n - (fp)_{j-1/2}^n + f_{j-1/2}^{n-1} p_{j-1/2}^n - p_{j-1/2}^{n-1} f_{j-1/2}^n\right] = T_{j-1/2}^{n-1}, \tag{14.36b}$$

where α_n is given by

$$\alpha_n = \frac{3\xi_{n-1/2}}{\xi_n - \xi_{n-1}} \tag{14.37}$$

and $R^{n-1}_{j-1/2}$, $L^{n-1}_{j-1/2}$, $T^{n-1}_{j-1/2}$, and $M^{n-1}_{j-1/2}$ are given by

$$R^{n-1}_{j-1/2} = -L^{n-1}_{j-1/2} + \alpha_n \left[(fv)^{n-1}_{j-1/2} - (u^2)^{n-1}_{j-1/2} \right], \qquad (14.38a)$$

$$L^{n-1}_{j-1/2} = h_j^{-1} \left(b_j^{n-1} v_j^{n-1} - b_{j-1}^{n-1} v_{j-1}^{n-1} \right) + (u^2)^{n-1}_{j-1/2} + (fv)^{n-1}_{j-1/2}, \qquad (14.38b)$$

$$T^{n-1}_{j-1/2} = -M^{n-1}_{j-1/2} + \alpha_n \left[(fp)^{n-1}_{j-1/2} - (ug)^{n-1}_{j-1/2} \right], \qquad (14.38c)$$

$$M^{n-1}_{j-1/2} = h_j^{-1} \left(e_j^{n-1} p_j^{n-1} - e_{j-1}^{n-1} p_{j-1}^{n-1} \right) + (fp)^{n-1}_{j-1/2}. \qquad (14.38d)$$

We next linearize the nonlinear difference equations by Newton's method as described in Section 13.4 to yield equations that are identical to Eqs. (13.62) and (13.63) except that

$$(r_2)_j = R^{n-1}_{j-1/2} - \left[h_j^{-1} (b_j v_j - b_{j-1} v_{j-1}) + (1 + \alpha_n)(fv)_{j-1/2} \right.$$
$$+ (1 - \alpha_n)(u^2)_{j-1/2} + \alpha_n \left(v^{n-1}_{j-1/2} f_{j-1/2} - f^{n-1}_{j-1/2} v_{j-1/2} \right) \Big], \qquad (14.39a)$$

$$(r_3)_j = T^{n-1}_{j-1/2} - \left[h_j^{-1} (e_j p_j - e_{j-1} p_{j-1}) + (fp)_{j-1/2} - \alpha_n \{ (ug)_{j-1/2} \right.$$
$$+ u^{n-1}_{j-1/2} g_{j-1/2}$$
$$- g^{n-1}_{j-1/2} u_{j-1/2} - (fp)_{j-1/2} + f^{n-1}_{j-1/2} p_{j-1/2}$$
$$\left. - p^{n-1}_{j-1/2} f_{j-1/2} \} \right]. \qquad (14.39b)$$

The coefficients of the momentum equation $(s_k)_j$ are identical to those given by Eqs. (13.64), provided that we take $m_1 = 1$ and $m_2 = -1$. Except for $(\beta_1)_j$ and $(\beta_2)_j$, the coefficients of the energy equation $(\beta_k)_j (k = 3, \ldots, 10)$ are also identical to those given by Eqs. (13.65c)–(13.65j) provided that we take $m_1 = 0$ and $d_j = 0$. The coefficients $(\beta_1)_j$ and $(\beta_2)_j$ are given by

$$(\beta_1)_j = h_j^{-1} e_j^{(i)} + \tfrac{1}{2} f_j^{(i)} + \frac{\alpha_n}{2} \left(f_j^{(i)} - f^{n-1}_{j-1/2} \right), \qquad (14.40a)$$

$$(\beta_2)_j = -h_j^{-1} e_{j-1}^{(i)} + \tfrac{1}{2} f_{j-1}^{(i)} + \frac{\alpha_n}{2} \left(f_{j-1}^{(i)} - f^{n-1}_{j-1/2} \right). \qquad (14.40b)$$

The linearized boundary conditions become

$$\delta f_0 = \delta v_0 = 0, \qquad \delta p_0 = 0; \qquad \delta u_J = 0, \qquad \delta g_J = 0. \qquad (14.41)$$

Fortran Program

To solve the linear system given by Eqs. (13.62) and (13.63), subject to the modifications discussed above and subject to the boundary conditions given by Eq. (14.41) and the turbulence model given by Eq. (14.32), several changes to the computer program of Section 13.5 are required. We define the initial velocity and temperature profiles given by Eq. (14.31) in sub-routine IVPL, the coefficients of the linearized finite-difference equations,

$(s_k)_j$ and $(\beta_l)_j$, and the $(r_k)_j$ in subroutine COEF, and the eddy-viscosity formula, Eq. (14.32), in MAIN.

One of the centerline boundary conditions for the momentum equation in our problem is different from the one used in the computer program of Section 13.5. Rather than $\delta u_0 = 0$, we now have $\delta v_0 = 0$ in Eq. (14.41). For this reason, the second row of the A_0 matrix of Eq. (13.68a) must be changed to

$$0 \quad 0 \quad 1 \quad 0 \quad 0. \tag{14.42}$$

To incorporate the centerline boundary condition $\delta p_0 = 0$ for the energy equation, we set $\alpha_0 = 0$ and take $\alpha_1 = 1.0$. No changes are needed in the SOLV5 subroutine for the edge boundary conditions.

The revised computer program, which consists of MAIN and the three subroutines IVPL, COEF, and SOLV5, is given below.

```
C     MAIN
      COMMON /INPT1/ WW(60),ALFA0,ALFA1
      COMMON /BLC0/ NP,NPT,NX,NXT,NTR,IT
      COMMON /BLC1/ F(61,2),U(61,2),V(61,2),B(61,2),G(61,2),P(61,2),
     1              C(61,2),D(61,2),E(61,2),RMU(61),BC(61)
      COMMON /GRD/ X(60),ETA(61),DETA(61),A(61)
      COMMON /EDGE/ UE(60),TE(60),RHOE(60),RMUE(60),PE(60),P1(60),P2(60)
      COMMON /BLC3/ DELF(61),DELU(61),DELV(61),DELG(61),DELP(61)
      COMMON/AK1/GW,PR,PRT,RL
C - - - - - - - - - - - - - - - - - - - - - - - - - - - - - - - - -
      EPS = 0.0001
      ITMAX = 6
      NX    = 1
      NPT = 61
      PRT = 0.9
C  INPUT DATA
C     ADIABATIC         ALFA0 = 0.0      ALFA1 = 1.0
C     SPECIFIED WALL TEMP.    ALFA0 = 1.0      ALFA1 = 0.0
      READ(5,8000) NXT,NTR
      READ(5,8100) ETAE,DETA(1),VGP,ALFA0,ALFA1,PR,RL
      READ(5,8100) (X(I),I=1,NXT)
      SQRL = SQRT(RL)
      WRITE(6,7000) NXT,NTR,PR,ETAE,DETA(1),VGP,RL
C  GRID GENERATION
      IF((VGP-1.0) .LE. 0.001) GO TO 5
      NP    = ALOG((ETAE/DETA(1))*(VGP-1.0)+1.0)/ALOG(VGP) + 1.0001
      GO TO 10
    5 NP    = ETAE/DETA(1) + 1.0001
   10 IF(NP .LE. 61) GO TO 15
      WRITE(6,9000)
      STOP
   15 ETA(1)= 0.0
      DO 20 J=2,NPT
      DETA(J)=VGP*DETA(J-1)
      A(J)  = 0.5*DETA(J-1)
   20 ETA(J)= ETA(J-1)+DETA(J-1)
      CALL IVPL
   30 WRITE(6,9100) NX,X(NX)
      P1(NX) = 1.0
      P2(NX) =-1.0
      IF(NX .EQ. 1) GO TO 90
      IT    = 0
   40 IT    = IT+1
      IF(IT .LE. ITMAX) GO TO 60
      WRITE(6,2500)
      GO TO 90
```

```
      60 CONTINUE
         IF(NX .LT. NTR) GO TO 80
C    CALCULATE EDDY VISCOSITY BASED ON SCHLICHTING FORMULA
C    EDDY = 0.037*B*UMAX, WHERE B IS HALF WIDTH
         EPS = 0.02
         UMAXH  = 0.5*U(1,2)
         DO 1 J=1,NP
         IF(U(J,2).LT.UMAXH) GOTO 2
       1 CONTINUE
         ETAB   = ETA(NP)
         GOTO 3
       2 ETAB   = ETA(J-1)+(ETA(J)-ETA(J-1))/(U(J,2)-U(J-1,2))*(UMAXH-
       1          U(J-1,2))
       3 EDV    = .037*ETAB*U(1,2)*SQRL*X(NX)**(1.0/3.0)
         DO 70 J=1,61
         E(J,2) = EDV/PRT
      70 B(J,2)= EDV
      80 CONTINUE
         CALL COEF
         CALL SOLV5
         WRITE(6,3000) U(1,2),DELU(1)
C    CHECK FOR CONVERGENCE
         IF(ABS(DELU(1)/U(1,2)) .GT. EPS) GO TO 40
         IF(U(1,2) .LT. 0.0) STOP
      90 CONTINUE
C    OUTPUT DATA
         WRITE(6,4400)
         WRITE(6,4500)  (J,ETA(J),F(J,2),U(J,2),V(J,2),G(J,2),P(J,2),
       1               B(J,2),J=1,NP,5)
         WRITE(6,4500) NP,ETA(NP),F(NP,2),U(NP,2),V(NP,2),G(NP,2),P(NP,2),
       1               B(NP,2)
         NX     = NX+1
         IF(NX .GT. NXT) STOP
C    SHIFT PROFILES
         DO 110 J=1,NP
         F(J,1)= F(J,2)
         U(J,1)= U(J,2)
         V(J,1)= V(J,2)
         G(J,1)= G(J,2)
         P(J,1)= P(J,2)
         E(J,1)= E(J,2)
     110 B(J,1)= B(J,2)
         GO TO 30
C - - - - - - - - - - - - - - - - - - - - - - - - - - - - - - - - - - -
    8000 FORMAT(8I3)
    8100 FORMAT(8F10.0)
    3000 FORMAT(1H ,7HU(1,2)=,E13.6,2X,6HDELU =,E13.6)
    4400 FORMAT(1H0,2X,1HJ,4X,3HETA,9X,1HF,13X,1HU,13X,1HV,13X,1HG,13X,1HP,
       1      13X,1HB)
    4500 FORMAT(1H ,I3,F10.3,6E14.6)
    7000 FORMAT(1H0,6HNXT  =,I3,14X,6HNTR  =,I3,14X,6HPR    =,F5.3/
       1      1H ,6HETAE =,E14.6,3X,6HDETA1=,E14.6,3X,6HVGP  =,E14.6,3X,
       2      6HRL   =,E14.6/)
    2500 FORMAT(1H0,16X,25HITERATIONS EXCEEDED ITMAX)
    9000 FORMAT(1H0,'NP EXCEEDED NPT -- PROGRAM TERMINATED')
    9100 FORMAT(1H0,4HNX =,I3,5X,3HX =,F10.3/)
    9400 FORMAT(1H ,79X,2E14.6)
         END




         SUBROUTINE IVPL
         COMMON /BLCO/ NP,NPT,NX,NXT,NTR,IT
         COMMON /BLC1/ F(61,2),U(61,2),V(61,2),B(61,2),G(61,2),P(61,2),
       1              C(61,2),D(61,2),E(61,2),RMU(61),BC(61)
         COMMON /GRD/ X(60),ETA(61),DETA(61),A(61)
         COMMON/AK1/GW,PR,PRT,RL
C - - - - - - - - - - - - - - - - - - - - - - - - - - - - - - - - - - -
C        G(J,2) = U(J,2)
C        P(J,2) = V(J,2)
         PY = ARCOS(-1.0)
         SQPY = SQRT(PY)
         AA = 25.0
         SQRL  = SQRT(RL)
```

```
      F(1,2)= 0.0
      TERM    = 27.855*X(NX)**(2.0/3.0)/SQRL
      DO 30 J=1,NP
      ETAH = ETA(J)*ETA(J)/AA
      ETAS = ETAH*ETAH
      TANF    = TANH( TERM*(ETA(J)-5.0) )
      U(J,2)= 0.5*X(NX)**(1.0/3.0)*3.0*(1.0-TANF)
      V(J,2)= -TERM*X(NX)**(1.0/3.0)*0.5*(1.0-TANF**2)*3.0
      G(J,2) = 1.0 - ERF(ETAH)
      P(J,2) = -4.0*EXP(-ETAS)/SQPY/AA*ETA(J)
      C(J,2) = 1.0
      B(J,2) = 1.0
      E(J,2) = 1.0/PR
      D(J,2) = 0.0
      BC(J) = 1.0
      C(J,1) = 1.0
      D(J,1) = 0.0
      IF(J .EQ. 1) GO TO 30
      F(J,2) = F(J-1,2) + A(J)*(U(J,2)+U(J-1,2))
   30 CONTINUE
      IF(NX .LT. NTR) RETURN
C     CALCULATE EDDY VISCOSITY BASED ON SCHLICHTING FORMULA
C     EDDY = 0.037*B*UMAX, WHERE B IS HALF WIDTH
      UMAXH   = 0.5*U(1,2)
      DO 1 J=1,NP
      IF(U(J,2).LT.UMAXH) GOTO 2
    1 CONTINUE
      ETAB    = ETA(NP)
      GOTO 3
    2 ETAB    = ETA(J-1)+(ETA(J)-ETA(J-1))/(U(J,2)-U(J-1,2))*(UMAXH-
     1          U(J-1,2))
    3 EDV     = .037*ETAB*U(1,2)*SQRL*X(NX)**(1.0/3.0)
      DO 70 J=1,NP
      E(J,2) = EDV/PRT
   70 B(J,2)= EDV
      RETURN
      END



      SUBROUTINE COEF
      COMMON /INPT1/ WW(60),ALFA0,ALFA1
      COMMON /BLC0/ NP,NPT,NX,NXT,NTR,IT
      COMMON /BLC1/ F(61,2),U(61,2),V(61,2),B(61,2),G(61,2),P(61,2),
     1       C(61,2),D(61,2),E(61,2),RMU(61),BC(61)
      COMMON /GRD/ X(60),ETA(61),DETA(61),A(61)
      COMMON /EDGE/ UE(60),TE(60),RHOE(60),RMUE(60),PE(60),P1(60),P2(60)
      COMMON/BLC6/ S1(61),S2(61),S3(61),S4(61),S5(61),S6(61),S7(61),
     1       S8(61),B1(61),B2(61),B3(61),B4(61),B5(61),B6(61),
     2       B7(61),B8(61),B9(61),B10(61),R(5,61)
C - - - - - - - - - - - - - - - - - - - - - - - - - - - - - - - -
      IF(IT .GT. 1) GO TO 5
      CEL = 1.5*(X(NX) + X(NX-1))/(X(NX) - X(NX-1))
      P1P = P1(NX) + CEL
      P2P = P2(NX) + CEL
    5 CONTINUE
      DO 100 J=2,NP
C   PRESENT STATION
      USB     = 0.5*(U(J,2)**2+U(J-1,2)**2)
      FVB     = 0.5*(F(J,2)*V(J,2)+F(J-1,2)*V(J-1,2))
      FPB     = 0.5*(F(J,2)*P(J,2)+F(J-1,2)*P(J-1,2))
      UGB     = 0.5*(U(J,2)*G(J,2)+U(J-1,2)*G(J-1,2))
      UB      = 0.5*(U(J,2)+U(J-1,2))
      VB      = 0.5*(V(J,2)+V(J-1,2))
      FB      = 0.5*(F(J,2)+F(J-1,2))
      GB      = 0.5*(G(J,2)+G(J-1,2))
      PB      = 0.5*(P(J,2)+P(J-1,2))
      CB      = 0.5*(C(J,2) + C(J-1,2))
      DERBV   = (B(J,2)*V(J,2)-B(J-1,2)*V(J-1,2))/DETA(J-1)
      DEREP   = (E(J,2)*P(J,2)-E(J-1,2)*P(J-1,2))/DETA(J-1)
      DRDUV   = (D(J,2)*U(J,2)*V(J,2) - D(J-1,2)*U(J-1,2)*V(J-1,2))/
     1          DETA(J-1)
      CFB     = 0.5*(F(J,1)+F(J-1,1))
      CVB     = 0.5*(V(J,1)+V(J-1,1))
      CPB     = 0.5*(P(J,1)+P(J-1,1))
      CUB     = 0.5*(U(J,1)+U(J-1,1))
```

```
          CGB    = 0.5*(G(J,1)+G(J-1,1))
          CFVB   = 0.5*(F(J,1)*V(J,1)+F(J-1,1)*V(J-1,1))
          CFPB   = 0.5*(F(J,1)*P(J,1)+F(J-1,1)*P(J-1,1))
          CUGB   = 0.5*(U(J,1)*G(J,1)+U(J-1,1)*G(J-1,1))
          CUSB   = 0.5*(U(J,1)**2+U(J-1,1)**2)
          CCB = 0.5*(C(J,1) + C(J-1,1))
          CDERBV= (B(J,1)*V(J,1)-B(J-1,1)*V(J-1,1))/DETA(J-1)
          CDEREP= (E(J,1)*P(J,1)-E(J-1,1)*P(J-1,1))/DETA(J-1)
          CDRDUV = (D(J,1)*U(J,1)*V(J,1) - D(J-1,1)*U(J-1,1)*V(J-1,1))/
         1         DETA(J-1)
          CLB = CDERBV + P1(NX-1)*CFVB - P2(NX-1)*CUSB
          CRB = -CLB  - CEL*CUSB + CEL*CFVB
          CMB = CDEREP + CDRDUV + P1(NX-1)*CFPB
          CTB = -CMB + CEL*(CFPB-CUGB)
C     COEFFICIENTS OF THE DIFFERENCED MOMENTUM EQ.
          S1(J) = B(J,2)/DETA(J-1) + 0.5*P1P*F(J,2) - 0.5*CEL*CFB
          S2(J) = -B(J-1,2)/DETA(J-1) + 0.5*P1P*F(J-1,2) - 0.5*CEL*CFB
          S3(J) = 0.5*(P1P*V(J,2) + CEL*CVB)
          S4(J) = 0.5*(P1P*V(J-1,2) + CEL*CVB)
          S5(J) = -P2P*U(J,2)
          S6(J) = -P2P*U(J-1,2)
          S7(J) = 0.0
          S8(J) = S7(J)
          R(2,J) = CRB - (DERBV + P1P*FVB - P2P*USB + CEL*(FB*CVB-VB*CFB))
C     COEFFICIENTS OF DIFFERENCED ENERGY EQ.
          B1(J) = E(J,2)/DETA(J-1) + 0.5*P1P*F(J,2) - 0.5*CEL*CFB
          B2(J) = -E(J-1,2)/DETA(J-1) + 0.5*P1P*F(J-1,2) - 0.5*CEL*CFB
          B3(J) = 0.5*(P1P*P(J,2) + CEL*CPB)
          B4(J) = 0.5*(P1P*P(J-1,2) + CEL*CPB)
          B5(J) = D(J,2)*V(J,2)/DETA(J-1) - 0.5*CEL*(G(J,2)-CGB)
          B6(J) = -D(J-1,2)*V(J-1,2)/DETA(J-1) - 0.5*CEL*(G(J-1,2)-CGB)
          B7(J) = -0.5*CEL*(U(J,2)+CUB)
          B8(J) = -0.5*CEL*(U(J-1,2)+CUB)
          B9(J) = D(J,2)*U(J,2)/DETA(J-1)
          B10(J) = -D(J-1,2)*U(J-1,2)/DETA(J-1)
          R(3,J) = CTB - (DEREP + DRDUV + P1P*FPB - CEL*(UGB-CGB*UB+CUB*GB)+


          SUBROUTINE SOLV5
          COMMON /INPT1/ WW(60),ALFAO,ALFA1
          COMMON /BLCO/ NP,NPT,NX,NXT,NTR,IT
          COMMON /GRD/ X(60),ETA(61),DETA(61),A(61)
          COMMON /BLC1/ F(61,2),U(61,2),V(61,2),B(61,2),G(61,2),P(61,2),
         1              C(61,2),D(61,2),E(61,2),RMU(61),BC(61)
          COMMON /BLC3/ DELF(61),DELU(61),DELV(61),DELG(61),DELP(61)
          COMMON/BLC6/ S1(61),S2(61),S3(61),S4(61),S5(61),S6(61),S7(61),
         1              S8(61),B1(61),B2(61),B3(61),B4(61),B5(61),B6(61),
         2              B7(61),B8(61),B9(61),B10(61),R(5,61)
          DIMENSION     A11(61),A12(61),A13(61),A14(61),A15(61),A21(61),
         1              A22(61),A23(61),A24(61),A25(61),A31(61),A32(61),
         2              A33(61),A34(61),A35(61),G11(61),G12(61),G13(61),
         3              G14(61),G15(61),G21(61),G22(61),G23(61),G24(61),
         4              G25(61),G31(61),G32(61),G33(61),G34(61),G35(61),
         5              W1(61),W2(61),W3(61),W4(61),W5(61)
C - - - - - - - - - - - - - - - - - - - - - - - - - - - - - - - - - - -
          A11(1)=1.0
          A12(1)=0.0
          A13(1)=0.0
          A14(1)=0.0
          A15(1)=0.0
          A21(1)=0.0
          A22(1)=0.0
          A23(1)=1.0
          A24(1)=0.0
          A25(1)=0.0
          A31(1)=0.0
          A32(1)=0.0
          A33(1)=0.0
          A34(1)=ALFAO
          A35(1)=ALFA1
          W1 (1)=R(1,1)
          W2 (1)=R(2,1)
          W3 (1)=R(3,1)
          W4 (1)=R(4,1)
          W5 (1)=R(5,1)
```

```
C - FORWARD SWEEP
      DO 30 J=2,NP
      AA1=A(J)*A24(J-1)-A25(J-1)
      AA2=A(J)*A34(J-1)-A35(J-1)
      AA3=A(J)*A12(J-1)-A13(J-1)
      AA4=A(J)*A22(J-1)-A23(J-1)
      AA5=A(J)*A32(J-1)-A33(J-1)
      AA6=A(J)*A14(J-1)-A15(J-1)
      AA7=A(J)*S6(J)-S2(J)
      AA8=S8(J)*A(J)
      AA9=A(J)*B6(J)-B10(J)
      AA10=A(J)*B8(J)-B2(J)
      DET=A11(J-1)*(AA4*AA2-AA1*AA5)-A21(J-1)*(AA3*AA2-AA5*AA6)+
     1      A31(J-1)*(AA3*AA1-AA4*AA6)
      G11(J)=(-(AA4*AA2-AA5*AA1)+A(J)**2*(A21(J-1)*AA2-A31(J-1)*AA1))/
     1      DET
      G12(J)=((AA3*AA2-AA5*AA6)-A(J)**2*(A11(J-1)*AA2-A31(J-1)*AA6))/DET
      G13(J)=(-(AA3*AA1-AA4*AA6)+A(J)**2*(A11(J-1)*AA1-A21(J-1)*AA6))/
     1      DET
      G14(J)=G11(J)*A12(J-1)+G12(J)*A22(J-1)+G13(J)*A32(J-1)+A(J)
      G15(J)=G11(J)*A14(J-1)+G12(J)*A24(J-1)+G13(J)*A34(J-1)
      G21(J)=(S4(J)*(AA2*AA4-AA1*AA5)+A31(J-1)*(AA1*AA7-AA4*AA8)+
     1      A21(J-1)*(AA5*AA8-AA7*AA2))/DET
      G22(J)=(A11(J-1)*(AA2*AA7-AA5*AA8)+A31(J-1)*(AA3*AA8-AA6*AA7)+
     1      S4(J)*(AA5*AA6-AA2*AA3))/DET
      G23(J)=(A11(J-1)*(AA4*AA8-AA1*AA7)+S4(J)*(AA3*AA1-AA4*AA6)+
     1      A21(J-1)*(AA7*AA6-AA3*AA8))/DET
      G24(J)=G21(J)*A12(J-1)+G22(J)*A22(J-1)+G23(J)*A32(J-1)-S6(J)
      G25(J)=G21(J)*A14(J-1)+G22(J)*A24(J-1)+G23(J)*A34(J-1)-S8(J)
      G31(J)=(B4(J)*(AA4*AA2-AA5*AA1)-AA9*(A21(J-1)*AA2-A31(J-1)*AA1)+
     1      AA10*(A21(J-1)*AA5-A31(J-1)*AA4)/DET
      G32(J)=(-B4(J)*(AA3*AA2-AA5*AA6)+AA9*(A11(J-1)*AA2-A31(J-1)*AA6)-
     1      AA10*(A11(J-1)*AA5-A31(J-1)*AA3))/DET
      G33(J)=(B4(J)*(AA3*AA1-AA4*AA6)-AA9*(A11(J-1)*AA1-A21(J-1)*AA6)+
     1      AA10*(A11(J-1)*AA4-A21(J-1)*AA3))/DET
      G34(J)=G31(J)*A12(J-1)+G32(J)*A22(J-1)+G33(J)*A32(J-1)-B6(J)
      G35(J)=G31(J)*A14(J-1)+G32(J)*A24(J-1)+G33(J)*A34(J-1)-B8(J)
     1              CEL*(CPB*FB-CFB*PB))
C  DEFINITIONS OF RJ
      R(1,J) = F(J-1,2)-F(J,2)+DETA(J-1)*UB
      R(4,J-1) = U(J-1,2)-U(J,2)+DETA(J-1)*VB
      R(5,J-1) = G(J-1,2)-G(J,2)+DETA(J-1)*PB
  100 CONTINUE
      R(1,1) = 0.0
      R(2,1) = 0.0
      R(3,1) = 0.0
      R(4,NP)=0.0
      R(5,NP)=0.0
      RETURN
      END


      A11(J)=1.0
      A12(J)=-A(J)-G14(J)
      A13(J)=A(J)*G14(J)
      A14(J)=-G15(J)
      A15(J)=A(J)*G15(J)
      A21(J)=S3(J)
      A22(J)=S5(J)-G24(J)
      A23(J)=S1(J)+A(J)*G24(J)
      A24(J)=-G25(J)+S7(J)
      A25(J)=A(J)*G25(J)
      A31(J)=B3(J)
      A32(J)=B5(J)-G34(J)
      A33(J)=B9(J)+A(J)*G34(J)
      A34(J)=B7(J)-G35(J)
      A35(J)=B1(J)+A(J)*G35(J)
      W1(J) =R(1,J)-G11(J)*W1(J-1)-G12(J)*W2(J-1)-G13(J)*W3(J-1)-
     1        G14(J)*W4(J-1)-G15(J)*W5(J-1)
      W2(J) =R(2,J)-G21(J)*W1(J-1)-G22(J)*W2(J-1)-G23(J)*W3(J-1)-
     1        G24(J)*W4(J-1)-G25(J)*W5(J-1)
      W3(J) =R(3,J)-G31(J)*W1(J-1)-G32(J)*W2(J-1)-G33(J)*W3(J-1)-
     1        G34(J)*W4(J-1)-G35(J)*W5(J-1)
      W4(J) =R(4,J)
      W5(J) =R(5,J)
   30 CONTINUE
C - BACKWARD SWEEP
```

```
      J      =NP
      DP =      -(A31(J)*(A13(J)*W2(J)-W1(J)*A23(J))-A32(J)*(A11(J)*
     1         W2(J)-W1(J)*A21(J)))  +  W3(J)*(A11(J)*A23(J)-A13(J)*A21(J)))
      DV =      -(A31(J)*(W1(J)*A25(J)-W2(J)*A15(J))-W3(J)*(A11(J)*A25(J)
     1         -A15(J)*A21(J))+A35(J)*(A11(J)*W2(J)-W1(J)*A21(J)))
      DF =      -(W3(J)*(A13(J)*A25(J)-A23(J)*A15(J))-A33(J)*(W1(J)*A25(J)
     1         -A15(J)*W2(J)))  +  A35(J)*(W1(J)*A23(J)-A13(J)*W2(J)))
      D1 =      -(A31(J)*(A13(J)*A25(J)-A23(J)*A15(J))-A33(J)*(A11(J)*
     1         A25(J)-A21(J)*A15(J))+A35(J)*(A11(J)*A23(J)-A21(J)*A13(J)))
      DELP(J) = DP/D1
      DELV(J) = DV/D1
      DELF(J) = DF/D1
      DELG(J) = 0.0
      DELU(J) = 0.0
   40 J = J-1
      BB1=DELU(J+1)-A(J+1)*DELV(J+1)-W4(J)
      BB2=DELG(J+1)-A(J+1)*DELP(J+1)-W5(J)
      CC1=W1(J)-A12(J)*BB1-A14(J)*BB2
      CC2=W2(J)-A22(J)*BB1-A24(J)*BB2
      CC3=W3(J)-A32(J)*BB1-A34(J)*BB2
      DD1=A13(J)-A12(J)*A(J+1)
      DD2=A23(J)-A22(J)*A(J+1)
      DD3=A33(J)-A32(J)*A(J+1)
      EE1=A15(J)-A14(J)*A(J+1)
      EE2=A25(J)-A24(J)*A(J+1)
      EE3=A35(J)-A34(J)*A(J+1)
      DETT=A11(J)*DD2*EE3+A21(J)*DD3*EE1+A31(J)*DD1*EE2
     1     -A31(J)*DD2*EE1-A21(J)*DD1*EE3-A11(J)*DD3*EE2
      DELF(J)=(CC1*DD2*EE3+CC2*DD3*EE1+CC3*DD1*EE2-CC3*DD2*EE1
     1        -CC2*DD1*EE3-CC1*DD3*EE2)/DETT
      DELV(J)=(A11(J)*CC2*EE3+A21(J)*CC3*EE1+A31(J)*CC1*EE2-
     1        A31(J)*CC2*EE1-A21(J)*CC1*EE3-A11(J)*CC3*EE2)/DETT
      DELP(J)=(A11(J)*CC3*DD2+A21(J)*CC1*DD3+A31(J)*CC2*DD1-
     1        A31(J)*CC1*DD2-A21(J)*CC3*DD1-A11(J)*CC2*DD3)/DETT
      DELU(J)=BB1-A(J+1)*DELV(J)
      DELG(J)= BB2-A(J+1)*DELP(J)
      IF(J .GT. 1) GO TO 40
      DO 50 J=1,NP
      F(J,2)=F(J,2)+DELF(J)
      U(J,2)=U(J,2)+DELU(J)
      V(J,2)=V(J,2)+DELV(J)
      G(J,2)=G(J,2)+DELG(J)
      P(J,2)=P(J,2)+DELP(J)
   50 CONTINUE
      U(NP,2) = 0.0
      RETURN
      END
```

14.4 Mixing Layer between Two Uniform Streams at Different Temperatures

As a fourth example to illustrate the application of Fortran program of Section 13.5, we now discuss the solution of the mixing layer between two uniform streams at different temperatures that was discussed in Sections 8.2 and 8.4. Here, for simplicity, we consider a laminar similar flow.

The transformed momentum and energy equations and their boundary conditions for this flow are given by Eqs. (8.48), (8.49), and (8.44) and are repeated below for convenience.

$$f''' + \tfrac{1}{2}ff'' = 0. \tag{8.48}$$

$$\frac{g''}{\mathrm{Pr}} + \frac{1}{2}fg' = 0. \tag{8.49}$$

$$\eta = -\eta_e, \quad f' = \lambda, \quad g = 0; \quad \eta = \eta_e, \quad f' = 1, \quad g = 1. \tag{8.44a}$$

$$\eta = 0, \quad f = 0. \tag{8.44b}$$

We note that these equations, though not the boundary conditions, are identical to those for a uniformly heated flat-plate flow [$m = n = 0$ in Eqs. (4.31) and (4.32)]. If it were not for the middle boundary condition at $\eta = 0$ given by Eq. (8.44b), their solution would be rather straightforward with the Fortran program of Section 13.5. The presence of this boundary condition, however, complicates the situation and requires changes in the solution procedure used for external-flow problems.

To use the Fortran program of Section 13.5 with as few changes as possible, it is best to employ the nonlinear eigenvalue approach described in Section 14.1. In this approach, we first solve the *standard problem* given by Eqs. (8.48) and (9.49) subject to

$$\eta = -\eta_e, \qquad f' = \lambda, \qquad g = 0, \qquad f = s, \qquad (14.43a)$$

$$\eta = \eta_e, \qquad f' = 1, \qquad g = 1, \qquad (14.43b)$$

where s is an assumed constant. From the solution of the standard problem, we find the value of $f \, (\equiv f_0)$ at $\eta = 0$ to see whether the calculated value satisfies the boundary condition given by Eq. (8.44b). If not, since the solution is a function of the parameter s, we use Newton's method in order to find another value of s so that the newly computed f_0 will satisfy Eq. (8.44b). Specifically, if we denote

$$\phi(s^\nu) = f_0, \qquad (14.44)$$

then we compute

$$s^{\nu+1} = s^\nu - \frac{\phi}{\partial \phi / \partial s} \equiv s^\nu - \frac{f_0}{\partial f_0 / \partial s}, \qquad \nu = 0, 1, 2, \ldots. \quad (14.45)$$

As in Section 14.1, $\partial f_0 / \partial s$ is obtained by solving the variational equations subject to

$$\eta = -\eta_e, \qquad f = 1, \qquad f' = 0, \qquad g = 0, \qquad (14.46a)$$

$$\eta = \eta_e, \qquad f' = 0, \qquad g = 0. \qquad (14.46b)$$

With the newly computed value of s, we again solve the standard problem to compute whether

$$|\phi(s^\nu)| < \varepsilon, \qquad (14.47)$$

where ε is a tolerance parameter, say, equal to 10^{-3}. If not, we solve the variational equations again to get a new estimate of s, and we repeat the above procedure until Eq. (8.44b) is satisfied.

For the standard problem, the boundary conditions have the same form as a boundary-layer flow. Since the finite-difference coefficients of the momentum and energy equations correspond to those of a uniformly heated flat-plate flow; essentially the COEF subroutine does not need substantial changes. All we need to do is set $b_j = 1$, $e_j = 1/\text{Pr}$, $d = 0$, and $\alpha_n = 0$ and take $\alpha_0 = 1$ and $\alpha_1 = 0$ to handle the boundary conditions of the energy equation.

The coefficients of the variational equations also do not require substantial changes. Except for the $(r_1)_0$ term, which is equal to 1, the right-hand side of the variational equations $(r_k)_j$ are all zero, and their coefficients $(s_k)_j$ and $(\beta_l)_j$ are the same as those in the standard problem. As a result, to solve the variational equations, all we need to do is recompute $\mathbf{r}_j$, and since the $\mathbf{A}$ matrix in Eq. (13.28) is already known from the standard problem, solve for $\boldsymbol{\delta}_j$, which is now defined by Eq. (14.15). Unlike the problem of Section 14.1, however, the calculations in SOLV5 must continue from the "outer" edge $\eta = \eta_j$ up to $\eta = 0$ to compute $(f_1)_{\eta=0}$. All these modifications are rather straightforward; so we do not present sample calculations or a revised listing of the computer program.

References

[1] Bradshaw, P., Cebeci, T., and Whitelaw, J. H.: *Engineering Calculation Methods for Turbulent Flows*. Academic, London, 1981.

[2] Cebeci, T. and Smith A. M. O.: *Analysis of Turbulent Boundary Layers*. Academic, New York, 1974.

[3] Cebeci, T., Khattab, A. A., and Lamont, R.: Combined natural and forced convection in vertical ducts. Proc. Seventh International Heat Transfer Conf., Munich, Germany, Hemisphere Publishing Co., Washington, 1982.

[4] Isaacson, E. and Keller, H. B.: *Analysis of Numerical Methods*. Wiley, New York, 1966, p. 192.

APPENDIX A
Conversion Factors

1. Acceleration
$1 \text{ ft s}^{-2} = 0.3048 \text{ m s}^{-2}$
$1 \text{ m s}^{-2} = 3.2808 \text{ ft s}^{-2}$

2. Area
$1 \text{ in}^2 = 6.4516 \text{ cm}^2$
$1 \text{ ft}^2 = 0.0929 \text{ m}^2$
$1 \text{ m}^2 = 10.764 \text{ ft}^2$

3. Density
$1 \text{ lb in}^{-3} = 27.680 \text{ g cm}^{-3}$
$1 \text{ lb ft}^{-3} = 16.019 \text{ kg m}^{-3}$
$1 \text{ kg m}^{-3} = 0.06243 \text{ lb ft}^{-3}$
$1 \text{ slug ft}^{-3} = 515.38 \text{ kg m}^{-3}$

4. Diffusivity (heat, mass, momentum)
$1 \text{ ft}^2 \text{ s}^{-1} = 0.0929 \text{ m}^2 \text{ s}^{-1}$
$1 \text{ ft}^2 \text{ h}^{-1} = 0.2581 \times 10^{-4} \text{ m}^2 \text{ s}^{-1}$
$1 \text{ m}^2 \text{ s}^{-1} = 10.7639 \text{ ft}^2 \text{ s}^{-1}$
$1 \text{ cm}^2 \text{ s}^{-1} = 3.8745 \text{ ft}^2 \text{ h}^{-1}$

5. Energy, heat, power
$1 \text{ J} = 1 \text{ W s} = 1 \text{ N m}$
$1 \text{ J} = 10^7 \text{ erg}$

$1 \text{ Btu} = 1055.04 \text{ J}$
$1 \text{ Btu} = 1055.04 \text{ W s}$
$1 \text{ Btu} = 252 \text{ cal}$
$1 \text{ Btu} = 778.161 \text{ ft lb}_f$
$1 \text{ Btu h}^{-1} = 0.2931 \text{ W}$
$1 \text{ Btu h}^{-1} = 3.93 \times 10^{-4} \text{ hp}$
$1 \text{ cal} = 4.1868 \text{ J (or W s or N m)}$
$1 \text{ cal} = 3.968 \times 10^{-3} \text{ Btu}$
$1 \text{ hp} = 550 \text{ ft lb}_f \text{ s}^{-1}$
$1 \text{ hp} = 745.7 \text{ W} = 745.7 \text{ N m s}^{-1}$
$1 \text{ Wh} = 3.413 \text{ Btu}$

6. Heat capacity, heat per unit mass, specific heat
$1 \text{ Btu h}^{-1} \text{ °F}^{-1} = 0.5274 \text{ W °C}^{-1}$
$1 \text{ W °C}^{-1} = 1.8961 \text{ Btu h}^{-1} \text{ °F}^{-1}$
$1 \text{ Btu lb}^{-1} = 2325.9 \text{ J kg}^{-1}$
$1 \text{ Btu lb}^{-1} \text{ °F}^{-1} = 4.18669 \text{ kJ kg}^{-1}$
$\text{°C}^{-1} \text{ (or J g}^{-1} \text{ °C}^{-1})$
$1 \text{ Btu lb}^{-1} \text{ °F}^{-1} = 1 \text{ cal g}^{-1}$
$\text{°C}^{-1} = 1 \text{ kcal kg}^{-1} \text{ °C}^{-1}$

456

7. Heat flux

$1 \text{ Btu h}^{-1} \text{ ft}^{-2} = 3.1537 \times 10^{-3} \text{ kW m}^{-2}$

$1 \text{ W m}^{-2} = 0.31709 \text{ Btu h}^{-1} \text{ ft}^{-2}$

8. Heat-generation rate

$1 \text{ Btu h}^{-1} \text{ ft}^{-3} = 10.35 \text{ W m}^{-3}$

$1 \text{ W m}^{-3} = 0.0966 \text{ Btu h}^{-1} \text{ ft}^{-3}$

9. Heat-transfer coefficient

$1 \text{ Btu h}^{-1} \text{ ft}^{-2} \,^\circ\text{F}^{-1} = 5.677 \text{ W m}^{-2} \,^\circ\text{C}^{-1}$

$1 \text{ W m}^{-2} \,^\circ\text{C}^{-1} = 0.1761 \text{ Btu h}^{-1} \text{ ft}^{-2} \,^\circ\text{F}^{-1}$

$1 \text{ Btu h}^{-1} \text{ ft}^{-2} \,^\circ\text{F}^{-1} = 4.882 \text{ kcal h}^{-1} \text{ m}^{-2} \,^\circ\text{C}^{-1}$

10. Length

$1 \text{ in} = 2.54 \text{ cm}$

$1 \text{ in} = 2.54 \times 10^{-2} \text{ m}$

$1 \text{ ft} = 0.3048 \text{ m}$

$1 \text{ m} = 3.2808 \text{ ft}$

$1 \text{ mile} = 1609.34 \text{ m}$

$1 \text{ mile} = 5280 \text{ ft}$

11. Mass

$1 \text{ oz} = 28.35 \text{ g}$

$1 \text{ lb} = 16 \text{ oz}$

$1 \text{ lb} = 453.6 \text{ g}$

$1 \text{ lb} = 0.4536 \text{ kg}$

$1 \text{ kg} = 2.2046 \text{ lb}$

$1 \text{ g} = 15.432 \text{ grains}$

$1 \text{ slug} = 32.1739 \text{ lb}$

12. Mass flux

$1 \text{ lb ft}^{-2} \text{ h}^{-1} = 1.3563 \times 10^{-3} \text{ kg m}^{-2} \text{ s}^{-1}$

$1 \text{ lb ft}^{-2} \text{ s}^{-1} = 4.882 \text{ kg m}^{-2} \text{ s}^{-1}$

$1 \text{ kg m}^{-2} \text{ s}^{-1} = 737.3 \text{ lb ft}^{-2} \text{ h}^{-1}$

$1 \text{ kg m}^{-2} \text{ s}^{-1} = 0.2048 \text{ lb ft}^{-2} \text{ s}^{-1}$

13. Pressure, force

$1 \text{ N} = 1 \text{ kg m s}^{-2}$

$1 \text{ N} = 0.22481 \text{ lb}_f$

$1 \text{ N} = 7.2333 \text{ poundals}$

$1 \text{ N} = 10^5 \text{ dyn}$

$1 \text{ N m}^{-2} = 1 \text{ Pa}$

$1 \text{ lb}_f = 32.174 \text{ ft lb s}^{-2}$

$1 \text{ lb}_f = 4.4482 \text{ N}$

$1 \text{ lb}_f = 32.1739 \text{ poundals}$

$1 \text{ lb}_f \text{ in}^{-2} \equiv (1 \text{ psi}) = 6894.76 \text{ Pa}$

$1 \text{ lb}_f \text{ ft}^{-2} = 47.880 \text{ Pa}$

$1 \text{ bar} = 10^5 \text{ Pa}$

$1 \text{ atm} = 14.696 \text{ lb}_f \text{ in}^{-2}$

$1 \text{ atm} = 2116.2 \text{ lb}_f \text{ ft}^{-2}$

$1 \text{ atm} = 1.0132 \times 10^5 \text{ Pa}$

14. Specific heat

$1 \text{ Btu lb}^{-1} \,^\circ\text{F}^{-1} = 1 \text{ kcal kg}^{-1} \,^\circ\text{C}^{-1} = 1 \text{ cal g}^{-1} \,^\circ\text{C}^{-1}$

$1 \text{ Btu lb}^{-1} \,^\circ\text{F}^{-1} = 4.18669 \text{ J g}^{-1} \,^\circ\text{K}^{-1} \text{ (or W s g}^{-1} \,^\circ\text{C}^{-1})$

$1 \text{ J g}^{-1} \,^\circ\text{C}^{-1} = 0.23885 \text{ Btu lb}^{-1} \,^\circ\text{F}^{-1}$

$\quad \text{(cal g}^{-1} \,^\circ\text{C}^{-1} \text{ or kcal kg}^{-1} \,^\circ\text{C}^{-1})$

15. Speed

$1 \text{ ft s}^{-1} = 0.3048 \text{ m s}^{-1}$

$1 \text{ m s}^{-1} = 3.2808 \text{ ft s}^{-1}$

$1 \text{ mile h}^{-1} = 1.4667 \text{ ft s}^{-1}$

$1 \text{ mile h}^{-1} = 0.44704 \text{ m s}^{-1}$

16. Temperature

$1^\circ \text{ K} = 1.8^\circ \text{R}$

$T(^\circ\text{F}) = 1.8(^\circ\text{K} - 273) + 32$

$T(^\circ\text{K}) = \dfrac{1}{1.8}(^\circ\text{F} - 32) + 273$

$T(^\circ\text{C}) = \dfrac{1}{1.8}(^\circ\text{R} - 492)$

17. Thermal conductivity

$1 \text{ Btu h}^{-1} \text{ ft}^{-1} \,^\circ\text{F}^{-1} = 1.7303 \text{ W m}^{-1} \,^\circ\text{C}^{-1}$

$1 \text{ Btu h}^{-1} \text{ ft}^{-1} \,^\circ\text{F}^{-1} = 0.4132 \text{ cal s}^{-1} \text{ m}^{-1} \,^\circ\text{C}^{-1}$

$1 \text{ W m}^{-1} \,^\circ\text{C}^{-1} = 0.5779 \text{ Btu h}^{-1} \text{ ft}^{-1} \,^\circ\text{F}^{-1}$

18. Thermal resistance

$1 \text{ h}^{-1} \,^\circ\text{F}^{-1} \text{ Btu}^{-1} = 1.896 \,^\circ\text{C W}^{-1}$

$1 \,^\circ\text{C W}^{-1} = 0.528 \text{ h }^\circ\text{F Btu}^{-1}$

19. Viscosity

$1 \text{ poise} = 1 \text{ g cm}^{-1} \text{ s}^{-1}$

$1 \text{ poise} = 10^2 \text{ centipoise}$

$1 \text{ poise} = 241.9 \text{ lb ft}^{-1} \text{ h}^{-1}$

$1 \text{ lb ft}^{-1} \text{ s}^{-1} = 1.4882 \text{ kg m}^{-1} \text{ s}^{-1}$

$1 \text{ lb ft}^{-1} \text{ s}^{-1} = 14.882 \text{ poises}$

$1 \text{ lb ft}^{-1} \text{ h}^{-1} = 0.4134 \times 10^{-3} \text{ kg m}^{-1} \text{ s}^{-1}$

$1 \text{ lb ft}^{-1} \text{ h}^{-1} = 0.4134 \times 10^{-2} \text{ poise}$

20. Volume

$1 \text{ in}^3 = 16.387 \text{ cm}^3$

$1 \text{ cm}^3 = 0.06102 \text{ in}^3$

$1 \text{ oz (U.S. fluid)} = 29.573 \text{ cm}^3$

1 ft^3 = 0.0283168 m^3
1 ft^3 = 28.3168 liters
1 ft^3 = 7.4805 gal (U.S.)
1 m^3 = 35.315 ft^3
1 gal (U.S.) = 3.7854 liters
1 gal (U.S.) = 3.7854×10^{-3} m^3
1 gal (U.S.) = 0.13368 ft^3

Constants

g_c = gravitational acceleration
conversion factor   = 32.1739 ft lb lb$_f^{-1}$ s^{-2}

= 4.1697×10^8 ft lb lb$_f^{-1}$ h^{-2}
= 1 g cm dyn^{-1} s^{-2}
= 1 kg m N^{-1} s^{-2}
= 1 lb ft poundal^{-1} s^{-2}
= 1 slug ft lb$_f^{-1}$ s^{-2}

J = mechanical equivalent of heat = 778.16 ft lb$_f$ Btu^{-1}
$\mathscr{R}$ = gas constant   = 1544 ft lb$_f$ lb^{-1} mol^{-1} °R^{-1}
= 8.314 N m g^{-1} mol^{-1} °K^{-1}
= 1.987 cal g^{-1} mol^{-1} °K^{-1}

Physical Properties of Gases, Liquids, Liquid Metals, and Metals

Table B-1 Physical properties of gases at atmospheric pressure

T, °K	ρ kg m^{-3}	c_p, kJ kg^{-1} °K^{-1}	μ, kg m^{-1}s^{-1}	ν, m^2 s^{-1} $\times 10^6$	k, W m^{-1} °K^{-1}	κ, m^2 s^{-1} $\times 10^4$	Pr
Air							
100	3.6010	1.0266	0.6924×10^{-5}	1.923	0.009246	0.02501	0.770
150	2.3675	1.0099	1.0283	4.343	0.013735	0.05745	0.753
200	1.7684	1.0061	1.3289	7.490	0.01809	0.10165	0.739
250	1.4128	1.0053	1.488	9.49	0.02227	0.13161	0.722
300	1.1774	1.0057	1.983	15.68	0.02624	0.22160	0.708
350	0.9980	1.0090	2.075	20.76	0.03003	0.2983	0.697
400	0.8826	1.0140	2.286	25.90	0.03365	0.3760	0.689
450	0.7833	1.0207	2.484	28.86	0.03707	0.4222	0.683
500	0.7048	1.0295	2.671	37.90	0.04038	0.5564	0.680
550	0.6423	1.0392	2.848	44.34	0.04360	0.6532	0.680
600	0.5870	1.0551	3.018	51.34	0.04659	0.7512	0.680
650	0.5430	1.0635	3.177	58.51	0.04953	0.8578	0.682
700	0.5030	1.0752	3.332	66.25	0.05230	0.9672	0.684
750	0.4709	1.0856	3.481	73.91	0.05509	1.0774	0.686
800	0.4405	1.0978	3.625	82.29	0.05779	1.1951	0.689

Table B-1 (continued)

T, °K	ρ kg m^{-3}	c_p, kJ kg^{-1} °K^{-1}	μ, kg m^{-1}s^{-1}	ν, m^2 s^{-1} $\times 10^6$	k, W m^{-1} °K^{-1}	κ, m^2 s^{-1} $\times 10^4$	Pr
Air							
850	0.4149	1.1095	3.765	90.75	0.06028	1.3097	0.692
900	0.3925	1.1212	3.899	99.3	0.06279	1.4271	0.696
950	0.3716	1.1321	4.023	108.2	0.06525	1.5510	0.699
1000	0.3524	1.1417	4.152	117.8	0.06752	1.6779	0.702
1100	0.3204	1.160	4.44	138.6	0.0732	1.969	0.704
1200	0.2947	1.179	4.69	159.1	0.0782	2.251	0.707
1300	0.2707	1.197	4.93	182.1	0.0837	2.583	0.705
1400	0.2515	1.214	5.17	205.5	0.0891	2.920	0.705
1500	0.2355	1.230	5.40	229.1	0.0946	3.262	0.705
1600	0.2211	1.248	5.63	254.5	0.100	3.609	0.705
1700	0.2082	1.267	5.85	280.5	0.105	3.977	0.705
1800	0.1970	1.287	6.07	308.1	0.111	4.379	0.704
1900	0.1858	1.309	6.29	338.5	0.117	4.811	0.704
2000	0.1762	1.338	6.50	369.0	0.124	5.260	0.702
2100	0.1682	1.372	6.72	399.6	0.131	5.715	0.700
2200	0.1602	1.419	6.93	432.6	0.139	6.120	0.707
2300	0.1538	1.482	7.14	464.0	0.149	6.540	0.710
2400	0.1458	1.574	7.35	504.0	0.161	7.020	0.718
2500	0.1394	1.688	7.57	543.5	0.175	7.441	0.730
Helium							
3		5.200	8.42×10^{-7}		0.0106		
33	1.4657	5.200	50.2	3.42	0.0353	0.04625	0.74
144	3.3799	5.200	125.5	37.11	0.0928	0.5275	0.70
200	0.2435	5.200	156.6	64.38	0.1177	0.9288	0.694
255	0.1906	5.200	181.7	95.50	0.1357	1.3675	0.70
366	0.13280	5.200	230.5	173.6	0.1691	2.449	0.71
477	0.10204	5.200	275.0	269.3	0.197	3.716	0.72
589	0.08282	5.200	311.3	375.8	0.225	5.215	0.72
700	0.07032	5.200	347.5	494.2	0.251	6.661	0.72
800	0.06023	5.200	381.7	634.1	0.275	8.774	0.72
900	0.05286	5.200	413.6	781.3	0.298	10.834	0.72
Carbon dioxide							
220	2.4733	0.783	11.105×10^{-6}	4.490	0.010805	0.05920	0.818
250	2.1657	0.804	12.590	5.813	0.012884	0.07401	0.793
300	1.7973	0.871	14.958	8.321	0.016572	0.10588	0.770
350	1.5362	0.900	17.205	11.19	0.02047	0.14808	0.755
400	1.3424	0.942	19.32	14.39	0.02461	0.19463	0.738
450	1.1918	0.980	21.34	17.90	0.02897	0.24813	0.721
500	1.0732	1.013	23.26	21.67	0.03352	0.3084	0.702
550	0.9739	1.047	25.08	25.74	0.03821	0.3750	0.685
600	0.8938	1.076	26.83	30.02	0.04311	0.4483	0.668

Table B-1 (continued)

T, °K	ρ, kg m^{-3}	c_p, kJ kg^{-1} °K^{-1}	μ, kg m^{-1}s^{-1}	ν, m^2 s^{-1} $\times 10^6$	k, W m^{-1} °K^{-1}	κ, m^2 s^{-1} $\times 10^4$	Pr
Carbon monoxide							
220	1.55363	1.0429	13.832×10^{-6}	8.903	0.01906	0.11760	0.758
250	0.8410	1.0425	15.40	11.28	0.02144	0.15063	0.750
300	1.13876	1.0421	17.843	15.67	0.02525	0.21280	0.737
350	0.97425	1.0434	20.09	20.62	0.02883	0.2836	0.728
400	0.85363	1.0484	22.19	25.99	0.03226	0.3605	0.722
450	0.75848	1.0551	24.18	31.88	0.0436	0.4439	0.718
500	0.68223	1.0635	26.06	38.19	0.03863	0.5324	0.718
550	0.62024	1.0756	27.89	44.97	0.04162	0.6240	0.721
600	0.56850	1.0877	29.60	52.06	0.04446	0.7190	0.724
Ammonia, NH$_3$							
220	0.3828	2.198	7.255×10^{-6}	19.0	0.0171	0.2054	0.93
273	0.7929	2.177	9.353	11.8	0.0220	0.1308	0.90
323	0.6487	2.177	11.035	17.0	0.0270	0.1920	0.88
373	0.5590	2.236	12.886	23.0	0.0327	0.2619	0.87
423	0.4934	2.315	14.672	29.7	0.0391	0.3432	0.87
473	0.4405	2.395	16.49	37.4	0.0467	0.4421	0.84
Steam (H$_2$O vapor)							
380	0.5863	2.060	12.71×10^{-6}	21.6	0.0246	0.2036	1.060
400	0.5542	2.014	13.44	24.2	0.0261	0.2338	1.040
450	0.4902	1.980	15.25	31.1	0.0299	0.307	1.010
500	0.4405	1.985	17.04	38.6	0.0339	0.387	0.996
550	0.4005	1.997	18.84	47.0	0.0379	0.475	0.991
600	0.3652	2.026	20.67	56.6	0.0422	0.573	0.986
650	0.3380	2.056	22.47	64.4	0.0464	0.666	0.995
700	0.3140	2.085	24.26	77.2	0.0505	0.772	1.000
750	0.2931	2.119	26.04	88.8	0.0549	0.883	1.005
800	0.2739	2.152	27.86	102.0	0.0592	1.001	1.010
850	0.2579	2.186	29.69	115.2	0.0637	1.130	1.019
Hydrogen							
30	0.84722	10.840	1.606×10^{-6}	1.895	0.0228	0.02493	0.759
50	0.50955	10.501	2.516	4.880	0.0362	0.0676	0.721
100	0.24572	11.229	4.212	17.14	0.0665	0.2408	0.712
150	0.16371	12.602	5.595	34.18	0.0981	0.475	0.718
200	0.12270	13.540	6.813	55.53	0.1282	0.772	0.719
250	0.09819	14.059	7.919	80.64	0.1561	1.130	0.713
300	0.08185	14.314	8.963	109.5	0.182	1.554	0.706
350	0.07016	14.436	9.954	141.9	0.206	2.031	0.697
400	0.06135	14.491	10.864	177.1	0.228	2.568	0.690
450	0.05462	14.499	11.779	215.6	0.251	3.164	0.682

Table B-1 (continued)

T, °K	ρ, kg m^{-3}	c_p, kJ kg^{-1} °K^{-1}	μ, kg m^{-1}s^{-1}	ν, m^2 s^{-1} $\times 10^6$	k, W m^{-1} °K^{-1}	κ, m^2 s^{-1} $\times 10^4$	Pr
Hydrogen							
500	0.04918	14.507	12.636	257.0	0.272	3.817	0.675
550	0.04469	14.532	13.475	301.6	0.292	4.516	0.668
600	0.04085	14.537	14.285	349.7	0.315	5.306	0.664
700	0.03492	14.574	15.89	455.1	0.351	6.903	0.659
800	0.03060	14.675	17.40	569	0.384	8.563	0.664
900	0.02723	14.821	18.78	690	0.412	10.217	0.676
1000	0.02451	14.968	20.16	822	0.440	11.997	0.686
1100	0.02227	15.165	21.46	965	0.464	13.726	0.703
1200	0.02050	15.366	22.75	1107	0.488	15.484	0.715
1300	0.01890	15.575	24.08	1273	0.512	17.394	0.733
1333	0.01842	15.638	24.44	1328	0.519	18.013	0.736
Oxygen							
100	3.9918	0.9479	7.768×10^{-6}	1.946	0.00903	0.023876	0.815
150	2.6190	0.9178	11.490	4.387	0.01367	0.05688	0.773
200	1.9559	0.9131	14.850	7.593	0.01824	0.10214	0.745
250	1.5618	0.9157	17.87	11.45	0.02259	0.15794	0.725
300	1.3007	0.9203	20.63	15.86	0.02676	0.22353	0.709
350	1.1133	0.9291	23.16	20.80	0.03070	0.2968	0.702
400	0.9755	0.9420	25.54	26.18	0.03461	0.3768	0.695
450	0.8682	0.9567	27.77	31.99	0.03828	0.4609	0.694
500	0.7801	0.9722	29.91	38.34	0.04173	0.5502	0.697
550	0.7096	0.9881	31.97	45.05	0.04517	0.6441	0.700
600	0.6504	1.0044	33.92	52.15	0.04832	0.7399	0.704
Nitrogen							
100	3.4808	1.0722	6.862×10^{-6}	1.971	0.009450	0.025319	0.786
200	1.7108	1.0429	12.947	7.568	0.01824	0.10224	0.747
300	1.1421	1.0408	17.84	15.63	0.02620	0.22044	0.713
400	0.8538	1.0459	21.98	25.74	0.03335	0.3734	0.691
500	0.6824	1.0555	25.70	37.66	0.03984	0.5530	0.684
600	0.5687	1.0756	29.11	51.19	0.04580	0.7486	0.686
700	0.4934	1.0969	32.13	65.13	0.05123	0.9466	0.691
800	0.4277	1.1225	34.84	81.46	0.05609	1.1685	0.700
900	0.3796	1.1464	37.49	91.06	0.06070	1.3946	0.711
1000	0.3412	1.1677	40.00	117.2	0.06475	1.6250	0.724
1100	0.3108	1.1857	42.28	136.0	0.06850	1.8591	0.736
1200	0.2851	1.2037	44.50	156.1	0.07184	2.0932	0.748

Table B-2 Physical properties of saturated liquids

t, °C	ρ, kg m^{-3}	c_p, kJ kg^{-1}°K^{-1}	ν, m^2 s^{-1}	k, W m^{-1}°K^{-1}	κ, m^2 s^{-1} $\times 10^7$	Pr	β, °K^{-1}
Ammonia, NH$_3$							
-50	703.69	4.463	0.435×10^{-6}	0.547	1.742	2.60	
-40	691.68	4.467	0.406	0.547	1.775	2.28	
-30	679.34	4.476	0.387	0.549	1.801	2.15	
-20	666.69	4.509	0.381	0.547	1.819	2.09	
-10	653.55	4.564	0.378	0.543	1.825	2.07	
0	640.10	4.635	0.373	0.540	1.819	2.05	
10	626.16	4.714	0.368	0.531	1.801	2.04	
20	611.75	4.798	0.359	0.521	1.775	2.02	2.45×10^{-3}
30	596.37	4.890	0.349	0.507	1.742	2.01	
40	580.99	4.999	0.340	0.493	1.701	2.00	
50	564.33	5.116	0.330	0.476	1.654	1.99	
Carbon dioxide, CO$_2$							
-50	1,156.34	1.84	0.119×10^{-6}	0.0855	0.4021	2.96	
-40	1,117.77	1.88	0.118	0.1011	0.4810	2.46	
-30	1,076.76	1.97	0.117	0.1116	0.5272	2.22	
-20	1,032.39	2.05	0.115	0.1151	0.5445	2.12	
-10	983.38	2.18	0.113	0.1099	0.5133	2.20	
0	926.99	2.47	0.108	0.1045	0.4578	2.38	
10	860.03	3.14	0.101	0.0971	0.3608	2.80	
20	772.57	5.0	0.091	0.0872	0.2219	4.10	14.00×10^{-3}
30	597.81	36.4	0.080	0.0703	0.0279	28.7	

Table B-2 (Continued)

t, °C	ρ, kg m^{-3}	c_p, kJ kg^{-1} °K^{-1}	ν, m^2 s^{-1}	k, W m^{-1} °K^{-1}	κ, m^2 s^{-1} $\times 10^7$	Pr	β, °K^{-1}
Dichlorodifluoromethane (Freon), CCl$_2$F$_2$							
-50	1,546.75	0.8750	0.310×10^{-6}	0.067	0.501	6.2	2.63×10^{-3}
-40	1,518.71	0.8847	0.279	0.069	0.514	5.4	
-30	1,489.56	0.8956	0.253	0.069	0.526	4.8	
-20	1,460.57	0.9073	0.235	0.071	0.539	4.4	
-10	1,429.49	0.9203	0.221	0.073	0.550	4.0	
0	1,397.45	0.9345	0.214	0.073	0.557	3.8	
10	1,364.30	0.9496	0.203	0.073	0.560	3.6	
20	1,330.18	0.9659	0.198	0.073	0.560	3.5	
30	1,295.10	0.9835	0.194	0.071	0.560	3.5	
40	1,257.13	1.0019	0.191	0.069	0.555	3.5	
50	1,215.96	1.0216	0.190	0.067	0.545	3.5	
Engine oil (unused)							
0	899.12	1.796	0.00428	0.147	0.911	47,100	0.70×10^{-3}
20	888.23	1.880	0.00090	0.145	0.872	10,400	
40	876.05	1.964	0.00024	0.144	0.834	2,870	
60	864.04	2.047	0.839×10^{-4}	0.140	0.800	1,050	
80	852.02	2.131	0.375	0.138	0.769	490	
100	840.01	2.219	0.203	0.137	0.738	276	
120	828.96	2.307	0.124	0.135	0.710	175	
140	816.94	2.395	0.080	0.133	0.686	116	
160	805.89	2.483	0.056	0.132	0.663	84	
Ethylene glycol, C$_2$H$_4$(OH$_2$)							
0	1,130.75	2.294	57.53×10^{-6}	0.242	0.934	615	0.65×10^{-3}
20	1,116.65	2.382	19.18	0.249	0.939	204	

t (°C)	ρ	c_p	ν	k	α	Pr	β
40	1,101.43	2.474	8.69	0.256	0.939	93	
60	1,087.66	2.562	4.75	0.260	0.932	51	
80	1,077.56	2.650	2.98	0.261	0.921	32.4	
100	1,058.50	2.742	2.03	0.263	0.908	22.4	
Eutectic calcium chloride solution, 29.9% CaCl₂							
−50	1,319.76	2.608	36.35×10^{-6}	0.402	1.166	312	
−40	1,314.96	2.6356	24.97	0.415	1.200	208	
−30	1,310.15	2.6611	17.18	0.429	1.234	139	
−20	1,305.51	2.688	11.04	0.445	1.267	87.1	
−10	1,300.70	2.713	6.96	0.459	1.300	53.6	
0	1,296.06	2.738	4.39	0.472	1.332	33.0	
10	1,291.41	2.763	3.35	0.485	1.363	24.6	
20	1,286.61	2.788	2.72	0.498	1.394	19.6	
30	1,281.96	2.814	2.27	0.511	1.419	16.0	
40	1,277.16	2.839	1.92	0.523	1.445	13.3	
50	1,272.51	2.868	1.65	0.535	1.468	11.3	
Glycerin, C₃H₆(OH)₃							
0	1,276.03	2.261	0.00831	0.282	0.983	84.7×10^{3}	
10	1,270.11	2.319	0.00300	0.284	0.965	31.0	
20	1,264.02	2.386	0.00118	0.286	0.947	12.5	0.50×10^{-3}
30	1,258.09	2.445	0.00050	0.286	0.929	5.38	
40	1,252.01	2.512	0.00022	0.286	0.914	2.45	
50	1,244.96	2.583	0.00015	0.287	0.893	1.63	
Mercury, Hg							
0	13,628.22	0.1403	0.124×10^{-6}	8.20	42.99	0.0288	
20	13,579.04	0.1394	0.114	8.69	46.06	0.0249	
50	13,505.84	0.1386	0.104	9.40	50.22	0.0207	
100	13,384.58	0.1373	0.0928	10.51	57.16	0.0162	1.82×10^{-4}
150	13,264.28	0.1365	0.0853	11.49	63.54	0.0134	
200	13,144.94	0.1570	0.0802	12.34	69.08	0.0116	

Table B-2 (Continued)

t, °C	ρ, kg m^{-3}	c_p, kJ kg^{-1} °K^{-1}	ν, m^2 s^{-1}	k, W m^{-1} °K^{-1}	κ, m^2 s^{-1} $\times 10^7$	Pr	β, °K^{-1}
250	13,025.60	0.1357	0.0765	13.07	74.06	0.0103	
315.5	12,847	0.134	0.0673	14.02	81.5	0.0083	

Methyl chloride, CH$_3$Cl

t, °C	ρ, kg m^{-3}	c_p, kJ kg^{-1} °K^{-1}	ν, m^2 s^{-1}	k, W m^{-1} °K^{-1}	κ, m^2 s^{-1} $\times 10^7$	Pr	β, °K^{-1}
-50	1,052.58	1.4759	0.320×10^{-6}	0.215	1.388	2.31	
-40	1,033.35	1.4826	0.318	0.209	1.368	2.32	
-30	1,016.53	1.4922	0.314	0.202	1.337	2.35	
-20	999.39	1.5043	0.309	0.196	1.301	2.38	
-10	981.45	1.5194	0.306	0.187	1.257	2.43	
0	962.39	1.5378	0.302	0.178	1.213	2.49	
10	942.36	1.5600	0.297	0.171	1.166	2.55	
20	923.31	1.5860	0.293	0.163	1.112	2.63	
30	903.12	1.6161	0.288	0.154	1.058	2.72	
40	883.10	1.6504	0.281	0.144	0.996	2.83	
50	861.15	1.6890	0.274	0.133	0.921	2.97	

Sulfur Dioxide, SO$_2$

t, °C	ρ, kg m^{-3}	c_p, kJ kg^{-1} °K^{-1}	ν, m^2 s^{-1}	k, W m^{-1} °K^{-1}	κ, m^2 s^{-1} $\times 10^7$	Pr	β, °K^{-1}
-50	1,560.84	1.3595	0.484×10^{-6}	0.242	1.141	4.24	
-40	1,536.81	1.3607	0.424	0.235	1.130	3.74	
-30	1,520.64	1.3616	0.371	0.230	1.117	3.31	
-20	1,488.60	1.3624	0.324	0.225	1.107	2.93	
-10	1,463.61	1.3628	0.288	0.218	1.097	2.62	
0	1,438.46	1.3636	0.257	0.211	1.081	2.38	
10	1,412.51	1.3645	0.232	0.204	1.066	2.18	
20	1,386.40	1.3653	0.210	0.199	1.050	2.00	1.94×10^{-3}
30	1,359.33	1.3662	0.190	0.192	1.035	1.83	
40	1,329.22	1.3674	0.173	0.185	1.019	1.70	
50	1,299.10	1.3683	0.162	0.177	0.999	1.61	

Water, H_2O

			1.788×10^{-6}			0.18×10^{-3}
0	1,002.28	4.2178	1.788 $\times 10^{-6}$	0.552	1.308	13.6
20	1,000.52	4.1818	1.006	0.597	1.430	7.02
40	994.59	4.1784	0.658	0.628	1.512	4.34
60	985.46	4.1843	0.478	0.651	1.554	3.02
80	974.08	4.1964	0.364	0.668	1.636	2.22
100	960.63	4.2161	0.294	0.680	1.680	1.74
120	945.25	4.250	0.247	0.685	1.708	1.446
140	928.27	4.283	0.214	0.684	1.724	1.241
160	909.69	4.342	0.190	0.680	1.729	1.099
180	889.03	4.417	0.173	0.675	1.724	1.004
200	866.76	4.505	0.160	0.665	1.706	0.937
220	842.41	4.610	0.150	0.652	1.680	0.891
240	815.66	4.756	0.143	0.635	1.639	0.871
260	785.87	4.949	0.137	0.611	1.577	0.874
280.6	752.55	5.208	0.135	0.580	1.481	0.910
300	714.26	5.728	0.135	0.540	1.324	1.019

Table B-3 Physical properties of liquid metals

Metal	Melting Point °C	Boiling Point °C	T °C	ρ kg m⁻³	c_p kJ kg⁻¹ °C⁻¹	$\mu \times 10^4$ kg m⁻¹ s⁻¹	$\nu \times 10^6$ m² s⁻¹	k W m⁻¹ °C⁻¹	$\kappa \times 10^6$ m² s⁻¹	Pr
Bismuth	271	1477	315	10,011	0.144	16.2	0.160	16.4	11.25	0.0142
			538	9739	0.155	11.0	0.113	15.6	10.34	0.0110
			760	9467	0.165	7.9	0.083	15.6	9.98	0.0083
Lead	327	1737	371	10,540	0.159	2.40	0.023	16.1	9.61	0.024
			704	10,140	0.155	1.37	0.014	14.9	9.48	0.0143
Lithium	179	1317	204.4	509.2	4.365	5.416	1.1098	46.37	20.96	0.051
			315.6	498.8	4.270	4.465	0.8982	43.08	20.32	0.0443
			426.7	489.1	4.211	3.927	0.8053	38.24	18.65	0.0432
			537.8	476.3	4.171	3.473	0.7304	30.45	15.40	0.0476
Mercury	−38.9	357	−17.8	13,707.1	0.1415	18.334	0.1342	9.76	5.038	0.0266
			93.3	13,409.4	0.1365	12.224	0.0903	10.38	5.619	0.0161
			204.4	13,168.1	0.1356	10.046	0.0748	12.63	7.087	0.0108
Sodium	97.8	883	93.3	931.6	1.384	7.131	0.7689	84.96	56.29	0.0116
			204.4	907.5	1.339	4.521	0.5010	80.81	66.80	0.0075
			315.6	878.5	1.304	3.294	0.3766	75.78	66.47	0.00567
			426.7	852.8	1.277	2.522	0.2968	69.39	64.05	0.00464
			537.8	823.8	1.264	2.315	0.2821	64.37	62.09	0.00455
			648.9	790.0	1.261	1.964	0.2496	60.56	61.10	0.00408
			760.0	767.5	1.270	1.716	0.2245	56.58	58.34	0.00385
Potassium	63.9	760	426.7	741.7	0.766	2.108	0.2839	39.45	69.74	0.0041
			537.8	714.4	0.762	1.711	0.2400	36.51	67.39	0.0036
			648.9	690.3	0.766	1.463	0.2116	33.74	64.10	0.0033
			760.0	667.7	0.783	1.331	0.1987	31.15	59.86	0.0033
NaK (56% Na, 44% K)	−11.1	784	93.3	889.8	1.130	5.622	0.6347	25.78	25.76	0.0246
			204.4	865.6	1.089	3.803	0.4414	26.47	28.23	0.0155
			315.6	838.3	1.068	2.935	0.3515	27.17	30.50	0.0115
			426.7	814.2	1.051	2.150	0.2652	27.68	32.52	0.0081
			537.8	788.4	1.047	2.026	0.2581	27.68	33.71	0.0076
			648.9	759.5	1.051	1.695	0.2240	27.68	34.86	0.0064

Table B-4 Physical properties of metals

| Metal | Melting Point °C | Properties at 20°C | | | | Thermal conductivity k, W m^{-1} °C^{-1} | | | | | | | | |
		ρ, kg m^{-3}	c_p, kJ kg^{-1} °C^{-1}	k, W m^{-1} °C^{-1}	κ, m^2 s^{-1} ×10^5	-100°C	0°C	100°C	200°C	300°C	400°C	600°C	800°C	1000°C
Aluminum:														
Pure	660	2,707	0.896	204	8.418	215	202	206	215	228	249			
Al-Cu (Duralumin), 94–96% Al, 3–5% Cu, trace Mg		2,787	0.883	164	6.676	126	159	182	194					
Al-Si (Silumin, copper-bearing), 86.5% Al, 1% Cu		2,659	0.867	137	5.933	119	137	144	152	161				
Al-Si (Alusil), 78–80% Al, 20–22% Si		2,627	0.854	161	7.172	144	157	168	175	178				
Al-Mg-Si, 97% Al, 1% Mg, 1% Si, 1% Mn		2,707	0.892	177	7.311	•	175	189	204					
Beryllium	1277	1,850	1.825	200	5.92									
Bismuth	272	9,780	0.122	7.86	0.66									
Cadmium	321	8,650	0.231	96.8	4.84									
Copper:														
Pure	1085	8,954	0.3831	386	11.234	407	386	379	374	369	363	353		
Aluminum bronze 95% Cu, 5% Al		8,666	0.410	83	2.330									
Bronze 75% Cu, 25% Sn		8,666	0.343	26	0.859									
Red brass 85% Cu, 9% Sn, 6% Zn		8,714	0.385	61	1.804		59	71						
Brass 70% Cu, 30% Zn		8,522	0.385	111	3.412	88	128	144	147	147	147			

Table B-4 (continued)

Metal	Melting Point, °C	Properties at 20°C				Thermal conductivity k, W m^{-1} °C^{-1}								
		ρ, kg m^{-3}	c_p, kJ kg^{-1} °C^{-1}	k, W m^{-1} °C^{-1}	κ, m^2 s^{-1} ×10^5	−100°C	0°C	100°C	200°C	300°C	400°C	600°C	800°C	1000°C
German silver 62% Cu, 15% Ni, 22% Zn		8,618	0.394	24.9	0.733	19.2		31	40	45	48			
Constantan 60% Cu, 40% Ni		8,922	0.410	22.7	0.612	21	22.2	26						
Iron:														
Pure	1537	7,897	0.452	73	2.034	87	73	67	62	55	48	40	36	35
Wrought iron, 0.5% C		7,849	0.46	59	1.626		59	57	52	48	45	36	33	33
Steel (C max ≈1.5%):														
Carbon steel C ≈ 0.5%		7,833	0.465	54	1.474		55	52	48	45	42	35	31	29
1.0%		7,801	0.473	43	1.172		43	43	42	40	36	33	29	28
1.5%		7,753	0.486	36	0.970		36	36	36	35	33	31	28	28
Nickel steel Ni ≈ 0%		7,897	0.452	73	2.026									
20%		7,933	0.46	19	0.526									
40%		8,169	0.46	10	0.279									
80%		8,618	0.46	35	0.872									
Invar 36% Ni		8,137	0.46	10.7	0.286									
Chrome steel Cr = 0%		7,897	0.452	73	2.026	87	73	67	62	55	48	40	36	35
1%		7,865	0.46	61	1.665		62	55	52	47	42	36	33	33
5%		7,833	0.46	40	1.110		40	38	36	36	33	29	29	29
20%		7,689	0.46	22	0.635		22	22	22	22	24	24	26	29
Cr-Ni (chrome-nickel): 15% Cr, 10% Ni		7,865	0.46	19	0.527									
18% Cr, 8% Ni (V2A)		7,817	0.46	16.3	0.444	16.3	17	17	19	19	19	22	27	31
20% Cr, 15% Ni		7,833	0.46	15.1	0.415									
25% Cr, 20% Ni		7,865	0.46	12.8	0.361									

Material	Melting point	Density	Specific heat			Thermal conductivity (at various temperatures)
Tungsten steel						
W = 0%		7,897	0.452	73	2.026	
1%		7,913	0.448	66	1.858	
5%		8,073	0.435	54	1.525	
10%		8,314	0.419	48	1.391	
Lead	328	11,373	0.130	35	2.343	36.9, 35.1, 33.4, 31.5, 29.8
Magnesium:						
Pure	650	1,746	1.013	171	9.708	178, 171, 168, 163, 157
Mg-Al (electrolytic) 6–8% Al, 1–2% Zn		1,810	1.00	66	3.605	52, 62, 74, 83
Molybdenum	2,621	10,220	0.251	123	4.790	138, 125, 118, 114, 111, 109, 106, 102, 99
Nickel:						
Pure (99.9%)	1,455	8,906	0.4459	90	2.266	104, 93, 83, 73, 64, 59
Ni-Cr 90% Ni, 10% Cr		8,666	0.444	17	0.444	17.1, 18.9, 20.9, 22.8, 24.6
80% Ni, 20% Cr		8,314	0.444	12.6	0.343	12.3, 13.8, 15.6, 17.1, 18.0, 22.5
Silver:						
Purest	962	10,524	0.2340	419	17.004	419, 417, 415, 412
Pure (99.9%)		10,525	0.2340	407	16.563	419, 415, 415, 374, 362, 360
Tin, pure	232	7,304	0.2265	64	3.884	74, 65.9, 59, 57
Tungsten	3,387	19,350	0.1344	163	6.271	166, 151, 142, 133, 126, 112
Uranium	1,133	19,070	0.116	27.6	1.25	
Zinc, pure	420	7,144	0.3843	112.2	4.106	112, 109, 106, 100, 93

Gamma, Beta and Incomplete Beta Functions

Gamma function definition

$$\Gamma(\alpha) = \int_0^\infty t^{\alpha-1} e^{-t} dt$$

Recursion formula:

$$\Gamma(\alpha+1) = \alpha\Gamma(\alpha)$$

α	$\Gamma(\alpha)$	α	$\Gamma(\alpha)$	α	$\Gamma(\alpha)$
1.00	1.0000	1.35	0.8912	1.70	0.9086
1.05	0.9735	1.40	0.8873	1.75	0.9191
1.10	0.9514	1.45	0.8857	1.80	0.9314
1.15	0.9330	1.50	0.8862	1.85	0.9456
1.20	0.9182	1.55	0.8889	1.90	0.9618
1.25	0.9064	1.60	0.8935	1.95	0.9799
1.30	0.8975	1.65	0.9001	2.00	1.0000

Beta function definition:

$$B_1(\alpha, \beta) = \int_0^1 t^{\alpha-1}(1-t)^{\beta-1} dt = \frac{\Gamma(\alpha)\Gamma(\beta)}{\Gamma(\alpha+\beta)} = B_1(\beta, \alpha)$$

Incomplete Beta function definition:

$$B_x(\alpha, \beta) = \int_0^x t^{\alpha-1}(1-t)^{\beta-1}\, dt$$

Recursion formula:

$$B_x(\alpha, \beta) = B_1(\alpha, \beta) - B_{1-x}(\alpha, \beta)$$

The following table [1] gives the functional ratios $I_x(\alpha, \beta) = B_x(\alpha, \beta)/B_1(\alpha, \beta)$ for typical combinations of α and β:

Incomplete beta function ratios $I_x(\alpha, \beta)$

x	$\alpha=1/3$ $\beta=2/3$	$\alpha=1/3$ $\beta=4/3$	$\alpha=1/3$ $\beta=8/3$	$\alpha=2/3$ $\beta=4/3$	$\alpha=1/9$ $\beta=8/9$	$\alpha=1/9$ $\beta=10/9$	$\alpha=1/9$ $\beta=20/9$	$\alpha=8/9$ $\beta=10/9$
0	0	0	0	0	0	0	0	0
0.02	0.2249	0.3068	0.4007	0.0912	0.6346	0.6588	0.7281	0.0342
0.04	0.2838	0.3859	0.5007	0.1443	0.6856	0.7113	0.7845	0.0628
0.06	0.3254	0.4410	0.5684	0.1886	0.7173	0.7439	0.8186	0.0917
0.08	0.3588	0.4845	0.6204	0.2278	0.7407	0.7679	0.8431	0.1174
0.10	0.3872	0.5210	0.6627	0.2636	0.7595	0.7870	0.8622	0.1416
0.20	0.4924	0.6506	0.8008	0.4124	0.8213	0.8490	0.9199	0.2607
0.30	0.5694	0.7377	0.8793	0.5321	0.8603	0.8870	0.9506	0.3715
0.40	0.6337	0.8038	0.9284	0.6339	0.8895	0.9146	0.9696	0.4765
0.50	0.6911	0.8566	0.9599	0.7225	0.9133	0.9362	0.9820	0.5767
0.60	0.7448	0.8998	0.9796	0.7999	0.9335	0.9538	0.9901	0.6725
0.70	0.7970	0.9352	0.9912	0.8671	0.9515	0.9686	0.9952	0.7640
0.80	0.8501	0.9640	0.9972	0.9244	0.9679	0.9812	0.9982	0.8507
0.90	0.9084	0.9863	0.9996	0.9706	0.9835	0.9917	0.9996	0.9313
1.00	1.0000	1.0000	1.0000	1.0000	1.0000	1.0000	1.0000	1.0000
$B_1(\alpha, \beta)$	3.6275	2.6499	2.0153	1.2092	9.1853	8.8439	7.9839	1.0206

References

[1] Baxter, D. C., and Reynolds, W. C.: Fundamental solutions for heat transfer from nonisothermal flat plates. *J. Aero. Sci.* **25**:403 (1958).

APPENDIX D

Fortran Program for Head's Method

In this Appendix we present a computer program for predicting the turbulent boundary-layer development on two-dimensional bodies by Head's method. The code uses FORTRAN IV and is based on the integration of Eqs. (3.68) and (6.117) by a fourth-order Runge-Kutta method [1], which requires the specification of the following arguments to solve the equations:

X The dependent variable x. The initial value of x must be input.

B The number of dependent variables to be computed. For example,

$$\begin{bmatrix} \theta \\ u_e \theta H_1 \end{bmatrix} = \begin{bmatrix} B(1) \\ B(2) \end{bmatrix}$$

Initial values of θ, $(u_e \theta H_1)$ must be input.

C Denotes the right-hand sides of the differential equations. For example

$$\begin{bmatrix} \dfrac{c_f}{2} - (H+2)\dfrac{\theta}{u_e}\dfrac{du_e}{dx} \\ u_e F \end{bmatrix} = \begin{bmatrix} C(1) \\ C(2) \end{bmatrix}$$

DX The increment in x. DX must be input.

N The number of simultaneous equations to be integrated.

F	The array used by the subroutine (RKM) to store values of the array *B*. It is dimensioned *N*.
G	An array that contains intermediate values computed by the subroutine. Four entries of *G* are used to compute one entry of *G*. *G* is dimensioned 4*N*.
IS	A code variable that must be set to zero to initialize the subroutine. It is automatically stepped through the values 1, 2, 3 and 4 and is reset to zero by the subroutine after the variables for *X* and the array *B* are computed.

Since the solution of Eq. (3.68) requires the specification of the external velocity distribution, we input $u_e(x)/u_\infty$, $UE(I)$, as a function of surface distance x/L, $X(I)$, with u_∞ denoting the reference freestream velocity and *L* a reference length. The initial conditions consist of a dimensionless momentum thickness, θ/L, $T(1)$, and shape factor *H*, $H(1)$, at the first station. In addition, we specify a reference Reynolds number $R_L = u_\infty L/\nu$, *RL*, and the total number of *x*-stations, *NXT*. In the code, the derivative of external velocity du_e/dx is computed by using a three-point Lagrange-interpolation formula.

```
      COMMON/SHARE/ NXT,RL,X(41),UE(41),DUEDX(41),T(41),S(41),H(41),
     1              RTH(41),CF(41)
C - - - - - - - - - - - - - - - - - - - - - - - - - - - - - - - - - - -
      RGNG(X1,X2,X3,Y1,Y2,Y3,Z) = Y1*(2.0*Z-X2-X3)/(X1-X2)/(X1-X3)+
     1              Y2*(2.0*Z-X1-X3)/(X2-X1)/(X2-X3)+Y3*(2.0*Z-X1-X2)/(X3-X1)
     1              /(X3-X2)
C
    1 READ(5,2,END=20) NXT,RL,T(1),H(1)
      READ(5,3) (X(I),I=1,NXT)
      READ(5,3) (UE(I),I=1,NXT)
      X(NXT+1) =2.*X(NXT)-X(NXT-1)
C  CALCULATE THE VELOCITY GRADIENT BY THREE POINT LAGRANGIAN FORMULA
      NXTM1 = NXT-1
      DUEDX(1)   = RGNG(X(1),X(2),X(3),UE(1),UE(2),UE(3),X(1))
      DUEDX(NXT)= RGNG(X(NXT-2),X(NXT-1),X(NXT),UE(NXT-2),UE(NXT-1),
     1              UE(NXT),X(NXT))
      DO 10 I=2,NXTM1
   10 DUEDX(I)   = RGNG(X(I-1),X(I),X(I+1),UE(I-1),UE(I),UE(I+1),X(I))
C
      S(1)   = UE(1)*T(1)*HOFH1(-H(1))
      CALL STNDRD
C
      WRITE(6,4)
      DO 15 I=1,NXT
      H1     = S(I)/UE(I)/T(I)
      DELST = T(I)*H(I)
      DELTA = T(I)*H1+DELST
      WRITE(6,5) I,X(I),UE(I),DUEDX(I),T(I),H(I),DELST,DELTA,CF(I),
     1              RTH(I)
   15 CONTINUE
      GOTO 1
   20 STOP
C - - - - - - - - - - - - - - - - - - - - - - - - - - - - - - - - - - -
    2 FORMAT(I5,3F10.0)
    3 FORMAT(6F10.0)
    4 FORMAT(1H1,1X,2HNX,6X,1HX,12X,2HUE,10X,5HDUEDX,9X,5HTHETA,11X,
     1              1HH,11X,5HDELST,9X,5HDELTA,11X,2HCF,10X,6HRTHETA,/)
    5 FORMAT(1H ,I3,9E14.6)
      END

      SUBROUTINE STNDRD
      COMMON/SHARE/ NXT,RL,X(41),UE(41),DUEDX(41),T(41),S(41),H(41),
     1              RTH(41),CF(41)
      DIMENSION C(2),B(2),Z(2),G(8)
```

```
C - - - - - - - - - - - - - - - - - - - - - - - - - - - - - - - - - - -
      N       = 2
      IS      = 0
      XX      = X(1)
      B(1)    = T(1)
      B(2)    = S(1)
      UI      = UE(1)
      UP      = DUEDX(1)
      DX      = X(2)-X(1)
      NXTP1 = NXT+1
      DO 100 I=2,NXTP1
      DO 110 LL=1,4
      GO TO (18,16,18,17), LL
   16 UI      = (UI+UE(I))/2.0+DX*(UP-DUEDX(I))/8.0
      UP      = (UP+DUEDX(I))/2.0
      GO TO 18
   17 UI      = UE(I)
      UP      = DUEDX(I)
   18 CONTINUE
      H1      = B(2)/B(1)/UI
      IF ( H1 .LE. 3.0 ) GO TO 120
      HB      = HOFH1(H1)
      RTHE    = UI*B(1)*RL
      CF02    = 0.123/10.0**(0.678*HB)/RTHE**0.268
      C(1)    = -(HB+2.0)*B(1)/UI*UP+CF02
      C(2)    = UI*0.0306/(H1-3.0)**0.6169
      IF(LL .GT. 1) GO TO 19
      H(I-1)= HB
      RTH(I-1) = RTHE
      CF(I-1)  = 2.0*CF02
      IF(I .GT. NXT) GO TO 100
   19 CONTINUE
  110 CALL RKM(XX,B,C,DX,N,Z,G,IS)
      T(I)    = B(1)
      S(I)    = B(2)
      DX      = X(I+1)-X(I)
  100 CONTINUE
      RETURN
  120 WRITE(6,10)I
   10 FORMAT(1H0,'** WARNING : PROGRAM FAILS AT NX=',I3,
     1       ' INDICATING POSSIBILITY OF FLOW SEPARATION **',/)
      NXT = I-1
      RETURN
      END
      FUNCTION HOFH1(A)
C  H = FUNCT(H1=A), INVERSE IF A NEGATIVE,
C  H1= FUNCT(H=-A)
      REAL C1/1.5501/,C2/0.6778/,C3/-3.064/,C4/3.3/,C5/0.8234/,
     1     C6/1.1/,C7/-1.287/
C - - - - - - - - - - - - - - - - - - - - - - - - - - - - - - - - - - -
      HOFH1 = 0.0
      IF(A .LT. -C6) GO TO 2
      IF(A .LE. C4) RETURN
      IF(A .LT. 5.3) GO TO 3
      HOFH1 = ((A-C4)/C5)**(1.0/C7)+C6
      RETURN
    3 HOFH1 = ((A-C4)/C1)**(1.0/C3)+C2
      RETURN
    2 IF(A .LT. -1.6) GO TO 4
      HOFH1 = C5*(-A-C6)**C7+C4
      RETURN
    4 HOFH1 = C1*(-A-C2)**C3+C4
      RETURN
      END

      SUBROUTINE RKM(A,B,C,DX,N,F,G,IS)
      DIMENSION B(1),C(1),F(1),G(1)
C - - - - - - - - - - - - - - - - - - - - - - - - - - - - - - - - - - -
      IS      = IS+1
      GO TO (10,30,60,80), IS
C  FIRST ENTRY
   10 E       = A
      DO 20 I=1,N
      F(I)    = B(I)
      G(4*I-3) = C(I)*DX
   20 B(I)    = F(I)+G(4*I-3)/2.0
      GO TO 50
```

```
C    SECOND ENTRY
     30 DO 40 I=1,N
        G(4*I-2) = C(I)*DX
     40 B(I)    = F(I)+G(4*I-2)/2.0
     50 A       = E+DX/2.0
        GO TO 100
C    THIRD ENTRY
     60 DO 70 I=1,N
        G(4*I-1) = C(I)*DX
     70 B(I)    = F(I)+G(4*I-1)
        A       = E+DX
        GO TO 100
C    FOURTH ENTRY
     80 DO 90 I=1,N
        G(4*I)= C(I)*DX
        B(I)    = G(4*I-3)+2.0*(G(4*I-2)+G(4*I-1))
     90 B(I)    = (B(I)+G(4*I))/6.0+F(I)
        IS      = 0
    100 RETURN
        END
```

References

[1] Hildebrand, F. B.: *Advanced Calculus for Application*. Prentice-Hall, NJ, 1962.

Index